Trends in Nonlinear Analysis

Springer
Berlin
Heidelberg
New York
Hong Kong
London
Milan
Paris
Tokyo

Markus Kirkilionis Susanne Krömker
Rolf Rannacher Friedrich Tomi (Eds.)

Trends in Nonlinear Analysis

With 109 Figures, Including 14 Color Plates

Markus Kirkilionis
Mathematics Department
and Center for Scientific Computing
University of Warwick
CV4 7AL Coventry, UK
E-mail: mak@maths.warwick.ac.uk

Susanne Krömker
Institute of Scientific Computing
University of Heidelberg
Im Neuheimer Feld 368
69120 Heidelberg, Germany
E-mail: kroemker@iwr.uni-heidelberg.de

Rolf Rannacher
Institute of Applied Mathematics
University of Heidelberg
Im Neuheimer Feld 293
69120 Heidelberg, Germany
E-mail: rannacher@iwr.uni-heidelberg.de

Friedrich Tomi
Department of Mathematics
University of Heidelberg
Im Neuheimer Feld 288
69120 Heidelberg, Germany
E-mail: tomi@mathi.uni-heidelberg.de

The cover design is based on a painting by Raquel Diaz, Madrid.

Mathematics Subject Classification (2000): 35-O2, 37-O2, 65-O2, 78-O2, 92-O2

Cataloging-in-Publication Data applied for
A catalog record for this book is available from the Library of Congress.

Bibliographic information published by Die Deutsche Bibliothek
Die Deutsche Bibliothek lists this publication in the Deutsche Nationalbibliografie;
detailed bibliographic data is available in the Internet at http://dnb.ddb.de

ISBN 3-540-44198-0 Springer-Verlag Berlin Heidelberg New York

This work is subject to copyright. All rights are reserved, whether the whole or part of the material is concerned, specifically the rights of translation, reprinting, reuse of illustrations, recitation, broadcasting, reproduction on microfilm or in any other way, and storage in data banks. Duplication of this publication or parts thereof is permitted only under the provisions of the German Copyright Law of September 9, 1965, in its current version, and permission for use must always be obtained from Springer-Verlag. Violations are liable for prosecution under the German Copyright Law.

Springer-Verlag Berlin Heidelberg New York
a member of BertelsmannSpringer Science+Business Media GmbH

http://www.springer.de

© Springer-Verlag Berlin Heidelberg 2003
Printed in Germany

The use of general descriptive names, registered names, trademarks, etc. in this publication does not imply, even in the absence of a specific statement, that such names are exempt from the relevant protective laws and regulations and therefore free for general use.

Typesetting: LaTeX-data from the authors
Cover design: *design & production*, Heidelberg

Printed on acid-free paper   SPIN 10778079   46/3142/yl   5 4 3 2 1 0

On occasion of the 60th birthday of Professor Dr. Dr. h. c. mult. Willi Jäger.

Preface

The idea for this book originated from a symposium with the same title "Trends in Nonlinear Analysis" held at Heidelberg, October 7–12, 2000, in honour of our colleague Willi Jäger on the occasion of his 60th birthday. The symposium, moreover, was the contribution of Heidelberg to the "World Mathematical Year 2000".

This book, which we dedicate to Willi Jäger, is not meant as a mere proceedings volume of this conference. Rather, we asked some of the participants, each one a renowned expert in his field, to contribute chapters illustrating the state of the art of nonlinear analysis in a number of selected areas. This list of fields being selected consisted of the following subjects:

- Dynamical Systems
- Elliptic and Parabolic Equations
- Geometrical Partial Differential Equations
- Multiscale Problems
- Numerical Methods
- Mathematical Biology
- Material Sciences

The choices made certainly reflect specific research interests of the mathematical community at Heidelberg and of course does not comprise the complete field of nonlinear analysis. Nevertheless, we are convinced that the chosen subjects currently act as driving forces for the field as a whole.

We feel that nonlinear analysis has gained its current strength from the fruitful interplay between pure and applied mathematics. This interplay is not only reflected in the selection of the above themes, but also in each of the individual contributions. As the reader will notice, there are references to applications in the more theoretically oriented articles and, conversely, references to theoretical background when the applications are in the foreground of the discussion.

The use of modern computers has become an important tool in mathematics, changing many of the above fields in a fundamental way. Therefore we have chosen several contributions involving numerical simulations. The growing importance of this technique is also reflected in the increasing number of research centres for scientific computing all over the world. The University of Heidelberg has fostered this development from an early stage, as is documented by several "Sonderforschungsbereiche" (special research units of the German science foundation), and above all, by the *Interdisziplinäres Zentrum für Wissenschaftliches Rechnen (IWR)*.

This leads us to the person to whom we dedicate this book, to Willi Jäger. He has shaped all these institutions in a decisive way and his unremitting efforts during the past years have greatly contributed to making Heidelberg an attractive place for mathematical research. For all this we wish to say a deepfelt "thank you". Our TiNA conference confirmed once more Willi Jäger's eminent role in applied mathematics on the international level, documented by his many scientific contacts and cooperations as well as by the impressive number of his pupils, many of whom have meanwhile become established scientists themselves at various places around the globe.

Questions of science policy and academic as well as general education have always been issues of major concern for Willi Jäger. We invite the reader to learn about his personal views on these and some other subjects in the first chapter of this book which contains the transcript of an interview the editors carried out with him recently. We wish Willi Jäger many more creative and fruitful years!

Heidelberg, Warwick, *Markus Kirkilionis*
October 2002 *Susanne Krömker*
Rolf Rannacher
Friedrich Tomi

Table of Contents

1 Interview with Willi Jäger
*Willi Jäger in discussion with Markus Kirkilionis, Susanne Krömker,
Rolf Rannacher, and Friedrich Tomi* 1

**2 Spatio-Temporal Dynamics
of Reaction-Diffusion Patterns**
Bernold Fiedler, Arnd Scheel 23
2.1   Introduction and Overview 23
2.2   One Space Dimension: Global Attractors 27
    2.2.1 Lyapunov Functions, Comparison Principles,
         and Sturm Property.. 27
         Lyapunov functions 27
         Comparison principles 29
         Sturm property, revisited 30
    2.2.2 Sturm Attractors on the Interval.......................... 33
         Global attractors 33
         Sturm attractors and Sturm permutations 34
         Sturm permutations and heteroclinics 38
         Combinatorics of Sturm attractors 41
    2.2.3 Sturm Attractors on the Circle 45
         Poincaré-Bendixson theory 45
         Heteroclinic connections of rotating waves 47
2.3   One Unbounded Space-Dimension: Travelling Waves 51
    2.3.1 Unbounded Domains and Essential Spectra 51
         From bounded to unbounded domains 51
         Spectra of travelling waves: group velocities and Fredholm
            indices 53
    2.3.2 Instabilities of Travelling Waves 60
         Instability of a front caused by point spectrum 61
         The Turing instability 61
         Essential Hopf instability of a front 64
         Instability of a pulse caused by the essential spectrum 67
         Fredholm indices and essential instabilities 67
         Spatial dynamics and essential instabilities 70

- 2.3.3 From Unbounded to Large Domains: Absolute Versus Essential Spectra 74
- 2.4 Two Space Dimensions: Existence of Spiral Waves 80
  - 2.4.1 Kinematics and its Defects 80
    - Curvature flow of Archimedean spirals 82
    - The front-back matching problem 83
  - 2.4.2 Archimedean Spiral Waves in Radial Dynamics 86
    - Rigid rotation and asymptotic wavetrains 86
    - Linear and nonlinear group velocities 88
    - Characterizing Archimedean spirals 89
  - 2.4.3 Bifurcation to Spiral Waves 91
- 2.5 Two Space Dimensions: Bifurcations from Spiral Waves 93
  - 2.5.1 Phenomenology of Spiral Instabilities 93
  - 2.5.2 Meandering Spirals and Euclidean Symmetry 95
    - Euclidean equivariance 96
    - Relative center manifolds 97
    - Palais coordinates 98
    - Spiral tip motion, Hopf meandering, and drift resonance 99
    - Relative normal forms 102
    - Relative Hopf resonance 104
    - Relative Takens-Bogdanov bifurcation 105
  - 2.5.3 Spectra of Spiral Waves 107
    - The eigenvalue problem for spiral waves: core versus farfield .. 107
    - Spatial Floquet theory and the dispersion relation of wavetrains 109
    - Relative Morse indices and essential spectra of spiral waves .. 111
    - Absolute spectra of spiral waves 114
    - Point spectrum and the shape of eigenfunctions 116
  - 2.5.4 Comparison with Experiments 118
    - Meander instabilities 118
    - Farfield and core breakup 120
- 2.6 Three Space Dimensions: Scroll Waves 123
  - 2.6.1 Filaments, Scrolls, and Twists 123
    - Spirals, tips, and Brouwer degree 124
    - Scroll waves, filaments, and twists 124
  - 2.6.2 Generic Changes of Scroll Filament Topology 127
    - Generic level sets 128
    - Sturm property, revisited 129
    - Comparison principle and nodal domains 130
    - Annihilation of spiral tips 131
    - Collisions of scroll wave filaments 131
  - 2.6.3 Numerical Simulations 134
- References 140

3 Some Nonclassical Trends in Parabolic and Parabolic-like Evolutions

Paul Fife 153
3.1   Introduction 153
3.2   The Simplest Nonlocal Parabolic-like Evolution
      and its Relatives 154
      3.2.1   Comparison Between the Local and Nonlocal Equations 156
      3.2.2   Models from Statistical Mechanics 157
      3.2.3   Related Nonlocal Evolutions 158
      3.2.4   Digression on the Role of Gradient Flows in Modeling 158
      3.2.5   The Issue of Discontinuous Profiles in the Nonlocal Problem 160
3.3   The Simplest Pattern-Forming Parabolic Equation............. 160
      3.3.1   Overview 160
      3.3.2   Spinodal Decomposition in Higher Dimensions.............. 162
3.4   Layer Phenomena Related
      to the Cahn-Hilliard Equation 164
      3.4.1   The Slowness of Some Motions 164
              Phenomena in 1D...................................... 164
              Bubbles and such 165
      3.4.2   Reduction to the Mullins-Sekerka Problem 165
      3.4.3   Further Reductions: Ripening 166
3.5   Patterning Due to Competition
      in General Gradient Systems................................ 168
      3.5.1   An Abstract Setting................................... 168
              Examples 169
              Threshold results 170
              Properties of the minimizers 170
              Restriction to real-valued functions 170
      3.5.2   Conserved Evolutions 171
      3.5.3   A Paradigm 171
3.6   Ginzburg-Landau Energies with Nonlocal Additions 171
      3.6.1   A Prototypical Inverse Elliptic Reduction 172
3.7   Free Boundary Reductions................................... 173
3.8   Another Kind of Competition 175
      3.8.1   Models for Copolymers 176
3.9   Conclusion 177
References 178

4 Mathematical Aspects of Design of Beam Shaping Surfaces in Geometrical Optics

Vladimir Oliker.. 193
4.1   Introduction 193
4.2   Creating a Prescribed Intensity Distribution
      in the Far-Field 195

        4.2.1 Statement of the Problem................................ 195
        4.2.2 Weak Formulation of the Problem 196
        4.2.3 Strong Solutions of the Reflector Problem................. 200
        4.2.4 Existence, Uniqueness and Regularity 200
        4.2.5 Computational Methods 203
              The method of supporting paraboloids (SP method)......... 203
        4.2.6 Open Problems 204
  4.3 Creating a Prescribed Intensity Distribution
      in the Near-Field 205
        4.3.1 Statement of the Near-Field (NF) Reflector Problem 206
        4.3.2 Weak Formulation and Solution
              of the NF Reflector Problem 207
        4.3.3 Some Open Problems 210
  4.4 Two-Reflector System for Transforming
      a Beam of Parallel Rays 210
        4.4.1 Statement of the Problem................................ 211
        4.4.2 Properties of Reflectors R_1 and R_2...................... 214
        4.4.3 Weak Formulation and Weak Solutions 215
        4.4.4 Regularity and Numerics 218
  4.5 Two-Reflector System with a Point Source.................... 219
  References 222

5 Recent Developments in Multiscale Problems Coming from Fluid Mechanics
Andro Mikelić 225
  5.1 Homogenization of Flow Problems in the Presence
      of Rough Boundaries and Interfaces............................. 226
        5.1.1 Wall Laws at Rough Boundaries 226
              Introduction 226
              Navier's boundary layer 228
              Justification of the Navier's slip condition for the laminar 3D
                    Couette flow 230
        5.1.2 Drag Reduction and Homogenization 236
        5.1.3 Law of Beavers and Joseph.............................. 238
              Introduction 238
              Modeling of the experiment by Beavers and Joseph 241
              Navier's boundary layer 243
              Justification of the law by Beavers and Joseph............. 245
  5.2 Interactions Flow-Structures 249
        5.2.1 Introduction 249
        5.2.2 Biot's Model Without Dissipation........................ 252
        5.2.3 Biot's Model with Dissipation 259
  References 264

6 From Molecular Dynamics to Conformation Dynamics in Drug Design
Peter Deuflhard 269
6.1   Introduction 269
6.2   Classical Molecular Dynamics 270
     6.2.1   Hamiltonian Differential Equations 270
     6.2.2   Condition of Molecular Initial Value Problems 271
            Example: Trinucleotide ACC. 272
6.3   Metastable Conformations as Almost Invariant Sets 272
     6.3.1   Perron–Frobenius Operator 274
     6.3.2   Stochastic Transition Operator 274
     6.3.3   Perron Cluster Analysis (PCCA) 276
6.4   Approximation of the Transition Operator 280
            Example: HIV protease inhibitor VX-478. 284
6.5   Perspectives 286
References 286

7 A Posteriori Error Estimates and Adaptive Methods for Hyperbolic and Convection Dominated Parabolic Conservation Laws
Dietmar Kröner, Marc Küther, Mario Ohlberger, Christian Rohde 289
7.1   Introduction 289
7.2   A Posteriori Error Estimates
     for Scalar Hyperbolic Conservation Laws 291
     7.2.1   Cell Centered Finite Volume Approximations 292
     7.2.2   Staggered Lax-Friedrichs Approximations 295
7.3   A Posteriori Error Estimates for Weakly Coupled Systems 298
            The finite volume scheme. 299
7.4   Numerical Experiments 302
     7.4.1   Transport of Contaminants with Degradation 302
7.5   Conclusion 304
References 305

8 On Anisotropic Geometric Diffusion in 3D Image Processing and Image Sequence Analysis
Karol Mikula, Tobias Preußer, Martin Rumpf, Fiorella Sgallari 307
8.1   Introduction 307
8.2   Review of Related Work 308
8.3   Anisotropic Geometric Diffusion on Still Images 311
8.4   Processing Image Sequences
     via Coupled Anisotropic Geometric Diffusion 315
8.5   Local Curvature and Motion Evaluation 316
8.6   Finite Element Discretization 317
References 320

9 Population Dynamics: A Mathematical Bird's Eye View
Odo Diekmann, Markus Kirkilionis 323
9.1 The Chemostat 323
9.2 Consumer-Resource Interaction 324
9.3 Competition for Substrate in the Chemostat 327
9.4 A Chemostat Containing a Food-Chain 328
9.5 Infectious Agents and the Art of Averaging 331
9.6 Heterogeneity 333
    9.6.1 Heterogeneity Deriving from Physiological Differences...... 333
    9.6.2 Heterogeneity Deriving from Spatial Position 334
    The gradostat and the creation of niches 335
9.7 The Pecularities of Semelparity 335
9.8 Concluding Sermon 336
References 337

10 Did Something Change? Thresholds in Population Models
Frank Hoppensteadt, Paul Waltman 341
10.1 Introduction 341
10.2 Mathematical Background on Bifurcations 343
10.3 Disease Thresholds 347
    10.3.1 Kermack-McKendrick 347
    10.3.2 Schistosomiasis 350
10.4 Predator-Prey Systems 352
    10.4.1 The Basic Model 353
    10.4.2 Subcritical Bifurcation 356
    10.4.3 Bifurcation from a Limit Cycle 359
10.5 Chaos 361
    10.5.1 Iterating Reproduction Curves 361
10.6 Random Perturbations of Ecological Systems 364
    10.6.1 Lotka-Volterra Model with Random Perturbations 365
    10.6.2 The Basic Model with Random Perturbations 371
10.7 Summary 372
References 373

11 Multiscale Modeling of Materials – the Role of Analysis
Sergio Conti, Antonio DeSimone, Georg Dolzmann, Stefan Müller, Felix Otto 375
11.1 Introduction 375
11.2 Soft Magnetic Films 377
    11.2.1 Micromagnetics 378
    11.2.2 Thin Film Limit 380
    11.2.3 Numerical Results and Comparison with Experiment 383

11.2.4 Discussion 385
11.3　Nematic Elastomers 386
　　11.3.1 Microscopic Model 387
　　11.3.2 Quasiconvexification 391
　　11.3.3 Finite–Element Computations 396
　　11.3.4 Attainment Results 399
　　　　Attainment and non–attainment for Dirichlet boundary
　　　　　conditions. 399
　　　　Attainment for a Dirichlet–Neumann problem............... 401
　　11.3.5 Discussion and Perspectives 404
References 406

Appendix. Color Plates... 409

1 Interview with Willi Jäger

Willi Jäger[1] in discussion with Markus Kirkilionis, Susanne Krömker, Rolf Rannacher, and Friedrich Tomi

[1]Interdisziplinäres Zentrum für Wissenschaftliches Rechnen der Universität Heidelberg, Im Neuenheimer Feld 368, D-69120 Heidelberg, Germany – jaeger@iwr.uni-heidelberg.de

Translated by Aminia Brueggemann

Summary. A while ago, we had the idea of publishing a book about our conference "Trends in Nonlinear Analysis, TiNA 2000". As the book will be published in honor of the 60th birthday of Willi Jäger, we decided to include an interview that would highlight his biography as well as his view of mathematics. We tried to find good questions. On April 30, 2000, the interview was conducted spontaneously and without prior knowledge of the questions. The interview is roughly organized as follows: His own biography, the current situation of mathematics, science organization, and the representation of mathematics in the public eye, and finally, perspectives, and visions. The questions were posed in the following order: Rolf Rannacher (RR), Susanne Krömker (SK), Friedrich Tomi (FT), Markus Kirkilionis (MK).

RR: It is a long road from minimal surfaces and harmonic mappings to chemical reactions and porous media. How did this come about?

Of course, it has been a long road. However, I did not begin with minimal surfaces but rather with the Schrödinger equations. For a long time, I had been working on scattering theory for the Schrödinger equation. Because I wished to work on a physical problem, my teacher Erhard Heinz simultaneously offered me both the topic and a job. This is how I changed from physics to mathematics, where I was able to study the Schrödinger equations, as I would have done within the theoretical physics.

Discussions with Erhard Heinz and Friedrich Tomi in Göttingen brought me to minimal surfaces. In those days, we lived through a time during which a real breakthrough in the analysis of minimal surfaces occurred. In 1969, when I went to the Courant Institute, I was of the opinion that I would occupy my time with minimal surfaces. I was very aware that – for example – Cathleen Morawetz or Ludwig worked on scattering theory. However, when I arrived at the Courant Institute, I was the only one studying minimal surfaces. Later, Bombieri, who just had discovered excellent results for graphs of minimal surfaces, arrived as a visiting scientist. Minimal surfaces are related to harmonic mappings, which are described by partial differential equations with strong nonlinearities. Similarly, a variational problem leads to such equations.

It is a variational problem that yields surfaces with prescribed non-vanishing average curvature as solutions. Before I moved from Münster to Heidelberg, I worked mainly on that topic.

When I came to Heidelberg, I immediately noticed that there was a special research area (SFB[1] 123) *Stochastic Mathematical Models*, but only as a corps which needed to be resuscitated. Together with Wilhelm von Waldenfels and Hermann Rost the resuscitation was a success! One day, I received a phone call from DFG[2] with the following question: "Would you be prepared to file an SFB-application, consisting of a concrete scientific program and its budget? If you do this, DFG will be prepared to accept the proposal and to fund it. If this is not possible, the already designated, but deferred SFB 123 will be eliminated." At that time, I was the managing director of the institute for applied mathematics. What could be done in such a situation? I certainly could not include the research of minimal surfaces or the solution of variational problems, leading to harmonic mappings, in this SFB, because those were subjects of the SFB in Bonn. We had to figure out which concrete problems offered themselves. In those days, we had already begun to get interested in variational problems; in problems with nonlinearities, which did not necessarily came from the minimal surfaces, and in geometric problems, all of which lead to systems of partial differential equations with similar strong nonlinearities. One example is the model equations by Keller-Segel for chemotaxis, which also have square terms in the derivations and likewise were outside the framework of common theory. If one only thought about the applications, then it became a natural step to go from systems of elliptic equations to parabolic equations. The applications – which we finally intended to consider in Heidelberg – were generally dynamic. In brief, my research was influenced – among other factors – by the necessity to define a new goal. We could not simply duplicate what was happening in Bonn, or we would not have been able to push the SFB through. Many methodical aspects remained the same. But because of the changing applications – especially from chemistry and biology – new ground was broken on the methodological level.

SK: Which teachers, scientists, and colleagues have mainly influenced you? Who are your role models?

Without doubt, Erhard Heinz's influence was extensive. I would like to say the same for Fritz Tomi. Through his work, his mathematical sharpness and strength, his content-rich lectures, Erhard Heinz had a decisive affect on his co-workers and students. Perhaps, less in his decisions of what one should do: we as his Assistants could decide about our own research topic – a big advantage. Of course, his advice and criticism were very important. His mathematical comments were always appropriate and right on target. He also impressed

---

[1] Sonderforschungsbereich – SFB special research area
[2] Deutsche Forschungsgemeinschaft – DFG is the central public funding organization for academic research in Germany

Fig. 1.1. From left to right: Rolf Rannacher, Markus Kirkilionis, Willi Jäger, Susanne Krömker, Friedrich Tomi

though the rigor of his work: He was very accurate and very meticulous – which was, apart from his discipline, quite important. So, Heinz has influenced me very much. I would also like to mention Martin Kneser. I did not attend the introductory lectures by Erhard Heinz, but rather the ones given by Martin Kneser, who then was a young man. Already during my studies, his lectures strongly influenced me. Naturally, among my teachers were also physicists who influenced me, for example the Sommerfeld disciple Bopp, who gave very mathematical lectures, but also other scientists from mathematics and physics have influenced me during my studies. Later, I aligned myself with the famous analysts at the Courant Institute. Although I worked there only for one year, and returned – as originally planned – to Göttingen. Exceptional scientists such as Courant, Friedrichs, John, Lax, Morawetz, Moser, Nirenberg and their younger colleagues at the Institute influenced me deeply. For example, Friedrich's lectures were very important to me. I had a close personal contact to Richard Courant, whom I frequently visited in New Rochelle during the weekends. During our many walks, he made many valuable comments and gave tips, which especially influenced me in the manner of research and teaching.

I am very pleased that I later had the opportunity to built personal relationships with exceptional mathematicians. Among the older generation, two people come to mind, Heinrich Behnke and Herbert Seifert, who also exerted a strong influence on me, although they worked in different mathematical areas. I was lucky to meet such important mathematicians, to have lengthy discussions – of course not discussions between specialists but talks with more seasoned colleagues, who had lengthy scientific experience. Later I met Olga Oleinik and Olga Ladyzhenskaya, whose research influenced my group and me very much. Especially important for my development was the cooperation with Stefan Hildebrandt and his group. This cooperation, ongoing since 1968, is the basis for many contributions to the progress of nonlinear analysis in Germany. It also included the exceptional Italian analysts. The cooperation was mathematically characterized by calculus of variation, the theory of minimal surfaces, and the existence and regularity theory of elliptic problems. The *Mathematische Forschungsinstitut Oberwolfach* played an important role for all of us, providing the ideal atmosphere and the necessary support.

I could mention an entire list of mathematical friends and colleagues who have influenced me through our cooperation. I am glad that so many of them contributed to TiNA's success with presentations or by acting as session chairs. Several of these friendships have their roots in the years I spent in New York, Minneapolis or Salt Lake City. The relationships with some Eastern Europeans took decades to develop, before the Iron Curtain opened. For many years, Prague as a meeting place had an important function.

FT: After you came to Heidelberg you have remained here until today. What keeps you here?

Well, let's first talk about all the reasons, which brought me to Heidelberg.

My first full professorship was in Münster, a city that in fact we liked very much. For personal reasons, however, we wanted to move to the south in the long run. I received offers from Marburg and Heidelberg. I made my decision for Heidelberg, because there existed a scientific environment, which fit me better. In addition, another factor was the pure mathematics – colleagues such as Dold, Puppe and Roquette, and the applied mathematics with colleagues such as von Waldenfels and Romberg as well as the younger members Rost, Krieger and Müller, whom I personally knew from my studies in Munich.

Twice there was a concrete temptation, to move to Bavaria. In fact, it had always been my dream to return to Bavaria, especially to Munich. When I received the call to the TU Munich, I compared the scientific situation, and found positive arguments for remaining in Heidelberg as well as for leaving Heidelberg. Arguments for Munich were the chance of a new beginning with all its challenges and the very pleasant attitude of all partners in Munich, especially the cooperation with the colleagues Bulirsch and Hoffmann. Arguments for Heidelberg were the proven cooperation, especially in the new center IWR[3]. I had the impression that IWR had gone beyond the growing

---

[3] Interdisziplinäres Zentrum für Wissenschaftliches Rechnen der Universität Heidelberg – IWR Interdisciplinary Center for Scientific Computing

pains, so that my departure would not have caused any damage. In hindsight, this might not have been completely accurate, because in those days the financing and the equipment of IWR had only been partially secured, und the subsequent support were essentially tied to negotiations with me remaining there. I did not find this quite fair, as this connection did not allow me to feel completely free in some decision-making processes. The decision was finally made for other reasons: The move to Munich would have been too expensive, at least for me.

We also had learned to appreciate that in Heidelberg, a much more accessible city, personal contact can be easier realized. Heidelberg possesses research institutions with a high, scientific status, Munich has them as well of course. But when there are obstacles such as the purchase of a home, then it becomes difficult to give up the familiar. If those obstacles had not existed, I would probably have succumbed to the temptation.

MK: For every scientist is it is quite important to have foreign contacts, as we already mentioned. Which foreign contacts are nowadays very important for you?

For years, the contact to American scientists has been good. The scientific cooperation within Europe has become stronger over the years. Because of European cooperation, the contact to France became more intense, especially to Lyon and Paris, and to Italy, especially Milan and Pisa. We are participating in European networks such as the network *Homogenization and Multiscale Systems 2000*. We are in the process of organizing educational graduate programs with Warsaw and Milan. We are also planning *Euronet-Bioquant*, in which the application of mathematical methods in the biosciences will be promoted throughout Europe.

For me personally, the scientific contact to Eastern Europe is always important, among others the contact to Prague, to Warsaw, to Moscow, to St. Petersburg, and Bratislava. Why? First, there are reasons of quality, interesting high-quality mathematics happens there. Of course, exciting mathematic also happens in e.g. Western Europe, the United States and in Japan. The second reason is that I personally feel committed to go to the East, and to foster or initiate contacts. For this reason, I have often regretfully declined an invitation from a Western colleague.

RR: How is your self-perception as a scientist? Do you have a motto for your scientific work?

First of all, I believe science should be fun. If you do not enjoy something you might as well stop it. Furthermore, it is very important for me not to view mathematics as being isolated. I try to view mathematics within the scientific context, i.e. mathematics in relation to physics, chemistry, biology, and other fields. It is an important for me to encourage these contacts. Originally, I wanted to study philosophy of nature. But soon I realized that this was not a beneficial undertaking because the discrepancy between the cognitive methods in philosophy and the natural sciences seemed unbridgeable. In addition, I have always tried not to loose my ties to the humanities. So, number one: Science must be fun, and mathematics is fun if you have time for it. One has to take this time if needed.

Fig. 1.2. Willi Jäger

SK: Which scientific results make you proud?

If one asks me spontaneously, and I view the different scientific areas in front of my eyes in fast motion, then perhaps the result that makes me most proud is a maximum principle for harmonic mappings. As everyone knows it is not easy to get good estimates of the maximum norm of solutions to systems of elliptic or parabolic partial differential equations. Together with Helmut Kaul, I have completely adapted such a principle to the underlying geometry. It truly was a breakthrough. When the idea turned into a proof, it was as if everything became clear. But also what we have accomplished in cooperation with Andro Mikelić in the area of homogenization, the study of transition conditions between different scales, for example the Navier-Stokes equation, satisfies me thoroughly. I believe the final results surpass in their importance several things that were accomplished in different parts of my work. Naturally, the judgment of quality is also time-depended.

Earlier I would have said that the results of the scattering theory would be very important. But these results were gained a long time ago. I believe that I was able to contribute with those studies to the inclusion of appropriate methods from functional analysis. In those days, one used integral equations and not the methods of Sobolev-spaces and imbedding theorems.

FT: After our look back into the past let's look towards the future. Which scientific goals are you trying to further and which results would you like to see in the near future?

I would like to gain a better understanding of the manner one can move from micro scales of processes such as flow, reactions, transport to a macroscopic description. One controls some of these processes well in media, which have a periodic structure. But this is not the case, if for example one has complex geometries such as branching, geometries, as come into existence through iterations of mappings. If one considers how heterogeneous media really are built, one notices very quickly that they have a stochastic geometry. I would like to understand the way stochastic, or even only deterministic, processes should be macroscopically described in stochastic geometry. For instance, how does one describe the flow in a capillary network?

I would thoroughly enjoy it if someone came to me and explained what a turbulent flow is. For a long time, I have attempted to understand what happens in a flow over a rough surface. All specialists say that the flow is turbulent in the most interesting and applicably relevant cases. If one only has results for the laminar situation, one always feels a little bit sniggered at. Under certain assumptions of fluid mechanics, we deducted recently the boundary layer equations, and the approximate boundary conditions for the flow at the rough surface including a turbulent situation. One assumes the essentials of Schlichting' boundary layer theory, that every turbulent flow develops a laminar boundary layer at the walls. Even its thickness is quantitatively determined. If the roughness is small in relation to its thickness, then we can show a Navier-type boundary condition for a smoother boundary. This yields a good approximation under the above stated assumptions. We reach Reynold numbers of several hundred thousands, as occur realistically, for example for swimming fishes such as a shark. Understanding the observed reduction of the drag by appropriate roughness poses a challenge. Indeed, we find a situation that we frequently encounter in theoretical mathematics: Under the assumption of a hypothesis another one is proven. In this case we begin with the assumption, that the boundary layer theory is valid, and prove under this assumption that the claimed law is valid.

MK: Science consists of different aspects: Our area of research, our committees – all this is important. How would you define your role in science? How do you understand your role in science?

One is not always free in making decisions. First of all, I understand my role as being an active scientist by gaining new insights, by deriving mathematical statements, opening new fields by mathematical modeling, by formulating the analysis of the models in such a way that the analysis can be numerically converted and the modeled processes can be simulated on the computer. For all that, I would like to have more time. Unfortunately no one can freely implement his goals as desired. Today, a single person can hardly solve scientifically demanding problems.

The moment one wants to have a team that is integrated in an adequate scientific environment, then, as a scientist, one has to deal with scientific management. I would like to add, that design of scientific tasks and projects is the fun part of scientific management, but the rest is everything but enjoyable – as we all found out just recently. The financial supporters should be included more closely. Without this the costly processes of filing an application, the review, reporting, and the administration are taken away from the resources of the productive scientific work.

Every responsible scientist has to be able to forgo something – as long as it remains reasonable. I have learnt the following: Often, it is necessary to pass on the chance to prove one or another mathematical theorem – a chance which I would have liked to take – to a younger co-worker, a student, a PhD candidate, in order to focus more on the big picture of research. That is not always easy but is a conscious sacrifice. It was my goal to promote science also with the young co-workers in the group, diploma and PhD candidates, to provide enough stimulation and problem sets in order for something new to develop. Through that, I hope science will progress. Perhaps the obtained results show that this is the right attitude. But I have to admit that sacrifice does not always come easy.

I never avoided committee work – I could give you many examples. However, it is certainly a desirable goal to limit committee work to a minimum.

Unfortunately, many so-called reforms often take us further away from this goal.

RR: In the next part, we remove ourselves from your person and turn to the general role of applied mathematics in sciences. How do you define the role of nonlinear analysis in the sciences and did this role become visible at the TiNA conference?

Perhaps I may answer the last question first: My view of the role of nonlinear analysis was quite visible at the TiNA conference. The conference covered different aspects of nonlinear analysis, its applied and theoritical aspects. Perhaps the applied aspects were more emphasized. Many scientists, all originating in the same essence and later evolving in different directions, have rarely met as they, for instance, have attended different conferences. I noticed that at this conference they began talking to each other again – also about scientific subjects.

But let me emphasize, that one should not forget the linear analysis that is contained in nonlinear analysis. Linear analysis is important in mathematical theory as well as in applications. In the mathematical theory, there is no doubt about the relevance of nonlinear analysis. A commentary is necessary for the area of applications. If one seriously pursues applications, as an analyst one quickly realizes that one enters into a complexity, which is not directly accessible with analytical methods. So one might question whether one needs analytical methods, if one cannot analytically treat high dimensional nonlinear systems – as they appear in applications. Already with systems of ordinary differential equations of first order, we see that everything causes problems when going beyond dimension two. Just bear in mind conservation laws: In nonlinear analysis, how far are we apart with our knowledge from that which we should in fact be able to control?

Here is my argument against it: It is indeed necessary to explain – at least in partial models – questions which arise in nonlinear problems, and the treatment of complex systems of equation, and to deduce the answers from the treatment of simple systems. If one has not understood the underlying phenomena of a system, one can evaluate the prescribed model equations with computers, but one cannot be sure, if the outcome really describes what is contained in these equations and what they pretend to describe. Naturally, I also think of error estimations, if one wishes to approximate solutions. But equally important is the analysis of the mathematical models by themselves. If one presupposes a model equation, one should know what is the nature of the solutions of the equations, and one should examine through analysis, if they describe the essential expected structure and characteristics, at least qualitatively. In this way, the quantitative calculation is not the first step. The interplay between the model formation, the analysis, and the numerics is of crucial importance.

Of course, the pure existence assertions remain important to theory, whereby the algorithmic access to the solution is preferred. Qualitative analysis of complex system is of essential importance. Real non-linear systems are generally controlled though many parameters. The high dimension of the parameter space requires preliminary analytical investigations, in order to compute solutions of partial differential equations in the decisive parameter area. One has to know where the interesting phenomena occur. The nonlinear analysis contributes with preliminary investigation to this.

SK: The title of our conference was "Trends in Nonlinear Analysis (TiNA)." Which trends are recognizable, or what do you consider to be milestones?

This question reminds me of a conversation with Charles Conley, during his last visit to Heidelberg. While leisurely walking to the Heidelberg castle, he asked me what would be, in my opinion, the biggest challenge of mathematics in the next years. I thought about that for a long time. Together we arrived at the same observation: The treatment of complex systems. Meanwhile, complex dynamics has become a central topic of research. Now, there even exists a Max-Planck-Institute for complex dynamics where, however, the processes in natural sciences is more emphasized than their mathematical analysis. New ideas and methods have to be developed in order to get a grip on this complexity, entering through nonlinearity, number of system variables or stochastics.

Multi scale approaches follow an important trend. We, in Heidelberg, belonged to the first mathematicians in Germany, who dealt with questions of multi scale analysis. In the meantime, multi scale approaches assert themselves everywhere. The trend is not limited to analysis, but it also gains importance in numerical methods. This is a development I regard as being positive. In the coming years, I expect a strong interaction between the modeling of systems, the analysis and their numerical simulation that will contribute to fundamental advancement. I have to add: Attention! While the analyst quickly writes down an epsilon, the computation has to find the epsilon as an intrinsic scale, to which we should help more.

It is a significant challenge and chance for mathematics to build bridges between the different scales as in the biosciences, from molecules and molecule complexes, from genes and proteins to cells, cell connections, tissues, organs, and organisms, to ecological system and finally to biospheres. Methods of computer science do not suffice for this goal. In Germany, if one takes a closer look at the departments and funding institutions one notices the need to overcome ignorance and prejudice, and to expose the life sciences to mathematical methods. The same is true for economics and finance, although there the development is more developed.

FT: What should a prospective scientist tackle in the area that we discuss? Is the time of scientific individualists over?

I do not believe that at all. I hope that scientific individualists are always to be found, because they are the icing on the cake. Original mathematicians, orginal scientists, create important impulses because of their surpassing current fashion trends. Of course, the selections of research areas are subjectively determined and their identification is not simple and demands experience and vision. While I believe that the time of *isolated* scientists is over, this does not mean that individuality has to be abandoned. Scientific cooperation is more and more project-oriented. Naturally everyone should bring to the project one's own profile; otherwise no real valuable cooperation is possible.

How should a future scientist behave? First, he should receive a solid mathematical training. The requirements will be different in the future. He should certainly be interdisciplinary, as once he realizes that he needs a basic knowledge not only in almost all mathematical areas, but also in computer science, physics, chemistry, biosciences, and economics.

When someone approaches me as a candidate for a thesis, my first question is: What do you dream about? What would you like to do? The answer to this question is important. For a diploma thesis it is fine not to have any concrete ideas, because generally he is still searching. If someone, who wishes to get a PhD, has no concept and limits himself to the following request: "Please give me a topic" he has already disqualified himself. Recently a student with the right attitude approached me and wanted *pattern formation in biology* as a topic for his master thesis.

If I should assign a project in my own field, I would currently choose from two areas of application-oriented analysis for which I foresee a big future: One is the connection between analysis and computation of multi-scale problems, and the second area is the investigation of diffusion and reaction processes, in flow in complex media, or in media with complex geometry. The stochastic geometry has been well developed. However, if one considers processes on such geometric objects, very few results are available.

If one focuses on the applications, then one notices that the cooperation with mathematics has already been accepted in physics. Since 1978, we were able to strengthen cooperation between mathematics and chemistry in Heidelberg. The big challenge now lies with the biosciences. There I envision major challenges for mathematicians, especially young mathematicians. An important subject is the derivation and simulation of model equations in nano- and microstructures. The mathematical formulation of models for such processes is essentially still in the works. This is not only true for the biosciences but also for micro technology.

MK: The next question fits well into this context: Discrete models versus differential equations. Which approach withstands the demands of reality better?

I do not see the opposition formulated in this question. I consider – if such a comparison is at all meaningful – a discussion of "discrete" versus "continuous" for more appropriate.

In continuous analysis, we often use also integral and functional equations, not only differential equations. In my view, in the applied analysis, but also in many other areas, one should focus first on the problem and the solution of the problem and then on the methods. This is more valid for the applied analyst than for the numerical analyst, who develops numeric algorithms for a class of problems. This may sound trivial, but it is not trivial. One has to establish which question should be answered for which problem. The answer determines the methods to be used. It may be that one has to work in a discrete setting, but it can also turn out to be more appropriate to work continuously or even using a mixed description. Finally, when we want to compute something, we will work discrete – if I can say it like that – more algebraic than geometric.

RR: In future, will computer simulation dominate in modeling?

Probably not. Because each computer simulation is built on a modeling, it automata its simulation. Computer simulations are important. Quantitative statements usually require numerical solutions of model equations. Therefore, I understand simulation rather as the computation of model equations of a process. There are also computer simulations, where no well-formulated model exists. Doing so, one has to pay attention to the understanding of the causal contexts and not only the production of pictures. A computer simulation is included in most of my projects that investigate models, because the complexity of the result demands it. The concept of *Scientific Computing* represents both areas, the development of algorithms as well as the assessment, by means of experiments, of the results.

MK: It is a question of future trend. Will it still increase?

Yes, with certainty, the use of computers will still increase. I would like to formulate the question differently. One could also think about describing processes though algorithms, i.e. using discrete rules. In future, it could be that one does not insist so much on describing the problems in basic equations. Let's take the engineer who deals with the question, of how an implantation affects the human jawbone in the long run. If one asks him about the model, one may receive the following answer: I use a finite-element-model. He refers to a specific algorithm, which is formulated in finite elements. But he does not say – what a mechanical engineer would have to say – which laws of elasticity he assumes. In the future, it could be that algorithms rightly will move more into the foreground. Cellular automates are based on algorithms, which create certain patterns. Another example is the so-called L-Systems[4], with which one can create plants on the computer. No modeling of the growth has been successful for the creation of geometric patterns.

The description of processes by means of algorithms will become stronger; so far it has been neglected.

SK: You have mentioned many examples of interdisciplinary work. People often claim that interdisciplinary research is so soft because one can always move on to the discipline that is unknown to the other. Is there some validity to this criticism?

If one looks around, then one can partially become critical. In an area, which I wish would be more represented in Germany – mathematical biology – this often happened. When one asked about the relationship between an analysis and an experiment, one often received this answer: I cannot say anything about that, I am a mathematician. If one looked at the mathematical part just by itself, then one often did not find the standard of mathematical theory, as it should have been, had the biological context not remained in the foreground. The reaction to the criticism was a referral to the biological application. The criticism of the continuous jumping back and forth between two excuses is valid. This criticism does not relate to interdisciplinary research but rather to individual scientists, who do not deal with it in a competent manner.

For interdisciplinary research, the same remains valid that we already have established for scientific computing: Quality standards have to be developed. These standards cannot be the same as the standards in theory or in experiment. In some areas it would be impossible to accomplish interdisciplinary work, if one were to apply the standards of mathematical theories. How far would we have come in interdisciplinary research in the area of flows if we had waited until the important questions would have been solved by proofs, e.g. questions for Navier-Stokes in three dimensions or for the Euler equations in several dimensions. If a proof is not being offered or cannot be offered, then this has to be clearly explained, omissions are to be shown clearly and hypotheses have to be obvious.

FT: After these more critical comments how would you define good applied mathematics?

Application-oriented mathematics differs from a more theoretical mathematics as follows: It usually begins with an application, a problem area, and it uses this problem area as a test case for the development of a theory, an approach to solve the problem. A good applied mathematician should not limit

---

[4] Lindenmayer-Systeme – Lindenmayer systems

himself by picking up a model equation from somewhere and by working on it, but rather the application by itself should remain in the center of one's vision. I do not expect that someone has to perform his own experiments. For example, if he works on a specific chemical reaction, he should integrate real data as far as they are available, and should they be missing, he should encourage additional experiments. The application-oriented mathematician should not redefine problems, as Goethe claims: *The mathematicians are some kind of French: If one talks to them, they translate it into their own language, and then it is something completely different.*[5]

He should at least do justice to the applications by giving himself a test problem. He should not evade but rather commit himself. Otherwise, he should rather focus on theoretical mathematics, adapt to its criterions and let himself be measured by them.

Application-oriented mathematics, if it pursues real applications, mostly encounters situations where the necessary theory is missing. For example, take the transport of particles in a flow. The necessary regularity theory has not yet been developed for the case of colliding particles.

MK: In your long experience as representative of SFB 123 *Stochastic Mathematical Models* you have strongly promoted the dialog between different sciences and mathematics and finally you put your energy into the creation of the Interdisciplinary Center for Scientific Computing (IWR). Nowadays, how do you perceive the mutual fertilization?

Between whom? Between natural science and mathematics? In my opinion, the reciprocal interaction is very strong. Certainly, Heidelberg's landscape has been essentially changed through SFB 123 and the current SFB 359[6], in physical chemistry. Wolfrum's or Grunze's experimental groups have substantially contributed to the mathematical modeling and the simulation of combustion processes or rather catalytic surfaces. Many things would not have flourished as well in Heidelberg, if the close cooperation between mathematics, chemistry, and physics had not been in existence.

The cooperation has always existed between physics and mathematics. Unfortunately, I have to say that mathematical physics, as it was generally carried out within mathematics for many years, had nothing to do with physics. There existed investigations dealing with some special functions, whose importance I do not underestimate, or some specific differential equations. However, during the last years the necessary interaction between the disciplines has been reestablished.

---

[5] Johann Wolfgang von Goethe: *Die Mathematiker sind eine Art Franzosen: redet man zu ihnen, so übersetzen sie es in ihre eigene Sprache, und dann ist es alsobald ganz etwas anderes.*

[6] Sonderforschungsbereich 359: *Reactive flow, diffusion, and transport*

With building the Interdisciplinary Center for Scientific Computing, while developing scientific computing as its own interdisciplinary field, it was our goal to combine strengths from different sciences, with the use of mathematical methods and the use of computer in sciences. One had to seek synergetic effects that accelerate scientific and technological progress. I believe, through such initiatives and research work Heidelberg contributed substantially to the development of new fields such as computational science and scientific computing. Several times, an international committee of experts has acknowledged IWR's international leading role in interdisciplinary research. The importance of mathematization and of quantative descriptions of biosciences is not yet generally appreciated. This is not true in Heidelberg. Just recently, the planning of BIOQUANT has begun, a central building for the research of stronger quantitatively oriented biosciences. With this undertaking, a strong basis was created for collaborations between mathematics, scientific computing, and biosciences. I view this development as very positive and visionary.

Years ago at IWR, we did not dare to use the term *scientific and technical computing*, because the interaction with industry was viewed with suspicion. Meanwhile many things have happened: The relationship of mathematics with industry has improved through the BMBF[7] support program in mathematics and industry, by all means to the advantage of mathematics.

At this point, I would like to emphasize what I have already mentioned: One has to take great care that quality does not suffer. Quality will have to be assessed differently, because the demands are different, but the desire for quality has to remain.

RR: I would like to insert a question. What is mathematical quality? I am not quite sure about that. How can that be measured?

Mathematical quality is certainly not related to how complicated the mathematical methods are. Even here, the faculty has quarreled about this. A measurement could be whether a difficult problem can be answered well and appropriately by the mathematical method. Naturally, one has to agree on the definition of a difficult problem. The easier the method of solution the better is the treatment of the problem. Simplicity is a fundamental principle in the description of nature. We prefer the laws, which are easier; we also prefer the aesthetically better descriptions.

Undoubtedly, criterions are simplicity, a certain mathematical aesthetics, and of course scientific rigor. Also in the area of interdisciplinary research, one has to point out existing gaps or used hypothesis – if one works mathematically, yet cannot work accurately enough.

---

[7] Bundesministerium für Bildung und Forschung – BMBF Federal Ministry of Education and Research

RR: On several occasions, we have used the term modeling. What in fact is modeling? Is modeling already an own area within mathematics or the natural sciences. Finally, can modeling be taught?

Modeling is – if I may formulate it a bit more abstract – the representation of a real process or a situation in mathematical abstract structure with the goal of describing the occurring processes. That means at the same time: If I formulate an abstract model I will have to explain how I will test it. The representations are not always reversible. The task to repeatedly compare the model, one has set up, with nature, with the real processes, is an essential component.

There are different points of views about the way to develop such models. Certainly, once again, simplicity of the model is guiding. If one aims to answer a question, one includes only what is needed in a model for answering the question. An important fact is often being overlooked: At some places granularity is resolved too finely, although it is not necessary. Modeling is a discipline where one has to have a lot of mathematical experience, but also experience in application. In any case, let me say: It is not so easy to teach. There is hardly a textbook about mathematical modeling. There are certain techniques, which one can teach such as perturbation methods and asymptotic expansions. Not before dealing with modeling did I encounter these areas. Before then, they certainly did not belong to my favorite areas. It became necessary that I taught it to myself. Model development has to be a component of the education, but it is not easy to teach because several scientific areas have to be familiar, in order to bring across the necessary information.

SK: The next question links up very well with the previous discussion: How much that is non mathematical a mathematician should know?

If one performs model development, I believe it is important to have knowledge of physics. The reason is simple: The fundamental principles are taken from physics, i.e. physical chemistry uses the model techniques from physics. This is my advice to anyone who wishes to do modeling, even if they the do financial or economic models: You have to know what speed, and acceleration is, or you will have no clue about the procedures. You also have to know what a diffusion process is. One can only comprehend this by using physical notions. The most simple model developments come from physics.

SK: How should interdisciplinary work be organized?

I do not believe that organization alone is sufficient for successful science. One cannot expect that. The close collaboration between partners is crucial in interdisciplinary research. It is essential that it is fun to work with neurobiologists or with biophysicists. And to recognize their way of thinking and the way they approach problems in neuroscience.

One should organize opportunities for interdisciplinary dialog and create workshops that bring scientists together e.g. in a center for scientific computing. In this way, they are encouraged to talk with each other and to initiate common projects. When they sit in their individual departments, this is more complicated. Establishing such centers, which focus on interdisciplinary research, can lead to extensive progress.

RR: Money, does money play a role?

Of course, money plays an important role. If the SFB 123 had not been financed, there would be no IWR. The *Deutsche Forschungsgemeinschaft* (DFG) deserves a lot of credit for the development of interdisciplinary research in Heidelberg. There would be no IWR, if the DFG had not helped with the establishment of IWR. The representatives of DFG immediately understood the underlying concept and its impact. I hope this will happen again if modeling and simulation should be increasingly used in biosciences. Without the strong support of DFG it is likely that it will not be possible to realize the underlying concept of BIOQUANT. The support of mathematical methods in the biosciences is urgently necessary. For this we need financial support for personnel as well as for equipment. Outside pressure is necessary to implement the appropriate work at the college level. The view, that this already has happened with the support of bioinformatics, is wrong.

FT: If one wants to promote interdisciplinary work, what are the consequences for the education of future scientific generations?

The combination of mathematics and computer science during the studies does not suffice, because both fields are basic sciences. I always advised students to combine their mathematical education with a chosen field of application. It is very important to have a solid education in at least one of the subjects. Good knowledge of computer science, and, especially, of the application area is very important for the coming mathematical generation. Fairly early on, students should be guided towards non-mathematical problem sets in seminars, diploma thesis and in internships. This is a big advantage for anyone, who plans to leave the university, because at the university they already learn what they hopefully will need later on. I say hopefully, because that what they have to work on outside of the university may not correspond to their expectations during their university education.

MK: And should departments remain organized in that way?

There is a big need for reform. If one looks at research, it becomes clear that research does not limit itself to subjects. It is a fact that departments cannot direct research. Research follows its own paths. Scientists do not keep to subject borders. This has to be taken into consideration when restructuring scientific institutions and financing the universities. No obstacle to interdisciplinary research should be built.

Departments are institutions that are important for teaching, while doing that they should not erect narrow barriers. We need new structures for interdisciplinary studies. The tendency for specialization will have to be reversed on the departmental level. The entire college of natural sciences again gains in importance through new tasks. These new structures must remain flexible.

MK: Fine, let's move to a different area. It is often assumed that mathematics will perish due to its own complexity. It is increasingly difficult to gain an overview even over a small part. Will mathematics reach a standstill because of that? How do we cut this Gordian knot?

I do not believe mathematics will perish due to its own complexity. Perhaps it demands too much from some mathematicians. In any case, one has to distinguish between mathematics and the mathematician. For sure there is a danger that too little exchange happens between mathematical disciplines, because too complex languages have been developed. Of course, we wish these languages were simpler; in order to increase communication and to avoid the situation where someone cannot follow the lecture of a potential future colleague. Mathematics as a science might be in danger if it remained this way. The complexity of mathematical problems poses challenges and impulses for further development of mathematics, but no danger for its existence.

RR: Now, let's look at a broader picture. The image of the German university is based in a large part on Humboldt's ideal of combining research and teaching. In the context of the reform of higher education, his idea is again on the test bench. How can today's combination – as we experience it daily – of research, teaching and fundraising be viable?

First of all, I am still a proponent of Humboldt's idea. Once, I listened to a presentation by Ralf Dahrendorf, in which he argued against this university ideal. In this presentation, he stated his error in his estimation of the university reforms in 1968 – reforms, which he strongly influenced. I told him after the lecture: "You will err a second time, if you regard this idea as antiquated." By all means, there is a large discrepancy between the demands: The split between teaching, research, and fund raising – as it is nowadays demanded – cannot be maintained in the long run.

There is a simple approach to address this – of course it is costly. It consists of creating jobs for scientific management, in my view in every department or specific research units. The scientists could be relieved of several managerial tasks. Long nightshifts would not have to be used for the completion of applications for research projects, so that I do not even realize that the high frequency in my ear comes from a tinitus and not from a broken hard drive. A change of the current funding situation is necessary.

The combination of teaching and research has to remain. In my opinion, researchers, especially excellent researchers, should give lectures for beginning students. Who else could transfer the enthusiasm of science and the correct image of science to the students? They can give such lectures in simplified and not highly complex versions. If one leaves these lectures to scientifically secondary people, the impetus is often missing – an impetus which is perhaps more important than a faultless blackboard. I do not agree with the objection that not all students wish a scientific career. It is my experience – once again confirmed at the presentation of the *Alwin Walther Medaille*[8] – that the influence of university professors is tremendous, it can last for decades.

SK: An increasing lack of studying abilities of the beginning students is being noticed, especially with missing mathematical and natural scientific knowledge, as became clear in the most recent studies. What is you diagnosis and what are your proposals for a cure?

Certainly, the quality of the beginning students in the area of natural sciences and mathematics leaves much to be desired. It appears to me this is worse in physics than in mathematics. One cannot simply expect anymore from a student to comprehend mathematics from a physical example. The high school education in physics is lacking. The curriculum does not correspond to the requirements expected by a university professor. Unfortunately we have to acknowledge that the politicians who are responsible for the education do not consult with the scientists. They do consider the advice of pedagogues, but the expertise of the scientists is not utilized. Scientific representatives should take part in a required discussion of the content of the curriculum and teaching methods. To fill the gaps between schools and the universities is a task, which we in Heidelberg have also consequently tried to address.

It is noticeable that many teachers are needed as social workers as well as mathematics teachers. Teachers have to be better trained – in the basics of their subject as well as in pedagogy. This education is not finished with the examination but rather must be continuously supplemented. At the moment, the discussion about schools is on the agenda. I have my doubts about many diagnoses but even more about some proposals for the cure. The education discussion has a central role and we as university professors have to be a part of it.

---

[8] Alwin Walther Medaille der Technischen Universität Darmstadt – Scientific award in the fields of Informatics and Applied Mathematics of the Technical University of Darmstadt, Germany

FT: As a final question to this subject, let's look at an even broader picture: Mathematics and public. The writer Enzensberger compared mathematics with a fortress whose drawbridge is closed. Is this a correct comparison?

The drawbridge has opened a little bit. Hans Magnus Enzensberger has a very positive attitude towards mathematics, yet he does not spare mathematicians in his criticism. It was necessary to let down the drawbridge, because a mentality of the ivory tower was cultivated to some degree. The editing of mathematical statements and results, so that others could understand them, was neglected.

In his lecture on the occasion of the *International Congress of Mathematicians*[9] in Berlin, Enzensberger described a discussion between a mathematician and his friend, who wished to understand his work. The mathematician regretted, that he could not explain to him what he does, because this would be too complex and could not be illustrated in a simple and correct fashion. His Friend answered: Just lie a little bit.

I do not believe, one should lie but one should attempt to explain it in a more simplified version. Mathematicians have to take greater care in allowing their work and results to be understood by a broader public. In addition, they have to penetrate better into the areas where mathematics is in demand. A lot has happened to improve the relationship between mathematics and industry, and we – in Heidelberg – have been able to aid in this endeavor. Within the universities, the drawbridge should be let down. As mathematicians, we are not permitted to think of ourselves as being so exceptional, that everyone completely accepts our reason for existence. Only when we are able to present mathematics openly, can we secure its importance.

MK: What kind of answers can mathematics give to the modern world and its problems? Can mathematics actually provide such answers or will mathematics pose new questions?

Thank God, mathematics always poses new questions. Mathematics also has the ability to help uncover still unsolved problems, and to formulate problems correctly, e.g. to describe the biosciences not only qualitatively but also quantitatively. When we see how the mathematization grows in all areas, for example in economics, we notice the growing need for mathematics. When I use the term mathematics, then I do not necessarily mean mathematicians, because non-mathematicians can also do mathematics. Much inspiration for the development of mathematics came from physics, where mathematics was at first practiced without worrying too much about a theoretical background. Here discoveries have been made which mathematicians might not have made, had

---

[9] International Congress of Mathematicians – ICM 98, Berlin, August 18–27, 1998

the discoveries not been in the direct context of concrete problems. "Mathematics, key technology for the future[10]", that is the title that we chose for a volume about BMBF supported scientific projects in industry. This title expresses my conviction: Nowadays, less and less works without mathematics, what has to get into the computer needs mathematics. One might call it something different, but it is mathematics. Many things from computer science, if it is done well and correctly, are mathematics. And I believe more and more fields will profit from mathematical methods. However, we do need a core mathematics that works in areas, where it might not be so foreseeable what the final result are and how they can be used in specific disciplines. One should not forget: Mathematics delivers essential contribution to culture through its insights and discoveries. Mathematics should not only be measured using a utilitarian point of view. Mathematics is also part of culture.

RR: With this background in mind, in which direction should mathematics develop in the 21. Century?

This question has implicitly already been answered: It should not loose its identity. We should not be afraid of it. Among theoretically oriented, so-called pure mathematicians, I sometimes feel this fear, which I consider incorrect. If the theoretical contribution is really good, it will become clear that it is needed. I always defend myself against a differentiation between pure and applied mathematics, because it suggests that theoretical mathematics is isolated and has little or no consequences for applications. This is nonsense. Theoretical mathematics cares more about theory development. As mathematicians, one of our future tasks is to open up even more, and to open up mathematics without loosing its identity. If we look at the job opportunities, it becomes clear that mathematicians hold top positions. We will see that tendency also in the future, because they are needed, since mathematics is needed.

FT: One last question, which will be answered, very quickly – I believe. Today, would you still study mathematics?

I believe I would study mathematics again. Perhaps something different, but I would certainly study mathematics again.

---

[10] K.-H. Hoffmann, W. Jäger, T. Lohmann, H. Schunk (Eds.): Mathematik, Schlüsseltechnologie für die Zukunft, Springer Verlag, Heidelberg, 1997

2 Spatio-Temporal Dynamics of Reaction-Diffusion Patterns

Bernold Fiedler[1] and Arnd Scheel[2]

[1] Freie Universität Berlin, FB Mathematik I, Arnimallee 2–6, 14195 Berlin, Germany – *fiedler@math.fu-berlin.de*

[2] University of Minnesota, School of Mathematics, 206 Church St. S.E., Minneapolis, MN 55455, USA – *scheel@math.umn.edu*

2.1 Introduction and Overview

In this survey we look at parabolic partial differential equations from a dynamical systems point of view. With origins deeply rooted in celestial mechanics, and many modern aspects traceable to the monumental influence of Poincaré, dynamical systems theory is mainly concerned with the global time evolution $\mathcal{T}(t)u_0$ of points u_0 — and of sets of such points — in a more or less abstract phase space X. The success of dynamical concepts such as gradient flows, invariant manifolds, ergodicity, shift dynamics, etc. during the past century has been enormous — both as measured by achievement, and by vitality in terms of newly emerging questions and long-standing open problems.

In parallel to this development, the applied horizon now reaches far beyond the classical sources of celestial and Hamiltonian mechanics. Applications areas today include physics, many branches of engineering, economy models, and mathematical biology, to name just a few. This influence can certainly be felt in several articles of this volume and cannot possibly be adequately summarized in our survey.

Some resources on recent activities in the area of dynamics are the book series Dynamical Systems I - X of the Encyclopedia of Mathematical Sciences [AnAr88, Si89, Ar&al88, ArNo90, Ar94a, Ar93, ArNo94, Ar94b, An91, Ko02], the Handbook of Dynamical System [BrTa02, Fi02, KaHa02], the Proceedings [Fi&al00], and the fundamental books [ChHa82, GuHo83, Mo73, KaHa95].

In the context of partial differential equations, the phase space X of solutions $u = u(t, x)$ becomes infinite-dimensional: typically a Sobolev space of spatial profiles $u(t) = u(t, \cdot)$. More specifically, the evolution of $u(t) = \mathcal{T}(t)u_0 \in X$ with time t is complemented by the behavior of the x-profiles $x \mapsto u(t,x)$ of solutions u. Such spatial profiles could be monotone or oscillatory; in dim $x = 1$ they could define sharp fronts or peaks moving at constant or variable speeds, with possible collision or mutual repulsion. Target patterns or spirals can emerge in $\dim x = 2$. Stacks of spirals, which Winfree called scroll waves, are possible in $\dim x = 3$. For a first cursory orientation in such phenomena and their mathematical treatment we again refer to [Fi02] and the article of Fife in the present volume, as well as the many references there.

For our present survey, we focus on the spatio-temporal dynamics of the following rather "simple", prototypical parabolic partial differential equation:

$$u_t = D\Delta u + f(x, u, \nabla u). \tag{2.1}$$

Here $t \geq 0$ denotes time, $x \in \Omega \subseteq \mathbb{R}^m$ is space and $u = u(t, x) \in \mathbb{R}^N$ is the solution vector. We consider C^1-nonlinearities f and constant, positive, diagonal diffusion matrices D. This eliminates the beautiful pattern formation processes due to chemotaxis; see [JäLu92, St00], for example. Note that we will not always restrict f to be a pure reaction term, like $f = f(u)$ or $f = f(x, u)$. More general than that, we sometimes allow for a dependence of f on ∇u to include advection effects. The domain Ω will be assumed smooth and bounded, typically with Dirichlet or Neumann boundary conditions, or else unbounded, typically $\Omega = \mathbb{R}^m$, an unbounded cylinder, or a half-space. At any rate, assumptions will be such that questions like global forward existence, uniqueness, and smoothness of weak solutions $u(t, x)$ for prescribed initial conditions $u_0(x) = u(t = 0, x) \in X$ will not constitute a problem; see [He81, Pa83]. To emphasize the dynamical systems aspect of dependence on initial conditions $u_0(x) = u(t = 0, x)$, we frequently write $\mathcal{T}(t)u_0 = u(t) = u(t, \cdot)$ for the strongly continuous solution semigroup $\mathcal{T}(t)$ of (2.1) on X.

Within this general setting, we devote the following chapters to a discussion of increasingly specialized spatio-temporal patterns in increasing space dimensions $m = \dim x = 1, 2, 3$. Our approach is motivated to some extent by classical dynamics in finite-dimensional spaces: while flows in $\dim X = 2$ admit the beautifully simple Poincaré-Bendixson theory, present in any good textbook on ordinary differential equations, some of the complications arising in $\dim X = 3$ are illustrated by the Lorenz attractor. Chapter 2.2 is devoted to single equations, $N = \dim u = 1$, in one space dimension, $m = \dim x = 1$. On bounded interval domains Ω, the gradient-like dynamics is governed by a decreasing Lyapunov-functional, and by additional nodal properties, which were first discussed by Sturm; see Section 2.2.1. In Section 2.2.2 these Sturm properties provide a rather complete combinatorial characterization of the global PDE attractors $\mathcal{A}_f$ associated to (2.1). The case of a circle domain $\Omega = S^1$, in some sense an amphibium like $m = \dim x = 1.5$, destroys the gradient-like Lyapunov structure, in Section 2.2.3, but preserves the Sturm structure. We sketch a new description of the global attractors for this case, which involves rotating waves on the circle and their heteroclinic connections. It is worth noting that individual orbits, but not the arbitrarily high-dimensional global attractor $\mathcal{A}_f$, do satisfy a Poincaré-Bendixson theorem in this case. Adding a linear, but nonlocal rank-1 perturbation can destroy this simple Morse-Smale structure completely, allowing for embeddings of arbitrarily complicated dynamics. Chapter 2.2 can also be read as a very classically-minded warm-up for the much more advanced discussion of gradient flows in the article of Fife in the present volume.

In the unbounded limit $\Omega = \mathbb{R}$, travelling waves can arise along with various instabilities; see Chapter 2.3. Stability considerations are greatly complicated — and phenomena as well as their mathematics greatly enriched — by the presence of continuous spectrum. As a prelude to our treatment of spiral wave bifurcations, in Chapter 2.5, we discuss dispersion relations and characterize Fredholm properties of linearizations at fronts in one space dimension in Section 2.3.1. The role of group velocities is then illustrated in a Hopf bifurcation from travelling waves caused by continuous spectrum in Section 2.3.2. We return to bounded but large domains $\Omega = (-L, L)$ in Section 2.3.3, where we investigate linear and nonlinear stability of fronts in the limit $L \to \infty$.

Chapter 2.4, dwelling on planar domains $\Omega \subseteq \mathbb{R}^2$, addresses the issue of spiral waves. Following [WiRo46], spiral patterns arise naturally in systems ($N = 2$) of excitable media type. As a link to geometric dynamics and mean curvature flow, we begin in Section 2.4.1 with a negative result: Archimedean rotating spirals cannot be described by a reduced model of curve-shortening type. Some nonlocal effects or extra input describing the tip motion, or other relics of the system-character $N = \dim n = 2$ of the viscous approximation to the geometric model, have to be added. In Section 2.4.2, we develop a characterization of Archimedean spiral wave solutions to reaction-diffusion systems of N species, valid for excitable and oscillatory media, but independent of singular limiting regimes such as fast relaxation kinetics or small amplitude oscillations. Our characterization ensures robustness of spiral waves when system parameters are varied. A bifurcation result in Section 2.4.3 provides us with an open class of reaction diffusion systems, where spiral waves are actually proved to exist.

Beyond rigidly rotating Archimedean spiral patterns, spiral waves generate a variety of spatio-temporal structures. In Chapter 2.5 we attempt a mathematical analysis based on bifurcation theory. In Section 2.5.1, we phenomenologically describe some of the patterns that may arise, building on experimental observations. We address in some detail meandering and drifting spirals, Doppler-induced super-spirals, and spiral breakup in the core and farfield. In Section 2.5.2 we outline a symmetry approach, which accurately describes observed meandering cycloid bifurcations from the unperturbed circular tip motion of rigidly rotating spiral wave patterns. The elementary Euclidean group $SE(2)$ of rotations and translations in the x-plane causes these motions by Hopf bifurcation transverse to the group orbit: a prime example of symmetry breaking under noncompact Lie groups. However, this simplistic analysis ignores the possible presence of continuous spectrum. In Section 2.5.3 we characterize spectral properties of spiral waves in large and unbounded domains. We focus on two major consequences of the presence of continuous spectrum in the unbounded domain. First, the shape of eigenfunctions to isolated eigenvalues is largely governed by the complex dispersion relation of the wavetrains in the farfield. Second, instabilities arising through the continuous spectrum come in two steps. At onset, the instability is convective in nature: perturbations decay pointwise, and the instability disappears after a possibly long transient. Only after driving the instability further by increasing the bifurcation parameter beyond threshold, the instability manifests itself in bounded domains. The discussion of spectra, here, is in strong analogy to the simpler setup in one space-dimension, Chapter 2.3. We conclude Chapter 2.5 relating spectral theory to the experimentally and numerically observed spiral wave patterns.

Chapter 2.6, where $m = \dim x = 3$, discusses scroll wave dynamics. Following [Wi73], scroll waves can be viewed as stacks of rotating spirals with tip positions $x = z^t(\tau)$ aligned along filament curves $\tau \mapsto z^t(\tau) \in \Omega \subseteq \mathbb{R}^3$, for (most) fixed times t. Filaments z^t can be knots with parameter $\tau \in S^1$. Moreover the spirals, which appear in spatial sections transverse to filaments $z^t(\tau)$ at τ, define a local phase angle $\varphi^t(\tau) \in S^1$ there. We relate knotting orientations of filaments z^t, and "twist", alias the winding of φ^t, in Section 2.6.1. For generic initial conditions $u_0(x) = u(t = 0, x)$, it is possible to describe all scenarios changing the knotting or linking topology of scroll wave filaments z^t; see section 2.6.2. We conclude with a discussion of some simulations of spatio-temporal scroll wave dynamics, in Section 2.6.3.

Acknowledgments. We are grateful for the hospitality of the program of the Twente Summer School and of the Seminars of DMV (German Mathematical Society) at Oberwolfach, where these results have first been compiled. This work was substantially supported by the Deutsche Forschungsgemeinschaft, in particular by the Sonderforschungsbereich "Complex Nonlinear Processes", DFG Priority Research Programs "Ergodic Theory, Analysis, and Efficient Simulation of Dynamical Systems" and "Analysis, Modeling, and Simulation of Multiscale Problems". For generous support of several research visits both authors are indebted to the Institute of Mathematics and its Applications (IMA) at Minneapolis. The first author would like to thank Regina Löhr for patiently typesetting quite a few versions of this manuscript. We are also indebted to many colleagues who have helped us along the way, most notably Sigurd Angenent, Pavol Brunovský, Klaus Ecker, Giorgio Fusco, Jack Hale, Jim Keener, Yannis Kevrekidis, John Mallet-Paret, Alexander Mikhailov, Waldyr Oliva, Peter Poláčik, Carlos Rocha, Björn Sandstede, Art Winfree, and Matthias Wolfrum.

Most of all, we feel deeply indebted to Willi Jäger, who untiringly created and maintained an inspiring work atmosphere at Heidelberg with such lasting influence on his scientific children and grandchildren.

2.2 One Space Dimension: Global Attractors

2.2.1 Lyapunov Functions, Comparison Principles, and Sturm Property

In this chapter, we consider various special cases of reaction-advection-diffusion systems of the general form

$$u_t = D\Delta u + f(x, u, \nabla u), \qquad (2.2)$$

$\Omega \subseteq \mathbb{R}^m$, with $u \in \mathbb{R}^N$, $f \in C^1$, and Dirichlet, Neumann, or mixed boundary conditions; see (2.1). Mostly we will restrict our attention to the special case of a scalar equation, $N = 1$, in one space dimension $m = 1$. We begin our discussion, however, by recalling the larger, overlapping domains of validity for the concepts of gradient flows or gradient-like dynamics, as given by decreasing Lyapunov functions, and of monotonicity or comparison principles.

Lyapunov functions We first consider the variational structure which makes system (2.2) an L^2-gradient flow. We assume the vector nonlinearity $f = f(x, u)$ to be independent of the term ∇u. In addition, we assume $f \in \mathbb{R}^N$ to be the gradient of a scalar potential $F = F(x, u)$ with respect to $u \in \mathbb{R}^N$,

$$F_u(x, u) = f(x, u). \qquad (2.3)$$

Then the energy functional

$$\mathcal{V}(u) := \int_\Omega \big(\frac{1}{2}(\sum_{j,k} \nabla u_j \cdot D_{jk} \nabla u_k) - F(x, u)\big)\, dx \qquad (2.4)$$

is a *Lyapunov function* for system (2.2), if the positive definite diffusion matrix D is symmetric. Indeed we observe the Lyapunov decay property:

$$\frac{d}{dt}\mathcal{V}(u(t)) = -\int_\Omega u_t^2\, dx \qquad (2.5)$$

is strictly negative, except at equilibrium solutions of the elliptic system $u_t = 0$. Of course we have to impose appropriate growth conditions on f and choose an appropriate Sobolev space X containing functions in $H^1(\Omega)$ with the chosen boundary conditions. Critical points of the energy functional $\mathcal{V}$ provide weak solutions of the associated elliptic equilibrium system $u_t = 0$ and are amply studied; see for example the books [GiHi96a, GiHi96b, MaWi89, St90, Ze85], and the many references there.

By LaSalles invariance principle, the monotone decay (2.5) of the Lyapunov functional $\mathcal{V}$ implies $\mathcal{V} \equiv const.$ on the ω-limit set

$$\omega(u_0) := \{v \in X;\ v = \lim u(t_n),\ \text{for some}\ t_n \to \infty\}, \qquad (2.6)$$

of any bounded solution $u(t) = T(t)u_0 \in X$ with initial condition $u = u_0$ at $t = 0$.

If we substitute the term u_t in (2.2) by Au_t, with uniformly positive definite $N \times N$-matrices $A = A(x, u, \nabla u)$, and substitute $D\Delta u$ by an x-dependent matrix diffusion term div $(D(x)$ grad $u)$ in divergence form, the same functional $\mathcal{V}(u)$ works as a valid Lyapunov function. Indeed, the Lyapunov decay property (2.5) remains valid with u_t^2 replaced by the nonnegative integrand $u_t^T A u_t$.

Another generalization of (2.4), due to [Al79], relaxes the very restrictive gradient condition (2.3) under Neumann boundary conditions and for x-independent $f = f(u)$. Since homogeneous x-independent solutions then satisfy the ODE system $\dot{u} = f(u)$, a gradient-like condition

$$F_u(u) \cdot f(u) < 0 \tag{2.7}$$

has to be imposed, except when $f(u) = 0$ provides a spatially homogeneous equilibrium already. An additional convexity condition then ensures that $\mathcal{V}$ becomes a Lyapunov function, decreasing strictly along any non-equilibrium solutions.

In the case of scalar equations, $N = 1$, the gradient condition (2.3) is of course trivially satisfied. Even in the one-dimensional case $m = 1$, however, with periodic boundary conditions $x \in S^1 = \mathbb{R}/\mathbb{Z}$, Lyapunov functions may fail to exist, if we let the nonlinearity $f = f(x, u, u_x)$ depend on the gradient term u_x:

$$u_t := u_{xx} + f(x, u, u_x). \tag{2.8}$$

In fact a particularly simple example arises for x-independent nonlinearities $f = f(u, u_x)$. Note S^1-*equivariance* with respect to x-shift in that case: with $u(t, x)$, also $u(t, x - x_0)$ is a solution of (2.8), for any fixed x-shift by $x_0 \in S^1$. We can then find *rotating wave solutions* of the form

$$u(t, x) = U(x - ct), \tag{2.9}$$

$c \neq 0$, which are provided by any nonstationary 1-periodic solution of the second order ODE

$$U'' + cU' + f(U, U') = 0. \tag{2.10}$$

Such solutions U are easily constructed, for suitable nonlinearities f; for example the class (2.10) contains the van der Pol oscillator. See [AnFi88] and Section 2.2.3 for more details. Clearly the existence of rotating waves (2.9) with time period $1/c$ contradicts the existence of a Lyapunov function $\mathcal{V}$ which would have to strictly decrease with time. Similar constructions work for thin annulus domains $\Omega \subset \mathbb{R}^2$ and for disks. In fact we could just transform any system with $f = f(u)$ to coordinates rotating at uniform angular speed $c \neq 0$. Any equilibrium solution which is not rotationally symmetric will then provide a rotating wave solution for suitable $f = f(x, u, \nabla u)$.

In view of such examples it is therefore perhaps surprising, and not widely known, that a Lyapunov functional of the form

$$\mathcal{V}(u) = \int_\Omega \gamma(x, u, u_x) \mathrm{d}x \qquad (2.11)$$

does exist for (2.8), in the case of an interval domain $\Omega = (0, 1)$ with separated boundary conditions of Dirichlet, Neumann, or Robin type; see [Ze68, Ma78]. See also the beautiful construction in [Ma88]. Again the Lyapunov decay property (2.5) remains valid with u_t replaced by Au_t, for some uniformly positive $A = A(x, u, u_x)$.

The construction of γ by Matano in [Ma88] is particularly appealing: γ is constructed via solutions of subsidiary semilinear hyperbolic equations, which defy periodicity conditions in x and thus quite appropriately fail to exist under periodic boundary conditions $\Omega = S^1$. For analogous constructions of Lyapunov functions in the spatially discrete case of Jacobi systems and for certain graphs of neural networks see [FiGe98, FiGe99].

Comparison principles *Monotonicity properties*, alias *comparison principles*, are the second major structural property of reaction-advection-diffusion equations (2.2). We refer to [Po02] for a recent survey and only sketch the main line of thought, briefly. Let $u_0(x), \tilde{u}_0(x)$ be two initial conditions in our phase space $X \subseteq C^0(\overline{\Omega})$, with associated (global) solutions $u(t) = T(t)u_0$, $\tilde{u}(t) = T(t)\tilde{u}_0 \in X$. The (strong) monotonicity property, or parabolic comparison principle, states that

$$u_0 \leq \tilde{u}_0 \Rightarrow u(t) < \tilde{u}(t), \text{ for all } t > 0. \qquad (2.12)$$

Here $\leq$ indicates that $u_0(x) \leq \tilde{u}_0(x)$, for all $x \in \Omega$, whereas $<$ indicates pointwise strict ordering inside Ω. See [Sm83, Wa70, PrWe67] for proofs of this property in a scalar parabolic context $N = \dim n = 1$, $m = \dim x \geq 1$. The proof is based on the elementary observation that the difference $w := \tilde{u} - u$ of the two solutions satisfies a nonautonomous linear parabolic equation of the form

$$w_t = D\Delta w + a(t, x)w + b(t, x)\nabla w \qquad (2.13)$$

for suitable coefficients a, b which depend on the solutions $u, \tilde{u}$. For example

$$a = \int_0^1 f_u(x, u + \vartheta(\tilde{u} - u), \nabla u + \vartheta(\nabla \tilde{u} - \nabla u)) \cdot (\tilde{u} - u) \mathrm{d}\vartheta \qquad (2.14)$$

where $\tilde{u}, u$ have to be evaluated at (t, x). Note $w_0 \geq 0$. From (2.13) it is immediately clear that any nondegenerate local x-minimum $w(t_0, x_0) = 0$, should it ever develop at some time $t = t_0$, immediately retracts to locally strictly positive values. Indeed (2.13) then implies $w_t = \Delta w > 0$ at (t_0, x_0). The rest is parabolic technique.

A very successful general framework for monotonicity properties has been initiated by Hirsch and Matano; see [Hi83, Hi85, Hi88, Ma86, Ma87]. For example, Hirsch has proved that most initial conditions, in the Baire sense, give rise to solutions $u(t)$ which, if bounded, converge to some equilibrium in X for $t \to +\infty$. In particular this implies that any hyperbolic periodic solution of a (strongly) monotone system is necessarily unstable. Let us briefly indicate a proof of this latter fact. Indeed the compact Floquet-operator is given by $w_0 \mapsto w(p) \in X$, with linearization along the periodic orbit of the form (2.13) and with p denoting the time period. The compact Floquet-operator therefore preserves positivity. By Perron-Frobenius theory, any eigenvalue of maximal modulus possesses a positive eigenfunction. For a periodic solution $u(t)$ with trivial Floquet multiplier 1, the trivial Floquet eigenfunction $w := u_t$ cannot possibly be positive. By hyperbolicity, the periodic solution therefore possesses another Floquet multiplier outside the unit circle and with positive eigenfunction. Instability ensues. For an excellent survey of more recent, deeper results in this context, in particular concerning generic convergence to *subharmonicity* for time-periodically forced monotone systems, see [Po02].

From a pure dynamics point of view, monotonicity is not a serious restriction. As Smale has pointed out, any Lipschitz flow in $\mathbb{R}^n$ can be embedded into $u \in \mathbb{R}^{n+1}$ — normally unstable, of course — such that the resulting system is monotone. Just view $\mathbb{R}^n$ as the hyperplane $u_1 + \cdots + u_{n+1} = 0$ and add the fast scalar equation $\varepsilon \dot{v} = v$ in the normal direction, for sufficiently small $\varepsilon > 0$. The resulting system will be monotone with respect to the componentwise order $\leq$ on $\mathbb{R}^{n+1}$. Of course, this argument somewhat neglects the spatial aspects of dynamics which are so relevant for PDEs. Moreover, instability does not necessarily indicate invisibility or irrelevance. Unstable solutions characterize basin boundaries of stable objects. And sometimes their time scales of instability, which are not at all affected by the monotonicity requirement, may comfortably exceed the life times of their observers.

Sturm property, revisited We now return to the scalar case, $N = \dim u = 1$, in one space-dimension, $m = \dim x = 1$ with nonlinearity $f = f(x, u, u_x)$, as specified in (2.8) above. We admit separated boundary conditions, $x \in (0, 1)$, with a resulting gradient-like flow, as well as periodic boundary boundary conditions, $x \in S^1 = \mathbb{R}/\mathbb{Z}$, which defy Lyapunov functionals and admit nontrivial time-periodic solutions. In either case, there exists a very powerful refinement of the above monotonicity structure: the *Sturm property*. For any continuous function $w : [0, 1] \to \mathbb{R}$ let the *zero number* $z(w) \leq \infty$ denote the number of strict sign changes of w. In other words z is the supremum of all n, for which we can find $x_1 < \cdots x_n$ such that the nonzero signs of $w(x_i)$ alternate. Let $w = w(t, x)$ denote a solution (2.13) in one space dimension

$$w_t = w_{xx} + b(t, x) w_x + a(t, x) w. \tag{2.15}$$

Clearly $z = z(w(t, \cdot))$, the number of sign changes of the x-profiles, may then depend on t. As far back as 1836, Sturm has observed that the zero number

$$z(w(t, \cdot)) \searrow_t \qquad (2.16)$$

is *nonincreasing* with t, at least for time-independent a, b. See [St36] and also [Po33].

Sturm has used his beautiful observation to prove that any nontrivial linear combination φ of eigenfunctions φ_j, $k \leq j \leq \ell$, of any Sturm-Liouville problem

$$\lambda \varphi = \varphi_{xx} + b(x)\varphi_x + a(x)\varphi \qquad (2.17)$$

on the interval $0 \leq x \leq 1$ possesses at least k and at most ℓ sign changes:

$$k \leq z(\varphi) \leq \ell. \qquad (2.18)$$

Here we number the eigenfunctions φ_j such that $z(\varphi_j) = j$. Defining $w(t, \cdot)$ to be the explicit exponential solution of (2.15), for all $t \in \mathbb{R}$, with initial condition $w = \varphi$ at $t = 0$, we indeed see how $w/|w|$ approaches eigenfunctions $\varphi_{j\pm}$, for $t \to \pm\infty$. Invoking (2.16), we obtain for some large enough $t > 0$ that

$$\ell \geq j_- = z(w(-t, \cdot)) \geq z(w(0, \cdot)) = z(\varphi) \geq z(w(t, \cdot)) = j_+ \geq k. \quad (2.19)$$

This single line proves (2.18).

It is easy to prove that (2.16) indeed holds, locally, at a nondegenerate double zero

$$w = w_x = 0 \neq w_{xx}, \quad \text{at } (t_0, x_0). \qquad (2.20)$$

Then (2.15) implies sign $w_t =$ sign w_{xx}, and an elementary argument shows that the zero number $z(w(t, \cdot))$ drops by 2, locally at $(t_0, x_0) \in (0, \infty) \times [0, 1]$. See also Figure 2.1(a). It is less obvious that z also drops *strictly* at a triple, or multiple zero, where only $w = w_x = 0$ is required; see Figure 2.1(b).

Proposition 2.2.1 *Let $w(t, x) \not\equiv 0$ solve the linear, nonautonomous equation (2.15) in one space dimension and with separated or periodic boundary conditions. Let the zero number $z(w(t, \cdot))$ denote the number of strict sign changes of the x-profiles $x \mapsto w(t, x)$, $x \in [0, 1]$.*

Then w satisfies the Sturm property, *that is*

(i) $z(w(t, \cdot))$ is nonincreasing with time t;
(ii) $z(w(t, \cdot))$ drops strictly, whenever $w(t, \cdot)$ possesses any multiple zero at any $x \in [0, 1]$;
(iii) $z(w(t, \cdot))$ is finite, for any positive t.

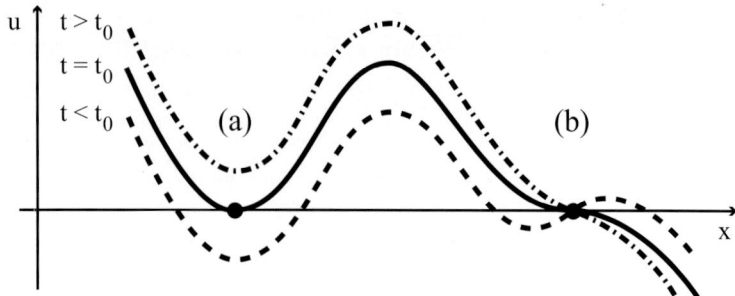

Fig. 2.1. Local dropping of $z(w(t,\cdot))$ at (a) a double zero, (b) a triple zero of $z(w(t_0,\cdot))$.

Note how the trivial case $z = 0$ of sign definite w corresponds to the monotonicity or comparison principle (2.12) discussed above. For solutions $w(t, x)$ which are analytic in (t, x), the proposition can be proved by Newton's polygon. In the autonomous case, this was Sturm's proof. For a modern version see [AnFi88]. The definitive version of this result, under rather mild regularity assumptions on a, b, was given by [An88]. We have explained above the renewed interest in the Sturm property for *nonautonomous* coefficients a, b: any difference $w = \tilde{u} - u$ of two solutions $u, \tilde{u}$ of the semilinear reaction-advection-diffusion equation (2.8) satisfies (2.15) and thus possesses the Sturm property.

We will see in the next two sections, how this simple additional Sturm property has enormous consequences for the description of the global spatio-temporal dynamics of (2.8). It has definitely been due to the deeply inspiring insight of Matano [Ma82], that these enormous nonlinear consequences have been first realized. We just illustrate the beginnings of his contributions with two further remarks here.

First, Matano's original idea was based on using the *lap number*, $z(u_x(t,\cdot))$, to characterize properties of solutions $u(t,\cdot)$ for nonlinearities $f = f(u, u_x)$ independent of x. Clearly this amounts to a choice $w = u_x$ and of course w satisfies the linearized equation (2.8), which takes the form (2.15). In particular, the Sturm property implies that the x-profile $u(t,\cdot)$ decomposes into finitely many intervals of strict monotonicity, for any positive t.

Second, we can use the zero number z and the Sturm property to prove convergence of any bounded solution $u(t, x)$ to a single equilibrium, for $t \to \infty$, for $x \in [0, 1]$, and under Neumann boundary conditions. Since $w = u_t$, like u_x, satisfies the linearized equation (2.15), even for nonlinearities $f = f(x, u, u_x)$ which do depend on x, the Sturm property implies that all zeros of u_t must be simple, eventually, for large $t \geq t_0$. But $u_{tx} \equiv 0$ at $x = 0$, due to the Neumann boundary condition. Therefore $u(t, x = 0)$ must be eventually monotone, hence convergent, and hence constant on the ω-limit set $\omega(u) \subseteq X$. Therefore $u_t \equiv u_{tx} \equiv 0$ on $\omega(u)$ and $z(u_t)$ would have to drop there, all the time. Well, can it? No way: by the Sturm property, again, we see that $u_t \equiv 0$ on $\omega(u)$. Hence $\omega(u)$ consists entirely of equilibria $v = v(x)$. But $v(x = 0) = \lim_{t\to\infty} u(t, x = 0)$ is fixed on $\omega(u)$. Together with the Neumann boundary condition $v_x(x = 0) = 0$, this specifies $v \in \omega(u)$ uniquely: the ω-limit set always consists of just a *single* equilibrium. Note how this argument provides a second proof of the gradient-like behavior of the scalar, one-dimensional reaction-advection-diffusion equation (2.8), which is completely independent of the approach by Zelenyak and Matano [Ze68, Ma78, Ma88] via Lyapunov functions (2.11).

2.2.2 Sturm Attractors on the Interval

In this section and the following one we describe some progress during the last 15 years in our understanding of global attractors of the scalar, one-dimensional reaction-advection-diffusion equation

$$u_t = u_{xx} + f(x, u, u_x). \tag{2.21}$$

The present section addresses the interval case $0 < x < 1$ under separated boundary conditions; to be completely specific we consider Neumann boundary

$$u_x = 0 \quad \text{at} \quad x = 0, 1. \tag{2.22}$$

For periodic boundary conditions $x \in S^1 = \mathbb{R}/\mathbb{Z}$ and nonlinearities $f = f(u, u_x)$ see Section 2.2.3.

Global attractors For an excellent recent survey on the theory of global attractors, including topics like dimension estimates, inertial manifolds, determining modes and applications to retarded functional differential equations, damped hyperbolic wave equations, and the gradient-like parabolic equations studied here, we refer to the book by Chepyzhov and Vishik [ChVi02], which includes the nonautonomous case, and to the article by Raugel [Ra02]. For earlier work on global attractors see for example the books [Ha88, Te88, La91, BaVi92] and the many references there.

Abstractly, the concept is the following. Recall that $\mathcal{T}(t)u_0 := u(t)$ denotes the solution semigroup of (2.21) on a suitable Banach space X of x-profiles $u(t) = u(t, \cdot) \in X$. We require $f \in C^2$ to be *point dissipative*, that is, there exists a (large) ball in X, in which any solution $\mathcal{T}(t)u_0$ stays eventually, for all $t \geq t_0 = t_0(u_0)$. Explicit sufficient conditions on f require, for example,

$$f(x, u, 0) \cdot u < 0 \tag{2.23}$$

for all sufficiently large $|u|$, along with uniformly subquadratic growth of f in the variable u_x. Condition (2.23) provides an L^∞-bound for u, whereas subquadratic growth prevents blow-up of u_x. Since $\mathcal{T}(t)$ is a compact, point dissipative semigroup, in the sense of the references quoted above, the *global attractor* $\mathcal{A} = \mathcal{A}_f$ of (2.21) can be equivalently characterized as follows:

$$\begin{aligned}\mathcal{A} &= \text{the smallest set attracting all bounded sets} \\ &= \text{the largest compact invariant set} \\ &= \text{the set of all bounded solutions } \mathcal{T}(t)u_0 \in X,\ t \in \mathbb{R}\end{aligned} \qquad (2.24)$$

Here *attractivity* is understood in the sense that $\mathcal{T}(t)\mathcal{B}$ stays in any arbitrarily small δ-neighborhood of $\mathcal{A}$, for any bounded set $\mathcal{B} \subset X$ of initial conditions and for all $t \geq t_0(\mathcal{B}, \delta)$. *Invariance* is understood in both forward and backward time. Negative time invariance of $\mathcal{A}$ requires the existence of a *past history* $u(-t) \in \mathcal{A}$, $t \geq 0$, of u_0, such that $\mathcal{T}(t)u(-t) = u_0$ for all $t > 0$. Similarly, the set of all bounded solutions is understood to consist of precisely those u_0 which possess a bounded past history, in addition to $\mathcal{T}(t)u_0$ being bounded uniformly for $t \geq 0$. We need not really bother about the mind-boggle of huge fans of past histories of u_0 here, because backwards uniqueness for parabolic equations implies that past histories are uniquely determined.

Under suitable dissipativeness conditions on the nonlinearity f, the above statements hold true in complete generality (2.1), including systems $u \in \mathbb{R}^N$, $N \geq 1$, higher-dimensional smoothly bounded $\Omega \subseteq \mathbb{R}^m$, $m \geq 1$, or the circle case $\Omega = S^1$. The gradient-flow property of (2.21) with Neumann boundary conditions (2.22), however, allows for an even more specific characterization of the global attractor:

$$\mathcal{A} = \mathcal{E} \cup \{\text{heteroclinics}\}. \qquad (2.25)$$

Here $\mathcal{E} = \{u_t = 0\} \subset X$ denotes the set of equilibria. We call $u_0 \in X$ *heteroclinic*, if there exists a solution $u(t), t \in \mathbb{R}$, including a past history, such that the distinct α- and ω-limit sets of u_0 for $t \to \pm\infty$ are both equilibria. We have already seen in section 2.1 how these sets will consist of a single equilibrium $U_\pm$, respectively, as $t \to \pm\infty$. The Lyapunov function $\mathcal{V}$, which only exists under separated boundary conditions, implies $\mathcal{V}(U_+) < \mathcal{V}(U_-)$ and hence $U_+ \neq U_-$: heteroclinicity. In fact *homoclinic* solutions with $U_+ \equiv U_- \not\equiv u(t)$ are possible for $x \in S^1$; see [SaFi92] and the remarks at the end of Section 2.2.3.

Sturm attractors and Sturm permutations For simplicity of presentation, we henceforth assume that all equilibria $U \in \mathcal{E}$ are *hyperbolic*: all eigenvalues λ of the Sturm-Liouville eigenvalue problem (2.17), arising by linearization at U, are nonzero. Such an assumption holds for most nonlinearities f, in a Baire sense: zero has to be a regular value of $u \mapsto u_{xx} + f$. Hyperbolicity implies that equilibria U are locally isolated and, by compactness of $\mathcal{E} \subseteq \mathcal{A}$, finite in number,

$$\mathcal{E} = \{U_1, \cdots, U_n\}. \qquad (2.26)$$

Moreover, they are accompanied by a local saddle structure, much in the geometric spirit of ODEs, characterized by local stable and unstable manifolds W_j^s, W_j^u at each equilibrium U_j. These invariant manifolds consist of those $u_0 \in X$ which remain in a neighborhood of U_j and converge to that same equilibrium in forward resp. backward time. Their tangent spaces at U_j are of course spanned by the Sturm-Liouville eigenfunctions φ of (2.17) with negative resp. positive eigenvalues λ. The unstable manifolds W_j^u in fact extend forward, globally, by the solution semigroup $\mathcal{T}(t), t \geq 0$. Noting that trivially $U_j \in W_j^u$, even if U_j is stable, the above characterization (2.25) of the global attractor therefore takes the equivalent form

$$\mathcal{A} = \bigcup_{j=1}^{n} W_j^u. \tag{2.27}$$

So much for generalities. As to the specific determination of global attractors $\mathcal{A}_f$ associated to nonlinearities f of the parabolic equation (2.21), and in fact for any gradient-like system, several basic questions arise:

Q0: What are the equilibria $U_1, \cdots, U_n$?
Q1: Given all equilibria, which pairs possess a heteroclinic orbit?
Q2: Given dissipative nonlinearities $f, \tilde{f}$, how can we determine whether or not their global attractors $\mathcal{A}_f, \mathcal{A}_{\tilde{f}}$ coincide?
Q3: How many different global attractors with n equilibria are there?
Q4: What is the geometry of the global attractors?

Note that **Q0** is an ODE question concerning the scalar second order equation

$$0 = U_{xx} + f(x, U, U_x) \tag{2.28}$$

with Neumann boundary conditions. See [Sc90] for a detailed analysis of the case $f = f(U)$. In contrast, question **Q1** – **Q4** address the PDE (2.2). We will now summarize some results which show how some essentially combinatorial information on the ODE question **Q0** is sufficient to answer the PDE questions **Q1** and **Q2**. In particular we will make precise what we mean by "coinciding" attractors. The challenging questions **Q3**, **Q4**, must remain open at this time.

Since our answers to **Q0** – **Q2** for the parabolic equation (2.2) are crucially determined by the Sturm property of solutions, Proposition 2.2.1, we call the associated global attractors *Sturm attractors*.

The combinatorial ODE information on the equilibria $U_1, \cdots, U_n$ has been ingeniously distilled by [FuRo91]. Under Neumann boundary conditions, let the equilibria $U_1, \cdots, U_n$ be numbered according to the strict ordering

$$U_1 < U_2 < \cdots < U_n, \quad \text{at } x = 0. \tag{2.29}$$

Indeed $U_x = 0$ at $x = 0$, together with the second order type of the equilibrium equation (2.28), implies that $i = j$, if $U_i = U_j$ at $x = 0$. Now define the *Sturm permutation* $\pi = \pi_f \in S_n$ by the corresponding, possibly different ordering at the other boundary point $x = 1$:

$$U_{\pi(1)} < U_{\pi(2)} < \cdots U_{\pi(n)}, \quad \text{at } x = 1 \tag{2.30}$$

To get used to Sturm permutations, let us quickly prove that $\pi(1) = 1$, $\pi(n) = n$. Indeed choose a very large positive or negative initial condition $u_0^\pm$: pointwise above resp. below any equilibrium U_j. By dissipativeness, the solutions $T(t)u_0^\pm$ must then each converge to some equilibrium $U^\pm$. By monotonicity (2.12), alias nonincrease of $z(u(t) - U_j) = 0$, that equilibrium must still lie above, resp. below, all other equilibria, pointwise. Hence $U^+ = U_n$ and $U^- = U_1$. This proves $\pi(1) = 1$, $\pi(n) = n$.

Question **Q2** was answered by [FiRo00] as follows.

Theorem 2.2.2 *Let $f, \tilde{f}$ be dissipative nonlinearities in (2.21), (2.22) such that their Sturm permutations π_f and $\pi_{\tilde{f}}$ coincide. Then the associated global attractors, the Sturm attractors $\mathcal{A}_f$ and $\mathcal{A}_{\tilde{f}}$, are C^0 orbit equivalent. In symbols*

$$\pi_f = \pi_{\tilde{f}} \quad \Rightarrow \quad \mathcal{A}_f \cong \mathcal{A}_{\tilde{f}} \tag{2.31}$$

We recall that C^0 orbit equivalence is given by a homeomorphism $H : \mathcal{A}_f \to \mathcal{A}_{\tilde{f}}$, in the topology of X, which maps f-orbits in $\mathcal{A}_f$ onto $\tilde{f}$-orbits in $\mathcal{A}_{\tilde{f}}$, preserving time direction.

We sketch some of the ingredients to the proof of this theorem. Of central importance in the *Morse-Smale property* on attractors $\mathcal{A}$ with hyperbolic equilibria: all intersections $W_i^u \cap W_j^s$ of stable and unstable manifolds are transverse, automatically, due to the Sturm property of the semiflow. This striking fact was discovered independently by [An86] and [He85]. An indirect zero number argument runs as follows. Consider a heteroclinic orbit $u(t) \in W_k^s$ from U_j to U_k. If transversality failed, we could choose a nontrivial solution $\psi(t)$ of the adjoint linearization, along $u(t)$, perpendicular to the tangent spaces

$$T_{u(t)} := T_{u(t)} W_j^u + T_{u(t)} W_k^s. \tag{2.32}$$

Letting $t \to -\infty$, we then see

$$u(\psi(t)) \geq i_j, \quad \text{for all } t. \tag{2.33}$$

Here i_j denotes the unstable dimension, alias the *Morse index* at the hyperbolic equilibrium U_j. Indeed the adjoint linearization is parabolic in backwards time and hence $z(\psi(t))$ increases with t. Sturm-Liouville theory then shows the inequality (2.33) "at" $t = -\infty$, because $\psi(t) \perp T_{u(t)} W_j^u$. A similar argument at $t = +\infty$ proves

$$z(\psi(t)) < i_k, \quad \text{for all } t. \tag{2.34}$$

Together, this implies $i_j < i_k$ for the Morse indices of the source U_j and the target U_k. Arguing for the heteroclinic tangent u_t itself, on the other hand, and using that $u_t/|u_t|$ converges to eigenfunctions for $t \to \pm\infty$, by [BrFi86], we see that the Sturm property for the linearization along $u(t)$ implies

$$i_k \leq z(u_t(t)) < i_j, \quad \text{for all } t. \tag{2.35}$$

This is a clear contradiction to $i_j < i_k$, and hence proves the Morse-Smale property.

For dynamical systems on compact manifolds, Palis and Smale have proved *structural stability* of Morse-Smale systems; see [Pa69, PaSm70]. Infinite-dimensional versions are due to Oliva; see [Ol02]. In our setting this implies the claim of Theorem 2.2.2, provided that $\tilde{f}$ is near f in the C^2-topology. It also proves the claim, if we can find a homotopy f^τ from $f^0 = f$ to $f^1 = \tilde{f}$, in the class of dissipative C^2-nonlinearities and parabolic equations (2.21), such that all equilibria in $\mathcal{E}_\tau = \mathcal{E}_{f^\tau}$ remain hyperbolic throughout this homotopy. Unfortunately, however, it is still not clear to us whether or not such a homotopy f^τ exists.

We have circumvented this difficulty by discretization within the class of *Jacobi systems*, also studied by Oliva, see [Ol02]. These are finite tridiagonal nonlinear systems

$$\dot{u}_i = f_i(u_{i-1}, u_i, u_{i+1}), \tag{2.36}$$

$i = 1, \cdots, \ell$, such that the off-diagonal partial derivatives of all f_i are strictly positive everywhere. Boundary conditions like $u_0 := u_1$, $u_{\ell+1} := u_\ell$ have to be imposed. Counting strict sign changes along the discrete positions $1 \leq i \leq \ell$, instead of $0 \leq x \leq 1$, provides a Sturm property which is completely analogous to the continuous parabolic case. The Morse-Smale property follows, likewise, from hyperbolicity of equilibria. Finite difference discretization of (2.21) for f and $\tilde{f}$, for example, leads to Jacobi systems ℓ large enough to ensure structural stability of the respective finite-dimensional global attractors under discretization.

But not all Jacobi systems are discretizations of parabolic equations. The additional freedom thus gained, adorned with an artificial unstable suspension trick which quadruples the dimension of the system, is sufficient to find a homotopy f^τ as described above, but on the discrete level. This completes our sketch of the proof of Theorem 2.2.2.

As an aside we note that the class $\mathbf{A}^J$ of (Morse-Smale) Sturm attractors of Jacobi systems coincides with the same class $\mathbf{A}^P$ for one-dimensional parabolic equations. Discretization shows $\mathbf{A}^J \subseteq \mathbf{A}^P$. The combinatorial characterization of possible Sturm permutations, given in Theorem 2.2.4 below, shows that $\mathbf{A}^J \subseteq \mathbf{A}^P$; see [FiRo00]. Similarly, the class $\mathbf{A}^P$ is independent of the choice of separated boundary conditions; see [Fi96]. Moreover, this suggests extensions to quasilinear or nonlinear parabolic equations: as long as finite-difference discretization to Jacobi systems prevails, no new attractors are to be expected. An open question, however, is realization of Sturm attractors $\mathcal{A}$ by Jacobi systems of minimal dimension. Let

$$\ell := \dim \mathcal{A} = \max_{1 \leq j \leq n} i_j \qquad (2.37)$$

denote the maximal Morse-index of equilibria $U_j \in \mathcal{A}$. Is it then possible to realize that same attractor $\mathcal{A}$, up to C^0 orbit equivalence, by a Jacobi system (2.36) of dimension only ℓ? A positive answer would provide the ultimate "qualitative" discretization.

Similar realization questions in the class A^P of Sturm attractors immediately arise when we restrict the class of admissible nonlinearities $f = f(x, u, u_x)$. For example it seems possible to exhaust the class A^P of Sturm attractors by "Hamiltonian" nonlinearities $f = f(x, u)$ alone [Wo02a]. On the other hand, consider nonlinearities with $f(-x, u, -p) = f(x, u, p)$, for all x, u, p. Then the reflection symmetry $x \mapsto 1-x$ immediately implies that π_f is an involution: $\pi_f = \pi_f^{-1}$. In particular, nonlinearities $f = f(u)$ which depend neither on x nor on u_x, explicitly, do not generate the class A^P completely. Indeed there exist Sturm permutations π_f with $n = 9$ equilibria, which contain 3-cycles and hence fail to be involutions; see Figure 2.2 below. The minimal such example requires $n = 7$ equilibria and is given by $\pi_f = (2, 4, 6)(3, 5)$.

Sturm permutations and heteroclinics We address question **Q1** next: how to determine all heteroclinics between equilibria? A first answer is given by the following theorem of [FiRo96, Wo02b].

Theorem 2.2.3 *The Sturm permutation $\pi = \pi_f$, defined by (2.29), (2.30) above, determines explicitly and constructively for all equilibria U_j, U_k, $1 \leq j, k \leq n$ of (2.21), (2.22)*

(i) the Morse indices (unstable dimensions) i_j, i_k of U_j, U_k
(ii) the zero numbers $z(U_k - U_j)$
(iii) whether or not there exists a heteroclinic solution $u(t)$ from U_j to U_k.

The "explicit and constructive" algorithm asserted by the theorem is slightly involved. For example

$$i_k = \sum_{j=1}^{k-1} (-1)^{j+1} \operatorname{sign}\left(\pi^{-1}(j+1) - \pi^{-1}(j)\right), \qquad (2.38)$$

and a similar formula holds for the zero numbers $z(U_k - U_j)$; see [FiRo96]. Note that indeed $i_1 = 0$ indicates stability of the minimal equilibrium U^- as was to be expected from dissipativeness and the monotone convergence to $U^- = U_1$ mentioned above. Similarly, $i_n = 0$. In particular, this implies that the total number n of equilibria is odd: just regard the $k-1$ entries ± 1 in the sum (2.38) which must add up to zero, for $k = n$. Alternatively, this oddness can easily be derived from Leray-Schauder degree theory of the equilibria, or from the classical Morse inequalities.

As for assertion (iii) of Theorem 2.2.3, we mention a *cascading principle* which is peculiar to Sturm attractors. Whenever there is a heteroclinic orbit from U_j to U_k, there exists a cascade of heteroclinic orbits from $U_{j_0} = U_j$ through $U_{j_1}, \cdots, U_{j_{\ell-1}}$ to $U_{j_\ell} = U_k$ such that the Morse indices drop by 1 along each heteroclinic in the cascade:

$$i_{j_\ell} = i_{j_{\ell+1}} + 1. \tag{2.39}$$

By transversality, each heteroclinic orbit in the cascade is just an isolated orbit, locally. It is in fact unique. Already the simple example of a gradient flow with respect to the height function on the standard Euclidean 2-sphere shows that the cascading principle fails for Morse-Smale flows, in general.

Transitivity, in contrast, is a general principle for Morse-Smale flows: if there are heteroclinics from U_{j_1} to U_{j_2} and from U_{j_2} to U_{j_3}, then there also exists a direct heteroclinic from U_{j_1} to U_{j_3}. The proof is based on the λ-Lemma.

Combined, cascading and transitivity reduce the problem of finding heteroclinics from U_j to U_k to the case $i_j = i_k + 1$. Here the zero number and the Sturm property strike again. Such a heteroclinic exists if, and only if, there does not exist an equilibrium U, between U_j and U_k at $x = 0$, such that

$$z(U_j - U) \leq z(U_k - U). \tag{2.40}$$

The "only if" part indeed follows indirectly from Proposition 2.2.1 Just consider

$$w(t) := u(t) - U \tag{2.41}$$

along a hypothetical heteroclinic $u(t)$ from U_j to U_k. Since U lies between U_j and U_k, at $x = 0$, and since $w_x \equiv 0$ there by the Neumann boundary conditions for $u(t)$ and U, a multiple zero of w has to arise at $x = 0$, for some $t = t_0$. Therefore $z(w(t))$ must drop strictly at least once, in contradiction to (2.40). Thus U as in (2.40) *blocks* any heteroclinic from U_j to U_k.

The "if" part is proved by a Conley index argument and by a suitable homotopy to a saddle-node bifurcation situation for U_j and U_k. By the explicit formulae for the Morse indices and zero numbers of parts (i) and (ii) of Theorem 2.2.3, this procedure enables us to recursively determine all heteroclinics in the Sturm attractor $\mathcal{A}_f$ from the combinatorial ODE information on the Sturm permutation $\pi = \pi_f$ alone.

To illustrate Theorems 2.2.2, 2.2.3 we ask for the Sturm attractor $\mathcal{A}$ of *maximal dimension* $\ell = \max_k i_k$, with a fixed (odd) number n of equilibria. By our explicit expression (2.38) for the Morse indices i_k in terms of the Sturm permutation π, we first observe $i_{k+1} = i_k \pm 1$. Since $i_0 = i_n = 0$ are stable, this implies a bound $2\ell + 1 \leq n$. Indeed i_k must ascend from $i_0 = 0$

to the maximal value ℓ, and also descend back to $i_n = 0$. A combinatorial exercise then shows that the maximality requirement $2\ell + 1 = n$ for ℓ is satisfied if, and only if, the permutation π is given by

$$\pi = \pi_{CI} := (2,\, n-1)(4,\, n-3)\cdots(\ell',\, n+1-\ell') \qquad (2.42)$$

where ℓ' denotes the largest even integer not exceeding $\ell = (n-1)/2$. By elementary ODE phase plane methods ("time map", see for example [ChIn74, Sm83, BrCh84, Sc90, BiRo62]), the permutation π_{CI} arises in the *Chafee-Infante problem*

$$f = \mu u(1 - u^2). \qquad (2.43)$$

Indeed $\pi_f = \pi_{CI}$ for $\lambda_{\ell-1} < \mu < \lambda_\ell$, where $\lambda_k := (k\pi)^2$ and $\ell = \dim \mathcal{A}_{CI}$ is the Morse-index of the trivial equilibrium $U \equiv 0$. The maximal Morse index ℓ is indeed attained at $U \equiv 0$, only. By Theorem t-2.2.1, we can now conclude that *the* Sturm attractor $\mathcal{A}$ of maximal dimension ℓ coincides with the *Chafee-Infante attractor* $\mathcal{A}_{CI} := \mathcal{A}_f$, for f as in (2.43), up to C^0 orbit equivalence. For an interesting geometric description of the Chafee-Infante attractor see [HaMi91].

Wolfrum has developed a much more elegant approach to Theorem 2.2.3, based on his notions of *z-ordering* and *z-adjacency*. For $z = 0, 1, 2, \cdots$ define partial orders $<_z$ on the equilibria $\mathcal{E} = \{U_1, \cdots, U_n\}$ as follows: $U_j <_z U_k$ if, and only if,

$$z(U_j - U_k) = z \quad \text{and} \quad U_j < U_k \quad \text{at} \quad x = 0. \qquad (2.44)$$

Clearly heteroclinics between such z-ordered U_j, U_k are blocked in the sense of (2.40), by the Sturm property, unless U_j and U_k are adjacent in this order. Of course z-adjacency means that there does not exist another equilibrium U such that $U_j <_z U <_z U_k$. We caution the reader that the z-order $<_z$ is not, in general, transitive.

Wolfrum has proved the following; see [Wo02b].

Theorem 2.2.4 *Let U_j, U_k be distinct equilibria with Morse indices i_j, i_k in the Sturm attractor $\mathcal{A}_f$ of (2.21), (2.22) on the interval. Let $z := z(U_j - U_k)$. Then there exists a heteroclinic orbit from U_j to U_k if, and only if, $i_j > i_k$ and U_j, U_k are z-adjacent.*

This very concise and direct description of all heteroclinic orbits is again purely combinatorial in terms of the Sturm permutation $\pi = \pi_f$. The necessary data $i_j, i_k, z(U_j - U_k)$ are provided by the explicit formulae of parts (i), (ii) of theorem 2.2; see for example (2.38). The result was not, however, derived from that theorem. Instead of cascading, transitivity, and blocking, it is based on a refined geometric analysis of the transversality and zero-number properties along the hierarchies of strong unstable and stable manifolds of all equilibria U_j. These strong manifolds are submanifolds of W_j^u, W_j^s characterized by the faster exponential convergence rates associated to the higher unstable and stable eigenvalues; see for example [BrFi86].

Combinatorics of Sturm attractors Theorems 2.2.2 – 2.2.4 highlight the central importance of the Sturm permutation π_f for the characterization of Sturm attractors $\mathcal{A}_f$. It is therefore worthwhile to note that there exists a purely combinatorial characterization of the Sturm permutations, which does not recur to any specific dissipative nonlinearities $f = f(x, u, u_x)$. See [FiRo99] for the following result.

Theorem 2.2.5 *Let $\pi \in S_n$ be any permutation. Then π is a Sturm permutation, that is, $\pi = \pi_f$ for some dissipative nonlinearity $f = f(x, u, u_x)$ in (2.21), if, and only if, n is odd and π is a dissipative Morse meander.*

Here a permutation $\pi \in S_n$ is called *dissipative*, if $\pi(1) = 1$ and $\pi(n) = n$. The permutation π is called *Morse*, if $\iota_k \geq 0$, holds for all *Morse numbers*

$$\iota_k := \sum_{j=1}^{k-1} (-1)^{j+1} \operatorname{sign}\left(\pi^{-1}(j+1) - \pi^{-1}(j)\right), \qquad (2.45)$$

$k = 1, \cdots, n$. Of course, $\iota_1 := 0$ anyways. To define the *meander* property of a permutation choose any two distinct numbers $1 \leq j, k \leq n$ of the same even/odd parity, such that $\pi^{-1}(k)$ is between $\pi^{-1}(j)$ and $\pi^{-1}(j+1)$. Then π is called a meander permutation, if necessarily $\pi^{-1}(k+1)$ is also between $\pi^{-1}(j)$ and $\pi^{-1}(j+1)$, for any such choice of j, k.

We briefly indicate why Sturm permutations $\pi = \pi_f$ are necessarily dissipative Morse meanders. The converse direction, which is much more involved, requires the construction of a nonlinearity $f = f(x, u, u_x)$ realizing a prescribed permutation π as its Sturm permutation, and will be omitted.

Immediately following our definition (2.29), (2.30) of Sturm permutations π_f, we have already indicated how the monotonicity of the semiflow implies that π_f is dissipative. The Morse property follows because the Morse numbers ι_k defined in (2.45) coincide with the Morse indices i_k of the equilibria U_k for Sturm permutations $\pi = \pi_f$; see (2.38) and [Ro85, Ro91, FuRo91]. Of course $i_k \geq 0$ holds for the Morse indices, which are the dimensions of the unstable manifolds W_k^u.

To understand the meander property we recall Arnold's definition [ArVi89] of a *meander permutation*. Consider a connected oriented non-selfintersecting curve in the plane, intersecting a fixed oriented base line in n points. The intersections are assumed to be strict crossings. The permutation defined by ordering the intersection points, first along the base line and then along the curve, is called a meander permutation. Here we label the intersection points by $1, \cdots, n$, when ordered along the base line, and by $\pi^{-1}(1), \cdots, \pi^{-1}(n)$ when ordered along the curve. This geometric definition of a meander permutation π indeed coincides with ours, by the Jordan curve theorem.

We now show that Sturm permutations $\pi = \pi_f$ indeed possess the meander property. We first solve the Neumann boundary-value problem

$$0 := U_{xx} + f(x, U, U_x) \qquad (2.46)$$

for equilibria $U = U(x)$ by a "shooting method" in the phase plane (U, U_x). Consider the base line of initial conditions $(U, U_x) = (a, 0), a \in \mathbb{R}$, at $x = 0$, given by the U-axis. Solving the initial value problem (2.46) up to $x = 1$, for this line of initial conditions, we obtain a differentiable curve at $x = 1$ in the (U, U_x)-plane, which we call the *shooting curve*. The shooting curve is parameterized by a. Intersection points of the shooting curve with the base line $U_x = 0$ are equilibria: they correspond to the solutions $U_1, \cdots, U_n$ of the Neumann problem. The labeling by $1, \cdots, n$ along the base line corresponds to the ordering of these solutions at $x = 1$. The labeling by $\pi^{-1}(1), \cdots, \pi^{-1}(n)$ along the shooting curve corresponds to the ordering by a at $x = 0$. The shooting curve is a connected Jordan curve because the ODE-flow (2.46) from $x = 0$ to $x = 1$ defines a diffeomorphism of the (U, U_x)-plane. Hyperbolicity of the equilibria $U_1, \cdots, U_n$ corresponds to transverse intersections of the shooting curve with the base line $U_x = 0$; see [BrCh84, Ro85]. These observations clearly prove that Sturm permutations are indeed dissipative Morse meanders.

With the help of our combinatorial characterization of Sturm permutations it is now possible to characterize Sturm attractors. See Table 2.1 for numbers of Sturm permutations $\pi \in S_n, n = 1, \cdots, 17$. We note that the trivial transformation $x \mapsto -x$ changes Sturm permutations π to their inverse π^{-1}. Similarly $u \mapsto -u$ conjugates $\pi \in S_n$ with the involution $j \mapsto n+1-j$ in S_n. Neither symmetry operation changes the Sturm attractor, geometrically. See Figure 2.2 for all sixteen Sturm attractors with 9 equilibria together with their Sturm permutations, reduced by the above trivial symmetries [Fi94].

Observe that apparently unrelated Sturm permutations, of different cycle lengths and hence non-conjugate, may lead to the "same" Sturm attractor. Interestingly, these cases differ by the geometry of fast unstable manifolds, as was discovered by Wolfrum [Wo98]; see Figure 2.3. In general, it is still an open question whether this geometric distinction is sufficient to provide a 1-1 correspondence between Sturm attractors $\mathcal{A}_f$ and Sturm permutations π_f, up to the above trivial symmetries.

Table 2.1. Numbers of Sturm permutations $\pi = \pi_f \in S_n$ for odd $n = 1, \cdots, 17$

n	dissipative meanders	Sturm permutations	Sturm mod symmetry
1	1	1	1
3	1	1	1
5	2	2	2
7	8	7	5
9	42	32	18
11	262	175	75
13	1828	1083	383
15	13820	7342	2850
17	110954	53372	14984

2 Spatio-Temporal Dynamics of Reaction-Diffusion Patterns

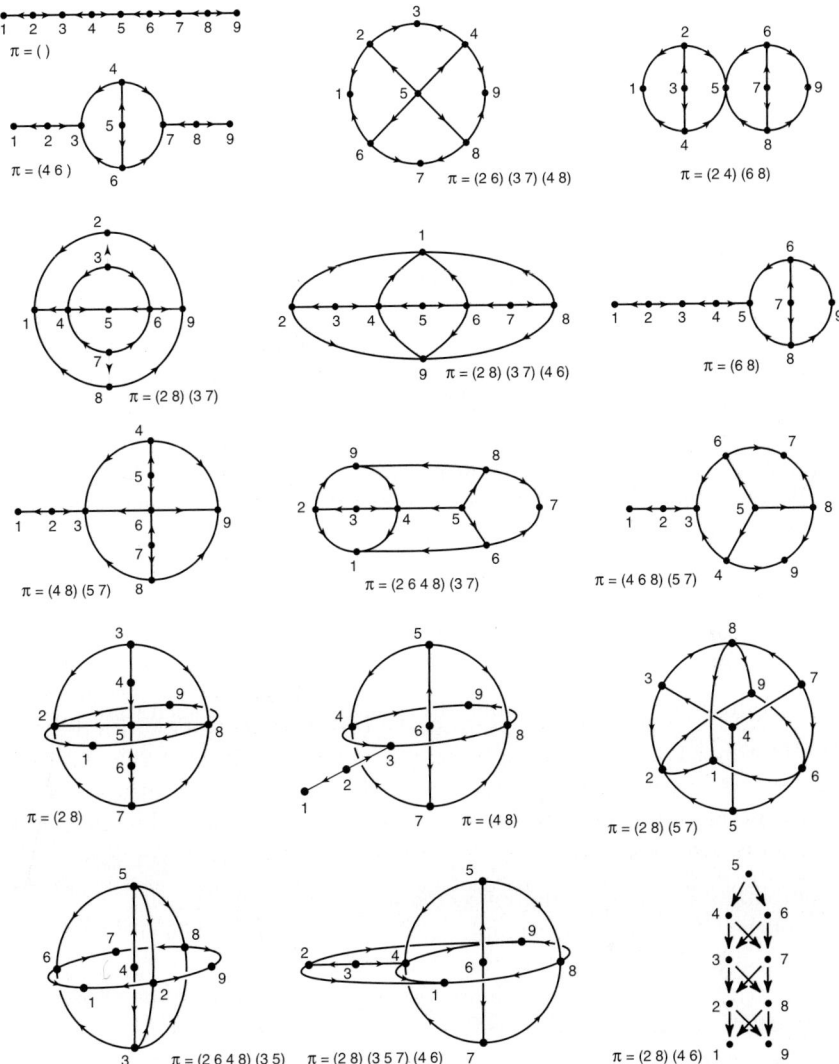

Fig. 2.2. All sixteen Sturm attractors $\mathcal{A}_f$ with $n = 9$ equilibria. The corresponding Sturm permutations π_f are indicated in cycle notation.

We conclude this section with a short summary of the intriguing results by Härterich on the Sturm attractors $\mathcal{A}_\varepsilon$ of viscous approximations to non-linear *hyperbolic balance laws*; see [Hä98]. For scalar equations in one space dimension such viscous approximations take the form

$$u_t + F(u)_x + G(u) = \varepsilon u_{xx}, \qquad (2.47)$$

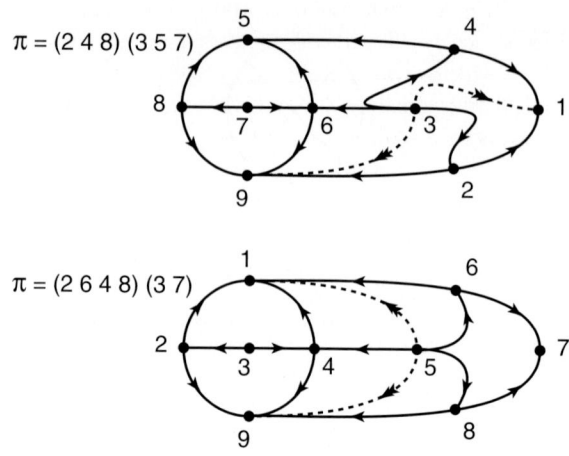

Fig. 2.3. C^0 orbit equivalent Sturm attractors $\mathcal{A}_f$ with nonconjugate permutations π_f. Note how the attractors differ in the geometry of the fast unstable manifolds, indicated by double arrows.

This fits in the framework (2.21), (2.22) with $f := F'(u)u_x + G(u)$, under Neumann boundary conditions on the unit interval. The object of interest is the limiting behavior of the Sturm attractor $\mathcal{A}_\varepsilon$, for vanishing viscosity $\varepsilon \searrow 0$. Härterich has derived sufficient conditions on F, G such that

(i) $\ell_\varepsilon := \dim \mathcal{A}_\varepsilon$
(ii) $n_\varepsilon :=$ the number of equilibria of $\mathcal{A}_\varepsilon$
(iii) $\pi_\varepsilon :=$ the Sturm permutations of $\mathcal{A}_\varepsilon$
(iv) $\mathcal{A}_\varepsilon$ up to orbit equivalence

all stabilize, that is, eventually become independent of ε for $0 < \varepsilon < \varepsilon_0$. In view of the Chafee-Infante example (2.43), which is equivalent to (2.47) with $F = 0$, cubic G, and $\varepsilon := 1/\mu$, this is surprising. Indeed $\ell_\varepsilon = 1 + [(\pi\sqrt{\varepsilon})^{-1}]$ and $n_\varepsilon = 2\ell_\varepsilon + 1$ grow unboundedly for $\varepsilon \searrow 0$, in this case.

In [Hä98] sufficient conditions for a uniform dimension bound $\ell_\varepsilon \leq \ell_0$ were derived, for $\varepsilon \searrow 0$. Specifically, if F possesses only nondegenerate critical points and G does not vanish at any critical point of F, then there exist numbers $\ell_0 \in \mathbb{N}$ and $\varepsilon_0 > 0$ such that $\ell_\varepsilon \leq \ell_0$ for all $0 < \varepsilon \leq \varepsilon_0$.

In [Hä97] these conditions were refined to imply stabilization of π_ε, (iii). Note that (iii) implies stabilization (ii) of n_ε, trivially, and stabilization (i) of $\ell_\varepsilon = \dim \mathcal{A}_\varepsilon$, by our explicit representation (2.38) for the Morse indices i_k. Specifically, there exists an open class of functions F and G (with respect to the strong Whitney topology) such that the following holds: There exist numbers $\ell_0 \in \mathbb{N}$ and $\varepsilon_0 > 0$ such that $\ell_\varepsilon \in \{\ell_0 - 1, \ell_0\}$ for all $0 < \varepsilon \leq \varepsilon_0$. Explicit conditions on F and G can also be stated which can be verified in concrete situations. Note that it is not clear whether n_ε stabilizes for $\varepsilon \to 0$.

If $F \in C^2$ is strictly convex and $G \in C^1$ possesses only simple zeros that do not coincide with the critical point of F, then all four stabilization properties (i)–(iv) hold, as was claimed above; see [Hä99].

The proofs of these results are based on a very careful singular perturbation analysis of the ODE equilibrium problem, for $\varepsilon \searrow 0$, including solutions of canard type.

For another example of a singular perturbation of Sturm permutations and Sturm attractors, involving x-dependent nonlinearities $f = f(x, u)$ of cubic Chafee-Infante type in u, see [Fi&al02a] and the references there.

2.2.3 Sturm Attractors on the Circle

In Section 2.2.1, (2.8)–(2.10) we have seen how the case $x \in S^1 = \mathbb{R}/\mathbb{Z}$ of periodic boundary conditions for the scalar reaction-advection-diffusion equation

$$u_t = u_{xx} + f(x, u, u_x) \tag{2.48}$$

is a hybrid between one- and two-dimensional domains Ω. The gradient-flow property fails, due to rotating waves which arise in the S^1-equivariant case $f = f(u, u_x)$. The Sturm property on the other hand prevails; see Proposition 2.2.1. At the end of Section 2.2.1 we have used the Sturm property, only, to prove a gradient-flow feature of (2.48) on the interval: convergence to equilibrium for any bounded solution. We will first indicate the fate of this convergence property on the circle domain. We will then present an analogue of Theorems 2.2.3, 2.2.4 for heteroclinic orbits between rotating waves.

Poincaré-Bendixson theory Intuitively we can say that the dynamics of (2.48) is "essentially one-dimensional", asymptotically for any single trajectory $u(t) = \mathcal{T}(t)u_0$ and under separated boundary conditions. Indeed, $u(t)$ converges to a single equilibrium as would be the case for any scalar ODE $\dot{u} = g(u)$, $u \in \mathbb{R}$. Still the global attractor $\mathcal{A}_f$ may be of arbitrarily large dimension ℓ, as we recall from the Chafee-Infante example (2.43).

The dynamics of (2.48) on the circle $x \in S^1$ is then "essentially two-dimensional", asymptotically for $u(t) = \mathcal{T}(t)u_0$. The analogous convergence statement is given by the Poincaré-Bendixson theorem, well known from ODE dynamics $\dot{\mathbf{u}} = g(\mathbf{u})$ in the phase plane $\mathbf{u} \in \mathbb{R}^2$. Of course, the Chafee-Infante example persists under periodic boundary conditions, if we reflect solutions through their Neumann boundary. Therefore, the global attractor may still be of arbitrarily large dimension.

For a precise statement of our Poincaré-Bendixson result we fix any $x_0 \in S^1$, arbitrarily, and consider the evaluation projection

$$\begin{aligned} P : X &\to \mathbb{R}^2 \\ u &\mapsto (u(x_0), u_x(x_0)). \end{aligned} \tag{2.49}$$

The following theorem has been proved in [FiMP89b]; for a related result obtained independently see [Na90]. Recall that $\mathcal{E} = \{u_t = 0\}$ denotes the set of equilibria.

Theorem 2.2.6 *Fix any $x_0 \in C^1$. Let $u_0 \in X \hookrightarrow C^1$ possess a uniformly bounded solution $u(t) = \mathcal{T}(t)u_0 \in X, t \geq 0$, of (2.48) with ω-limit set $\omega(u_0)$.*

Then the evaluation projection $P : \omega(u_0) \to \mathbb{R}^2$ is injective on the ω-limit set $\omega(u_0)$. Moreover the following alternative holds:

(i) either $\omega(u_0)$ is a single periodic orbit, or else
(ii) both the α-limit set and the ω-limit set of any nonequilibrium solution $\tilde{u}(t), t \in \mathbb{R}$, in $\omega(u_0)$ consist entirely of equilibria.

To indicate the relation of Theorem 2.2.6 with the Sturm property, Proposition 2.2.1, we show injectivity of the restricted evaluation projection

$$P : \mathrm{clos}\,(\mathrm{orb}(\tilde{u}_0)) \to \mathbb{R}^2 \tag{2.50}$$

only along the X-closure of any single trajectory $\mathrm{orb}(\tilde{u}_0) := \{\mathcal{T}(t)\tilde{u}_0;\ t \in \mathbb{R}\}$ within the original ω-limit set: $\tilde{u}_0 \in \omega(u_0)$. Invoking the Jordan curve theorem, it is then easy to show that $\mathrm{clos}\,(\mathrm{orb}(\tilde{u}_0))$ contains a periodic orbit, if it does not contain any equilibrium. See for example [Fi89]. The fine tuning which leads to Theorem 2.2.6 is much more involved, see [FiMP89b].

To prove restricted injectivity (2.50), choose any two distinct initial conditions $\tilde{u}_0^1, \tilde{u}_0^2$ in $\mathrm{clos}\,(\mathrm{orb}(\tilde{u}_0))$. Suppose that $P\tilde{u}_0^1 = P\tilde{u}_0^2$, proceeding indirectly. Denoting solution curves through $\tilde{u}_0^\iota$, $\iota = 1, 2$, by $\tilde{u}^\iota(t), t \in \mathbb{R}$, we then see that

$$t \mapsto z(\tilde{u}^1(t) - \tilde{u}^2(t)) \tag{2.51}$$

drops strictly at $t = 0$, by Sturm proposition 2.2.1(ii). Choosing t_0 and $t_0 + \vartheta$ large enough, (2.51) implies that

$$t \mapsto z(\tilde{u}(t + t_0 + \vartheta) - \tilde{u}(t + t_0)) \tag{2.52}$$

also drops, for some $|t| \leq 1$. Indeed, the x-profile $\tilde{u}^1(t) - \tilde{u}^2(t)$ possesses only simple zeros in x before and after dropping, and $\tilde{u}_0^1, \tilde{u}_0^2 \in \mathrm{clos}\,(\mathrm{orb}(\tilde{u}_0))$ in the topology of $X \subseteq C^1$. Since $\tilde{u}_0 \in \omega(u_0)$, the dropping in (2.52) implies that there also exists an increasing sequence $t_n \nearrow \infty$ of dropping times $t = t_n$ of

$$t \mapsto z(u(t + t_0 + \vartheta) - u(t + t_0)), \tag{2.53}$$

for the original solution u. This contradicts the finiteness property (iii) of the Sturm Proposition 2.2.1, and hence proves our injectivity claim (2.50) indirectly.

Similar Poincaré-Bendixson theorems have been proved to hold for *differential delay equations*

$$\dot u(t) = f(u(t), u(t-1)) \tag{2.54}$$

under a positive or negative monotonicity assumption on f in the delayed feedback argument $u(t-1)$; see [MP88, MPSm90]. For discrete analogues, the so-called *cyclic monotone feedback* systems, see [Sm95] as well as [MPSe96b] for combinations of both structures. The approach in [FiMP89b] is axiomatic, and includes Jacobi systems (2.36).

It is interesting to note that (2.54) under positive feedback, where $f(u, \cdot)$ is strictly increasing, does not exhibit asymptotically stable periodic solutions. This is reminiscent of the Hirsch theorem on the absence of stable periodic orbits for monotone dynamical systems, as discussed in Section 2.2.1. For negative feedback, in contrast, stable periodic solutions of (2.54) do arise; see for example [Di&al95] and the references there. As a curiosity, we note here the twisted periodic boundary condition

$$\begin{aligned} u(t,1) &= -u(t,0) \\ u_x(t,1) &= -u_x(t,0) \end{aligned} \tag{2.55}$$

for which all the above results remain valid, due to the Sturm structure. The monotonicity property fails, however: although formally in effect, it is prevented by the twisted boundary condition (2.55). Indeed $z(\tilde u - u) \geq 1$ for any two distinct solution profiles $\tilde u, u$ satisfying (2.55), as soon as $t > 0$. This phenomenon is closely related to the Smale objection to monotonicity, indicated in Section 2.2.1. Explicit examples of odd rotating wave solutions $u = u(x - ct)$ for suitable f show that asymptotically stable periodic orbits in fact do occur under the twisted boundary condition (2.55). Unfortunately this is a purely mathematical observation, at the moment: we are not aware of any applied relevance of our twisted periodic boundary condition.

Heteroclinic connections of rotating waves When we now attempt to describe the Sturm attractors $\mathcal{A}_f$ under the usual periodic boundary conditions $x \in S^1 = \mathbb{R}/\mathbb{Z}$, the Poincaré-Bendixson Theorem 2.2.6 encourages us to pursue the program outlined in questions **Q0** – **Q4** of Section 2.2.2. Of course we will have to replace the word "equilibria" in **Q0**, **Q1** and address "equilibria and time periodic orbits" as well as their heteroclinic orbits, instead. A fundamental obstacle, however, arises immediately: the loss of the Morse-Smale transversality property. In fact, nontransverse intersections of stable and unstable manifolds may arise. Specifically, any autonomous planar ODE vector field can be embedded into (2.48) for a suitable choice of the nonlinearity $f = f(x, u, u_x)$, $x \in S^1$; see [SaFi92] and [Br&al92]. The dynamics is supported, for example, on a time invariant linear subspace spanned by the spatial Fourier modes $\sin x$ and $\cos x$. Embedding any planar homoclinic orbit then immediately shows that the Morse-Smale property may fail, even when all equilibria and periodic orbits are hyperbolic. As an aside we mention that time periodic nonlinearities $f = f(t, x, u, u_x)$, $x \in S^1$, can give rise to transverse homoclinic orbits and hence to Smale horseshoes with Bernoulli type shift dynamics, by similar constructions; see [SaFi92].

A second difficulty of periodic boundary conditions lies in the combinatorial description of $\mathcal{A}_f$. The Fusco-Rocha definition (2.29), (2.30) of the Sturm permutation π_f, by ordering of the boundary values u, clearly depends on the separated nature of the boundary conditions at $x = 0, 1$. The Sturm permutation probably would have to be replaced by the braid type of the braid in $(x, u, u_x) \in S^1 \times \mathbb{R}^2$, which is defined by the x-profiles of all equilibria and time periodic orbits. Any periodic orbit $u(t)$ in X can in fact be represented by any snap-shot $u(t_0)$ without changing the braid type. In the case of Neumann boundary conditions the braid embeds into a surface: the shooting surface of solution of (2.46) with $U_x(0) = 0$. The braid type is then in fact determined by the Sturm permutation $\pi = \pi_f$.

The above remarks sufficiently demonstrate our lack of understanding of the general case $f = f(x, u, u_x)$. We therefore now restrict to the S^1-equivariant case of x-independent nonlinearities

$$u_t = u_{xx} + f(u, u_x), \quad x \in S^1. \tag{2.56}$$

From Section 2.2.1, (2.9), (2.10) we recall that rotating waves $u(t, x) = U(x - ct)$ arise in that case. More precisely it was proved in [AnFi88] that all periodic orbits $U(t, x)$ are rotating waves $U = U(x - ct)$. This follows from Sturm Proposition 2.2.1: the zero number z must be constant on the span of U_t, U_x, and hence U_t and U_x must indeed be linearly dependent. The Sturm attractor $\mathcal{A}_f$ then consists of equilibria $\mathcal{E}$, rotating waves $\mathcal{R}$, and their heteroclinic orbits $\mathcal{H}$:

$$\mathcal{A}_f = \mathcal{E} \cup \mathcal{R} \cup \mathcal{H}. \tag{2.57}$$

We assume $\mathcal{E}$ and $\mathcal{R}$ to be *normally hyperbolic*, admitting at most a single trivial eigenvalue (or Floquet exponent) resulting from x-shift.

To formulate our result on heteroclinic connections in $\mathcal{A}_f$, we slightly adapt Wolfrum's notation of z-adjacency, from (2.44). First we represent any nonhomogeneous equilibrium in $\mathcal{E}$ and any rotating wave in $\mathcal{R}$ by *two* elements $U \in X$ satisfying $U_x = 0$ at $x = 0$. This can always be achieved because $\mathcal{E}, \mathcal{R}$ are invariant under x-shift, by S^1-equivariance of (2.56). The *two* representatives are chosen to attain their maximum and their minimum at $x = 0$, respectively. We then define z-order and z-adjacency as in (2.44), based on the ordering and adjacency properties of these representatives, at $x = 0$.

We also adapt the Morse index i of $U \in \mathcal{E} \cup \mathcal{R}$ to indicate the strong unstable dimension. For equilibria, i counts the total algebraic multiplicity of complex eigenvalues λ, after linearization, with strictly positive real part. For rotating waves U we similarly count the nontrivial Floquet multipliers λ strictly outside the unit circle.

The following theorem is due to [Fi&al02b] and assumes normal hyperbolicity; see (2.57).

Theorem 2.2.7 *Let U_j, U_k be equilibria or rotating waves with strong unstable dimensions i_j, i_k in the Sturm attractor of the S^1-equivariant reaction-advection-diffusion equation (2.56) on the circle $x \in S^1$. Let $z := z(U_j - U_k)$. Then there exists a heteroclinic orbit from U_j to U_k if, and only if, $i_j > i_k$ and U_j and U_k are z-adjacent.*

The proof of this theorem at present involves four steps, which we call

(i) transversality
(ii) freezing
(iii) symmetrization
(iv) Neumann embedding

We only briefly mention the topics involved. In step (i), we slightly extend results by [Ol02] to show that the Morse-Smale property of transverse intersections of unstable and stable manifolds holds, in the S^1-equivariant case, for normally hyperbolic equilibria and rotating waves. This allows us, in step (ii), to construct a homotopy f^τ of $f = f^0$ which "freezes" all rotating waves:

$$f^\tau := f(u, u_x) + \tau c u_x \tag{2.58}$$

Here $c = c(u, u_x)$ is chosen to coincide with the constant wave speed along the x-profile $(u, u_x)(x)$ of any rotating wave of f, without introducing any additional rotating waves for any $0 \leq \tau \leq 1$. In effect, this homotopy reduces the speed c_0 of any rotating wave at $\tau = 0$ to

$$c_\tau = (1 - \tau) c_0. \tag{2.59}$$

By $\tau = 1$, any rotating wave speed has been tuned down to zero, and all periodic orbits have become spatially nonhomogeneous equilibria: $\mathcal{R} = \emptyset$ in (2.57). Step (iii) performs a second homotopy which keeps all (frozen) equilibria normally hyperbolic and reduces $f = f(u, u_x)$ to a nonlinearity which is even in u_x and hence commutes with the reflection $x \mapsto 1 - x$. In particular, any equilibrium solution U becomes reflection symmetric with respect to any of its local maxima or minima, by this x-reversibility. Shifting any equilibrium $U \in \mathcal{E}$ by some suitable value x_0, we may therefore assume $U_x = 0$ at $x = 0$ and $x = 1/2$. In step (iv) we observe that all heteroclinic orbits of Theorem 2.2.4 are then in fact already represented by reflection symmetric heteroclinic orbits. The latter are already known: we only have to apply Wolfrum's Theorem 2.2.4. to the resulting Neumann problem on the half-interval $0 \leq x \leq 1/2$. This completes our sketch of a proof of Theorem 2.2.7. For complete details see [Fi&al02b].

We conclude this section with a few remarks concerning related results. The results by Hale and Raugel concerning upper and lower semicontinuity of attractors $\mathcal{A}_f$ for *thin domains* [HaRa92] allow us to fatten the circle $\Omega_0 = S^1$ to thin annuli $\Omega_\varepsilon = S^1 \times (-\varepsilon, \varepsilon)$ and recover all dynamics of $\mathcal{A}_f$ on Ω_ε. More generally domains $\Omega_\varepsilon = S^1 \times \tilde{\Omega}_\varepsilon$ work, for suitably "small" cross sections $\tilde{\Omega}_\varepsilon$ in terms of a sufficiently large second eigenvalue of the Neumann problem on the cross section $\tilde{\Omega}_\varepsilon$. Similarly, nonlinearities $f + \varepsilon \tilde{f}$ can be allowed to depend on x, slightly, via $\tilde{f}(x, u, u_x)$ and small ε. In the spirit of [FiVi01a, FiVi01b], nonlinearities $f = f(\varepsilon, x/\varepsilon, u, u_x)$ are also admissible, under Diophantine quasiperiodicity conditions on the rapid spatial dependence x/ε of f.

Fast travelling waves in cylinder domains $x = (\xi, \eta) \in \Omega = S^1 \times \mathbb{R}$ provide another interpretation of our heteroclinic orbits. Consider travelling wave solutions

$$u(t, x) = U(t - \varepsilon \eta, x) \qquad (2.60)$$

of a scalar x-independent parabolic equation (2.1). For large wave speeds $1/\varepsilon$ in the unbounded η-direction, we obtain

$$U_\tau = U_{\xi\xi} + \varepsilon^2 U_{\tau\tau} + f(U, U_\xi, -\varepsilon U_\tau), \qquad (2.61)$$

abbreviating $\tau := t - \varepsilon \eta$. In [Sc96] it was shown how to omit both ε-terms, in the fast wave speed limit $\varepsilon \searrow 0$. See also [Ca&al93] for nonlinearities f which do not depend on the gradient term U_ξ. Invoking Theorem 2.2.7 then provides heteroclinic orbits $U(\tau, \xi)$ between rotating waves $U_1(\xi - c_1 \tau), U_2(\xi - c_2 \tau)$, typically rotating at different angular speeds c_1, c_2. Clearly $U(\tau, \xi)$ then describes a nonmonotone wave u propagating rapidly along the cylinder axis η of Ω, and connecting η-asymptotic states which rotate at different angular ξ-speeds c_1, c_2. The complexity of such wave profiles is in marked contrast with the monotone travelling waves found by Berestycki and Nirenberg using comparison methods [BeNi90]. See also the beautiful complementary approach by a variational characterization in [He89].

Returning to one-dimensional x we conclude this section by indicating why results like Theorems 2.2.2 – 2.2.7 do not carry over to systems $u \in \mathbb{R}^N, N = 2$, not even in a single space dimension. Following [FiPo90], we consider

$$u_t = u_{xx} + f(x, u) + c(x) \int_0^1 u \, dx, \qquad (2.62)$$

only for scalar $u \in \mathbb{R}$, but involving a linear nonlocal term $c(x) \int u \, dx$. We consider $0 < x < 1$ with Dirichlet boundary conditions. It has then been proved that very general finite-dimensional vector fields $\dot{\mathbf{u}} = g(\mathbf{u})$, $\mathbf{u} \in \mathbb{R}^n$ arise, on an invariant (center) manifold of (2.61) and up to any finite polynomial order, for suitable choices of f, c. The constraints on $g(\mathbf{u})$ are rather mild: $g(0) = 0$ and the eigenvalues of $g'(0)$ on the imaginary axis should be simple. In particular this includes the possibility of complicated dynamics of Bernoulli shift type. These results are very similar, in spirit, to earlier analogous observations by Hale for retarded functional differential equations; see [Ha85].

Similar remarks apply to the case of scalar equations (2.1), $u \in \mathbb{R}^N$, $N = 1$, in several space dimensions. Dancer and Poláčik [DaPo02] have proved that a C^1-dense set of finite-dimensional vector fields $\dot{\mathbf{u}} = g(\mathbf{u})$, $\mathbf{u} \in \mathbb{R}^\mathbf{n}$ can be realized on an invariant manifold, for suitable choices of x-independent reaction-advection nonlinearities $f = f(u, \nabla u)$ and of two-dimensional smooth bounded domains $\Omega \subset \mathbb{R}^2$. For suitable $f = f(x, u, \nabla u)$ with only linear dependence on ∇u and arbitrary, but fixed domains $\Omega \subset \mathbb{R}^2$, a similar result is due to [Po95], [PrRy98a, PrRy98b]. The gradient-like case $f = f(x, u)$ on a two-dimensional ball allows embeddings of C^1-dense sets of gradient vector fields $g(\mathbf{u}) = -\nabla_\mathbf{u} G(\mathbf{u})$. See also the survey [Po02].

These results certainly motivate a quest for more refined structural conditions on nonlinearities f, as well as domains Ω, which impose incisive restrictions on their resulting spatio-temporal dynamics.

2.3 One Unbounded Space-Dimension: Travelling Waves

In this chapter, we address some aspects of the dynamics of reaction-diffusion systems in large or unbounded, one-dimensional domains. Motivated by the motion of layers and fronts, we study essential spectra that arise in the linearization about travelling wave solutions, Section 2.3.1. In Section 2.3.2, we extend this linear analysis to a nonlinear bifurcation result. Depending on Fredholm indices in the essential spectrum, we find existence or nonexistence of periodic orbits in a Hopf bifurcation caused by the essential spectrum. We conclude this chapter with an investigation of the limiting behavior of the spectrum of the linearization about a travelling wave, when the domain size tends to infinity. Under separated boundary conditions, the limiting spectrum differs from the spectrum in the unbounded domain. The continuous parts of the limiting spectrum consist of a finite collection of curves which we call the absolute spectrum.

2.3.1 Unbounded Domains and Essential Spectra

From bounded to unbounded domains The results on global attractors in scalar reaction-diffusion equations reviewed in the preceding sections show that certain dynamical properties do not depend on boundary conditions. The Sturm nodal property prevails, for example, independently of the boundary conditions. This allows for homotopies between Neumann, Dirichlet, mixed, and sometimes even periodic boundary conditions. An extreme statement in this direction is that the class — in contrast to individual global attractors — of all Sturm attractors $\mathbf{A}^P$ does not depend on the type of (separated) boundary conditions; see Section 2.2.2.

In descriptions of phenomena in experiments, it is often quite desirable to separate the influence of the boundary from what we call the *inner dynamics* of a reaction-diffusion system. See for example [Bl&al00] for a critical view on the influence of boundary conditions in chemical experiments. Mathematical statements on the influence of boundaries can be found mainly in the context of singular perturbation theory. For an illustration, we rewrite the Chafee-Infante problem (2.43) as

$$u_t = \varepsilon^2 u_{xx} + u(1-u)(1+u), \qquad x \in (0,1),$$

say with Neumann boundary conditions. Rescaling x, the parameter ε can be interpreted as a measure for the size of the domain, $0 \le x \le L = 1/\varepsilon$. For small ε, the motion on the unstable manifolds of equilibria (which constitute the Chafee-Infante attractor, by (2.27)) becomes exponentially slow in the parameter ε; see [FuHa89, CaPe90]. "Most" of the time, solutions consist of a finite collection of transition layers at $x = x_j(t)$ of the approximate local form

$$u_j \sim \pm \tanh(\frac{x - x_j(t)}{\sqrt{2}\varepsilon}). \tag{2.63}$$

The exponentially slow motion $x_j(t)$ is driven by the inner dynamics, caused by interaction and annihilation of layers, and the influence of the boundary. The strength of the interaction and the influence of the boundary indeed decrease exponentially with distance:

$$|\dot{x}_j(t)| \le C \exp(-C'(\inf_{j' \ne j}\{|x_j - x_{j'}|/\varepsilon\} + \min\{|x_j|, |1 - x_j|\})).$$

For most initial conditions, the slow motion eventually leads to annihilation of all layers and we recover the Hirsch result [Hi88] on generic convergence to stable equilibria in monotone dynamical systems, $u_\pm(x) \equiv \pm 1$ in our case — however only after time spans which may well exceed realistic experimental conditions.

The limiting case $\varepsilon = 0$ is best described in the scaling

$$u_t = u_{xx} + u(1-u)(1+u), \qquad x \in (-L, L),$$

with $L = 1/\varepsilon$. In the limit $L = \infty$, we are led to consider the unbounded real line as an idealization describing the inner dynamics.

One important but largely unresolved question concerning the limit $L \to \infty$ of increasing domain size, is in how far energy considerations can be localized, separating effects of the boundary from the intrinsic dynamics. Energy, alias the Lyapunov function in (2.4), is a nonlocal function and need not even be finite in an unbounded domain. Gallay and Slijepčevič [GaSl02] show how this fact may lead to recurrent, quite unexpected behavior in "gradient" systems.

Unbounded domains feature yet another phenomenon: *non-compact symmetry*. Reaction-diffusion systems (2.1) with x-independent reaction term $f = f(u, \nabla u)$, are invariant under spatial translation $x \to x + g$: if $u(t, x)$ is a solution, so is $u(t, x+g)$ for any $g \in \mathbb{R}$, fixed. The slow motion $x_j(t)$ of layers in the Chafee-Infante problem reflects this translational symmetry. The time evolution is "composed" of translates of the single layer (2.63). Whereas the role of the noncompact symmetry group of translations $g \in \mathbb{R}$ is obvious, here, Euclidean symmetry $G = SE(2)$ will play a major and more subtle role in the discussion of spiral wave dynamics in two-dimensional domains; see Chapter 2.5.

Separated boundary conditions break this translational equivariance. Periodic boundary conditions, in contrast, preserve the action of at least a continuous, compact group $SO(2) = \mathbb{R}/\mathbb{Z}$ as a remaining symmetry, when the real line is truncated to a bounded interval; see Section 2.2.3. Still, even periodic boundary conditions fail to mimic the translational drift along noncompact group orbits like the observed motion of a single stable front: under periodic boundary conditions, fronts come in pairs and typically annihilate or strongly interact after some finite time interval. We will further investigate the role of boundary conditions in the presence of transport and drift in Section 2.3.3.

If the domain is the real line $x \in \mathbb{R}$, the discussion of equilibria in $u_t = u_{xx} + f(u, u_x)$ has at least to be augmented to include travelling waves, $u(t,x) = q(x - ct)$, $c \in \mathbb{R}$, with $\sup_\xi u(\xi) < \infty$, just like in the case of periodic boundary conditions, Section 2.2.3. Travelling waves are a special case of *relative equilibria*, where time evolution of a profile is described by motion along the group action, here, translated states. We discuss relative equilibria more generally in the context of spiral wave dynamics in Chapter 2.5. *Fronts* are special travelling waves, where $q(\xi)$ possesses a heteroclinic asymptotic behavior for $\xi \to \pm\infty$, for example $q(\xi) \to q_\pm$ for $\xi \to \pm\infty$. A particular case are the layers in the Chafee-Infante problem, where $c = 0$.

Spectra of travelling waves: group velocities and Fredholm indices
For a scalar equation $u_t = u_{xx} + f(u)$, travelling waves with $c \neq 0$ are either spatially constant, $u(\xi) \equiv u_0$, or heteroclinic front solution. This is due to the gradient-like structure of the travelling-wave ordinary differential equation

$$u_\xi = v, \quad v_\xi = -cv - f(u).$$

More generally, we address *reaction-diffusion systems*

$$u_t = Du_{xx} + f(u), \tag{2.64}$$

with $x \in \mathbb{R}$, $u \in \mathbb{R}^N$ and a positive, diagonal diffusion matrix $D = \text{diag}(d_j) > 0$. Systems of several reaction species, $N > 1$, in general, neither possess gradient-like structure, nor monotonicity properties — or even nodal properties — in the sense of Section 2.2.1. Still, heteroclinic travelling-wave solutions $u(t,x) = q_*(x - ct)$ remain a fundamental, and elementary, ingredient to their dynamics.

Given such a travelling wave, we ask for its stability. We therefore consider the reaction-diffusion system in a comoving frame $\xi = x - ct$,

$$u_t = Du_{\xi\xi} + cu_\xi + f(u), \quad x \in \mathbb{R}, \tag{2.65}$$

where the travelling wave becomes an equilibrium $q_*(\xi)$. Throughout, x always refers to the spatial coordinate in the steady frame, whereas ξ always refers to a comoving frame with speed $c \geq 0$. The case $c < 0$ is obtained reflecting $x \to -x$. The travelling wave moves towards $x = +\infty$ and, in the comoving frame, the reaction-diffusion system inherits a drift term cu_ξ, representing transport towards $x = -\infty$. In the comoving frame, we linearize at q_* to find

$$w_t = Dw_{\xi\xi} + cw_\xi + f'(q_*(\xi))w =: \mathcal{L}_*w. \tag{2.66}$$

For simplicity, we first consider the operator $\mathcal{L}_*$ as a closed, unbounded operator on $L^2(\mathbb{R})$. All results of this section remain valid if we replace the function space L^2 by L^p, or by $1 < p < \infty$, or bounded, uniformly continuous functions. The *spectrum* $\operatorname{spec} \mathcal{L}_* \subset \mathbb{C}$ is defined as the set of λ such that $\mathcal{L}_* - \lambda$ is not boundedly invertible. Various notions of *stability* for individual fronts q_* have been suggested. We call a travelling wave *spectrally stable*, if the spectrum of the linearized operator $\mathcal{L}_*$ is contained in $\{\operatorname{Re}\lambda \leq 0\}$. We caution the reader that spectral stability need not imply asymptotic stability on our unbounded domain $x \in \mathbb{R}$.

For a refined discussion of stability properties, we decompose the spectrum into the *essential spectrum*

$$\operatorname{spec}_{\operatorname{ess}} \mathcal{L}_* := \{\lambda \in \mathbb{C};\ \mathcal{L}_* - \lambda \text{ is not Fredholm of index } 0\}$$

and its complementary part, the *point spectrum*

$$\operatorname{spec}_{\operatorname{pt}} \mathcal{L}_* := \operatorname{spec} \mathcal{L}_* \setminus \operatorname{spec}_{\operatorname{ess}} \mathcal{L}_*.$$

The appearance of essential spectrum is caused by the non-compactness of the real line $x \in \mathbb{R}$, here. By robustness of Fredholm properties, the essential spectrum is closed. As usual, for $\lambda \in \operatorname{spec}_{\operatorname{pt}} \mathcal{L}_*$, nontrivial $u \in \operatorname{Ker}(\mathcal{L}_* - \lambda)$ are then called eigenfunctions and algebraic multiplicity is associated to Jordan blocks. Due to analyticity of the eigenvalue problem in λ, the only possible degeneracy in this description of point spectrum is characterized by connected components of the set of λ such that the Fredholm index of $\mathcal{L}_* - \lambda$ is zero and the kernel is nontrivial; see [Ka66, He81].

For example, consider the heteroclinic case of a travelling front q_*, where $q_*(\xi) \to q_\pm$ for $\xi \to \pm\infty$. In Proposition 2.3.1, below, we will characterize the essential spectrum of $\mathcal{L}_*$ on $L^2(\mathbb{R})$ in terms of the spectra of the asymptotic linearizations $\mathcal{L}_\pm$ at $q_\pm$:

$$\mathcal{L}_\pm w := Dw_{\xi\xi} + cw_\xi + f'(q_\pm)w.$$

Unlike $\mathcal{L}_*$, the differential operators $\mathcal{L}_\pm$ possess constant, x-independent coefficients. The spectra of $\mathcal{L}_\pm$ are therefore readily computed in terms of exponentials. Setting $w(\xi) = e^{-\nu\xi}w_0$, with $\nu \in \mathbb{C}$, the spectral problem $\mathcal{L}_\pm w = \lambda w$ is transformed into the *complex dispersion relation*

$$0 = d_\pm(\lambda, \nu) := \det(D\nu^2 - c\nu + f'(q_\pm) - \lambda), \tag{2.67}$$

which is polynomial in the complex variables $\lambda, \nu \in \mathbb{C}$. For fixed *spatial eigenvalue* ν, we find N roots $\lambda_\ell = \lambda_\ell(\nu)$. Conversely, fixing the *temporal eigenvalue* λ, we find $2N$ roots $\nu_j = \nu_j(\lambda)$. By Rouché's theorem, the roots depend continuously on ν or λ, respectively. We recover the Fourier transform in x when specializing to $\nu = ik$ with real spatial wavenumber $k \in \mathbb{R}$. We therefore conclude that

$$\lambda \in \operatorname{spec} \mathcal{L}_\pm \iff d_\pm(\lambda, ik) = 0 \text{ for some } k \in \mathbb{R}.$$

The spectra of $\mathcal{L}_\pm$ then each decompose into N algebraic curves

$$\operatorname{spec} \mathcal{L}_\pm = \bigcup_\ell \Gamma_\pm^\ell, \quad \Gamma_\pm^\ell = \{\lambda_\pm^\ell(ik);\ k \in \mathbb{R}\}. \tag{2.68}$$

The imaginary part of the tangent vector to any dispersion curve $\lambda_\pm^\ell(ik)$, $k \in \mathbb{R}$, is called its *group velocity* c_g:

$$c_{g,\pm}^\ell(k) = \operatorname{Im} \frac{d\lambda_\pm^\ell(ik)}{dk}. \tag{2.69}$$

From the dispersion relation (2.67), the sum $c_{g,\pm}^\ell(k) + c$ is independent of the speed c of the coordinate frame.

For an interpretation, first consider the special case of purely imaginary temporal eigenvalues $\lambda_\pm^\ell(ik)$. For classical linear conservative wave equations like $u_t = u_x$, $u_{tt} = u_{xx}$, the Schrödinger equation, etc., the dispersion curve Γ is indeed vertical. Then the group velocity c_g describes the speed and direction of x-propagation of an initial condition, which consists of a narrow Gaussian wave package of spatial oscillations e^{ikx}. The same observation holds true in the dissipative context, for $\lambda_\pm^\ell(ik)$ with nonvanishing real part. The real part $\operatorname{Re}\lambda_\pm^\ell(ik)$ then just produces a superimposed exponential temporal growth or decay of the wave packet, which moves along the x-axis with its group velocity c_g.

We orient the curves $\Gamma_\pm^\ell$ in the direction of increasing k. Upwards orientation of $\Gamma_\pm^\ell$, in the complex plane, therefore indicates a positive group velocity, and downwards orientation a negative group velocity. The following proposition identifies the curves $\Gamma_\pm^\ell$ as *Fredholm borders*, where the Fredholm index of $\mathcal{L}_*$ changes.

Proposition 2.3.1 *The operator $\mathcal{L}_* - \lambda$ on $L^2(\mathbb{R})$ is Fredholm for λ in the complement of $\operatorname{spec} \mathcal{L}_+ \cup \operatorname{spec} \mathcal{L}_-$. The Fredholm index increases by one upon crossing a curve $\lambda_+^\ell(ik)$ from left to right, with respect to its given orientation; it decreases by one upon crossing a curve $\lambda_-^\ell(ik)$ from left to right.*

For example, assume that the group velocity $c_{g,+}^\ell(k)$ or $c_{g,-}^\ell(k)$ is directed towards the front position, which is stationary in our comoving coordinates:

$\pm c^\ell_{g,\pm}(k) < 0$. Then the Fredholm index of $\mathcal{L}_* - \lambda$ *decreases* when $\operatorname{Re} \lambda$ increases across the spectral curve $\lambda^\ell_+(ik)$ or $\lambda^\ell_-(ik)$, respectively. Conversely, if the group velocity is directed away from the interface, $\pm c^\ell_{g,\pm}(k) > 0$, then the Fredholm index of $\mathcal{L}_* - \lambda$ *increases* when $\operatorname{Re} \lambda$ increases across the spectral curve $\lambda^\ell_+(ik)$ or $\lambda^\ell_-(ik)$, respectively; see Figure 2.4. In particular, the Fredholm index is $+1$ to the left of the most unstable curve $\Gamma^\ell_\pm$, with $\operatorname{Re} \lambda$ maximal for all choices of $\ell, \pm$, if the associated group velocity is directed towards the front. The Fredholm index is -1 if it is directed away from the front.

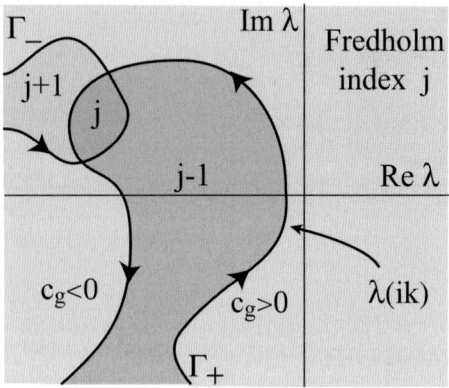

Fig. 2.4. A schematic plot of the oriented spectral curves $\Gamma^\ell_\pm$ and the Fredholm indices of $\mathcal{L}_* - \lambda$.

The proof of Propostion 2.3.1 is based on a formulation of the spectral problem $(\mathcal{L}_* - \lambda)u$ as a first-order differential equation

$$u_\xi = v, \qquad v_\xi = D^{-1}(-cv - f'(q_*(\xi))u + \lambda u). \tag{2.70}$$

Consider therefore a generalized eigenvalue problem for first-order differential operators

$$\mathcal{M}(\lambda)\mathbf{w} := \frac{d\mathbf{w}}{d\xi} - \mathbf{A}(\xi; \lambda)\mathbf{w} = 0. \tag{2.71}$$

Assume that the matrices $\mathbf{A}(x; \lambda)$ are asymptotically constant: $\mathbf{A}(x; \lambda) \to \mathbf{A}_\pm(\lambda)$ for $x \to \pm\infty$. Note that the asymptotic eigenvalues $\nu^\ell_\pm(\lambda)$ of $\mathcal{A}_\pm(\lambda)$ satisfy the dispersion relation

$$d_\pm(\lambda, \nu^\ell_\pm(\lambda)) := \det\left(-\nu^\ell_\pm - \mathbf{A}_\pm(\lambda)\right) = 0.$$

We define the asymptotic, *spatial Morse indices* $i_\pm(\lambda)$ as the dimensions of the unstable eigenspaces of $\mathbf{A}_\pm(\lambda)$, respectively. Note that from the Ansatz $\mathbf{w}(\xi) = e^{-\nu\xi}\mathbf{w}_0$, the spatial Morse index is given by the number of roots ν of the dispersion relation for given λ, with $\operatorname{Re} \nu < 0$.

Proposition 2.3.2 *The operator $\mathcal{M}(\lambda) = \frac{d}{d\xi} - \mathbf{A}(\xi; \lambda)$, considered as a closed, unbounded operator on $L^2(\mathbb{R})$, is Fredholm if, and only if, $\operatorname{Re} \nu_\pm^j \neq 0$ for all j. The Fredholm index is given by*

$$i(\mathcal{M}(\lambda)) = i_-(\lambda) - i_+(\lambda), \tag{2.72}$$

where the spatial Morse indices $i_\pm(\lambda) = \#\{\operatorname{Re} \nu_\pm^j < 0\}$ count the number of unstable eigenvalues of the asymptotic matrices $\mathbf{A}_\pm(\lambda)$, with algebraic multiplicity.

In the formulation as a first order, generalized eigenvalue problem, the proposition actually covers a much broader class of linear operators than linearizations about travelling waves in reaction-diffusion equations. We refer to [SaSc01d] for a still more general formulation, which allows to cover linearizations about time-periodic states as well as problems in several space dimensions $m > 1$.

Essential spectra do not depend on function spaces such as L^p, C^0, or H^k, as long as translation invariant norms are considered. However, the locations of the Fredholm borders $\Gamma_\pm^\ell$ do change when we consider spaces with exponential weights. For weights $\underline{\eta} = (\eta_+, \eta_-) \in \mathbb{R}^2$, define

$$L_{\underline{\eta}}^2 := \left\{ u \in L_{\text{loc}}^2;\ \|u\|_{L_{\underline{\eta}}^2} < \infty \right\}, \quad \|u\|_{L_{\underline{\eta}}^2}^2 := \int_{\mathbb{R}^+} |e^{\eta_+ \xi} u(\xi)|^2 + \int_{\mathbb{R}^-} |e^{\eta_- \xi} u(\xi)|^2. \tag{2.73}$$

Positive weights $\eta_\pm > 0$ stabilize transport to the left. For example, consider pure transport to the left by $u_t = u_\xi$, on the entire real line. The trivial solution $q_* \equiv 0$ is then stabilized by an exponential weight with $\eta := \eta_+ = \eta_- > 0$. Indeed, the exponentially weighted solution $v(t, \xi) = e^{\eta \xi} u(t, \xi)$ satisfies $\|u\|_{L_\eta^2} = \|v\|_{L^2}$, for all fixed t. Moreover, v solves the damped transport equation $v_t = v_\xi - \eta v$, and hence decays to zero exponentially, in $L^2(\mathbb{R})$, with rate $\eta > 0$.

More generally, transport is measured by the group velocity, as reflected by the following lemma.

Lemma 2.3.3 *The essential spectrum in $L_{\underline{\eta}}^2$ can be determined from Proposition 2.3.1, if we replace the dispersion relations (2.67) at the asymptotic states by the shifted dispersion relations*

$$d_\pm^\eta(\lambda, \nu) := d_\pm(\lambda, \nu - \eta_\pm). \tag{2.74}$$

In particular, small weights $\eta_\pm$ infinitesimally shift the Fredholm borders $\Gamma_\pm^\ell = \{\lambda_\pm^\ell(ik; \eta_\pm);\ k \in \mathbb{R}\}$ as follows:

$$\left. \frac{\partial \operatorname{Re} \lambda_\pm^\ell(ik; \eta)}{\partial \eta} \right|_{\eta_\pm = 0} = -c_{g,\pm}^\ell(k).$$

The key observation in the proof of Lemma 2.3.3 is that $\lambda_\pm^\ell(ik;\eta) = \lambda_\pm^\ell(ik-\eta;0)$, not only for real but also for complex k. Therefore

$$\left.\frac{\partial \operatorname{Re}\lambda_\pm^\ell(ik;\eta)}{\partial \eta}\right|_{\eta=0} =$$

$$\left.\frac{\partial \operatorname{Re}\lambda_\pm^\ell(ik-\eta;0)}{\partial \operatorname{Im} k}\right|_{\eta=0} = -\left.\frac{\partial \operatorname{Im}\lambda_\pm^\ell(ik-\eta;0)}{\partial \operatorname{Re} k}\right|_{\eta=0} = -c_{g,\pm}^\ell(k).$$

Here, we have used the Cauchy-Riemann equations for complex analytic functions in the second identity.

Proofs of Propositions 2.3.1 and 2.3.2 can be found, in a slightly different context, in [Pa88, He81]; see also [SaSc01d] for a more general setup. We conclude this section on essential spectra with several examples.

Consider the linear scalar reaction-advection-diffusion equation

$$w_t = w_{\xi\xi} + w_\xi + aw =: \mathcal{L}_* w \tag{2.75}$$

with positive drift velocity and constant reaction coefficient $a \in \mathbb{R}$. We first view $\mathcal{L}_*$ as an unbounded operator on $L^2(\mathbb{R})$, as in Propositions 2.3.1 and 2.3.2.

From (2.67) and (2.68), we obtain the dispersion relation and spectrum of $\mathcal{L}_*$ to be given by the parabola

$$0 = d(\lambda, ik) = -k^2 - ik + a - \lambda; \tag{2.76}$$
$$\operatorname{spec} \mathcal{L}_* = \{\lambda = -k^2 + a + ik;\ k \in \mathbb{R}\} = \{\lambda \in \mathbb{C};\ \operatorname{Re}\lambda = (\operatorname{Im}\lambda)^2 + a\} =: \Gamma.$$

Preparing for the more general case, where a depends on ξ, we now consider (2.75) on $\xi \in \mathbb{R}_+$ or $\mathbb{R}_-$ and equip the arising differential operators $\mathcal{L}_>$ and $\mathcal{L}_<$ with Dirichlet boundary conditions in $\xi = 0$. We claim that

$$\operatorname{spec}\mathcal{L}_> = \operatorname{spec}\mathcal{L}_< = \{\lambda \in \mathbb{C};\ \operatorname{Re}\lambda \le (\operatorname{Im}\lambda)^2 + a\}, \tag{2.77}$$

with Fredholm index $i(\mathcal{L}_> - \lambda) = 1$ and $i(\mathcal{L}_< - \lambda) = -1$ in the interior of the spectrum. Indeed, we follow (2.70) to rewrite $(\mathcal{L}_> - \lambda)w = 0$ as a first order system

$$\mathcal{M}(\lambda)\mathbf{w} = \mathbf{w}_\xi + \mathbf{A}(\lambda)\mathbf{w} = 0, \qquad \mathbf{A}(\lambda) = \begin{pmatrix} 0 & -1 \\ a-\lambda & 1 \end{pmatrix}, \tag{2.78}$$

for $\xi > 0$. For $\operatorname{Re}\lambda < (\operatorname{Im}\lambda)^2 + a$, to the left of the Fredholm border Γ, both eigenvalues $\nu^{1,2}(\lambda)$ of $\mathbf{A}(\lambda)$ possess strictly positive real part. The one-dimensional subspace in the $\mathbf{w}$-plane, selected by the Dirichlet boundary conditions at $\xi = 0$ therefore gives rise to a one-dimensional kernel of exponentially decaying functions. The operator $\mathcal{L}_>$ is surjective since solutions to the initial value problem

$$\mathbf{w}_\xi = -\mathbf{A}(\lambda)\mathbf{w} + \mathbf{h}(\xi)$$

with $\mathbf{w}(0) = 0$ for $\xi > 0$ belong to L^2. This shows that $\mathcal{L}_> - \lambda$ is Fredholm of index 1 to the left of the Fredholm border. The operator $\mathcal{L}_<$ is conjugate to the adjoint of $\mathcal{L}_>$ on $\mathbb{R}^+$, by the reflection $x \mapsto -x$, and therefore possesses Fredholm index -1. For λ to the right of the Fredholm border, both operators are invertible since the Dirichlet subspace does not coincide with an eigenspace of $\mathbf{A}(\lambda)$.

Next consider piecewise constant coefficients $a(\xi) = a_\pm$ for $\pm\xi > 0$ in the advection-diffusion equation (2.75). We then obtain corresponding spectral borders $\Gamma_\pm \subset \mathbb{C}$, which are parabolas shifted by $a_\pm$, respectively. Assume $a_+ > a_-$, first. We claim that

$$\operatorname{spec} \mathcal{L}_* = \{\lambda \in \mathbb{C};\ (\operatorname{Im}\lambda)^2 + a_- \leq \operatorname{Re}\lambda \leq (\operatorname{Im}\lambda)^2 + a_+\}, \quad (2.79)$$

with Fredholm index $i(\mathcal{L}_* - \lambda) = 1$ in the interior of the spectrum. Indeed, consider the first-order eigenvalue problem (2.78) with λ between the spectral borders $\Gamma_\pm$. The eigenvalues of the asymptotic matrices $\mathbf{A}_\pm$ then satisfy $\operatorname{Re}\nu_-^1 < 0 < \operatorname{Re}\nu_-^2$ and $\operatorname{Re}\nu_+^{1,2} > 0$. The differential equation therefore possesses a one-dimensional subspace of solutions $\mathbf{w}(\xi)$ which are continuous in $\xi = 0$ and decay for $\xi \to \pm\infty$. This contributes a one-dimensional kernel in the Fredholm index one region of $\mathcal{L}_*$. Surjectivity between the Fredholm borders, and invertibility in the Fredholm index zero region, follow just like for the operator $\mathcal{L}_>$.

In the opposite case $a_- > a_+$, the Fredholm index is -1 between the Fredholm borders $\Gamma_\pm$. In the interpretation of Proposition 2.3.1, the group velocity $c_g = -1$ is constant and negative on Γ_+, directed towards the "front" in $\xi = 0$. The Fredholm index therefore decreases from left to right. However, on Γ_- the group velocity $c_g = -1$ is directed away from the "front", and the Fredholm index increases from left to right.

Essential spectra for more general coefficients $a = a(\xi)$ with $a(\xi) \to a_\pm$ coincide with the essential spectra computed above, since localized ξ-dependence of $a(\xi)$ amounts to a relatively compact perturbation of $\mathcal{L}_*$, which does not change Fredholm properties [Ka66].

Just like in the example of simple translation $u_t = u_\xi$, exponential weights $\eta = (\eta_+, \eta_-)$ with $\eta_+, \eta_- > 0$ shift the Fredholm borders to the left, as can be readily seen from the complex dispersion relations, or Lemma 2.3.3.

As a final example, we comment on spectra in the case of spatially periodic coefficients $a = a(\xi) = a(\xi + p)$. This situation arises under linearization $a(\xi) = f'(q_{\mathrm{wt}}(\xi))$ along a periodic wavetrain with nonlinear spatial wave vector $k_{\mathrm{wt}} = 2\pi/p$, in comoving coordinates ξ; see for example the end of Section 2.3.2. In a more demanding context, the problem reappears in our stability analysis of Archimedean spiral waves, for radial spatial dynamics in the farfield limit of infinite radius; see Section 2.4.2, 2.5.3, and equations (2.123)-(2.125). Standard spatial Floquet theory for the eigenvalue problem

$$u_\xi = v, \quad v_\xi = -cv - a(\xi)u - \lambda u \tag{2.80}$$

will consider exponential functions $\mathbf{w} = (u, v)$,

$$\mathbf{w}(\xi) = e^{-\nu\xi}\tilde{\mathbf{w}}(\xi), \tag{2.81}$$

with spatially p-periodic $\tilde{\mathbf{w}}$ and spatial Floquet exponent $\nu \in \mathbb{C}$. As usual, ν is only determined up to integer multiples of $2\pi i/p = ik_{\text{wt}}$. Explicitly,

$$(\partial_\xi - \nu)^2 \tilde{\mathbf{w}} + (\partial_\xi - \nu)\tilde{\mathbf{w}} + a(\xi)\tilde{\mathbf{w}} = \lambda \tilde{\mathbf{w}} \tag{2.82}$$

Alternatively, we can work with $\mathbf{w}$ directly and impose the Floquet boundary condition

$$\mathbf{w}(p) = e^{-\nu p}\mathbf{w}(0).$$

The Floquet exponents ν play the role of the spatial wave numbers in the dispersion relation (2.67). Looking for solutions with neutral growth $\nu = ik$, $k \in \mathbb{R}$, in analogy to the spectral results of Propositions 2.3.1, 2.3.2, we obtain the dispersion relation

$$d(\lambda, ik) = \det\left(\Phi(\lambda) - e^{-ikp}\right) = 0, \tag{2.83}$$

where $\Phi(\lambda)$ denotes the period map to the linear differential equation (2.80). Note that the dispersion relation is invariant under the Floquet shift $k \to k + k_{\text{wt}}$. The values of k in the dispersion relation (2.83) are referred to as *Bloch wavenumbers*. The associated spectral values $\lambda^\ell(ik)$, $\ell = 1, 2, \ldots$ are the eigenvalues of the elliptic operator $\mathcal{L}_*$ on $L^2(\mathbb{R})$. The eigenfunctions $u(\xi)$ are called *Bloch waves* associated with the Bloch wavenumber k. We can also define group velocities

$$c_g^\ell(k) := \frac{d\,\text{Im}\,\lambda^\ell(ik)}{dk},$$

in analogy to (2.69). Analogous statements to Propositions 2.3.1, 2.3.2 then hold for equations with asymptotically periodic coefficients, with Fredholm borders Γ^ℓ defined through (2.83).

2.3.2 Instabilities of Travelling Waves

In this section, we investigate relative equilibria which lose stability for the PDE (2.64)

$$u_t = Du_{xx} + f(u; \mu),$$

on the unbounded real line $x \in \mathbb{R}$. Here, $\mu \in \mathbb{R}$ denotes a typical control parameter, driving the instability, which we assume to occur at the value $\mu = 0$. Instabilities can be caused either by point spectrum or by essential spectrum crossing the imaginary axis. Without striving for completeness, we present four different cases of Hopf bifurcation from travelling waves, here. Three of them are caused by essential spectrum crossing the imaginary axis.

Fredholm indices play a most prominent role when solving the resulting nonlinear equations, locally. Lyapunov-Schmidt reduction generally provides reduced equations, which map the kernel to the cokernel. Because the boundaries of the essential spectrum were characterized precisely by the loss of Fredholm property, in Section 2.3.1, we will rely on spatial dynamics ideas, in the spirit of [Ki82, Fi84] to discuss these latter bifurcations.

Instability of a front caused by point spectrum First consider a travelling wave $u = q_*(x - c_*t)$ with essential spectrum of the linearization $\mathcal{L}_*$ strictly contained in the open left half-plane, see (2.64),(2.68). We assume that there is a pair of simple, purely imaginary eigenvalues crossing the imaginary axis. In addition, there necessarily is a zero eigenvalue due to translation of the relative equilibrium q_*. This additional eigenvalue prevents the straight-forward application of standard Hopf bifurcation theorems. We choose to proceed by center manifold reduction, instead. With a careful choice of cut-off functions, we can construct a center manifold which is invariant under translations [He81, Sa&al97b]. With an appropriate choice of coordinates, we find a three-dimensional reduced system of ODEs on the center manifold, which is of skew-product form

$$\dot{v} = h(v; \mu), \qquad \dot{g} = c(v; \mu). \qquad (2.84)$$

Here, $v \in \mathcal{U} \subset \mathbb{C}$, $g \in \mathbb{R}$, $h(0; 0) = 0$ and $\operatorname{spec} h_v(0;0) = \{\pm i\omega\}$. The (small) v-variable parameterizes the Hopf eigenspace. The (global) variable g in the translation group $G = \mathbb{R}$ parameterizes the position of the front. Since we can preserve the translation equivariance of the original reaction-diffusion system (2.64) through the center manifold reduction, the Hopf bifurcation in the v-equation is independent of the translation component g. As a result, we typically encounter a generic Hopf bifurcation in the v-equation. Unique periodic orbits $v_*(t; \mu)$ arise super- or subcritically, depending only on the sign of a certain cubic normal form coefficient. The v-dynamics enter the g-equation, with the periodic solution $v_*(t; \mu)$ acting as a time-periodic forcing. Phenomenologically, this leads to a periodically oscillating speed $c = c(v_*(t; \mu); \mu)$ of the wave front position. The periodic variable $v_*(t; \mu)$ indicates small periodic shape fluctuations of the propagating wave.

We refer to Section 2.5.2 for a thorough discussion in the more complicated, but analogous situation of meandering spiral wave patterns.

The Turing instability When the essential spectrum of the linearization $\mathcal{L}_*$ at a front touches the imaginary axis, a naive finite-dimensional Lyapunov-Schmidt reduction fails, due to the absence of the Fredholm property. The dynamics, generated by a continuum of eigenmodes, cannot be represented by a finite collection of ordinary differential equations. We illustrate the effects of critical essential spectrum as opposed to critical point spectrum by means of an example: one of the asymptotic states of a propagating front experiences a Turing instability. We begin with some background on the nature of this pattern forming mechanism, predicted by Turing in 1952 [Tu52], and observed, only in 1989, in laboratory experiments [Ca&al90].

A *Turing instability* occurs at one of the asymptotic states $q_\pm$ if the spectrum of $\mathcal{L}_\pm$ in the steady frame touches the imaginary axis at $\lambda = 0$, with spatially inhomogeneous critical eigenfunction. To fix ideas, we assume that the Turing instability occurs at q_+ and we set $q_+(\mu) \equiv 0$. We consider the reaction-diffusion system (2.64) in the steady frame, first

$$u_t = Du_{xx} + f(u;\mu), \quad x \in \mathbb{R},$$

with linearization

$$\mathcal{L}_0(\mu) = Du_{xx} + f'(0;\mu)u.$$

In the notation of Section 2.3.1, one of the curves Γ_+^ℓ touches the imaginary axes with a zero eigenvalue $\lambda(ik_*) = 0$, for nonzero wavenumber, $k_* \neq 0$; see Figure 2.5. We assume that $\lambda^\ell(k_*;\mu = 0) = 0$ is a simple root of the

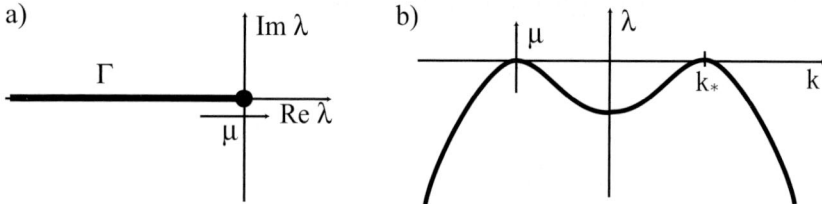

Fig. 2.5. The critical curve $\Gamma_+^\ell \subset \mathbb{C}$, **(a)**, and the temporal eigenvalue λ_+^ℓ as a function of the wavenumber k, **(b)**, are shown at criticality.

dispersion relation (2.67), which crosses the imaginary axis with nonvanishing speed. After a suitable scaling, we find

$$\lambda^\ell(k;\mu_0) = \mu - (k - k_*)^2 + O((k - k_*)^3), \tag{2.85}$$

for $k \sim k_*$. Since $\mathcal{L}_0$ possesses real coefficients, the curve Γ_+^ℓ is necessarily contained in the real axis close to the origin.

It was Turing's original discovery that this instability can actually occur in reaction-diffusion systems, although the spatially homogenizing effect of diffusion suggests that the most unstable eigenfunction ought to be spatially homogeneous. In case of a Turing instability, a spatially structured eigenfunction $u_* \cos(k_* x)$ is amplified by the time evolution for the linearization of the reaction-diffusion system (2.64) about the critical state $q_+ = 0$

$$u_t = Du_{xx} + f'(0;0)u. \tag{2.86}$$

The resulting spatially periodic, stationary pattern for the nonlinear equation (2.64) near 0 is generally referred to as a *Turing pattern* $T(x)$.

One possible theoretical approach to this instability imposes spatial periodicity with period $L \sim 2\pi/k_*$ in the function space, restricting the allowed values of the wavenumber of k to multiples of $2\pi/L$; see [Go&al88]. Then the Turing bifurcation reduces to a pitchfork bifurcation in the space of even functions with prescribed period L, with reduced equation on a center manifold given by

$$\dot{z} = (\mu - (k - k_*)^2)z + \beta(k;\mu)z^3 + O(z^5) \in \mathbb{R}. \tag{2.87}$$

We will assume throughout that $\beta(k_*;0) < 0$, such that the bifurcating solutions are asymptotically stable — in the space of functions with the prescribed period $2\pi/k$.

In a slightly different spirit, Turing instabilities on unbounded domains, without imposing periodicity, have been described by a partial differential equation instead of our single ordinary differential equation. The dynamics of the reaction-diffusion system can be approximated by a slow, long-wavelength modulation of the complex amplitude of the critical eigenfunction:

$$u(t,x) \sim \sqrt{\mu}\left(A(\mu t, \sqrt{\mu}x)e^{ik_*x} + \bar{A}(\mu t, \sqrt{\mu}x)e^{-ik_*x}\right)u_*.$$

The time evolution of the complex amplitude $A(\tilde{t},\tilde{x}) \in \mathbb{C}$ is to leading order given by a Ginzburg-Landau equation

$$A_{\tilde{t}} = A_{\tilde{x}\tilde{x}} + A - A|A|^2, \tag{2.88}$$

on time scales $\tilde{t} = O(1)$. In the Ginzburg-Landau description (2.88), the stationary Turing patterns are solutions of the form $A(\tilde{x}) = r(\tilde{k})e^{i\tilde{k}\tilde{x}}$. Their linearized stability can be explicitly calculated. In $(\mu,\tilde{k})$-parameter space, only Turing patterns which are sufficiently close to the critical wavenumber, $k = k_*$, or $\tilde{k} = 0$, are stable — with respect to not necessarily spatially periodic perturbations. The boundary of stability is commonly referred to as the Eckhaus boundary; see Figure 2.6.

Turing patterns $T(x;\mu,k)$ exist for parameter values $\mu > \mu_{\text{ex}}(k-k_*)$, and are asymptotically stable for $\mu > \mu_{\text{st}}(k-k_*) \geq \mu_{\text{ex}}(k-k_*)$. The boundary

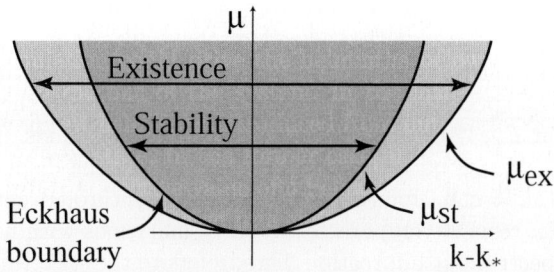

Fig. 2.6. Existence and stability of Turing patterns depending on spatial wavenumber k and bifurcation parameter k.

curves can be computed to leading order from the Ginzburg-Landau approximation (2.88):

$$\mu_{\text{ex}}(\tilde{k}) = \tilde{k}^2 + O(\tilde{k}^3), \quad \mu_{\text{st}}(\tilde{k}) = 3\tilde{k}^2 + O(\tilde{k}^3).$$

We refer to [CrHo93, MiSc96, Mi97, SaSc00b, Sc98c] for physics background and mathematical proofs in this context.

Essential Hopf instability of a front We are now going to describe a general bifurcation result for a front, where one of the asymptotic states undergoes a Turing instability. Reflecting the spatial variable x, if necessary, we may fix $c_* > 0$, that is, propagation towards $x = +\infty$. Then the direction of propagation of the front distinguishes between the asymptotic states $q_\pm$. We refer to q_+ as the state *ahead* of the front, and to q_- as the state which is *behind* the front.

Whereas the Turing bifurcation is stationary in the steady frame, the solution $u_* \exp i k_* x$ to the linearized equation (2.86) becomes time-periodic $u_* \exp(ik_* \xi + i\omega_* t)$, in a comoving frame with speed $c_* = \omega_*/k_* > 0$. Viewed in the comoving frame, the Turing bifurcation therefore becomes a Hopf bifurcation. Typically, in a Hopf bifurcation, we are interested in time-periodic solutions. More specifically, we shall be interested in the existence and stability of *modulated front solutions* to the reaction-diffusion system (2.64):

$$u(t,x) = q(t, x - ct), \quad \text{with } q(t, \xi) = q(t + P, \xi), \text{ for all } t, \xi \in \mathbb{R},$$

for some temporal period P, and an appropriate wave speed $c \sim c_*$.

Setting up a typical bifurcation result usually requires *existence* of a primary solution branch, *minimal critical spectrum*, and assumptions on *nonlinear terms* in the Taylor expansion. We are now going to make these assumptions precise for our set-up. Recall that x always refers to the spatial coordinate in the steady frame, whereas ξ always refers to some comoving frame.

At criticality $\mu = 0$, we assume *existence* of a uniformly translating front solution, $q(x - c_* t) \to q_\pm$ for $x \to \pm\infty$ for some strictly positive speed of propagation $c_* > 0$.

The *critical spectrum* is assumed to generate a Turing instability at one of the asymptotic states, either ahead of the front, at q_+, or behind the front, at q_-. In the comoving frame $\xi = x - c_* t$, this Turing instability corresponds to a spectral curve Γ_+ (or Γ_-, respectively) of the linearization about the front $\mathcal{L}_*$, (2.66) which touches the imaginary axis. The tangency at $\lambda(ik_*) = i\omega_* \neq 0$ is quadratic from the left. From the dispersion relation (2.67), we find $\omega_* = c_* k_*$. Increasing the parameter μ through zero, we assume that Γ_+ (or Γ_-, respectively) crosses the imaginary axis with nonzero speed.

Minimal spectrum at bifurcation, here, refers to the spectrum of the front in a comoving frame: we assume that the critical spectrum only consists of the part related to the Turing instability and the simple eigenvalue $\lambda = 0$, induced by translation. More precisely, we assume that $\lambda(ik_*) = \pm i\omega_*$ are the only roots of the dispersion relation (2.67) in the comoving frame $c = c_*$ on the imaginary axis $\lambda \in i\mathbb{R}$, and $\lambda = \pm i\omega_*$ are simple roots. Moreover, we require absence of critical point spectrum on the imaginary axis: in an exponentially weighted space $L^2_\eta(\mathbb{R})$ with norm

$$\|u\|^2_{L^2_\eta} := \int_\mathbb{R} |e^{\eta\xi} u(\xi)|^2 < \infty,$$

we assume that
$$\operatorname{spec}_{L^2_\eta} \mathcal{L}_* \cap i\mathbb{R} = \{0\},$$

is algebraically simple, for all $\eta > 0$ sufficiently small. Note that compared to Lemma 2.3.3, we used exponential weights $\eta_+ = \eta_- := \eta$.

Nonlinear terms for the Turing bifurcation in a neighborhood of the asymptotic state q_+ (or q_-, respectively) are most easily computed in the steady frame, restricting to even, $2\pi/k$-periodic functions. As mentioned above, the instability then induces a pitchfork bifurcation on a one-dimensional center manifold, with Taylor expansion

$$\dot{z} = \lambda(ik;\mu)z + \beta(k;\mu)z^3 + O(z^5).$$

We assume $\beta(k_*;0) < 0$ such that the bifurcating Turing patterns are stable within this class of periodic functions. From elementary bifurcation theory, we find for each $k \sim k_*$ and $\mu > 0$, $\mu \sim 0$, a periodic Turing pattern $T(x;\mu,k)$, which is time-independent in the original frame x and unique up to translation in x.

Theorem 2.3.4 *[SaSc01b] Assume existence of a front solution $q_*(x - c_*t)$, $c_* > 0$, with minimal, critical essential spectrum causing a supercritical Turing instability, as described above.*

(I) Assume the Turing instability occurs ahead of the front, at q_+, only. Then, for each Turing pattern $T(x;\mu,k)$ with (k,μ) near $(k_*,0)$, there exists a modulated front solution of the reaction-diffusion system (2.64), $q(t, x - c(\mu,k)t;\mu,k)$, invading the Turing pattern

$$q(t,\xi;\mu,k) \to T(\xi + c(\mu,k)t;\mu,k), \quad \text{for } \xi \to \infty,$$

uniformly in time t. The modulated front is $\sqrt{\mu}$-close to the primary front q_*, uniformly in ξ and t, and is $P(\mu,k)$-periodic in its first argument, with $P(0,k_*) = 2\pi/(c_*k_*)$ and $c(0,k_*) = c_*$. See Figure 2.7 for a schematic plot. The bifurcating modulated front is unique in the class of modulated fronts, close to the primary front $q_*(\xi)$, invading a fixed Turing pattern, up to spatial and temporal translations.

If the bifurcating Turing pattern is stable, that is, if it lies inside the Eckhaus boundary, then the modulated front is spectrally stable.

(II) Assume the Turing instability occurs behind the front, at q_-, only. Then for any parameter value $\mu \sim 0$, any speed $c \sim c_*$, any wavenumber $k \sim k_*$, and any temporal period $P \sim 2\pi/(c_* k_*)$, there does not exist a modulated front solution $q(t, x - ct)$ to (2.64), which is P-periodic in the first argument, and which leaves a Turing pattern behind

$$q(t, \xi) \to T(\xi + ct; \mu, k), \quad \text{for } \xi \to -\infty;$$

see Figure 2.8.

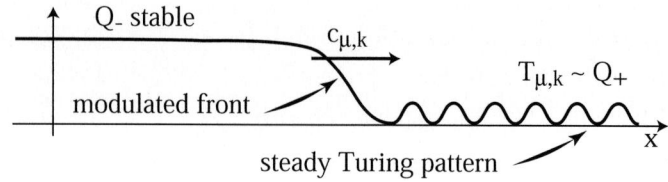

Fig. 2.7. A modulated front invading a Turing pattern, case I

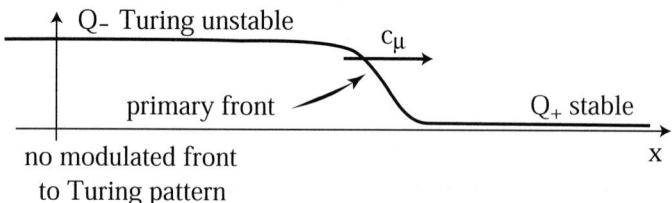

Fig. 2.8. Non-existence of modulated fronts leaving a Turing pattern behind, case II.

Stability in the theorem refers to the spectrum of the temporal period map of the reaction-diffusion system, linearized at the (time-periodic) modulated front in a comoving coordinate frame.

If the front moves to the left, $c_* < 0$, we may reflect x and find bifurcating modulated waves for an instability at q_-. In this sense, bifurcation, case I, occurs whenever the front is invading the Turing unstable state, and bifurcation failure, case II, occurs when the front leaves the Turing unstable state behind.

We emphasize that bifurcation failure, case II, is in contrast to finite-dimensional bifurcation theory, where generically periodic solutions arise in Hopf bifurcation.

Instability of a pulse caused by the essential spectrum If $q_+ = q_- = 0$, bifurcation is in conflict with bifurcation failure in Theorem 2.3.4, above. The Turing instability then occurs ahead of the front and behind the front, simultaneously. This scenario naturally arises if we consider pulses

$$0 \not\equiv q_*(\xi) \to 0, \text{ for } |\xi| \to \infty.$$

Theorem 2.3.5 *[SaSc99] Assume existence of a pulse solution $q_*(x - c_*t)$, with minimal, critical essential spectrum causing a supercritical Turing instability as described in Section 2.3.2.*

Then, for each Turing pattern $T(x; \mu, k)$ with $(\mu, k) \sim (0, k_*)$, there exists a modulated pulse solution of the reaction-diffusion system (2.64), $q(t, x - c(\mu, k)t; \mu, k)$, travelling through the Turing pattern. More precisely, there are (μ, k)-independent constants $C, \eta, \eta' > 0$, and phases $\theta_\pm(\mu, k)$ such that

$$|q(t, \xi) - T(\xi + ct + \theta_+)| \leq Ce^{-\eta|\xi|}, \text{ for all } \xi > 0,$$
$$|q(t, \xi) - T(\xi + ct + \theta_-)| \leq Ce^{-\eta'\mu|\xi|}, \text{ for all } \xi < 0.$$

We have suppressed dependence of q, T, c, and $\theta_\pm$ on (μ, k). The modulated pulse is $\sqrt{\mu}$-close to the primary pulse, uniformly in ξ and t, and periodic in its first argument t, with period $P = P(\mu, k)$, $P(0, k_*) = 2\pi/(c_* k_*)$ and $c(0, k_*) = c_*$. See Figure 2.9 for a schematic plot. The bifurcating modulated pulse is unique in the class of modulated pulses, (x, t)-uniformly close to the primary pulse $q_*(\xi)$, moving through a fixed Turing pattern, up to spatial and temporal translations.

If the bifurcating Turing pattern is stable, that is, if it lies inside the Eckhaus boundary, then the modulated pulse is spectrally stable.

Again, spectral stability refers to the linearized temporal period-P map along the time-periodic modulated pulse.

Summarizing, we find bifurcation of time-periodic solutions in this case of simultaneous instability ahead and behind a pulse, just like in the case of fronts invading the Turing pattern. Reminiscent of the nonexistence in case of an instability behind the front is the recovery zone of length $1/\mu$ behind the pulse, marked by the slow convergence to the Turing patterns for $\xi \to -\infty$.

Fredholm indices and essential instabilities There are several ways of understanding Theorems 2.3.4 and 2.3.5. We first appeal to intuition.

Consider bifurcation of fronts case I, first. Before the instability, $\mu < 0$, the homogeneous state q_- on the left invades the stable state q_+ on the right by means of the front moving with speed $c \sim c_* > 0$. For $\mu > 0$ the homogeneous state ahead of the front has become unstable and Turing

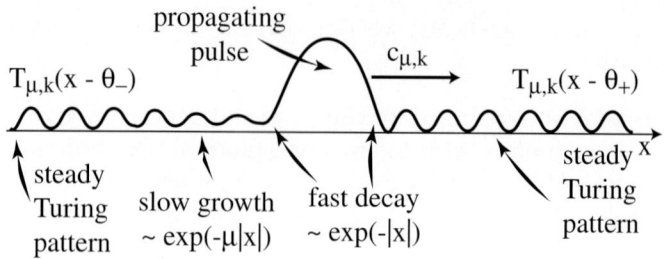

Fig. 2.9. A modulated pulse moving through a Turing pattern

patterns have formed. Intuition suggests that the front would invade these Turing patterns close to the homogeneous state q_+, which was invaded for $\mu < 0$.

Consider bifurcation failure, case II, next, with $\mu > 0$. We expect the dynamics near q_- behind the front to be described approximately by the Ginzburg-Landau equation (2.88). In this approximation, there exist travelling waves, where a Turing pattern invades the unstable homogeneous background $A = 0$. For example, we may consider the Turing pattern $A(x) \equiv 1$ constant. The spread of $A \equiv 1$ into the unstable background $A \equiv 0$ (alias q_-), is then described by the second order travelling wave equation

$$A'' + \tilde{c}A' + A - A^3 = 0,$$

for solutions $A = A(\tilde{x} - \tilde{c}\tilde{t})$ of (2.88). Monotone heteroclinic orbits, connecting $A = 1$ to $A = 0$ exist for all speeds $\tilde{c} \geq 2$. In the unscaled coordinates t, x, the speed $\tilde{c} = 2$ corresponds to a speed $c_{\text{Turing}} = O(\sqrt{\mu})$. The heteroclinic represents a small front of a Turing pattern to the left invading the unstable state q_-, to the right. At present, there is no proof, that Turing patterns actually spread with this (minimal) speed into the unstable homogeneous background state; see, however, [CoEc00] for evidence in this direction. Now observe that in this picture, the speed of a front between a Turing pattern and the homogeneous unstable state q_- is much slower than the speed of the primary front. Imagine an initial condition consisting of the Turing pattern on the left, followed to the right by the unstable state q_-, then the front, and finally the stable homogeneous state q_+ on the right. Then the interface between the Turing pattern and the unstable state q_- moves much slower than the primary front. This leads lead to a wedge opening between the two fronts; see Figure 2.10.

Another explanation would take a functional analytic point of view, still formal, and allude to the results on Fredholm properties of the linearization, Proposition 2.3.1.

Consider case I, first. Observe that the Fredholm index of the linearization $\mathcal{L}_* - i\omega_*$ along the front in a comoving frame changes at the bifurcation point

2 Spatio-Temporal Dynamics of Reaction-Diffusion Patterns   69

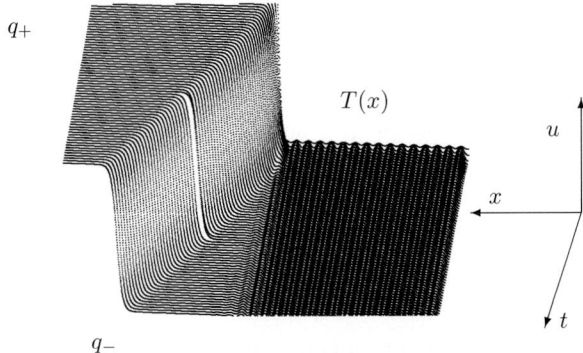

Fig. 2.10. The wedge of an unstable state formed between a small amplitude front of Turing patterns $T(x)$ and a large amplitude front connecting q_+ and q_-.

$\mu = 0$. The group velocity associated with the Turing instability in a steady frame is zero, since $\lambda(ik)$ is real for all $k \sim k_*$. In a comoving frame of speed $c_* > 0$, the group velocity is negative. Therefore, by Proposition 2.3.1, the Fredholm index changes from 0 to $+1$ when crossing the critical spectral curve $\lambda(ik)$ from the unstable to the stable complex half plane; see Figure 2.11.

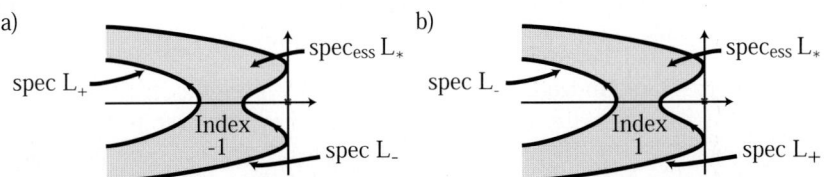

Fig. 2.11. Spectra, Fredholm indices, and oriented critical spectral curves of the linearization about a front, when a Turing instability occurs ahead of the front **(a)** or behind the front **(b)**.

On the linear level, this index computation suggests that after the instability, $\mu > 0$, the Fredholm index of the linearization at $i\omega_*$ is $+1$, such that the linearization possesses a one-dimensional kernel, and we should be able to find a family of solutions, parameterized by the wavenumber k.

In case II, the analogous argument shows that the Fredholm index is -1 after the instability, $\mu > 0$, since, again, the associated group velocity is negative, but now points away from the interface; see Proposition 2.3.1. We therefore do not expect to find any periodic solution since the cokernel is at least one-dimensional.

However, this reasoning remains formal, since right at the moment of bifurcation, $\mu = 0$, the linearization is not Fredholm and Lyapunov-Schmidt reduction does not apply. In fact, the reasoning fails in case of Theorem 2.3.5. The Turing instability occurs simultaneously ahead and behind the front. Therefore, two identical spectral curves touch the imaginary axis simultaneously, both with negative group velocity. One of the curves is associated with the linearization at q_+, one with q_-. Again, by Proposition 2.3.1, the Fredholm index to the left and to the right of these two identical critical curves is zero. Still, a family of solutions parameterized by the wavenumber k bifurcates, just like in the case of Fredholm index $+1$; see Theorem 2.3.5.

Spatial dynamics and essential instabilities We now outline our strategy for the proof of Theorems 2.3.4 and 2.3.5. The proofs rely on a spatial dynamics formulation of the parabolic equation (2.66) in the spirit of [Ki82]. Consider the system

$$U_\xi = V$$
$$V_\xi = D^{-1}(-cV + \omega \partial_t U - F(U; \mu)), \quad (2.89)$$

in a space $(U, V) \in X = H^{1/2}(S^1) \times L^2(S^1)$ of functions which are 2π-periodic in time $t \in S^1$. The wave speed c and the temporal period $P = 2\pi/\omega$ enter as additional parameters into the equation. We view system (2.89) as an abstract differential equation for $\mathbf{U} = (U, V)(\cdot)$

$$\mathbf{U}_\xi = \mathbf{F}(\mathbf{U}; \omega, c, \mu) \in X. \quad (2.90)$$

The goal is to find heteroclinic orbits to this abstract differential equation on a Hilbert space. Indeed, modulated fronts are periodic in time t for an appropriate wave speed c. In spatial "time" ξ, they converge to the asymptotic states $q_\pm$ or the Turing patterns. The Turing patterns are periodic in time t and spatial "time" ξ, in the comoving frame. They arise out of a Hopf bifurcation from the equilibrium which undergoes a Turing instability. The essential instability therefore reduces to a global heteroclinic bifurcation, where a heteroclinic orbit connects two equilibria, one of which undergoes a Hopf bifurcation.

Before we proceed to analyze this heteroclinic bifurcation, we want to point out the ill-posed character of (2.90). The initial value problem to this equation is ill-posed, even in the example of the simple heat equation

$$U_\xi = V, \quad V_\xi = \partial_t U. \quad (2.91)$$

Indeed, we have solutions $U(\xi, t) = e^{i\ell t + \nu_\ell \xi}$ to (2.91), with $\nu_\ell = \pm\sqrt{i\ell}$, for any $\ell \in \mathbb{Z}$. In particular, there is no a priori bound on the exponential growth of solutions, neither in forward, nor in backward spatial "time" ξ. However, solutions to the initial value problem are unique, provided they exist [SaSc01d, Ch98]. The situation is reminiscent of the elliptic boundary-value problem $u_{\xi\xi} + u_{yy} = f(u)$, posed on unbounded cylinders $x \in \mathbb{R}$, $y \in (0,1)$; see for example [Ki82, Fi84, Mi94, Fi&al98].

Still, we can formally linearize (2.90) at equilibria and compute the spectrum of the linearized operator. It then turns out that, just as in the case of the heat equation (2.91), most of the eigenvalues are bounded away from the imaginary axis. It is therefore possible to define stable, unstable, and center manifolds in certain regions of the phase space X. In particular, given a solution $\mathbf{U}_*(\xi)$ to the nonlinear equation, we can construct stable and unstable manifolds in the vicinity of this solution if the linearized equation

$$\mathbf{W}_\xi = \mathbf{F}'(\mathbf{U}_*(\xi); \omega, c, \mu)\mathbf{W}$$

possesses an *exponential dichotomy*; see [Pe&al97, SaSc01d] for a definition. Existence of exponential dichotomies on $\mathbb{R}_+$ is guaranteed, if, $\mathbf{U}_*(\xi) \to \mathbf{U}_{*,+}$ for $\xi \to \infty$ and the linearized equation about $\mathbf{U}_{*,+}$ does not possess solutions of the form $\mathbf{W}(t)e^{ik\xi}$ for any $k \in \mathbb{R}$; see [Pe&al97, SaSc01d]. Similarly, center, center-unstable, and center-stable manifolds can be constructed, if the weighted linearized equation

$$\mathbf{W}_\xi = \mathbf{F}'(\mathbf{U}_*(\xi); \omega, c, \mu)\mathbf{W} - \eta_+ \mathbf{W},$$

possesses an exponential dichotomy; see again [Pe&al97, SaSc01d].

The main advantage of our spatial dynamics formulation is that the independent variable t lives on a compact domain, the circle. Time shift provides an S^1-equivariance in phase space $\mathbf{U} \in X$

$$\mathbf{U}(\cdot) \mapsto \mathbf{U}(\cdot + \theta) \in X, \quad \theta \in S^1; \qquad (2.92)$$

see also [Va82] for the role of S^1-symmetry in Hopf bifurcation. The subspace of functions (U, V) which are fixed under this action of the circle group consists of the time-independent functions. In this subspace, the abstract differential equation (2.90) reduces to the usual travelling wave ODE

$$U_\xi = V, \quad V_\xi = D^{-1}(-cV - F(U; \mu)).$$

For $\mu = 0$, there is a heteroclinic orbit $\mathbf{U} = Q(\xi) = (q(\xi), q'(\xi))$ to this differential equation, converging to the "equilibria" $Q_\pm = (q_\pm, 0)$. Recall again, that the term "equilibria", here, refers to spatial dynamics: "equilibria" are spatially homogeneous solutions.

Consider first the instability ahead of the front, case I in Theorem 2.3.4. We fix the temporal frequency to $\omega_* = c_*/k_*$. The Turing instability turns into a Hopf bifurcation, reflecting spatial periodicity, with (spatial) eigenvalues $\pm ik_*$ from the dispersion relation (2.85). Turing patterns lie in a small neighborhood of Q_+, but outside of the subspace of time-independent functions. Simplicity of the critical spectral curve, which induces the Turing instability guarantees that the linearization at Q_+ possesses precisely two simple imaginary eigenvalues $\nu_H = ik_*$ and $\bar{\nu}_H$, constituting the center eigenspace. After reduction to a 2-dimensional center manifold in a neighborhood of the equilibrium Q_+, we find a periodic orbit arising through the Hopf bifurcation. Temporal time shift acts nontrivially on the center-manifold such that the periodic orbits actually are relative equilibria, with spatial time evolution given by the temporal time-shift symmetry.

We next argue that the Turing patterns are unstable with respect to the spatial ξ-dynamics inside the spatial center manifold. Since λ is real in the steady frame, the group velocity associated with the critical spectral curve in the comoving frame is negative; see (2.69) and the following remark. By Lemma 2.3.3, the temporal time-t-instability disappears in a space of functions with weight $e^{\eta\xi}$, $\eta > 0$ small; see Lemma 2.3.3. In our spatial dynamics picture, this weight shifts the spatial eigenvalues ν of the linearization at the equilibrium Q_+, adding $\eta > 0$. The effect is that the spatial eigenvalue $\nu_H(\mu)$ stays on the side of the imaginary axis, where it was located for $\mu < 0$, suppressing neutral spatial eigenvalues and failure of the Fredholm property throughout the bifurcation. This is only possible, if $\mathrm{Re}\,\nu_H(\mu) < 0$ for $\mu > 0$, that is, if the equilibrium Q_+ is stable inside the center manifold, after bifurcation. By exchange of stability, the periodic solution is therefore unstable inside the center manifold.

We turn to the global bifurcation along the heteroclinic orbit $Q(\xi)$, next; compare also Figure 2.12. We argue with the spatial ξ-dynamics naively, using objects like (infinite-dimensional) stable and unstable manifolds of the "equilibria" $Q_\pm$ freely, as is usual for ODEs; see [SaSc99, SaSc01b] for a technical justification. Just before bifurcation, $\mu < 0$, the only bounded solution of equation (2.89), linearized along the front $Q(\xi)$, is $Q'(\xi)$. Indeed, any other bounded solution would correspond to a time-periodic solution of the linearized parabolic equation and hence would yield an eigenvalue $2\pi i \ell/\omega$ of the linearization $\mathcal{L}_*$, for some $\ell \in \mathbb{Z}$. Therefore, the intersection of the global stable manifold of Q_+ and the global unstable manifold of Q_- consists of precisely the heteroclinic $Q(\xi)$. Upon varying c, stable and unstable manifold cross with nonzero speed, since the zero eigenvalue was assumed to be simple. This *transverse crossing* is best viewed in the extended phase space, where the equation (2.89) is augmented with an equation for the parameter, $\mu_\xi = 0$. In this extended phase space, transverse crossing can be defined as a transverse intersection of the μ-family of stable manifolds with the μ-family of unstable manifolds. Through the bifurcation point $\mu = 0$, the stable manifold of Q_+ can be continued smoothly as the strong stable manifold of Q_+. Indeed, considering the strong stable manifold eliminates the additional weakly stable center direction, which appears at the bifurcation point. Transverse crossing persists as transverse crossing of the strong stable manifold of Q_+ and the unstable manifold of Q_-, throughout the bifurcation $\mu \sim 0$. The central observation now is that, by continuity of the stable fibration of the center manifold, the strong stable manifold of Q_+ is close to any of the strong stable fibers of the periodic orbit in the center manifold! Therefore, the strong stable fibers of the periodic orbit cross the unstable manifold of Q_- transversely, which yields the desired heteroclinic orbit. If we choose a different strong stable fiber of the periodic orbit, we find a different heteroclinic connection. However, any of those connections can be obtained from the first heteroclinic through the time-t-shift symmetry. Uniqueness of the heteroclinic implies uniqueness of the modulated front in the class of functions with prescribed temporal frequency ω. Varying the temporal time-t-frequency ω amounts to a different choice of the wavenumber of the Turing pattern through the relation $k = c/\omega$. In other words, ω selects the spatial wavenumber of the Turing pattern that we choose to be placed ahead of the front.

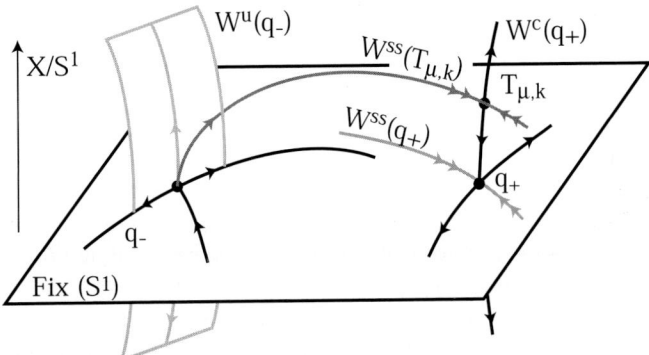

Fig. 2.12. The heteroclinic bifurcation creating a modulated front which invades a Turing pattern. See Plate 1 in the Appendix for a version of this figure in colour.

The picture for the instability behind the front, bifurcation failure case II, is very similar; compare Figure 2.13. The local bifurcation near Q_- is now the same as the one discussed for Q_+ above. Again, we find a periodic orbit, unstable for the spatial ξ-dynamics inside the center-manifold, representing Turing patterns behind the front, in the center manifold close to Q_-. Before bifurcation, $\mu < 0$, the unstable manifold of Q_- and the stable manifold of Q_+ cross transversely. This time however, we have to continue the unstable manifold of Q_- through the bifurcation point $\mu = 0$. Since the Hopf eigenvalue ν_H is unstable before the bifurcation, $\mu < 0$, the unstable manifold of Q_- continues smoothly to the center-unstable manifold of Q_- after bifurcation, $\mu > 0$. Transversality yields a unique intersection of the center-unstable manifold of Q_- with the stable manifold of Q_+ after bifurcation. However, this unique intersection, is already present in the $2N$-dimensional subspace of time-independent solutions: it is given by the trivial continuation of the robust primary front. This proves non-existence of modulated waves, leaving a Turing pattern behind, in the case of an instability behind the front.

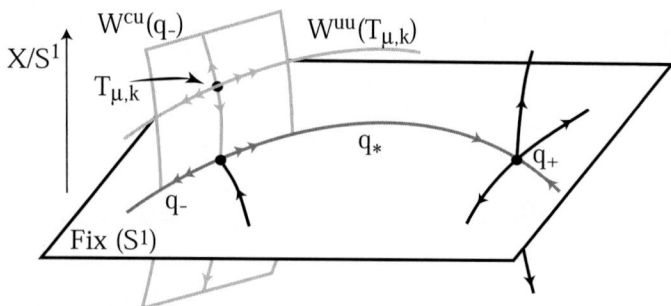

Fig. 2.13. The heteroclinic bifurcation failure of a front leaving a Turing pattern behind. See Plate 2 in the Appendix for a version of this figure in colour.

For a pulse, we combine the two arguments above; see also Figure 2.14. Throughout the bifurcation, the strong stable manifold of the origin crosses the center-unstable manifold of the origin transversely, along $Q(\xi)$. Again by continuity of the strong stable fibration of the center manifold the strong stable fiber of the periodic orbit, representing the Turing pattern ahead of the front, is close to the strong stable manifold of the origin. This implies transverse heteroclinic crossing of this strong stable fiber with the center-unstable manifold of the origin, at an $O(\mu)$-distance of the periodic orbit. As a consequence, the intersection is not contained in the stable manifold of the origin, which is distance $O(\sqrt{\mu})$ from the periodic Turing orbit. In backward spatial time $\xi \to -\infty$, the homoclinic orbit slowly approaches the periodic orbit: it creeps along the center manifold with rate $O(\exp(\eta'\mu\xi))$, as given by the linearization of the periodic orbit inside the center manifold.

For the technically more involved stability considerations, we refer to [SaSc01b, SaSc00b, SaSc01d].

2.3.3 From Unbounded to Large Domains: Absolute Versus Essential Spectra

In view of the bifurcation results in Theorems 2.3.4 and 2.3.5, we might try to describe the instability of fronts and pulses in a large, but bounded, comoving domain. Essential spectra disappear since the linearized operator $\mathcal{L}_*$ from

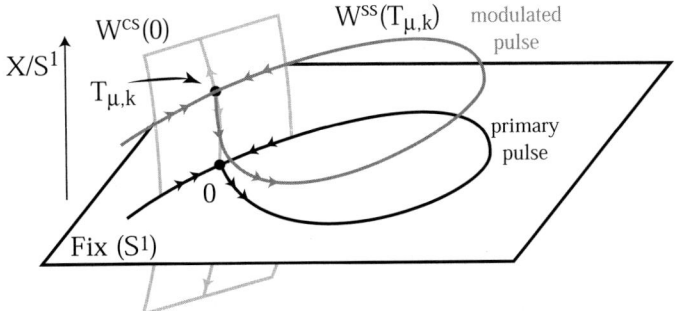

Fig. 2.14. The homoclinic bifurcation creating a modulated pulse travelling through a Turing pattern. See Plate 3 in the Appendix for a version of this figure in colour.

(2.66) possesses a compact resolvent on any finite domain. Therefore any instability in a bounded domain is necessarily due to point spectrum crossing the imaginary axis. In case of a pair of complex conjugate eigenvalues crossing the axis, a general Hopf bifurcation theorem guarantees bifurcation of periodic orbits — under mild nondegeneracy conditions. In consequence, there is an obvious difficulty trying to approximate instabilities in unbounded domains by instabilities in bounded domains. In particular the bifurcation failure in case II of Theorem 2.3.4, cannot be reproduced in any approximating bounded domain.

For a better understanding of this difficulty, we return to the simple example of the advection-diffusion problem

$$w_t = w_{\xi\xi} + w_\xi + \mu w =: \mathcal{L}_\mu w. \tag{2.93}$$

Compared to (2.75), we have added a constant linear driving term μw. In the unbounded domain $x \in \mathbb{R}$, the instability threshold is $\mu = 0$, where the essential spectrum,

$$\mathrm{spec}_{\mathrm{ess}} \mathcal{L}_\mu = \{\lambda = -k^2 - \mathrm{i}k + \mu;\ k \in \mathbb{R}\},$$

crosses the imaginary axis.

Truncating to $x \in (-L, L)$ and imposing periodic boundary conditions causes the curves of essential spectra to break up into point spectrum. The eigenvalues are still located in a neighborhood of the curve of essential spectrum, and can be computed explicitly:

$$\mathrm{spec}_{\mathrm{per},(-L,L)} \mathcal{L}_\mu = \left\{\lambda_\ell = -\left(\frac{\pi\ell}{2L}\right)^2 + \mathrm{i}\frac{\pi\ell}{2L} + \mu;\ \ell \in \mathbb{Z}\right\}.$$

On bounded subsets of the complex plane, $\mathrm{spec}_{\mathrm{per},(-L,L)} \mathcal{L}_\mu$ converges to the essential spectrum $\mathrm{spec}_{\mathrm{ess}} \mathcal{L}_\mu$, in the symmetric Hausdorff distance, for $L \to \infty$.

Imposing separated boundary conditions, for example Dirichlet boundary conditions $u(-L) = u(L) = 0$, the spectrum of $\mathcal{L}_\mu$ again consists of point spectrum. However, the point spectrum is now located on the real axis, with eigenvalues given explicitly by

$$\mathrm{spec}_{\mathrm{Dir},(-L,L)} \mathcal{L}_\mu = \{\lambda_\ell = -\left(\frac{\pi\ell}{2L}\right)^2 + \mu - \frac{1}{4}; \; \ell \in \mathbb{Z}\}.$$

The eigenfunctions are exponentially localized at the left boundary of the interval $(-L, L)$. On bounded subsets of the complex plane, the point spectrum converges to $(-\infty, \mu - \frac{1}{4}]$ in the symmetric Hausdorff distance, for $L \to \infty$. We will later identify this half line as the *absolute spectrum* of the operator $\mathcal{L}_\mu$, posed on $x \in \mathbb{R}$.

Note that the half line $\lambda \leq \mu - 1/4$ is not part of the spectrum of $\mathcal{L}_\mu$ on $x \in \mathbb{R}$, but merely arises as a set-wise limit of the spectrum if we truncate to finite intervals and let L tend to infinity. It is not difficult, but tedious, to check that almost all separated mixed boundary conditions induce a similar limiting behavior of the spectrum on large intervals $(-L, L)$, as the length of the interval tends to infinity. For example, in the case of Neumann boundary conditions, the spectrum converges

$$\mathrm{spec}_{(-L,L),\mathrm{sep}} \mathcal{L}_\mu \longrightarrow (-\infty, \mu - \frac{1}{4}] \cup \{\mu\} \quad \text{for } L \to \infty,$$

on bounded subsets of the complex plane $\mathbb{C}$, in the symmetric Hausdorff distance. It is an interesting exercise to follow the eigenvalue $\lambda = \mu$ through a homotopy from Neumann to Dirichlet through mixed boundary conditions.

For Dirichlet boundary conditions and $0 < \mu < 1/4$, the system is stable for any size of the domain, while it is unstable on the entire real axis. In this parameter regime, we call the system *convectively unstable*. Indeed, the instability on the unbounded domain is convective in nature: starting with initial conditions $u_0(x)$ having compact support on $\mathbb{R}$, the absolute value $|u(t, x_0)|$ converges to zero for every fixed point x_0 on the real axis, although the norm $\|u(t, \cdot)\|_{L^2}$ grows exponentially. The pointwise decay can be extracted from the explicit representation of solutions by the Green's function for (2.93).

This general phenomenon of pointwise decay of solutions, which nevertheless grow in norm, was first emphasized in the context of plasma physics and fluid flow instability [LaLi59, Br64]. We will now study the behavior of spectra of pulses or fronts in reaction-diffusion systems, in the limit $L \to \infty$; see [SaSc00c] for a more general result.

Consider the linear reaction-advection-diffusion system

$$\mathcal{L}u = Du_{\xi\xi} + cu_\xi + a(\xi)u, \tag{2.94}$$

where the $N \times N$-matrix $a(\xi)$ converges exponentially to $a_\pm$ for $\xi \to \pm\infty$. At $\xi = \pm\infty$, we encounter the asymptotic linearized operators

$$\mathcal{L}_\pm u = Du_{\xi\xi} + cu_\xi + a_\pm u.$$

Associated with $\mathcal{L}_\pm$ are the dispersion relations

$$d_\pm(\lambda, \nu) = \det(D\nu^2 - c\nu + a_\pm - \lambda) = 0,$$

which are polynomials of degree $2N$ in ν. Let $\nu^j_\pm(\lambda)$, $j = 1, \ldots, 2N$, denote the complex roots, ordered by decreasing real part

$$\operatorname{Re}\nu^1_+ \geq \ldots \geq \operatorname{Re}\nu^{2N}_+, \quad \operatorname{Re}\nu^1_- \geq \ldots \geq \operatorname{Re}\nu^{2N}_-.$$

Note that this numbering might depend discontinuously on the spectral parameter λ.

In the homoclinic case $a_+ = a_-$, the *essential spectrum*, $\operatorname{spec}_{\mathrm{ess}}(\mathcal{L})$, is given by the set of λ such that $\operatorname{Re}\nu^j_+(\lambda) = 0$ or $\operatorname{Re}\nu^j_-(\lambda) = 0$, for some j; see Proposition 2.3.1.

In the general case of a pulse or a front, we define the *absolute spectrum*, $\operatorname{spec}_{\mathrm{abs}}(\mathcal{L})$, as the set of $\lambda \in \mathbb{C}$ such that $\operatorname{Re}\nu^N_+(\lambda) = \operatorname{Re}\nu^{N+1}_+(\lambda)$ or $\operatorname{Re}\nu^N_-(\lambda) = \operatorname{Re}\nu^{N+1}_-(\lambda)$. Note that the absolute spectrum is not necessarily part of the essential spectrum. Still, from this definition, the absolute spectrum is defined by the operator posed on the unbounded domain $\xi \in \mathbb{R}$. The reader is invited to check that the absolute spectrum of the advection-diffusion problem (2.94) is actually given by the half line $(-\infty, \mu - \frac{1}{4}]$. In general, the absolute spectrum is a closed set and consists of a finite collection of curves.

An important property of the absolute spectrum is that in its complement, there exist exponential weights $\eta(\lambda) = (\eta_-(\lambda), \eta_+(\lambda)) \in \mathbb{R}^2$ such that $\mathcal{L} - \lambda$ is Fredholm with index zero when considered as a closed operator on $L^2_{\eta(\lambda)}$, the space of functions in L^2_{loc} with

$$\|u\|^2_{L^2_\eta} := \int_{\mathbb{R}^+} |e^{\eta_+\xi}u(\xi)|^2 + \int_{\mathbb{R}^-} |e^{\eta_-\xi}u(\xi)|^2 < \infty.$$

In Theorem 2.3.6 below we characterize the limiting behavior of spectra, as the domain size tends to infinity. We will consider periodic boundary conditions, $u(-L) = u(L)$, $u_\xi(-L) = u_\xi(L)$, as well as separated boundary conditions, $(u(\pm L), u_\xi(\pm L)) \in E_\pm$ with given, N-dimensional subspaces $E_\pm$ of $\mathbb{C}^{2N}$. Periodic boundary conditions are understood with the restriction $a_+ = a_-$, that is, we are interested in linearizations about pulses.

The *point spectrum*, $\operatorname{spec}_{\mathrm{pt}}(\mathcal{L})$, refers to the point spectrum in $L^2(\mathbb{R})$. For the case of separated boundary conditions, we distinguish two types of point spectrum. Fix $\lambda_* \notin \operatorname{spec}_{\mathrm{abs}}(\mathcal{L})$, and the weight $(\eta^*_-, \eta^*_+) = \eta_* = \eta_*(\lambda_*)$ such that $\mathcal{L}$ is Fredholm of index zero for λ near λ_* in $L^2_{\eta_*}$. We say λ_* belongs to the *extended point spectrum* $\operatorname{spec}_{\mathrm{expt}}(\mathcal{L})$, if $\mathcal{L} - \lambda$ is not invertible in $L^2_{\underline{\eta}_*}$. We say λ_* belongs to the *right boundary spectrum*, spec$_{\text{bdy},>}(\mathcal{L})$, if the asymptotic, constant coefficient operator $\mathcal{L}_+ - \lambda_*$ possesses nontrivial kernel when considered on $\xi > 0$, in $L^2_{\eta_+^*}(\mathbb{R}_+)$ with boundary condition E_+ in $\xi = 0$. Similarly, the *left boundary spectrum* is defined via the kernel of $\mathcal{L}_- - \lambda_*$ in $L^2_{\eta_-^*}(\mathbb{R}_-)$ with boundary condition E_- in $\xi = 0$. The *boundary spectrum* spec$_{\text{bdy}}\mathcal{L}$ is defined as the union of left and right boundary spectra. We refer to [SaSc00c] for a more geometric and constructive definition of boundary spectra, involving winding numbers or, alternatively, Evans function constructions.

Several assumptions are needed in the proof of the following theorem. Most of them are technical in nature and satisfied for "typical" systems. We therefore refer the reader to [SaSc00c] for precise hypotheses.

Theorem 2.3.6 *[SaSc00c] Under typical assumptions on the spectrum on the real line, we find that the spectrum of $\mathcal{L}$ on $L^2(-L, L)$ converges, as $L \to \infty$:*

- *under periodic boundary conditions,*

$$\text{spec}\,\mathcal{L}_{\text{per},(-L,L)} \longrightarrow \text{spec}_{\text{ess}}(\mathcal{L}) \cup \text{spec}_{\text{pt}}(\mathcal{L});$$

- *under typical, separated boundary conditions,*

$$\text{spec}\,\mathcal{L}_{\text{sep},(-L,L)} \longrightarrow \text{spec}_{\text{abs}}(\mathcal{L}) \cup \text{spec}_{\text{expt}}(\mathcal{L}) \cup \text{spec}_{\text{bdy}}(\mathcal{L}).$$

Convergence is understood on bounded subsets of the complex plane in the symmetric Hausdorff distance.

Moreover, multiplicity of the eigenvalues is preserved. The number of eigenvalues, counted with multiplicity, in any fixed open neighborhood of a point $\lambda_* \in \text{spec}_{\text{ess}}(\mathcal{L})$ (or $\text{spec}_{\text{abs}}(\mathcal{L})$, respectively) converges to infinity as $L \to \infty$. Multiplicities in neighborhoods of points $\lambda_* \in \text{spec}_{\text{expt}}(\mathcal{L})$, $\text{spec}_{\text{pt}}(\mathcal{L})$, $\text{spec}_{\text{bdy}}(\mathcal{L})$ stabilize and convergence of eigenvalues is of exponential rate in L, there.

The most important, and at first sight surprising, part of the theorem is that spectra on large intervals with separated boundary conditions do not approximate essential, but merely absolute spectra. The latter are in general different from the essential spectrum. However, a reflection symmetry $x \to -x$ in the problem can force both spectra to coincide.

The theorem shows that the limit of spectra on bounded intervals is entirely determined by "spectral" information from the unbounded domain limit, together with information on the boundary conditions.

Let us return to the bifurcation problem in Theorem 2.3.4. An intriguing question is, what happens to the instability when restricting the domain to a bounded interval in an appropriately comoving frame. This restriction becomes necessary, for example, when trying to detect instabilities of fronts numerically. The critical spectrum in Theorem 2.3.4 consists of an isolated zero eigenvalue and of essential spectrum with non-zero group velocity. With a positive exponential weight $\exp(\eta\xi)$ with $\eta > 0$, on $\xi \in \mathbb{R}$, we can shift the essential spectrum into the stable complex half plane. By Theorem 2.3.6, the only critical eigenvalue in a large domain is going to be the zero eigenvalue. Its precise location depends on the actual choice of boundary conditions. In particular, there is no Hopf bifurcation in any, arbitrarily large but finite, domain caused by the essential spectrum. The only possible instability mechanism would be the creation of unstable spectrum through the boundary conditions, in $\text{spec}_{\text{bdy}}\mathcal{L}$.

Linearized stability, $\text{Re spec }\mathcal{L}_* < 0$, in large bounded domains $(-L, L)$ implies nonlinear stability, for reaction-diffusion systems. However, our above result on the spectrum for $L \to \infty$ makes extensive use of exponential weights, which might be incompatible with the nonlinearity. For example, with $\eta := \eta_+ = \eta_- > 0$, the nonlinearity $u \mapsto u^2$ is not defined on H^1_η, the space of functions u with $u, u' \in L^2_\eta$. This technical observation is closely related to limitations of the above theorem. For an illustration, let us return to

$$u_t = u_{xx} + u_x + \mu u + \kappa u^3, \quad u(-L) = u(L) = 0 \qquad (2.95)$$

with L large. For $0 < \mu < 1/4$, the absolute spectrum is contained in the open left half plane and $u(x) \equiv 0$ is asymptotically stable. Define the basin of attraction B_L as the set of initial values $u_0 \in H^1_{\text{Dir}}(-L, L)$ such that the solution of (2.95) with initial value u_0 converges to zero. As a measure for stability, we define the *instability threshold*

$$\delta^u(L) = \inf\{\|u_0\|_{H^1}; \; u_0 \notin B_L\}.$$

The instability threshold measures the minimal amount of a perturbation needed to "permanently" drive the system away from the equilibrium. The smaller $\delta^u(L)$, the more sensitive the equilibrium $u \equiv 0$ will be to perturbations.

Proposition 2.3.7 *[SaSc02a] Consider equation (2.95) with $0 < \mu < 1/4$. Then the instability threshold $\delta^u(L)$ satisfies*

(a) $\delta^u(L) \geq \delta_0 > 0$ if $\kappa < 0$, for all $L > 0$;
(b) $\delta^u(L) = O(1/L)$ if $\kappa > 0$.

In particular, for the unstable sign (b) of the nonlinearity $\kappa > 0$, the stability predicted from the spectral analysis, with uniform exponential decay rate $\mu - 1/4$, is valid only in a very small region of phase space, if the size L of the domain is large. Even small initial conditions may then lead to blow-up in finite time. We do not know if the upper estimate $O(1/L)$ on the instability threshold is optimal in case (b).

2.4 Two Space Dimensions: Existence of Spiral Waves

The spontaneous appearance of rotating spiral-shaped planar patterns is a most striking phenomenon, both from the experimental and the theoretical point of view. Spiral wave patterns have been observed for example in heart tissue [WiRo46], in Belousov-Zhabotinsky (BZ) reactive media [Wi72, BrEn93, Un&al93], in the CIMA-reaction [dK&al94], in fluid convection experiments [PlBo96], and in surface catalysis [Ne&al]. From a mathematical view point, they pose challenges because they arise in large — ideally unbounded — domains; see Section 2.4.2.

Pioneering work by Wiener, 1946, has been motivated by waves of electrical excitation in heart muscle tissue. It is well worth obtaining the somewhat obscure original reference [WiRo46]. See also the overwhelming material collected by [Wi01, Wi87], and the references there. One approach initiated by Wiener aims at a direct geometric description of the spatio-temporal dynamics of the appearing sharp wave fronts, which far from a core region take the form of Archimedean spirals. This approach is sometimes called "kinematic theory of spirals", and has been developed further on a mostly formal level; see for example [KeTy92, Ke92, MiZy91] and the references there. The underlying idea is closely related to the problem of "curve shortening", which has seen significant recent progress; see [An90, An91, GaHa86].

Spiral waves appear both in excitable and oscillatory media. In the latter, the interpretation changes from a source of excitation waves to a defect in the pattern of phase waves emitted by the spiral core. At the center, or core, the oscillation phase cannot be defined in a continuous way. Along a simple closed path around the core, the phase of the oscillations winds around the circle once. Small-amplitude oscillations are amenable to the powerful tool of modulation equations such as the complex Ginzburg-Landau equation or λ-ω-systems; see [KoHo73, Ha82, KoHo81].

In the following two sections, we expose two simple approaches to this problem. We first address excitable media and a kinematic description of the spiral arms in Section 2.4.1.

In Section 2.4.2, we then characterize spiral wave solution to reaction-diffusion systems more generally, aiming at a mathematical definition which incorporates the characteristic features of experimentally observed spirals. The formulation in this section will be the basis for the spectral discussion in Section 2.5.3.

Small-amplitude spiral waves arising near Hopf bifurcations in the reaction kinetics then provide us with specific examples, in Section 2.4.3.

2.4.1 Kinematics and its Defects

To state the problem in its most simplistic and, as we will see, yet inadequate form, we consider a differentiable planar curve $s \mapsto (r, \varphi)(s)$ in polar coordinates (r, φ). Arc length is denoted by s. We assume that the curve propagates in time, with scalar normal velocity $c_\perp$ which depends on the signed local curvature κ in a strictly monotone way. (We fix the normal direction to the left of the tangent vector here.) For later convenience, we write

$$c_\perp = \mathrm{v}^{-1}(\kappa) \tag{2.96}$$

to denote the inverse dependence. Let $\theta \in S^1$ denote the angle which the unit tangent to our curve forms with the horizontal axis; see Figure 2.15. Denoting the derivative of θ with respect to arc length s by $\dot\theta$, we obtain $\dot\theta = \kappa$ and, by (2.96),

$$\dot\theta = \mathrm{v}(c_\perp). \tag{2.97}$$

If the curve is to rotate left at constant angular velocity ω with respect to the center $r = 0$ of our polar coordinate system (r, φ), then

$$c_\perp = \omega r \cos(\theta - \varphi). \tag{2.98}$$

Combining (2.97), (2.98) with the definition of arc length parameterization we obtain the ODE system

$$\begin{aligned} \dot\theta &= \mathrm{v}(\omega r \cos(\theta - \varphi)) \\ \dot\varphi &= r^{-1} \sin(\theta - \varphi) \\ \dot r &= \cos(\theta - \varphi) \end{aligned} \tag{2.99}$$

for any rigidly rotating curve $(r(s), \varphi(s), \theta(s))$ under the geometric equation (2.96).

Curve shortening as mentioned above corresponds to the special case $\mathrm{v}(c_\perp) = \kappa$ of normal velocity proportional to κ. As for all v of positive slope, a Sturm property analogous to Proposition 2.2.1 holds here: the number of strict crossings of two distinct curves never increases under this curvature flow. We note here that positive v-slope, $\mathrm{v}' > 0$, corresponds to a nonlinear forward parabolic equation for the curve evolution; see [An91]. The existence of rotating wave "Ying-Yang" spiral curves seems to be folklore in the subject. The resulting spirals, however, are not Archimedean: they limit from inside onto a circle of finite radius, instead. Negative slope, in contrast, would correspond to a backwards parabolic equation which is ill-posed in forward time. Although this may suggest an incompleteness in the model derivation of the curve-"lengthening" equation, it does not cause difficulties for our rotating wave ODE (2.99). Variants involving a power dependence like $\mathrm{v}^{-1}(\kappa) = \kappa^\beta$ have also been considered: most notably the affine invariant case $\beta = 1/3$ for self-focusing edge enhancement in image processing. For propagation of Belousov-Zhabotinsky (BZ) fronts, a good model is generally believed to be given by

$$\mathrm{v}^{-1}(\kappa) = c_0 + \kappa \tag{2.100}$$

on suitable time scales, see [KeTy92, MiZy91] and the references there.

Curvature flow of Archimedean spirals Let us look for *Archimedean spirals* in system (2.99), as are observed in BZ media. We consider Archimedean spirals to be characterized by an asymptotically proportional dependence of radius r on angle φ, given by a limiting slope

$$\sigma_* := \lim_{r \to \infty} \frac{dr}{d\varphi} = \lim_{s \to \infty} \frac{r \cos \alpha}{\sin \alpha} \notin \{0, \pm\infty\}. \tag{2.101}$$

Here we have used l'Hospital in (2.99), with the abbreviation

$$\alpha := \theta - \varphi. \tag{2.102}$$

Our curve evolution is in fact rotation invariant, as is $\alpha = \theta - \varphi$ but neither θ nor φ individually. It is therefore convenient to rewrite (2.99) in terms of (r, α):

$$\begin{aligned}\dot\alpha &= \mathrm{v}(\omega r \cos \alpha) - r^{-1} \sin \alpha \\ \dot r &= \cos \alpha.\end{aligned} \tag{2.103}$$

From (2.101) in the limit $s \to \infty$, alias $r \to \infty$, we immediately deduce that $\cos \alpha \to 0$ and in fact

$$\begin{aligned}\alpha &\to \pm\pi/2 \\ \sigma := r \cos \alpha &\to \pm\sigma_*.\end{aligned} \tag{2.104}$$

with corresponding signs $\pm$ for either limit. To analyze these limits further, we rewrite (2.103) in terms of r and $\rho := 1/r \to 0$ as follows

$$\begin{aligned}\sigma' &= \pm\sqrt{1 - \rho^2 \sigma^2}\, \mathrm{v}(\omega \sigma) + \rho \\ \rho' &= -\rho^4 \sigma\end{aligned} \tag{2.105}$$

Here we have rescaled "time" s such that $\rho d/ds = \,'$. The vector field (2.105) is regular along the invariant line $\rho = 0$, where

$$\sigma' = \pm \mathrm{v}(\omega \sigma). \tag{2.106}$$

The limit slope (2.101) therefore requires the equilibrium condition

$$0 = \mathrm{v}(\omega \sigma_*) \tag{2.107}$$

to hold. This condition amounts to a dispersion relation between the rotation frequency ω and the wave length $2\pi\sigma_*$ of the asymptotically uncurved, planar fronts of our Archimedean spirals. For example $\sigma_* = -c_0/\omega$ in the BZ case mentioned above.

A center manifold analysis at the Archimedean spiral equilibrium $\rho = 0$, $\sigma = \sigma_*$ of (2.105) yields additional information on those Archimedean spiral trajectories. Consider the case $\alpha \to +\pi/2$, $\sigma_* > 0$ of left outward winding spirals, first. Then $\sigma = \sigma_*$ is hyperbolically stable, within the line $\rho = 0$, if

$$\omega v'(\omega \sigma_*) < 0. \tag{2.108}$$

Since $\rho \searrow 0$ inside the center manifold, any initial condition of (2.105) with $\rho > 0$ and (ρ, σ) near $(0, \sigma_*)$ then converges to this Archimedean spiral equilibrium. Most notably this includes the BZ example, where $v' > 0$, $\sigma^* > 0$, and $\omega < 0$. The opposite case $\omega v'(\omega \sigma^*) > 0$ arises for example when curves shorten due to $v' > 0$, $c_0 < 0 < \omega$. Then $\sigma = \sigma_*$ is hyperbolically unstable, within the line $\rho = 0$, and the asymptotically Archimedean spiral is uniquely defined. In either case, the center manifold — be it unique or not — provides convergence to the Archimedean spiral equilibrium along the same asymptote

$$\sigma = \sigma(r) \sim \sigma_* - (\omega v'(\omega \sigma_*))^{-1} r^{-1} + O(r^{-2}). \tag{2.109}$$

The r^{-1}-term measures the deviation from a precise Archimedean spiral. Similarly, polar coordinate asymptotics for $r = r(\varphi)$ can be derived.

The front-back matching problem The two sign-cases in (2.105) are related by a simple transformation. Indeed $(\rho, \sigma, s) \mapsto (-\rho, \sigma, s)$ reverses the sign $\pm$ as well as the sign of $'$. This means that the unique equilibrium $\sigma = \sigma_*$ on the line $\rho = 0$ possesses the same stability properties under the s-flow, in either case. In particular it is impossible to find a rotating wave solution as in Figure 2.15(a), which would be biasymptotic to an Archimedean spiral for both $s \to +\infty$ and $s \to -\infty$.

This impossibility indicates a first defect of our naive, inadequate approach based on motion by curvature (2.96) alone. Two appealing remedies have been proposed. The first derives curvature motion for fronts ($s \to +\infty$) and backs ($s \to -\infty$) separately, by singular perturbation arguments, and then attempts a formal matching in the "core" region where the arc length parameter $|s|$ is small. See for example [Ke92].

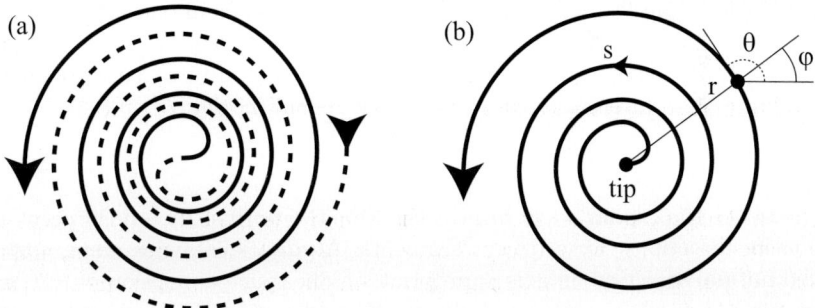

Fig. 2.15. Biasymptotic Archimedean spiral **(a)**, and spiral with spiral tip **(b)**.

More radically the second remedy terminates the spiral at $s = 0$ and proposes additional phenomenological equations of motion at this "tip" point, see for example [MiZy91], and Figure 2.15(b). Neither approach has been fully justified on a satisfactory mathematical basis, so far. For scalar equations, $u \in \mathbb{R}$, the singular perturbation limit $\varepsilon \searrow 0$ for reaction-diffusion equations

$$\varepsilon u_t = \varepsilon^2 \Delta u + f(u) \tag{2.110}$$

has been analyzed successfully for bounded domains $x \in \Omega \subset \mathbb{R}^N$, and (mostly) cubic nonlinearities

$$f(u) = (u-a)(1-u^2). \tag{2.111}$$

Concerning planar wave fronts, alias $N = 1$, between the stable equilibria $u \equiv \pm 1$ we have already mentioned the travelling waves $u = u((x - c_0 t)/\varepsilon)$ of wave speed $c_0 = c_0(a)$. Note that $c_0(-a) = -c_0(a)$, by the symmetry $u \mapsto -u$. Based on earlier formal expansions by Fife, an asymptotics

$$c_\perp = c_0(a) + \varepsilon \kappa + \cdots \tag{2.112}$$

has been proved to hold for $N = \dim x = 2$, over time spans $0 \le t \le T/\varepsilon$, for the interface curves of solutions $u = u^\varepsilon$ of (2.110). Mean curvature flow for $(N-1)$-dimensional interfaces has also been justified, for $N \ge 3$. See [Fi88, Ba&al93] and the references there.

Expansion (2.112) fits well with the BZ-Ansatz (2.100) for $c_\perp = \mathbf{v}^{-1}(\kappa)$. Choosing $a = a(\varepsilon) = \varepsilon a_1$ we can in fact scale ε out of (2.112) by rescaling time, because oddness of $c_0(a)$ then implies $c_0(a) = \varepsilon c_0'(0) a_1 + \cdots$. Then (2.112) takes precisely the form (2.100) discussed above.

However, rigorous analysis of (2.110) is heavily based on variational methods and hence fails, as it stands, in two important cases: in unbounded domains Ω because the Lyapunov functional $\mathcal{V} = \mathcal{V}_\varepsilon$ of (2.4) may become unbounded, and for systems $u \in \mathbb{R}^2$ because $\mathcal{V}$ may not exist.

The front-back difficulty, which causes the inexistence of biasymptotic Archimedean spirals as in Figure 2.15(a), already arises in one space dimension $N = 1$ and is due to the planar front wave speeds $c_0(a)$ in (2.100), (2.112). Indeed, adjacent curve segments in Figure 2.15(a) correspond to limits $s \to \pm\infty$, alternatingly, where curvature κ is negligible and

$$c_\perp \approx c_0(a) \tag{2.113}$$

approaches the planar wave front limit. But the normal directions point in opposite directions, for adjacent segments. Alternatively, with respect to a fixed normal direction or as is appropriate in one space dimension $N = 1$, we can invoke $c_0(-a) = -c_0(a)$ and observe that adjacent fronts and backs move in opposite directions, unless $c_0(a) = 0$. That latter possibility is excluded, both by experimental observation and by our constitutive assumption $\sigma_* \ne 0$ for Archimedean spirals, in (2.101).

As an alternative to this dilemma we might feel tempted to consider travelling pulses, for $N = 1$, instead of fronts and backs. Rescaling $u = $

$u((x - c_0 t)/\varepsilon)$ by $\varepsilon \searrow 0$, the pulse locations would contract to curves, for $N = 2$, which could move according to curvature. For scalar equations, $u \in \mathbb{R}$ however, this is prevented by the (damped) Hamiltonian structure of the travelling wave equation, which forces vanishing wave speed $c_0 = 0$ for any homoclinic pulse. Placing a single pulse on any large circle $r = const.$, moreover, the curve of pulse locations would have to terminate at a "tip" as in Figure 2.15(b).

Clearly this Hamiltonian difficulty disappears for non-variational systems $u \in \mathbb{R}^2$. We conclude with a standard example for excitable media systems: the FitzHugh-Nagumo system

$$\begin{aligned} u_t^1 &= \Delta u^1 + \varepsilon^{-1} u^1 (1 - u^1)(u^1 - (u^2 + b)/a) \\ u_t^2 &= d \Delta u^2 + u^1 - u^2. \end{aligned} \quad (2.114)$$

For given $u^2 = u^2(t, x)$, the singularly perturbed u^1-equation is cubic as discussed above, and u^2 enters as a parameter. Adjusting the constants a, b appropriately, homoclinic pulses arise which travel at nonzero speeds. This case is usually termed *excitable medium*; see Figure 2.16(a).

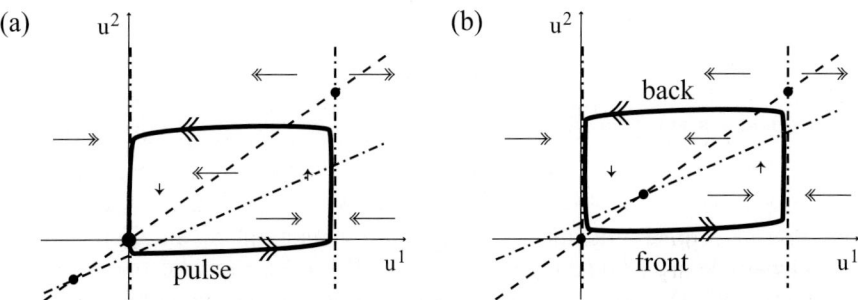

Fig. 2.16. Excitable (**a**) and oscillatory (**b**) media; homoclinic pulses and periodic travelling waves for $\varepsilon \searrow 0$. Isoclines in absence of diffusion. See also (2.114) for the excitable case (**a**).

The u^2-values are nearly constant through each u^1-jump, in the singular limit $\varepsilon \searrow 0$. The diffusive u^2-equation, however, constitutes a highly nonlocal and nonlinear coupling of the appropriate local wave speeds $c_0 = c_0(u^2)$ in the curvature motion (2.100) or (2.112). It seems tempting to derive an equation for the tip motion, in this case, involving local information on the "pulse" curve at the tip itself. See [MiZy91] for a phenomenological attempt.

Spatially periodic stable travelling waves can likewise be realized, for appropriate a, b and small ε, if we modify the equation u_t^2 by subtracting a suitable positive constant; see also Figure 2.16(b). This corresponds to an *oscillatory medium*.

In the following section we will take an entirely different approach, based on reaction-diffusion systems in the plane $x \in \Omega = \mathbb{R}^2$, but neither restricted to, nor complicated by, singular perturbations. In particular, in Section 2.4.3, we address the issue of primary bifurcation of spiral waves from spatially homogeneous equilibria in the presence of continuous spectrum. In Chapter 2.5, we will further pursue this approach to include secondary bifurcations from Archimedean rotating waves.

2.4.2 Archimedean Spiral Waves in Radial Dynamics

In Section 2.4.2, we collect some basic characteristics of spiral waves in reaction-diffusion systems: rigid rotation and Archimedean shape. Section 2.4.2 clarifies, in which sense spiral waves "emit" wave trains. We summarize some principal features of a spiral wave, in Section 2.4.2.

Rigid rotation and asymptotic wavetrains Proving existence of spiral waves in reaction-diffusion systems has been a long standing mathematical problem. In spite of their prevalence and robustness in experiments as well as in numerical simulations, proofs of existence for spiral waves are rare. Given the empirical evidence for existence, one might suspect that the existence problem is of purely mathematical interest. We believe, however, that a proper understanding of the existence problem also leads to a systematic approach to many observable stability and instability phenomena involving spiral wave dynamics.

Consider a reaction-diffusion system in $x \in \mathbb{R}^2$, the unbounded plane,

$$u_t = D\Delta_x u + f(u), \quad x \in \mathbb{R}^2. \tag{2.115}$$

We aim at a precise definition of what we understand by an Archimedean spiral wave solution to (2.115).

First of all, the simplest spiral waves observed in chemical experiments are rigidly rotating. Rigid rotation is easily described in polar coordinates (r, φ), as we saw in Section 2.4.1. The central point of our discussion of Archimedean spirals in Chapters 2.4 and 2.5, on top of that mere convenience, will be a particular view on the radial coordinate r: we will view radius r, rather than time t, as a new, spatial "time" variable — much in the same way as we have viewed the spatial variable x as "time" in our discussion of travelling waves in Chapter 2.3. We refer to this view point, which was first used in [Sc98] for the spiral wave problem and further exploited in [Sc01], as *radial dynamics*.

In polar coordinates (r, φ), rigid rotation allows us to write the spiral wave solution q_* in the form

$$u(t, r, \varphi) = q_*(r, \varphi + \omega_* t). \tag{2.116}$$

Here, ω_0 is the temporal frequency of rotation of the spiral. Inserting this Ansatz into the reaction-diffusion system, we find

$$D(u_{rr} + \frac{1}{r}u_r + \frac{1}{r^2}u_{\psi\psi}) - \omega_* u_\psi + f(u) = 0, \qquad (2.117)$$

a partial differential equation of elliptic type for $u = q_*(r,\psi)$, with $\psi = \varphi + \omega_* t$. Note, however, that the operator $\omega_* \partial_\psi$ is not relatively bounded with respect to the Laplacian. Indeed, the derivative ∂_ψ cannot be bounded by $r^{-2}\partial_{\psi\psi}$. As a consequence, the smooth function $\sin(x_1)$, for example, belongs to the domain of the Laplacian on spaces of bounded, uniformly continuous functions, but the angular derivative $x_2\partial_{x_1}\sin(x_1)$ is unbounded.

An additional characteristic of many observed spiral waves is their Archi"-medean shape. We require the profile to be Archimedean far away from the center of rotation,

$$|q_*(r,\psi) - q_\infty(r - (\psi/k_\infty))| \longrightarrow 0 \quad \text{for } r \to \infty, \qquad (2.118)$$

uniformly in ψ, for some function $q_\infty(\xi)$. It turns out that spiral waves typically do not possess an asymptotic phase as stated here. We stick to this simplifying assumption for clarity of presentation. The results on stability can be adapted to the more general case of logarithmic divergence of the phase. Note that q_∞ is constant along Archimedean spirals $\psi = k_\infty r + \psi_0$. Convergence (2.118) is the "boundary condition" that we will impose at $r = \infty$ for our spiral wave solution to the elliptic system (2.117). The asymptotic profile of the spiral wave, q_∞, has to satisfy certain compatibility conditions. First, q_∞ is $2\pi/k_\infty$-periodic in its argument by continuity. Second, if we formally let $r \to \infty$ in (2.117), we find the condition

$$Dq_\infty'' + \frac{\omega_*}{k_\infty}q_\infty' + f(q_\infty) = 0. \qquad (2.119)$$

This is nothing else but the travelling-wave equation for one-dimensional periodic wavetrain solutions $q(x - (\omega_*/k_{\text{wt}})t)$ of

$$u_t = Du_{xx} + f(u),$$

with speed ω_*/k_∞ and spatial period $2\pi/k_\infty$. We therefore call q_∞ the *asymptotic wavetrain* of the spiral wave solution. The constant k_∞ appearing in (2.118) is the wavenumber of the asymptotic wavetrain. The speed of the asymptotic wavetrain is computed from the standard relation

$$c_\infty = \omega_*/k_\infty. \qquad (2.120)$$

For fixed, arbitrary, speed c, periodic solutions to the travelling-wave equation (2.119) are typically isolated. The spatial period $2\pi/k$ depends nontrivially on c. Using (2.120) with $k_\infty = k$, $c_\infty = c$, we can turn this dependence in a relation between ω and k. We call this relation the *nonlinear dispersion relation* of the asymptotic wavetrain

$$d_{\text{nl}}(\omega, k) = 0. \qquad (2.121)$$

The spiral "selects" a particular solution $\omega = \omega_*$, $k = k_\infty$ of (2.121).

Linear and nonlinear group velocities Most spiral waves that have been observed in chemical experiments share an additional property: wavetrains are "emitted" by the center. Spirals with this property are sometimes referred to as active spirals. Mathematically, the emission of wavetrains is characterized by a positive group velocity $c_g > 0$ of the asymptotic wavetrains in the radial direction. The concept of group velocity we are referring to, here, is a slight generalization of the definition in equation (2.69). There are actually two equivalent definitions of the group velocity for wavetrains. The simplest definition would refer to the nonlinear dispersion relation (2.121). Assuming that we can solve for ω_∞ as a function of k_∞, we define the *nonlinear group velocity* through

$$c_{g,nl} = \frac{d\omega_\infty}{dk_\infty}, \tag{2.122}$$

evaluated at the wavenumber of the wavetrain, k_∞.

We say that a spiral waves *emits* the asymptotic wavetrain, if the nonlinear group velocity of the asymptotic wavetrain is positive.

Another definition of the group velocity would refer to the linearization about the wavetrain q_∞ in a comoving frame of speed ω_*/k_∞,

$$w_t = Dw_{\xi\xi} + \frac{\omega_*}{k_\infty} w_\xi + f'(q_\infty(\xi))w. \tag{2.123}$$

At the end of Section 2.3.1, (2.80)–(2.83), we explained how to solve this linear parabolic equation with spatially periodic coefficients by means of a generalized Fourier transformation. Indeed, the general solution can be decomposed into Bloch waves $w(t, \xi; k)$,

$$w(t, \xi; k) = e^{\lambda(ik)t - ik\xi} w_0(\xi; k), \tag{2.124}$$

where w_0 solves

$$(\lambda(ik) + c_\infty ik)w_0 = D\left(\frac{d}{d\xi} - ik\right)^2 w_0 + c_\infty \frac{d}{d\xi} w_0 + f'(q_\infty(\xi))w_0. \tag{2.125}$$

Here, $w_0(\xi; k) = w_0(\xi + (2\pi/k_\infty); k)$ is spatially periodic with the period of the wavetrain q_∞ and the Bloch wavenumber k introduces a second spatial period. The eigenvalue λ and the Bloch wavenumber k are related by a dispersion relation $d_{\text{lin}}(\lambda, ik) = 0$, see (2.83). Assuming that we can solve this equation for $\lambda = \lambda(k)$, with $\lambda(0) = 0$, the linear group velocity is computed in analogy to (2.69)

$$c_{g,\text{lin}} = c_\infty + \text{Im} \left.\frac{d\lambda}{dk}\right|_{k=0}. \tag{2.126}$$

The summand c_∞ accounts for the fact that we compute the dispersion relation in a comoving frame, whereas we are interested in the group velocity in the steady frame. Note that we can define a linear group velocity for different values of k, and actually on different solution branches $\lambda^\ell(ik)$ of the dispersion relation d_{lin}, again in analogy to (2.69).

Lemma 2.4.1 *Linear and nonlinear group velocities of periodic wavetrains, as defined in (2.122) and (2.126), coincide. In particular, whenever one of the two is well-defined, both are well-defined and*

$$c_{\mathrm{g,nl}} = c_{\mathrm{g,lin}} =: c_{\mathrm{g}}.$$

The proof of the lemma reduces the computation of both group velocities to a boundary-value problem, which can then readily be solved by Lyapunov-Schmidt reduction.

Characterizing Archimedean spirals In Sections 2.4.2 and 2.4.2 we have collected three properties of certain chemical spiral waves: rigid rotation, asymptotic Archimedean shape, and positive group velocity of the asymptotic wavetrains. Heading for an open class of "robust" solutions that resemble Archimedean spirals, we incorporate the apparent robustness of spiral waves with respect to changes in parameter values into our characterization. So far, we have set up a boundary-value problem for spiral wave solutions. Preferably, robustness proofs for spiral would rely on an implicit function theorem in a suitable function space. Unfortunately the properties of spiral waves, as collected above, prevent us from an application of the implicit function theorem, since the linearization of (2.117) lacks the Fredholm property.

More specifically, consider an Archimedean spiral wave solution $q_*(r, \psi)$ to (2.117) on $x \in \mathbb{R}^2$. Linearizing (2.115) at q_* in a corotating frame yields

$$w_t = D\Delta_{r,\psi} w - \omega_* \partial_\psi w + f'(q_*(r, \psi))w =: \mathcal{L}_* w. \qquad (2.127)$$

The elliptic operator $\mathcal{L}_*$ is not Fredholm on L^p-spaces or spaces of continuous functions; see [Sa&al99, Lemma 6.4], for example. We therefore cannot conclude persistence of a spiral wave solution under slight variation of diffusion coefficients or reaction kinetics, simply by an implicit function theorem.

In the remainder of this section, we aim at a definition of a spiral wave that will allow us to conclude robustness. The main idea is to invoke Lemma 2.3.3, relating Fredholm properties in exponentially weighted spaces to group velocities. Analogously to the weighted L^2-spaces in (2.73) on the real line $x \in \mathbb{R}$, we introduce the exponentially weighted spaces

$$L^2_\eta(\mathbb{R}^2) := \{ u \in L^2_{\mathrm{loc}}(\mathbb{R}^2);\ \|u\|^2_{L^2_\eta} := \int_{\mathbb{R}^2} |\mathrm{e}^{\eta|x|} u(x)|^2 < \infty \}, \qquad (2.128)$$

for weights $\eta \in \mathbb{R}$ in the radial direction $r = |x|$.

Definition 2.4.2 *An Archimedean spiral is a bounded rotating-wave solution $q_*(r, \varphi - \omega_* t)$ of the reaction-diffusion equation (2.115), with some nonzero rotation frequency ω_*, which converges to plane wavetrains in the farfield*

$$|q_*(r, \psi) - q_\infty(r - (\psi/k_\infty))| \longrightarrow 0, \quad \text{for } r \to \infty,$$

uniformly in $\psi \in [0, 2\pi]$. We call an Archimedean spiral transverse, *if the following conditions hold:*

(i) there exists $\eta_0 < 0$ such that the linearization $\mathcal{L}_*$, defined in (2.127), is Fredholm with index 0 in L_η^2, for all $\eta_0 < \eta < 0$, and the dimension of the generalized kernel is one;

(ii) there is $\eta_1 > 0$ such that the linearization $\mathcal{L}_*$ is Fredholm with index -1 for all η with $0 < \eta < \eta_1$.

The definition encodes positive group velocity of the wavetrain in the negative Fredholm index of the linearization; see Proposition 2.3.1, Lemma 2.3.3, and the discussion of the linearization in Section 2.5.3, below. It also reflects simplicity of the trivial zero eigenvalue $\lambda = 0$. For fronts and pulses, we have taken advantage of this property in order to conclude transverse crossing of stable and unstable manifolds in the travelling-wave equation, see Section 2.3.2. It is for this geometric interpretation that we refer to Archimedean spirals satisfying Definition 2.4.2(i),(ii) as transverse Archimedean spirals.

Theorem 2.4.3 *[SaSc00d] Assume the reaction-diffusion system (2.115) with analytic reaction kinetics $f = f(u)$ possesses a transverse Archimedean spiral wave solution. Then, for any reaction kinetics $\tilde{f}$, C^1-close to f, and any diffusion matrix $\tilde{D}$ close to D, the perturbed reaction-diffusion system*

$$u_t = \tilde{D}\Delta u + \tilde{f}(u)$$

also possesses a transverse Archimedean spiral wave.

For the proof of the theorem, we rewrite (2.117) in radial dynamics using $r = |x|$ as "time"; see [Sc98]. We then interpret a spiral wave as a heteroclinic front solution connecting the center $r = 0$ with the wavetrains in the farfield $r = \infty$. The radial dynamics formulation of (2.117) reads

$$U_r = V \qquad (2.129)$$
$$V_r = -\frac{1}{r}V - \frac{1}{r^2}\partial_{\psi\psi}U - D^{-1}(-\omega_*\partial_\psi U + f(U)).$$

Note how spiral waves $q_*(r, \psi)$ actually turn into heteroclinic orbits $(U, V)(r, \cdot)$, just like fronts in one space-dimension, Section 2.3.1, or the modulated fronts in Section 2.3.2. We view equation (2.129) as a non-autonomous differential equation on a Hilbert space $(U, V)(r, \cdot) \in H^1(S^1) \times L^2(S^1)$. Just like the travelling-wave ODE for modulated fronts (2.89), the initial value problem for the elliptic system (2.129) is ill-posed. Stable and unstable manifolds exist, but turn out to be infinite-dimensional. An additional complication arises since the term involving the highest derivative $\frac{1}{r^2}\partial_{\psi\psi}U$ is non-autonomous, singular at the origin $r = 0$, and degenerate in the farfield $r = \infty$.

The limit $r \to 0$ of the core region is easily understood in the scaling $r = e^\tau$, where we find

$$U_\tau = W \qquad (2.130)$$
$$W_\tau = \partial_{\psi\psi}U - e^{2\tau}D^{-1}(-\omega_*\partial_\psi U + f(U)).$$

In the farfield limit $r \to \infty$, we recover the equation for modulated travelling waves (2.89),

$$U_r = V \qquad (2.131)$$
$$V_r = -D^{-1}(-\omega_* \partial_\psi U + f(U)).$$

Spiral waves are solutions which converge to constant functions $U = U_0, V = 0$ in the center of rotation $r = 0$. For $r \to \infty$, they approach the asymptotic wavetrain $U = q_\infty(r - (\psi/k_\infty)), V = u'_\infty(r - (\psi/k_\infty))$. In this sense, spiral waves are heteroclinic orbits in the non-autonomous differential equation (2.129), connecting an equilibrium at $\tau = \log r = -\infty$ to a periodic orbit at $r = +\infty$.

Transversality of Archimedean spirals, in the sense of Definition 2.4.2, implies that the associated heteroclinic orbit in spatial dynamics consists of a *transverse crossing* of the stable manifold of the asymptotic periodic orbit, at $r = \infty$, and the unstable manifold at $r = 0$. Transverse crossing, here, again refers to a transverse intersection of stable and unstable manifolds in the extended phase space, augmented by the spiral frequency parameter ω_*; see also Section 2.3.2. We return to this geometric picture for the linearized equation in more detail in Section 2.5.3, below.

The proof of Theorem 2.4.3 is therefore based on smoothness of center-stable and center-unstable manifolds in the radial dynamics formulation, and a robustness argument for transverse intersections. Analyticity of f is used for uniqueness of solutions to the ill-posed initial-value problem in the radial dynamics formulation.

2.4.3 Bifurcation to Spiral Waves

Theorem 2.4.3 shows that transverse Archimedean spiral waves are robust. We now show that transverse Archimedean spiral waves actually exist. We construct these spirals through bifurcation from a spatially homogeneous equilibrium in the unbounded plane $x \in \mathbb{R}^2$, in the presence of critical continuous spectrum.

We consider reaction-diffusion equations with the pure reaction-kinetics undergoing a Hopf bifurcation. We assume that $f(u; \mu)$ depends on a parameter μ and $f(0; \mu) = 0$ for μ close to zero. The linearization $\partial_u f(0; 0)$ is assumed to possess a pair of purely imaginary eigenvalues $\pm i\omega_H$, with eigenvectors $u_*, \bar{u}_* \in \mathbb{C}^N$. We assume that all other eigenvalues are contained in the left complex half plane. Performing a center-manifold reduction and a subsequent normal-form transformation, the Hopf bifurcation in the pure kinetics $u_t = f(u; \mu)$ can be reduced to a two-dimensional ordinary differential equation of the form

$$\dot{z} = \lambda(\mu)z + \beta z|z|^2 + O(|z|)^5, \qquad (2.132)$$

with eigenvalue $\lambda(0) = i\omega_H$ and complex cubic normal form coefficient $\beta \in \mathbb{C}$. We assume that $\lambda'(0) > 0$, crossing of eigenvalues, and $\operatorname{Re}\beta < 0$. The linearization of the *spatially dependent* reaction-diffusion system in $u = 0$,

$$u_t = D\Delta u + \partial_u f(0;0)u,$$

can be analyzed after Fourier transform from the dispersion relation

$$d(\lambda, ik) = \det\left(-Dk^2 + \partial_u f(0;0) - \lambda\right) = 0, \quad k \in \mathbb{R}. \tag{2.133}$$

For wavenumbers $k \in \mathbb{R}$ near zero, the eigenvalue $\lambda = i\omega_H$ continues to a critical spectral curve $\lambda(ik;\mu)$, with $\lambda(0;0) = i\omega_H$ and expansion

$$\lambda(ik;0) = i\omega_H + \alpha k^2 + O(k^4). \tag{2.134}$$

Theorem 2.4.4 *[Sc98] There exists a nonempty open set $\mathcal{U} \subset C^3(\mathbb{R}^N)$ of nonlinearities $f = f(u)$, such that for all $f \in \mathcal{U}$ the reaction-diffusion system (2.115) possesses a transverse Archimedean spiral wave, in the sense of Definition 2.4.2.*

More specifically, assume that the reaction-diffusion system undergoes a Hopf bifurcation such that

(i) $\lambda(0;0) = i\omega$ *is a simple zero of (2.133) and the only purely imaginary solution to (2.133), for any $k \in \mathbb{R}$;*
(ii) $\partial_\mu \operatorname{Re}\lambda(0;0) > 0$;
(iii) *for α and β, as defined in (2.132) and (2.134), we have $\operatorname{Re}\beta < 0$, $\operatorname{Re}\beta < 0$ and $|\arg(\beta/\alpha)| < \delta$ sufficiently small.*

Then, for sufficiently small $\mu > 0$, there exists a transverse Archimedean spiral wave solution of (2.115).

One approach to Theorem 2.4.4 would be to approximate the reaction-diffusion system by a complex Ginzburg-Lan"-dau equation

$$A_{\tilde{t}} = \alpha \Delta_{\tilde{x}} A + A + \beta A|A|^2, \quad \tilde{x} \in \mathbb{R}^2, \ A, \alpha, \beta \in \mathbb{C}, \tag{2.135}$$

and then look for rigidly rotating spiral wave solutions. For persistence of spiral waves however, we might still have to rely on a spatial dynamics formulation as in Theorem 2.4.3.

The proof given in [Sc98] splits into three parts. The first part contains a reduction of the radial dynamics (2.89) to a finite-dimensional center manifold, which contains all solutions of (2.129) close to the equilibrium $(U,V) \equiv 0$ [Sc98, Theorem 1, Proposition 5]. On the center manifold, we find the bifurcation equation

$$A_{rr} = -\frac{1}{r}A_r - \frac{1}{r^2}A + A - i\tilde{\omega}A + (\beta/\alpha)A|A|^2, \tag{2.136}$$

for $A(r) \in \mathbb{C}$, to leading order in an appropriate scaling. The parameter $\tilde{\omega}$ depends on the frequency of rotation ω_*. In the second part, this reduced equation is shown to possess a transverse heteroclinic orbit, asymptotic to plane wave solutions with positive group velocity [Ha82, KoHo81, Sc98]. The proof is concluded with a persistence result under higher-order perturbations, similar to Theorem 2.4.3.

The results in [Sc97, Sc98] also show that spatially localized rotating wave solutions $u(r, \varphi - \omega t)$ with $u(r, \cdot) \to 0$ for $r \to \infty$ bifurcate if the Hopf bifurcation is subcritical. To our knowledge, these patterns have not been observed in experiments, yet.

2.5 Two Space Dimensions: Bifurcations from Spiral Waves

In this chapter we discuss several instability mechanisms for the spiral rotating wave solutions which are constructed in the previous chapter. We start with a biased review of experimentally observed instabilities, in Section 2.5.1. We then consider instabilities of Hopf type, caused by point spectrum. Such instabilities lead to meandering motions of the spiral tip and to periodic shape fluctuations. Equivariant bifurcations under the noncompact symmetry group $SE(m)$ are an appropriate framework here, as we will discuss in Section 2.5.2.

Beyond the two-frequency meandering motion, more complicated motions of the spatial tip have been observed [Un&al93]. Attempts to explain these motions strongly rely on a finite-dimensional description of the instability, coupled to the motion on the non-compact group $SE(2)$ [FiTu98, Go&al97, As&al01]. The motion of the tip, often referred to as hypermeander (although already meander originally refers to an irregular winding of river beds) can under certain assumptions be shown to trace Brownian motion paths; see Section 5.2.7.

In Section 2.5.3, we analyze the linearization about Archimedean spiral waves in a more systematic way. In particular, we classify continuous spectra and the super-spiral shapes of eigenfunctions. The results in this section are then compared with experiments and numerical simulations in Section 2.5.4: the spatio-temporal patterns arising in spiral breakup and meander instabilities are related to spectral properties of the primary Archimedean spiral.

The discussion of continuous spectra in Sections 2.5.3 and 2.5.4 is largely independent of Section 2.5.2 and closely follows the exposition on radial dynamics in Sections 2.4.2, 2.4.3 and on the one-dimensional case, Chapter 2.3.

2.5.1 Phenomenology of Spiral Instabilities

Beyond rigidly rotating spiral waves, destabilization of individual spiral waves and subsequent transition to still richer spatio-temporal patterns has been observed in chemical experiments. In this section, we review some experimentally and numerically observed instabilities.

Meandering of the spiral tip was observed in [Ja&al89, SkSw91, Ne&al, BrEn93, Un&al93], among others. Temporally, the spiral motion becomes a two-frequency motion. For example, in a two-species reaction-diffusion system $N = 2$, we may trace the evolution of a point $x = z_*(t)$ in physical space $x \in \mathbb{R}^2$, where the concentration vector possesses a fixed, prescribed value $u_* \in \mathbb{R}^2$. The curve $z_*(t)$ describes a one-frequency motion on a circle before the transition to meander movement. After transition the curve $x_0 = x_0(t)$ becomes an epicycle, as shown in Figure 2.17, 2.22. We refer to this (nonunique) point as the *tip* of the spiral; see Section 5.2.4. Beyond the two-frequency meandering motion. We recall more complicated motions of the spiral tip have been observed; see Section 5.2.7.

Explanations based on a curvature description of the wavetrain dynamics in the spirit of Section 2.4.1, together with an additional equation for the core of the spiral, were proposed in [MiZy91]. Barkley [Ba92] was the first to suggest simple Hopf bifurcation in the presence of Euclidean symmetry as the mechanism which leads to meandering. Barkley later discovered numerically that drift of the tip is related to a resonance between Hopf frequency and rotation frequency of the spiral wave [Ba93]. The numerical and theoretical predictions have been confirmed in careful experiments. In particular, the Hopf-typical square-root scaling of the amplitude of the instability as a function of the parameter driving the instability has been verified [Ou&al].

The patterns in the farfield appear as prominent super-imposed spiral-like regions, where the local wavelength is larger than the average. The numerical computation of critical eigenfunctions reveals an apparent dramatic change of the radial growth behavior of the eigenfunction when the bifurcation parameter is varied close to criticality: before instability, the amplitude of the eigenfunction appears to grow with the distance from the center of rotation, whereas after instability, the eigenfunction seems to be localized. Experimentally, the instability appears to be localized at the center of rotation in physical space. However, radial decay is very weak; see [Ou&al].

Spiral breakup is a phenomenologically different instability where, as in the meander instability prominent super-patterns occur in the farfield $|x| \to \infty$. This time, however, perturbations grow in amplitude with distance from the center; see [BäEi93, BäOr99, Be&al97, ZhOu00]. Before spiral breakup, the perturbations of the spiral wave again take the form of superimposed spirals of compression and expansion of the local wavelength of the wavetrains. Instability finally leads to visible collision of emitted pulses. Collision is generally referred to as breakup and eventually leads to the formation of new spiral cores, far away from the center of the primary spiral. The transition to instability is not sharp and depends on the size of the domain. We refer to this instability mechanism as *farfield breakup*.

Spiral breakup, i.e. collision of wavetrains, can also occur close to the core region, $r \to 0$, of the primary spiral. After collision, new spirals form close to the core of the primary spiral, which break up immediately. This mechanism of instability leads to very incoherent spatio-temporal states, with many small spiral domains; see [Ar&al94, BäOr99]. We refer to this scenario of instability as *core breakup*.

Both spiral breakup scenarios, farfield and core breakup, may occur as instabilities of rigidly rotating *or* meandering spirals [ZhOu00, BäOr99].

Aranson and Kramer [Ar&al92], considering the complex Ginzburg-Lan"-dau model equation, point out that the lack of a sharp transition to instability in spiral breakup is caused by transport of the wavetrains, rendering the instability convective: perturbations increase in amplitude, but decay at each fixed point in physical space. Sensitivity to instability can be increased by either further increasing the bifurcation parameter, or, by increasing the size of the domain. Baer et al. compare breakup instabilities with instabilities of one-dimensional wavetrains emitted by a Dirichlet source [BäOr99]. Striking qualitative and quantitative similarity suggests that spiral breakup is caused by an instability of the wavetrains in the farfield.

We briefly mention three other instability phenomena without attempting a complete explanation, here. The retracting wave bifurcation [MiZy91] manifests itself in a decrease of the spiral frequency ω_* to zero. The radius of the circle $x_0(t)$ that a fixed point u_0 on the spiral profile describes diverges until the spiral finally leaves the window of observation. The instability is partially explained in [As&al99] invoking a finite-dimensional reduction procedure. Two-dimensional, fingering-type instabilities of wavetrains, emitted by a spiral wave, have been studied in [MaPa97]. Period-doubling of a homogeneous oscillation which induced complicated bifurcations of the emitted wavetrains, has been observed in oscillatory media [Go&al98].

We do not attempt to list particular models giving rise to the plethora of phenomena around spiral waves. Most models center around two- or three-species reaction-diffusion equations with excitable or oscillatory kinetics. A different line of investigations concentrates on the complex Ginzburg-Lan"-dau equation, where in particular the wavetrain dynamics are amenable to explicit, algebraic computations.

Our emphasis is on the general framework, assuming existence of spiral waves without recurring to the precise kinetics. Instability scenarios will be explained from certain spectral assumptions on the linearization, rather than discussions of oscillatory versus excitable kinetics.

As a general guideline, we expect instabilities to be caused either by essential spectrum or by point spectrum crossing the imaginary axis. Essential spectrum crossing the axis will typically not lead to an instability in bounded domains, where the absolute spectrum, strictly to the left of the essential spectrum, decides upon stability.

2.5.2 Meandering Spirals and Euclidean Symmetry

In this section we consider a given rotating wave solution $u = q_*(r, \varphi - \omega_* t)$ of a reaction-diffusion system

$$u_t = D\Delta u + f(\mu, u). \tag{2.137}$$

We consider $u \in \mathbb{R}^N$, a positive diagonal diffusion matrix D, $x \in \mathbb{R}^m$, mostly $m = 2$, and a scalar bifurcation parameter μ. For $f \in C^2$, system (2.137) fits into the abstract framework of analytic semigroup theory on a Banach space X, [He81, Pa83], but we prefer to formulate results in the more specific setting (2.137). Specifically we may think of $X = L^2$, excluding the Archimedean spirals of Chapter 2.4, which do not decay in amplitude for $|x| \to \infty$. Suitable subspaces X of BUC, the space of bounded uniformly continuous functions, are another possible choice. Although BUC does accommodate Archimedean spirals, these may exhibit critical continuous spectrum — a complication which we postpone to Section 2.5.3.

Euclidean equivariance Since the Laplacian Δ as well as the point evaluation nonlinearity commute with translations $S \in \mathbb{R}^m$ and rotations $R \in SO(m)$ of spatial profiles $u \in X$, we observe equivariance of the semiflow generated by (2.137) under the Euclidean symmetry group

$$G = SE(m) = \{(R, S);\ (R \in SO(m),\ S \in \mathbb{R}^m\}. \tag{2.138}$$

The action on $g = (R, S) \in G$ on $u = u(x)$ is given explicitly by

$$((R, S)u)(x) := u((R, S)^{-1}x). \tag{2.139}$$

Note that the action on $x \in \mathbb{R}^m$ is affine,

$$(R, S)x := Rx + S, \tag{2.140}$$

whereas the action on $u \in X$ is linear. *Equivariance* of (2.137) under G means that $u(t) \in X$ is a solution if, and only if, $gu(t) \in X$ is a solution, for every fixed $g \in G$. For a general background of bifurcation theory in the presence of (mostly compact) symmetry groups we refer to [ChLa00, GoSc85, Go&al88, Va82].

The group $G = SE(m)$ can be viewed as a Lie subgroup of the special linear group $SL(m+1)$ of real $(m+1) \times (m+1)$ matrices with determinant one, via the embedding

$$\begin{aligned} SE(m) &\to SL(m+1) \\ (R, S) &\mapsto \begin{pmatrix} R & S \\ 0 & 1 \end{pmatrix} \end{aligned}. \tag{2.141}$$

For example, $(R, S)^{-1} = (R^{-1}, -R^{-1}S)$.

The embedding (2.141) is also useful for practical computations of the Lie algebra $se(m)$ and of the exponential map $\exp : se(m) \to SE(m)$. For example $\mathbf{a} := (\mathbf{r}, \mathbf{s}) \in se(m)$ is given by a skew symmetric matrix $\mathbf{r} \in so(m)$ and by $\mathbf{s} \in \mathbb{R}^m$ with explicit exponential

$$\exp(\mathbf{r}, \mathbf{s}) = (\exp \mathbf{r},\ \mathbf{r}^{-1}(\exp(\mathbf{r}) - \mathrm{id})\mathbf{s}). \tag{2.142}$$

For nonzero $\mathbf{r} \in so(2)$, we therefore see how any exponential $\exp(\mathbf{a}t)$ is conjugate to a pure rotation $\mathbf{s} = 0$, by a suitable translation. Identifying $\mathbf{r} \in so(2)$ with the purely imaginary number $i\omega$, we can also write rotations $R \in SO(2)$ as $R = \exp(\mathbf{r}t) = \exp(i\omega t)$. See [Fi&al96, Ch. 4] for many more examples.

In this abstract setting, a rotating wave solution $u(t) = u(t, \cdot) \in X$ of (2.137) takes the form

$$u(t) = \exp(\mathbf{a}_0 t) q_*, \qquad (2.143)$$

for some $q_* \in X$. The polar coordinate formulation (2.116) of Section 2.4.2 coincides with (2.143) for $\mathbf{a}_0 = (\mathbf{r}_0, 0)$ and $\mathbf{r}_0 = i\omega_*$. More generally, solutions of the form (2.143), with an arbitrary element $\mathbf{a}_0$ of the Lie algebra $\mathrm{alg}(G)$ of G, are called *relative equilibria:* the time orbit (2.143) is entirely contained in the group orbit Gq_* and hence does not exhibit any shape change. Travelling waves, where $G = (\mathbb{R}, +)$, are another simple example; see Chapter 2.3.

Relative center manifolds We now describe the center-manifold reduction near relative equilibria (2.143) due to [Sa&al97a, Sa&al97b, Sa&al99]. We restrict to the slightly simpler case of symmetry groups $G \leq SE(m)$. Fix the parameter $\mu = \mu_0$. Assume the reaction kinetics $f = f(u)$ are smooth, C^∞, and fix $k \in \mathbb{N}$ arbitrary. We first pass to a "rotating" or "comoving" coordinate frame $\tilde{u} := \exp(-\mathbf{a}_0 t) u$, where $u(t)$ becomes an equilibrium $\tilde{u} =: q_*$. By G-equivariance, this transformation leaves (2.137) autonomous. Assume that the group action is continuous on the relative equilibrium q_*. The central assumption to Theorem 2.5.1, below, is that the linearization at the equilibria Gq_* possess only point spectrum in $\{\operatorname{Re}\lambda \geq -\delta\} \subset \mathbb{C}$, for some $\delta > 0$. Let W denote the corresponding finite-dimensional generalized eigenspace.

Theorem 2.5.1 *[Sa&al97a, Sa&al97b, Sa&al99] Under the above assumptions there exists a G-invariant neighborhood $\mathcal{U}$ of the group orbit Gq_* of the relative equilibrium q_* and a G-invariant C^k center manifold $\mathcal{M} \subseteq \mathcal{U}$ with differentiable proper G-action on $\mathcal{M}$. The C^k-manifold $\mathcal{M}$ is time-invariant within $\mathcal{U}$, and tangent to the center eigenspace W at $q_* \in \mathcal{M}$. Moreover $\mathcal{M}$ contains any solution $u(t), t \in \mathbb{R}$, which remains in $\mathcal{U}$ for all positive and negative times t.*

Since the center manifold $\mathcal{M}$ is based on a parameterization over the group orbit Gq_* of the relative equilibrium q_*, we call $\mathcal{M}$ a *relative center manifold*.

Properness of the G-action on $\mathcal{M}$ is a technical property, which asserts that the map

$$\begin{aligned} G \times \mathcal{M} &\to \mathcal{M} \times \mathcal{M} \\ (g, u) &\mapsto (u, gu) \end{aligned} \qquad (2.144)$$

is *proper:* images of closed subsets are closed and preimages of points are compact. For example, consider the *isotropy* H of $q_* \in \mathcal{M}$:

$$H = G_{q_*} := \{h \in G;\ hq_* = q_*\}. \tag{2.145}$$

The isotropy H consists of the symmetries of the solution q_* in the group G. Note that isotropies are compact, for proper group actions. Indeed $H \times \{u\}$ is the preimage of (q_*, q_*) under the proper map (2.144), and hence compact.

We caution the reader that the eigenspace W is not G-invariant, in general. For example $G = SE(2)$ does not possess any nontrivial finite-dimensional G-invariant subspace in L^2. In fact, $g \in G$ will move W to contain the tangent space $T_{gq_*}Gq_*$ which may only intersect trivially with W. In particular, the center manifold $\mathcal{M}$ cannot be written as a graph of a globally G-equivariant map over the tangent space W. This situation differs markedly from the case of compact symmetries G, where X decomposes into finite-dimensional G-invariant subspaces.

The proof of theorem 2.5.1 first exploits finite-dimensionality of the center-eigenspace in order to prove smoothness of the group action on spectral projections associated with W and on individual elements of W. This allows for a smooth parameterization of the group orbit $G(q_* + W)$. The main difficulty is that the group action on the normal bundle, consisting of the strongly stable subspace, is not even strongly continuous, in general. Still, the normal bundle is smooth, as a fiber bundle, and discontinuity of the group action can be carefully circumvented in the construction of the center manifold as a section to the normal bundle. For details of the construction, we refer to [Sa&al99].

Palais coordinates Within the finite-dimensional relative center manifold $\mathcal{M}$, we now express the PDE dynamics (2.137) by an ODE, rather explicitly, using *Palais coordinates* (g, v). See [Fi&al96] for details. We choose a local Palais section V in $\mathcal{M}$, transverse at q_* to the group orbit Gq_*. Then

$$\begin{aligned} G \times V &\to \mathcal{M} \\ (g, v) &\mapsto g(q_* + v) \end{aligned} \tag{2.146}$$

identifies $\mathcal{M}$ as a principal fiber bundle over Gq_*, with fiber V and structure group H. Indeed (g, v) and (g_0, v_0) parameterize the same point in $\mathcal{M}$ if, and only if,

$$\begin{aligned} gq_* &= g_0 q_*, \quad \text{and} \\ gv &= g_0 v_0. \end{aligned} \tag{2.147}$$

Therefore $g = g_0 h$, for some $h \in H$, and thus $v_0 = hv$. Hence

$$(g_0, v_0) = (gh^{-1}, hv). \tag{2.148}$$

With respect to this free action of the isotropy H on $(g, v) \in \mathcal{M}$, we indeed see that

$$\mathcal{M} = (G \times V)/H \tag{2.149}$$

is the orbit space of H. The action of $g_0 \in G$ on $\mathcal{M}$ lifts to $(g, v) \in G \times V$ canonically:

$$g_0(g, v) := (g_0 g, v) \tag{2.150}$$

in the above notation. Combining the actions (2.148), (2.150) we see how $(g_0, h) \in G \times H$ act by

$$(g, h)(g, v) := (g_0 g h^{-1}, hv) \tag{2.151}$$

on $(g, v) \in G \times V$. We admit that our above presentation is simplistic due to the suggestive, but strictly speaking illegitimate, use of the notation $g(q_* + v)$ for an element in the transverse slice gV to the group orbit Gq_* at gq_*. See [Fi&al96] for full details. An alternative approach, focusing on the dynamics in the bundle $\mathcal{M}$ directly, was developed earlier by [Kr90] for compact groups G.

We now lift the flow from the relative center manifold $\mathcal{M}$ to the Palais coordinates $(g, v) \in G \times V$. By equivariance (2.150) alone, this flow must take the following form.

Theorem 2.5.2 *[Fi&al96] In Palais coordinates (g, v), the G-equivariant flow (2.137) in the relative center manifold $\mathcal{M}$ takes the skew product form*

$$\begin{aligned} \dot{g} &= g\mathbf{a}(v) \\ \dot{v} &= \Phi(v) \end{aligned} \tag{2.152}$$

for suitable vector fields $\Phi(v)$ on V and $\mathbf{a}(v)$ on the Lie algebra $\mathrm{alg}(G)$. With respect to the compact isotropy H of q_* defined in (2.145), these vector fields transform according to

$$\begin{aligned} \Phi(hv) &= h\Phi(v) \\ \mathbf{a}(hv) &= h\mathbf{a}(v)h^{-1} \end{aligned} \tag{2.153}$$

For analogues of Theorem 2.5.2 for nonautonomous, periodically forced systems see [Sa&al99]. For bifurcations from relatively periodic solutions, rather than relative equilibria, see [Sa&al99, Wu&al01].

The skew product form (2.152) is easily interpreted. Points $(g, v) \in G \times V$ with the same v-component mark points on the same G-orbit in $\mathcal{M}$, by the G-action (2.150). Therefore $v(t)$ indicates the *shape* of the original x-profile $u(t, \cdot) \in X$. The G-component $g(t)$, in contrast, indicates *position* $S(t)$ and *phase* $R(t)$, at least in the case $G = SE(m)$, $g = (R, S)$, which is of primary interest here.

Spiral tip motion, Hopf meandering, and drift resonance To understand meandering spiral patterns, we rewrite (2.152) explicitly in terms of $g = (e^{i\alpha}, z) \in SE(2)$, with obvious complex notation for phase $R = e^{i\alpha}$ and position $S = z \in \mathbb{C}$. We immediately obtain the general form

$$\begin{aligned}\dot{\alpha} &= \omega(v) \\ \dot{z} &= e^{i\alpha}\sigma(v) \\ \dot{v} &= \Phi(v).\end{aligned} \qquad (2.154)$$

We now argue that the position component $z = z(t)$ can be interpreted as *tip position* of a meandering spiral wave directly; see [FiTu98]. Let

$$z_* : \mathcal{M} \to \mathbb{C} \qquad (2.155)$$

be any function which associates to a (nearly) spiral profile $u \in \mathcal{M} \subseteq \mathcal{U}$ near Gq_* a position $z_*(u) \in \mathbb{C} = \mathbb{R}^2$ of its "tip". Several such "tip" functions z_* have been considered; see Figure 2.16. They all share the equivariance property

$$z_*(gu) = gz_*(u) \qquad (2.156)$$

and hence lift to functions

$$z_* : G \times V \to \mathbb{C}$$

in Palais coordinates, which we again denote by z_*. A simple calculation then shows that

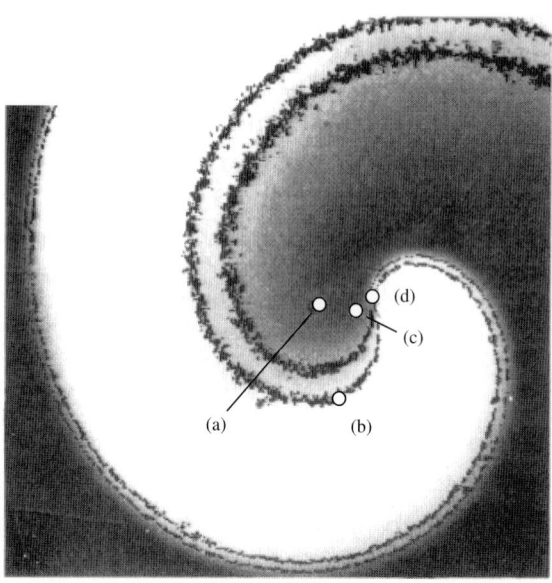

Fig. 2.17. "Tips" of a spiral wave pattern in the BZ reaction: (a) core center, (b) maximum curvature, (c) rotation center, (d) inflection point; courtesy of [MüZy94].

2 Spatio-Temporal Dynamics of Reaction-Diffusion Patterns

$$\dot{z} = e^{i\alpha}\sigma_*(v), \qquad (2.157)$$

which takes the same form as the positional z-equation in (2.153). Only the precise form of the nonlinearity σ depends on the precise form of the tip function z_*. Discussing the dynamics of skew product systems (2.154), in general, will therefore capture all the rather similar dynamics which are observed in the experimental literature — in spite of the varying definitions of a "tip". We therefore call $z(t)$ itself the tip of the spiral, henceforth.

Returning to a parameter dependent version of the skew product (2.154), we give a simple example. Consider shape variables $v \in \mathbb{R}^2$ undergoing Hopf bifurcation, due to a nontrivial pair of imaginary eigenvalues of the linearization at q_*, in corotating coordinates, crossing the imaginary axis transversely as the parameter μ increases. Then the shape variable $v = v(t)$ becomes periodic, say of Hopf frequency ω_H. Dividing (2.154) by the nonzero Euler multiplier $\omega(v)$ does not change the trajectory of the tip position $z \in \mathbb{C}$ and retains the abstract form of (2.154). We may therefore consider $\omega(v) \equiv \omega_*$ where ω_* indicates the normalized rotation frequency of the underlying rotating wave q_*, alias $v \equiv 0$.

The tip motion $z(t)$ is then obtained by simple integration. Fourier expansion

$$\sigma(v(t)) = \sum_{k \in \mathbb{Z}} \sigma_k e^{ik\omega_* t} \qquad (2.158)$$

provides the explicit expression

$$z(t) = z_0 + \int_0^t e^{i\omega_* t'} \sigma(v(t')) dt' =$$
$$= z_0 + \sum_{k \in \mathbb{Z}} \frac{\sigma_k}{i(\omega_* + k\omega_H)} (e^{i(\omega_* + k\omega_H)t} - 1) \qquad (2.159)$$

for Hopf frequencies ω_H which are not an integer fraction of the normalized rotation frequency ω_*. Clearly, (2.159) then describes a two-frequency quasi-periodic motion

$$z(t) = \tilde{z}_0 + \tilde{\sigma}(t) e^{i\omega_* t} \qquad (2.160)$$

with $\tilde{\sigma}$ periodic of frequency ω_H. For irrational frequency ratios ω_*/ω_H, the tip position $z(t)$ densely fills an annulus around $\tilde{z}_0$ of inner and outer radius $\min |\tilde{\sigma}|$ and $\max |\tilde{\sigma}|$, respectively. At integer *drift resonances*

$$k\omega_H = \omega_* \qquad (2.161)$$

the tip motion $z(t)$ typically becomes unbounded, due to a nonzero drift term $\sigma_{-k} t$.

Such meandering and drifting motions have been observed experimentally, and numerically; see the discussion in Section 2.5.1. In view of work by [Mi91] in the Hamiltonian context, even the "slinky" spiral springs described by Love as far back as 1892 [Lo04], can be considered as an example for meandering. A first mathematical analysis of meandering spirals is due to [Wu96].

Relative normal forms With these interpretations in mind, we now proceed to sketch a normal form analysis of skew products (2.152), due to [FiTu98]. For standard normal form procedures, going back essentially to Poincaré, see the modern account in [Va89]. For vector fields like the shape dynamics $\dot{v} = \Phi(v)$, these procedures successively simplify finite order terms in the Taylor expansion

$$\Phi(v) = Mv + \cdots \tag{2.162}$$

near the equilibrium $\Phi(0) = 0$, by transformations

$$\tilde{v} = \Psi(v) = v + \cdots . \tag{2.163}$$

The action of the transformations Ψ on Φ is given explicitly by the vector field pull-back

$$\tilde{\Phi}(v) = \Psi'(\Psi^{-1}(v))\Phi(\Psi^{-1}(v)). \tag{2.164}$$

Note linear dependence of $\tilde{\Phi}$ on Φ. The standard normal form theorem asserts that nonlinear, finite order terms of $\Phi(v)$ in the range of the linear map $\Phi \mapsto \mathrm{ad}(M)\Phi$ can be eliminated, by suitable normal form transformations Ψ. Here

$$((\mathrm{ad}M)\Phi)(v) := M\Phi(v) - \Phi'(v)Mv \tag{2.165}$$

preserves polynomial order. A normal form of Φ thus amounts to a nonunique choice of a complement of range(ad(M)). Nonremovable terms, in this procedure, are frequently called *resonant*.

For G-equivariant skew products (2.152) we consider such a normal form procedure as already applied to the shape equation $\dot{v} = \Phi(v)$. The only remaining transformation then takes the general form $\tilde{g} = g_0(g, v)$. The requirement of G-equivariance (2.150) enforces the particular form $\dot{g} = g\mathbf{a}(v)$ in (2.154), which our normal form transformation Ψ is supposed to preserve. Therefore our normal form transformation has to be of the form

$$\tilde{g} = gg_0(v); \tag{2.166}$$

with $g_0(0) = 0$. Geometrically, such a transformation simply corresponds to the choice of a different slice V to the group action in the center manifold $\mathcal{M}$; see (2.146).

The induced transformation of the Lie algebra factor $\mathbf{a}(v)$ is then given by

$$\tilde{\mathbf{a}}(v) = g_0^{-1}(v)\mathbf{a}(v)g_0(v) + g_0^{-1}(v)g_0'(v)\Phi(v) \tag{2.167}$$

with g_0^{-1} denoting the group-inverse of $g_0 \in G$. Again, $\tilde{\mathbf{a}}$ depends (affine) linearly on $\mathbf{a}$.

To formulate our normal form result, in this setting, we introduce some notation. Let p enumerate the Jordan blocks of (not necessarily distinct) eigenvalues λ_p of $M = \Phi'(0)$, and let $1 \leq q \leq d_p$ enumerate a basis for each block. Then the shape variable v possesses components v_{pq} such that

$$(Mv)_{pq} = \lambda_p v_{pq} + v_{p,q+1} \qquad (2.168)$$

with the convention $v_{p,d_p+1} := 0$. With integer nonnegative multi-indices $\mathbf{k} = (k_{pq})_{pq}$ we expand

$$\tilde{g} = \sum_{\mathbf{k}} \mathbf{a}_{\mathbf{k}} v^{\mathbf{k}}, \qquad (2.169)$$

with $\mathbf{a}_0 = \mathbf{a}(0)$ unaffected by our transformation. We further decompose each $\mathbf{a}_{\mathbf{k}}$ spectrally by the adjoint action

$$\begin{aligned} \mathrm{ad}(\mathbf{a}_0) : \mathrm{alg}(G) &\to \mathrm{alg}(G) \\ \mathbf{a} &\mapsto [\mathbf{a}_0, \mathbf{a}] \end{aligned} \qquad (2.170)$$

on the Lie algebra $\mathrm{alg}(G)$. Let η_j denote the distinct eigenvalues of this action, and spectrally decompose $\mathbf{a}_{\mathbf{k}} = \sum_j \mathbf{a}_{\mathbf{k}}^j$ accordingly.

Resonant terms, alias nonremovable terms, in our setting are those monomials $\mathbf{a}_{\mathbf{k}} v^{\mathbf{k}}$ in the expansion (2.169) which satisfy the *resonance condition*

$$0 = \eta_j + (\mathbf{k}, \lambda) := \eta_j + \sum_{p,q} k_{pq} \lambda_p. \qquad (2.171)$$

Note how this resonance condition captures an integer interaction between the eigenvalues λ_p of the linearized shape dynamics $\dot{v} = Mv$ and the eigenvalues of the unperturbed, exponential group dynamics $\dot{g} = g\mathbf{a}_0$ along the relative equilibrium $v = 0$. We have already encountered such a resonance, with $\eta = i\omega_*$ and $\lambda = i\omega_H$, as drift resonance of meandering spirals; see (2.161) and (2.143).

Theorem 2.5.3 *[FiTu98] With the above assumptions and notations, there exists a normal form transformation (2.166) which preserves the skew product structure (2.152) and removes all nonresonant terms $\mathbf{a}_{\mathbf{k}}^j$ from $\mathbf{a}(v)$, for any finite order $1 \leq |\mathbf{k}| = \sum k_{pq}$. The H-equivariance (2.148), (2.153) of Palais coordinates can also be preserved. In the parameter-dependent case $\mathbf{a} = \mathbf{a}(\mu, v)$, the same elimination holds with coefficients $\mathbf{a}_{\mathbf{k}}(\mu)$ and with resonance defined at $\mu = 0, v = 0$.*

We call any normal form reduction in the sense of Theorem 2.5.3 *relative normal form.*

Relative Hopf resonance Instead of giving a proof of relative normal form Theorem 2.5.3, we illustrate some consequences for the Euclidean group $G = SE(2)$, and for possible motions of spiral tips, by several examples. We begin with standard Hopf bifurcation at $\mu = \mu_0 = 0 \in \mathbb{R}$, as was already considered in (2.154), (2.158) – (2.161). Normal form theory for Hopf bifurcation in $\dot{v} = \Phi(\mu, v)$ implies $\Phi = \Phi_0(\mu, |v|^2)v$, in complex notation for $v \in \mathbb{C}$. Therefore $|v| \equiv const.$, for periodic solutions $v(t)$. The Hopf eigenvalue is $\lambda = i\omega_H \neq 0$, of course. In the notation (2.138) – (2.142) we have

$$\mathbf{a}_0 = (\mathbf{r}_0, \mathbf{s}_0) = (i\omega_*, 0) \qquad (2.172)$$

for the rotating wave at $v = 0$. Therefore $\eta \in \operatorname{spec} \mathbf{a}_0 = \{\pm i\omega_*, 0\}$ provides a resonance (2.171) if, and only if,

$$0 = -\omega_* + k\omega_H. \qquad (2.173)$$

This is precisely the condition for drift resonance encountered in (2.161) above. For noninteger ω_*/ω_H we can eliminate all v-terms in $\dot{g} = g\mathbf{a}(\mu, v)$, and the normal form

$$\dot{g} = \tilde{g}\mathbf{a}_0(\mu) \qquad (2.174)$$

yields purely exponential solutions $\tilde{g}(t)$ with frequency $\omega_*(\mu)$. Inverting the normal form transformation (2.166) then provides solutions

$$g(t) = \tilde{g}(t) g_0(v(t))^{-1} \qquad (2.175)$$

which are clearly quasiperiodic, with $g_0(v(t))^{-1}$ contributing the Hopf-frequency ω_H of the shape dynamics, and $\tilde{g}(t)$ contributing the rotation frequency ω_* of pure unperturbed rotation. Since $g = (e^{i\alpha}, z)$, the tip motion $z(t)$ is also quasiperiodic. Of course the same arguments apply to general finite-dimensional Lie groups G. In particular, the case $G = SE(3)$ provides fascinating motions $z(t) \in \mathbb{R}^3$ of scroll rings (see Chapter 2.6) and vibrating arrows characterized by periodically pulsating straight propagation and superimposed sidewards meandering; see also [Fi&al96].

Similar calculations allow us to analyze $k : 1$ drift resonance (2.173) of the frequencies $\omega_* : \omega_H$. The nonresonant normal form for the tip motion $\tilde{z}(t)$ associated to the dynamics of $\tilde{g}(t)$ then becomes

$$\dot{\tilde{z}} = e^{i\alpha} \sigma v^k \qquad (2.176)$$

where $\sigma = \sigma(\mu_1, \mu_2)$ depends on $\mu = (\mu_1, \mu_2)$ and $|v|^2$. For simplicity, we normalize $\mu_2 = |v|^2 \geq 0$ as the Hopf bifurcation parameter and let μ_1 measure the detuning of the linearized Hopf frequency $\omega_H(\mu_1) = \omega_H(0) + \mu_1$ at the bifurcation line $\mu_1 \in \mathbb{R}$, $\mu_2 = 0$, $v = 0$. Then (2.176) implies

$$\tilde{z}(t) = c - \sigma \mu_2^{k/2} (e^{-ik\mu_1 t} - 1)/ik\mu_1. \qquad (2.177)$$

For the tip motion $z(t)$ in original coordinates, we have to account for the transformation (2.166) by $g_0(v) = (e^{i\alpha_0(v)}, z_0(v))$ of period (nearly) $2\pi/\omega_H$. In total we obtain a meandering tip motion of $z(t)$ along large circles of radius proportional to

$$\mu_2^{k/2}/|\mu_1|, \tag{2.178}$$

for $\mu_2 > 0$, which blow up at the drift resonance $\mu_1 = 0$. Strictly speaking we have suppressed terms in (2.176) which may not be in normal form. These terms however can be assumed to be perturbations of arbitrarily high finite order.

The superposition $g(t) = \tilde{g}(t)g_0(v(t))^{-1}$ in (2.175) also leads to curious tip motions right at bifurcation $\mu = \mu_0 = 0$, in these resonant cases. In nondegenerate normal form and for stable $v \equiv 0$, the shape component $v(t)$ converges to zero with $|v(t)| \sim t^{-1/2}$ and asymptotic phase frequency ω_H. For the tip motion $\tilde{z}(t)$ in normal form coordinates, (2.176) provides a constant limit

$$\tilde{z}_\infty = \lim_{t\to\infty} z(t), \tag{2.179}$$

for orders $k \geq 3$ of the integer drift resonance $k = \omega_*/\omega_H$. In terms of the true tip motion $z(t)$ this corresponds to a rotating wave limit, because $\tilde{z}_\infty$ can be eliminated by conjugation in $G = SE(2)$.

For $k = \omega_* : \omega_H = 2 : 1$, however, $\tilde{z}(t)$ converges to a circular motion

$$\tilde{z}(t) \sim c_0 + c_1 \exp(-i\tau(t)) \tag{2.180}$$

with constants c_0, c_1, asymptotically logarithmic phase $\tau(t)$, and decreasing angular velocity $\dot{\tau}(t) \sim 1/t$. The tip motion $z(t)$ then corresponds to a meander with constant speed epicycles along a slowing base circle.

The 1:1 drift resonance $\omega_* = \omega_H$, finally, gives rise to an unbounded motion (2.180) with logarithmic phase and nonconstant $c_1 \sim \sqrt{t}$. All these results follow from just the normal form equation (2.176).

Clearly the analysis of relative mode interactions in the shape variable v becomes feasible now. For an example involving Hopf-Hopf mode interactions of two complex eigenvalues $i\omega_1, i\omega_2$ see [As&al01].

Relative Takens-Bogdanov bifurcation The case of nilpotent linearization $M = \Phi'(0)$ in the shape dynamics $\dot{v} = \Phi(v)$, with trivial (relative) equilibrium $v = 0$ contributes only eigenvalues $\lambda = 0$ to the resonance condition (2.171). In normal form it turns out that

$$\dot{\tilde{z}} = 0, \tag{2.181}$$

because all terms in the translational $\tilde{z}$-equation are nonresonant. Indeed $\eta = i\omega_*$ is generated only by the rotation frequency ω_* of the reference rotating wave; the zero eigenvalue of ad($\mathbf{a}_0$) does not appear in the component $\tilde{z}$. In particular, any drift terms vanish beyond finite order.

A prime example, not observed experimentally so far, are the effects of a relative Takens-Bogdanov bifurcation in the shape variable $v \in \mathbb{R}^2$. For classical analysis see [Ta74, Bo81a, Bo81b, GuHo83]. From (2.154) we recall the tip motion

$$\dot{z} = e^{i\omega_* t} \sigma(v(t)), \qquad (2.182)$$

again after rescaling time so that $\dot{\alpha} \equiv \omega_*$. If $v(t)$ is homoclinic to a hyperbolic relative equilibrium v_∞, the resulting tip motion $z(t)$ satisfies

$$z^h = [z(t)]_{-\infty}^{+\infty} = \int_{-\infty}^{+\infty} e^{i(\omega_* t + \alpha_0)} \sigma(v(t)) dt. \qquad (2.183)$$

The integral converges if $\sigma(v_\infty) = 0$ at the rotating wave v_∞. Again this can be assumed after a conjugation in the group $G = SE(2)$. By hyperbolicity this implies exponential convergence $\sigma(v(t)) \to 0$, for $t \to \pm\infty$. The homoclinic shape dynamics $\Gamma = \{v(t); t \in \mathbb{R}\}$ thus typically implies heteroclinic tip dynamics, with heteroclinic tip shift $z^h \neq 0$.

Next suppose that the homoclinic orbit $\Gamma \subseteq \mathbb{R}^2$ is attracting from inside, due to a negative saddle quantity $\lambda_1 + \lambda_2 < 0$ of the eigenvalues $\lambda_1 < 0 < \lambda_2$ at its equilibrium $v = v_\infty$. Let $v(t)$ converge to Γ from the inside. Then the passage times t_k near $v = v_\infty$ grow geometrically like

$$t_k \sim c_1 \nu^k, \qquad (2.184)$$

with $\nu = |\lambda_1/\lambda_2| > 1$. Therefore the respective tip positions behave like

$$z(t_{n+n_0}) \sim z(t_{n_0}) + \sum_{k=n_0+1}^{n} z^h \cdot \exp(i\omega_* c_1 \nu^k). \qquad (2.185)$$

With respect to Lebesgue measure on ω_*, this trigonometric power sum defines a random walk of $z(t_n)$ which can be rescaled to a Brownian motion of z. Indeed the increments are then only weakly coupled, for $k \to \infty$. See [BePh96] and, for explicit sample plots, Figure 2.18, with $\tilde{\omega} := \omega_* c_1$.

We caution our reader on two points. First, the waiting times for each step of our random heteroclinic tip walk increase exponentially at rate $\log \nu > 0$, with time. Observation of such a random walk may therefore take a *very* long time. Second, although the heteroclinic tip shift z^h in (2.183) will typically be nonzero along homoclinic orbits Γ, it will be small beyond finite order in the Takens-Bogdanov case. See (2.181). Indeed the rotating wave phase $\exp(i\alpha(t)) = \exp(i\omega_* t)$ enters as a rapid forcing into the slow dynamics generated by the zero eigenvalues of the shape dynamics. In spite of these subtleties, the example shows clearly how a planar — and therefore nonchaotic – Poincaré-Bendixson type shape dynamics v can cause complicated tip motions in the group component $\dot{g} = ga(v)$ of the skew product flow (2.152).

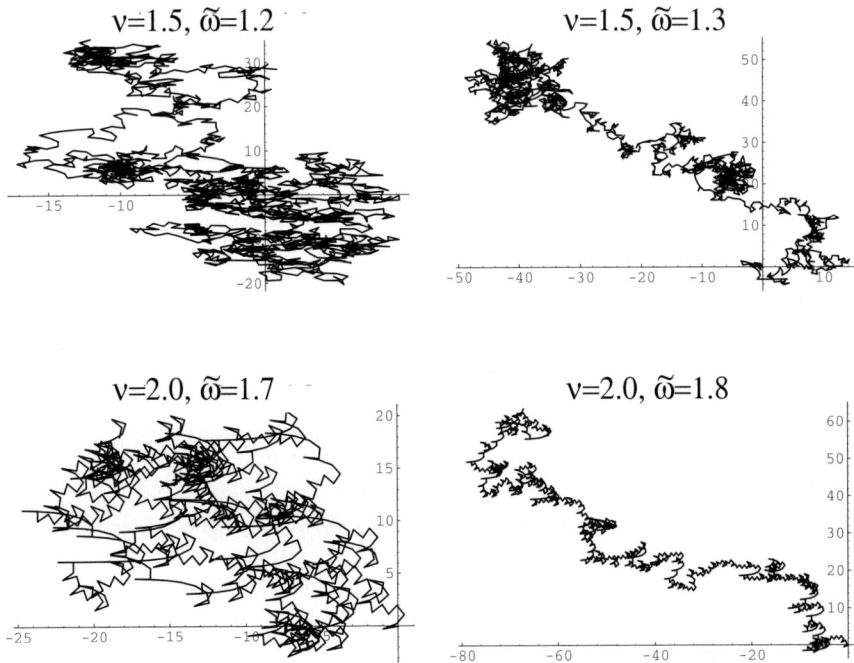

Fig. 2.18. Sample paths of Brownian tip motions due to asymptotically homoclinic shape dynamics.

2.5.3 Spectra of Spiral Waves

A major problem concerns a proper understanding of stability features of spiral waves. Even the notion of spectral stability turns out to be a very delicate problem. By means of a formal asymptotic matching procedure, Hagan [Ha82] showed that spiral waves in the complex Ginzburg-Lan"-dau equation (2.135) are likely to be stable. We give here a conceptual approach to the stability problem. We first characterize essential and absolute spectra of the spiral wave in terms of spectra of emitted wavetrains. We then describe the farfield shape of eigenfunctions to eigenvalues in the point spectrum in terms of the complex dispersion relation of the wavetrains.

The eigenvalue problem for spiral waves: core versus farfield As already pointed out in the discussion after Definition 2.4.2 a variant of Proposition 2.3.1 holds for spiral waves. Fredholm indices are again determined by group velocities on Fredholm borders. In order to make a precise statement, we consider the eigenvalue problem for the linearization $\mathcal{L}_*$ at an Archimedean spiral $q_*(r,\psi)$. In corotating polar coordinates, as defined in (2.127), the eigenvalue problem $(\mathcal{L}_* - \lambda)U = 0$ becomes

$$U_r = V \tag{2.186}$$
$$V_r = -\frac{1}{r}V - \frac{1}{r^2}\partial_{\psi\psi}U - D^{-1}\left(-\omega_*\partial_\psi U + f'(q_*(r,\psi))U - \lambda U\right).$$

As a first approach to the eigenvalue problem, we try to detect nontrivial bounded solutions of (2.186). An analogous consideration for the (formally) adjoint will then provide complete information about Fredholm properties such as the dimension of the kernel and the dimension of the cokernel.

The idea is to construct stable and unstable subspaces, containing solutions which are bounded in the farfield $r \to \infty$ and in the core $r \to 0$, respectively. From the dimension of these subspaces, we then read off the typical dimension of their intersection, which yields the dimension of the kernel. The dimension of the complement of their sum provides us with the dimension of the cokernel.

The core limit $r \to 0$ is again studied in the rescaled radial time $\tau = \log(r)$:
$$U_\tau = W \tag{2.187}$$
$$W_\tau = -\partial_{\psi\psi}U - e^{2\tau}D^{-1}\left(-\omega_*\partial_\psi U + f'(q_*(r,\psi))U - \lambda U\right).$$

In the limit $\tau = -\infty$, corresponding to $r = 0$, we formally obtain the autonomous, λ-independent system
$$U_\tau = V, \quad V_\tau = -\partial_{\psi\psi}U. \tag{2.188}$$

The equation is readily solved by angular Fourier decomposition. We find solutions to the spatial eigenvalues $\nu = \pm \ell$,
$$U(\tau,\psi) = U_+^\ell e^{i\ell\psi + \ell\tau}, \quad U(\tau,\psi) = U_-^\ell e^{i\ell\psi - \ell\tau},$$
for $\ell \neq 0$ and
$$U(\tau,\psi) = U_+^0, \quad U(\tau,\psi) = \tau U_-^0,$$
for the constant Fourier mode $\ell = 0$. The vectors $U_\pm^\ell \in \mathbb{C}^N$ are arbitrary. The condition $U_-^\ell = 0$ for all ℓ defines the (infinite-dimensional) *unstable subspace* E_-^u of solutions which stay bounded in the limit $\tau \to -\infty$. Note that these asymptotic eigenspaces are independent of the spectral parameter λ. A generalization of the concept of exponential dichotomies from ordinary differential equations [Pa88, Co78] to the ill-posed, elliptic problem (2.187) allows us to extend the unstable subspaces to the non-autonomous system (2.187) as a τ-dependent family of subspaces $E_-^u(\tau;\lambda)$; see [Pe&al97, Sc98].

In the farfield limit $r \to \infty$, we formally obtain
$$U_r = V \tag{2.189}$$
$$V_r = -D^{-1}(-\omega_*\partial_\psi U + f'(q_\infty(r - \frac{\psi}{k_\infty}))U - \lambda U).$$

If we formally replace $\psi/\omega_* \mapsto t$, $r \mapsto x$, and $U \mapsto Ue^{-\lambda t}$, we obtain the linearization about the one-dimensional wavetrains

$$U_t = DU_{xx} + f'(q_\infty(x - \frac{\omega}{k_\infty}t)U). \tag{2.190}$$

We have already discussed this equation in a comoving frame $\xi = x - \frac{\omega}{k_\infty}t$; see Section 2.3.1, (2.80).

Analogously to the unstable subspace E^u_- in the core region, we are going to determine the stable subspace E^s_+ in the farfield equation (2.189). Since the farfield system is periodic in spatial "time" r, Floquet theory (2.81)-(2.83) predicts that the stable subspaces are "time"-dependent and exponential growth or decay is governed by Floquet exponents ν. In the next section, we determine these Floquet exponents, depending on the spectral parameter λ.

Spatial Floquet theory and the dispersion relation of wavetrains
Our goal in this section is a characterization of exponential radial decay and growth properties of solutions to (2.189). The procedure is strongly reminiscent of Floquet theory (2.81)-(2.83) in Section 2.3.1, but the additional dependence on the angle ψ introduces some complications.

To eliminate explicit dependence on the angle ψ in (2.189), we exploit the Archimedean shape of the spiral by introducing "Archimedean coordinates". We replace ψ by $\xi = r - (\psi/k_\infty)$, and find

$$U_r = -\partial_\xi U + V \tag{2.191}$$
$$V_r = -\partial_\xi V - D^{-1}(\frac{\omega_*}{k_\infty}\partial_\xi U + f'(q_\infty(\xi))U - \lambda U).$$

Dependence of q_∞ on ξ is $2\pi/k_\infty$-periodic. The equation is now autonomous in r, and solutions are of the general form $U = e^{-\nu r}\hat{U}(\xi)$, where $\hat{U}$ solves the boundary-value problem

$$D(\frac{d}{d\xi} - \nu)^2\hat{U} + \frac{\omega_*}{k_\infty}\frac{d}{d\xi}\hat{U} + f'(q_\infty(\xi))\hat{U} = \lambda\hat{U}, \tag{2.192}$$

with periodic boundary conditions on $[0, 2\pi/k_\infty]$. Note that (2.192) is equivalent to the boundary-value problem (2.125) for the dispersion relation of the wavetrains. To simplify (2.192) further, we introduce $\tilde{U}(\xi) := e^{\nu\xi}\hat{U}(\xi)$. This yields

$$D\tilde{U}_{\xi\xi} + \frac{\omega_*}{k_\infty}\tilde{U}_\xi + f'(q_\infty(\xi))\tilde{U} = \tilde{\lambda}\tilde{U}, \tag{2.193}$$

with $\tilde{\lambda} := \lambda - (\omega_*\nu/k_\infty)$. Boundary conditions are now of Floquet type

$$\tilde{U}(0) = e^{2\pi\nu/k_\infty}\tilde{U}(2\pi/k_\infty) \tag{2.194}$$

induced from the $2\pi/k_\infty$-periodic boundary conditions for $\hat{U}$.

The dimension of the stable subspace E^s_+ for (2.189) changes whenever a Floquet exponent ν crosses the imaginary axis, as λ is varied. We therefore focus on solutions to (2.193), (2.194), with $\nu = ik$, purely imaginary, first.

In analogy to the one-dimensional discussion in Section 2.3.1, we call k the *Bloch wavenumber*.

Nontrivial solutions to the boundary-value problem (2.192),(2.194) can be found from the dispersion relation

$$d_{co}(\tilde{\lambda}, \nu) = \det\left(\Phi(\tilde{\lambda}) - e^{2\pi\nu/k_\infty}\right) = 0, \tag{2.195}$$

where $\Phi(\lambda)$ denotes the period-$2\pi/k_\infty$ map to (2.193), rewritten as a first-order differential equation; see (2.83). Substituting $\tilde{\lambda} = \lambda - (\omega_*\nu/k_\infty)$ we obtain

$$d_{st}(\tilde{\lambda}, \nu) = \det\left(\Phi(\lambda - (\omega_*\nu/k_\infty)) - e^{2\pi\nu/k_\infty}\right) = 0. \tag{2.196}$$

If we fix $\operatorname{Im}\nu \in [0, k_\infty)$, we find N solution $\lambda^\ell(\nu)$. The Floquet shift $\nu \mapsto \nu + ik_\infty$ gives us the vertically shifted spectral curves

$$\lambda^\ell(\nu) + i\omega_*\mathbb{Z}. \tag{2.197}$$

Note that the associated group velocity

$$c_{g,j}(ik) := \operatorname{Im}\left(\frac{d\lambda_j(ik)}{dk}\right) = \operatorname{Im}\left(\frac{d\tilde{\lambda}_j(ik)}{dk}\right) + c_\infty, \tag{2.198}$$

corresponds precisely to our previous definition (2.126) of the group velocity of the wavetrains.

Restricting to $\nu = ik$, $k \in \mathbb{R}$, we obtain the spectral curves $\Gamma^\ell = \{\lambda^\ell(ik); \; k \in \mathbb{R}\}$. We emphasize that together with a fixed spectral curve Γ^ℓ, there is the countable collection of vertically shifted (not necessarily disjoint) curves $\Gamma^\ell + i\omega_*\mathbb{Z}$. Comparing (2.190) with (2.193), we find that these curves consist of the continua of Floquet exponents for the linearized period map for the one-dimensional wavetrains.

For an illustration, we deviate from the general setup and discuss a specific example. Assume that the wavenumber k_∞ of the emitted wavetrains is small. More specifically, assume that on their very large interval of periodicity, $\xi \in [0, 2\pi/k_\infty]$, the wavetrains resemble a pulse solution $q_*(\xi)$, $|q_*(\xi)| \to 0$ for $|\xi| \to \infty$. We then refer to the wavetrains as pulse trains. Spectral properties of pulsetrains are determined by interaction of individual pulses, as we shall explain next. In particular, we can compute the Floquet spectrum of the pulse trains from certain geometric properties of a single pulse, alone. First consider the pulsetrains and the pulse in a comoving frame. Then the spectrum of the pulsetrains converges to the spectrum of the single pulse [SaSc01c]. Close to the (critical) zero-eigenvalue of the pulse with translation eigenfunction $q'_*(\xi)$, the spectrum of the pulsetrains consists of a small circle of essential spectrum

$$\tilde{\lambda}(\nu) = a(k_\infty)(e^{2\pi\nu/k_\infty} - 1) + o(|a(k_\infty)|), \quad \nu = ik \in i\mathbb{R}, \, ,$$

see Figure 2.19(a). The coefficient a is given as $a(k_\infty) = Me^{-\delta/k_\infty}$. The constant $\delta > 0$ is the weakest exponential decay rate at the tails of the individual pulse. The coefficient M measures the (signed) speed of crossing of stable and unstable manifolds of the origin in the travelling-wave equation

$$Dq'' + cq' + f(q) = 0,$$

when the speed c is varied. Note that $a(k_\infty) \to 0$, exponentially, for $k_\infty \to 0$. This reflects that the interaction strength between individual pulses decreases exponentially with their distance.

Spectral stability corresponds to $a(k_\infty) < 0$. The crucial coefficient M turns out to be positive in the classical FitzHugh-Nagumo equation.

In a comoving coordinate frame of speed c_∞, we have to add the speed of the frame to the group velocity and find

$$\lambda(\nu) = a(k)(e^{2\pi\nu/k_\infty} - 1) + c_\infty \nu + o(|a(k)|), \quad \nu = ik \in i\mathbb{R};$$

see Figure 2.19(b). The result is a vertically periodic epicycloid, winding along the imaginary axis, with tangencies at $i\omega_* \mathbb{Z}$.

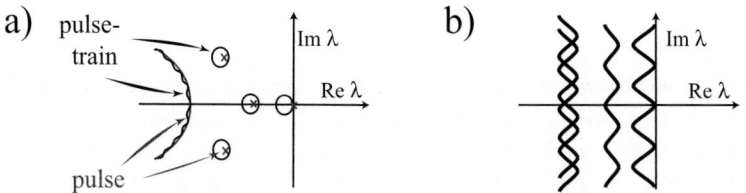

Fig. 2.19. The spectrum of wavetrains of weakly interacting pulses in the steady frame (**a**), and the Floquet exponents in the comoving frame (**b**).

Relative Morse indices and essential spectra of spiral waves In the spirit of Proposition 2.3.1, the oriented curves Γ^ℓ determine the Fredholm index of the linearization about a spiral wave, see (2.186). However, Proposition 2.3.2 has to be reformulated since both Morse indices $i_\pm$ at $\tau = -\infty$ (alias $r = 0$) and $r = \infty$ are infinite. Proposition 2.3.2 remains valid if both Morse indices $i_\pm$ are normalized to finite numbers, the *relative Morse indices*, which then enter formula (2.72) for the Fredholm index; see [SaSc01d] for details. A similar idea has been used repeatedly in the construction of Floer homology in various variational problems; see for example [RoSa95].

Since the asymptotic equation (2.188) in the core region, $\tau \to -\infty$, does not depend on λ, we can fix the Morse index in $-\infty$ and (arbitrarily) normalize to $i_- = 0$. We show next, how to define $i_+(\lambda) =: i_{\text{Morse}}(\lambda)$, such that Proposition 2.3.1 carries over to to linearizations at spiral waves: the Fredholm index of $\mathcal{L}_*$ is again given by the difference of the Morse indices

$$i(\mathcal{L}_* - \lambda) = i_- - i_+. \tag{2.199}$$

We solve the dispersion relation (2.196) for the radial Floquet exponent ν as a function of λ. For fixed λ, there exists an infinite number of spatial Floquet exponents ν_j, with $0 \leq \operatorname{Im} \nu < 2\pi/k_\infty$, which solve (2.196). Just like in the example of the heat equation (2.91), the real parts of the ν are unbounded on the positive and negative axis. Instead of counting unstable eigenvalues, we therefore count differences between numbers of unstable eigenvalues in different regions of the complex plane $\lambda \in \mathbb{C}$.

Consider the (open) complement of the spectral curves in the complex plane

$$\mathcal{U} = \mathbb{C} \setminus \bigcup_\ell (\Gamma^\ell + i\omega_* \mathbb{Z}).$$

For all $\lambda \in \mathcal{U}$, the asymptotic equation in the farfield, (2.189) is hyperbolic, that is, there are no purely imaginary solutions $\nu = ik$ to the dispersion relation (2.196). In consequence, the Morse index i_{Morse} is constant on connected components of $\mathcal{U}$.

In the connected component $\mathcal{U}_\infty$ of $\mathcal{U}$ containing $\operatorname{Re} \lambda \to +\infty$, the Fredholm index is zero. Indeed, the linearization $\mathcal{L}_* - \lambda$ is invertible, there, since the parabolic reaction-diffusion system is well-posed. In view of (2.199), we define $i_{\text{Morse}}(\lambda) \equiv 0$ in $\mathcal{U}_\infty$.

If we now move λ from one connected component of $\mathcal{U}$ to another, we cross a spectral curve $\Gamma^\ell + i\ell'\omega_*$. For $\lambda \in \Gamma^\ell + i\ell'\omega_*$, a Floquet exponent ν is located on the imaginary axis and hyperbolicity fails. The Morse index of $\mathcal{L}_* - \lambda$ changes by one when λ is varied across this single spectral curve. In order to determine the direction of change, we orient the curves Γ^ℓ such that vertical slope corresponds to positive group velocity, similarly to Proposition 2.3.1. With this orientation, the Floquet exponent $-\nu$ (recall the Ansatz $U \sim \exp(-\nu r)$) moves from $\operatorname{Re} \nu < 0$ to $\operatorname{Re} \nu > 0$ when λ crosses the spectral curve from left to right; see also Lemma 2.3.3. In particular, the (relative) number of Floquet exponents in $\operatorname{Re} \nu < 0$ decreases by one each time we cross spectral curves from left to right. We call this (relative) number of Floquet exponents in $\operatorname{Re} \nu < 0$ the *relative Morse index* i_{Morse}. This index is now defined in $\mathcal{U}$ via a λ-homotopy from $\mathcal{U}_\infty$ to any point in $\mathcal{U}$, counting signed crossings of the oriented curves $\Gamma^\ell + i\ell'\omega_*$ during the homotopy.

Proposition 2.5.4 *The operator $\mathcal{L}_* - \lambda$, defined in (2.127), considered as an unbounded operator on $L^2(\mathbb{R}^2)$ is Fredholm if, and only if $\lambda \notin \Gamma^\ell$, that is, if λ belongs to the complement $\mathcal{U}$ of the set of Floquet exponents of the emitted wavetrains. The Fredholm index is given by*

$$i(\mathcal{L}_* - \lambda) = -i_{\text{Morse}}(\lambda),$$

where the relative Morse index i_{Morse} was introduced above. In particular, the Fredholm index increases by one when crossing any of the oriented Fredholm borders $\Gamma^\ell + i\ell'\omega_*$ from left to right.

In the remainder of this section, we are going to derive a "typical" picture of the critical spectrum of a spiral wave. First notice that for $\lambda = 0$, the derivative of the wave train $U = q'_\infty$ provides a trivial solution to (2.193–2.194) with $\nu = 0$. If we assume that $\lambda = 0$ is simple as a root of the dispersion relation (2.196), then there is a unique spectral curve $\Gamma^0 = \lambda^0(ik)$, contained in the essential spectrum, that touches the imaginary axis in $\lambda = 0$. Since the positive group velocity of the emitted wavetrain does not vanish, the curve Γ^0 is not tangent to the real axis, see Definition 2.4.2. Because the spectrum of $\mathcal{L}_*$ is invariant under complex conjugation, the curve Γ^0 necessarily possesses a vertical tangent at $\lambda = 0$. Typically, the tangency will be quadratic. If the wave trains are spectrally stable, Γ^0 is the rightmost Fredholm border in the complex plane. In $\operatorname{Re} \lambda > 0$, the Fredholm index of $\mathcal{L}_*$ therefore is 0. Since the group velocity is positive, the region $\operatorname{Re} \lambda > 0$ is to the right of the Fredholm border Γ^0. To the left of this curve Γ^0, in $\operatorname{Re} \lambda < 0$, the Fredholm index of the linearization is -1, by Proposition 2.5.4. See Figure 2.20 for an illustration of the resulting spectral picture.

Positive group velocity was encoded in Definition 2.4.2 by means of exponential weights. Consider the linearization in L^2_η, (2.128), with radial weight $\eta < 0$ close to zero. Then the weight pushes the critical spectral curve Γ into the stable, left complex half plane. In other words, the exponentially decaying weight penalizes the outward transport; see Section 2.3.1 and Figure 2.20. Note that translation $(\partial_{x_1} + i\partial_{x_2})q_*$ and angular derivative $\partial_\psi q_*$ are bounded. They therefore contribute to the kernel in the space L^2_η, which allows for exponential growth.

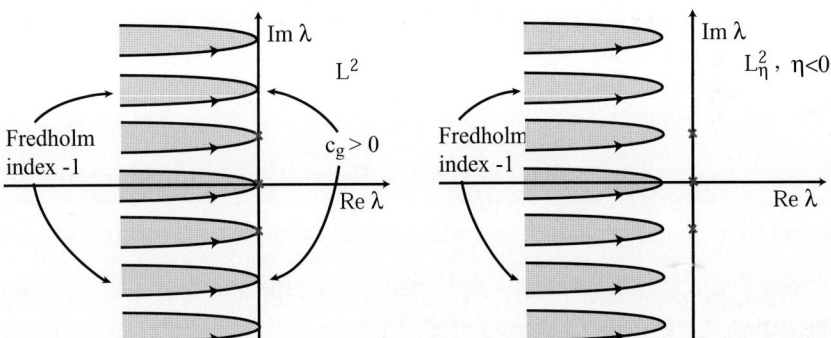

Fig. 2.20. Typical spectrum of the linearization about a spiral wave with Fredholm indices of the linearization in L^2 and in L^2_η, for $eta < 0$.

Along the imaginary axis, the spectral picture is periodic with period $i\omega_*$. The essential spectrum touches the imaginary axis at all points $i\omega_*\mathbb{Z}$. Since the wavetrains emitted from the center have a positive group velocity in the radial direction, this implies that the Fredholm index of $\mathcal{L}_* - \lambda$ is negative, $i(\mathcal{L}_* - \lambda) = -1$, for λ
to the left of all these critical spectral curves $\lambda^{\ell'}(ik)$, $\lambda^{\ell'}(0) = i\ell'\omega_*$, .

The abundance of critical spectrum, in the form of infinitely many tangencies of continuous spectrum at the imaginary axis, seems to suggest that spiral waves are extremely sensitive to perturbations. However, spiral waves in experiments appear to be very stable objects in many systems! In the next section, we will find a partial explanation for this apparent contradiction.

Absolute spectra of spiral waves We have seen in Section 2.3.3 how stability properties of waves on bounded domains are captured by the absolute spectrum rather than the essential spectrum. In this section, we carry out the analogous construction to define the absolute spectrum of a spiral wave. We then state a convergence theorem for spectra of linearizations along spiral waves on large bounded disks, as the diameter of the disk tends to infinity.

We return to the boundary-value problem (2.193),(2.194), with dispersion relation (2.196). For λ fixed, there is a countable family of spatial Floquet exponents $\nu_j = \nu_j(\lambda)$ with $0 \le \operatorname{Im} \nu_j(\lambda) < k_\infty$. The normalization eliminates trivial shifts by ik_∞. Each of these Floquet exponents can be seen to be of finite multiplicity as a root of the analytic dispersion relation (2.196). We order the Floquet exponents with increasing real part, repeated by multiplicity

$$\ldots \le \operatorname{Re} \nu_{-j-1} \le \operatorname{Re} \nu_{-j} \le \ldots \le \operatorname{Re} \nu_{-1} \le \operatorname{Re} \nu_1 \le \ldots \le \operatorname{Re} \nu_{j-1} \le \operatorname{Re} \nu_j \le \ldots. \quad (2.200)$$

Floquet exponents ν depend on λ and the labeling might be discontinuous in λ. We omitted the label 0 in the counting, for convenience. As already pointed out, the sequence is bi-infinite, that is, $\nu_j \to \pm\infty$ for $j \to \pm\infty$.

At this stage, we could shift all labels, $j \mapsto j + j'$ and find an equivalent labeling. However, the Morse index considerations in the Section 2.5.3 suggest a normalization at $\operatorname{Re} \lambda = +\infty$. Indeed, for $\operatorname{Re} \lambda$ positive and large, all $\nu_j(\lambda)$ have nonvanishing real part and we can write with a consistent splitting

$$\ldots \le \operatorname{Re} \nu_{-1}(\lambda) < 0 < \operatorname{Re} \nu_1(\lambda) \le \ldots. \quad (2.201)$$

Recall that the $\nu_j(\lambda)$ change sign when λ crosses one of the Fredholm borders Γ^ℓ.

With normalization (2.201), we continue the labeling for all values of λ. We define the *absolute spectrum* of $\mathcal{L}_*$ through

$$\operatorname{spec}_{\operatorname{abs}} \mathcal{L}_* := \{\lambda;\ \operatorname{Re} \nu_{-1}(\lambda) = \operatorname{Re} \nu_1(\lambda)\}. \quad (2.202)$$

As a first conclusion from the definition we note that the absolute spectrum is a subset of the essential spectrum of $\mathcal{L}_*$. Indeed, either $\operatorname{Re} \nu_{-1}(\lambda) = 0$

and the relative Morse index jumps at $\lambda \in \mathrm{spec}_{\mathrm{abs}}\mathcal{L}_*$, or else $\mathrm{Re}\,\nu_{-1}(\lambda) = \mathrm{Re}\,\nu_1(\lambda) \neq 0$ and $i_{\mathrm{Morse}}(\lambda) \neq 0$; see Proposition 2.5.4.

Note next that for λ in the complement of the absolute spectrum, $\mathcal{L}_* - \lambda$ is Fredholm of index zero in an exponentially weighted space L_η^2, with radial weight $\eta = \eta(\lambda)$ satisfying

$$\ldots \leq \mathrm{Re}\,\nu_{-1}(\lambda) < -\eta(\lambda) < \mathrm{Re}\,\nu_1(\lambda) \leq \ldots \quad (2.203)$$

Similar to the definition in Section 2.3.3, we can define the *extended point spectrum* $\mathrm{spec}_{\mathrm{expt}}\mathcal{L}_*$ and the *boundary spectrum* $\mathrm{spec}_{\mathrm{bdy}}\mathcal{L}_*$.

We now characterize the limiting behavior of the spectra on bounded disks. As in the case of Theorem 2.3.6, we omit a number of technical assumptions in the statement of the following theorem.

Theorem 2.5.5 *[SaSc00c, SaSc00d] Let $\mathcal{L}_*$ denote the linearization operator (2.127) considered on $L^2(\mathbb{R}^2)$. Let $\mathcal{L}_{*,R}$ denote the corresponding operator on a disk, $|x| \leq R$, equipped with typical mixed boundary conditions. Then the spectrum of $\mathcal{L}_{*,R}$ converges as $R \to \infty$,*

$$\mathrm{spec}\,\mathcal{L}_{*,R} \longrightarrow \mathrm{spec}_{\mathrm{abs}}(\mathcal{L}_*) \cup \mathrm{spec}_{\mathrm{expt}}(\mathcal{L}_*) \cup \mathrm{spec}_{\mathrm{bdy}}(\mathcal{L}_*).$$

Convergence is uniform on bounded subsets of the complex plane in the symmetric Hausdorff distance. Moreover, multiplicity of the eigenvalues is preserved: the number of eigenvalues, counted with multiplicity, in any fixed open neighborhood of any point $\lambda_* \in \mathrm{spec}_{\mathrm{abs}}(\mathcal{L}_*)$ converges to infinity as $R \to \infty$. Multiplicities in neighborhoods of points $\lambda_* \in \mathrm{spec}_{\mathrm{expt}}(\mathcal{L}_*)$, $\mathrm{spec}_{\mathrm{pt}}(\mathcal{L}_*)$, $\mathrm{spec}_{\mathrm{bdy}}(\mathcal{L}_*)$ stabilize and convergence of eigenvalues is of exponential rate in R, there.

For transverse Archimedean spiral waves, the rightmost curve $\Gamma^0 = \{\lambda^0(\mathrm{i}k)\}$ of the essential spectrum is simple, that is, only a single Floquet exponent $\nu = \nu_1(\lambda)$ is located on the imaginary axis. In particular, the absolute spectrum does not contain $\lambda = 0$. If $\Gamma^0 \cap \mathrm{i}\mathbb{R} = \{0\}$, and all other curves Γ^ℓ are contained in $\mathrm{Re}\,\lambda < 0$, then the absolute spectrum of the spiral wave is stable, i.e. contained in $\mathrm{Re}\,\lambda < 0$. Since the Fredholm borders Γ^ℓ consist of the Floquet exponents of the one-dimensional wavetrains, (2.190),(2.189), stability of the absolute spectrum is equivalent to one-dimensional stability of the wavetrains.

This explains, on a spectral level, why spirals waves appear to be very stable, despite the abundance of critical spectrum winding up the imaginary axis. If the spiral is considered as a solution in a bounded disk, only finitely many eigenvalues are close to criticality in a bounded domain — uniformly in the size R of the domain, if the absolute spectrum is stable.

We may say slightly more about the spectral properties of spirals. Rotation and translation of the spiral cause 0 and $\pm \mathrm{i}\omega_*$ to be in the spectrum: eigenfunctions are given explicitly by $\partial_\psi q_*(r, \psi)$, and $\mathrm{e}^{\pm \mathrm{i}\psi}(\partial_r \pm \frac{\mathrm{i}}{r}\partial_\psi)q_*(r, \psi)$, respectively. In particular, these eigenfunctions are bounded functions of

(r, ψ). Considering the linearization in L_η^2 with radial weight $\eta < 0$ small, the linearization $\mathcal{L}_*$ is Fredholm with index zero, by Definition 2.4.2, due to the positive group velocity of the wavetrains. Because L_η^2 allows for (small) exponential growth, radially, translation and rotation eigenfunctions contribute to the kernel. In other words, $\lambda = 0$ and $\lambda = \pm i\omega_*$ belong to the extended point spectrum and therefore, these eigenvalues stay exponentially close to the imaginary axis when truncating to a bounded domain.

Conversely, consider an Archimedean spirals with negative group velocity, violating Definition 2.4.2 of a *transverse* Archimedean spiral. Then the essential spectrum is pushed into the stable complex plane by an exponential weight $\exp(\eta r)$ with $\eta > 0$! As opposed to Figure 2.20, translation and rotation eigenvalues now do not belong to the extended point spectrum of $\mathcal{L}_*$, since the exponential weight requires exponential decrease of eigenfunctions. As a consequence, Euclidean equivariance does not imply critical spectrum of spirals with negative group velocity. Skew-product descriptions of tip motions as described in Section 2.5.2 is not possible, and even the abstract notion of a tip, (2.155) is not well defined in this case.

Point spectrum and the shape of eigenfunctions In the following, we will show how simple, temporal point spectrum λ_0 of the linearization $\mathcal{L}_*$ along an Archimedean spiral $q_*(r, \varphi - \omega_* t) \to q_\infty(k_\infty r - (\varphi - \omega_* t))$ gives rise to super-spiral patterns, near resonances $\lambda_0 \sim i\ell\omega_*$. See expansion (2.208), below. We prepare the resonance consideration with a more general result, which describes the radial asymptotics of the eigenfunction $u_0(r, \psi)$, related to any simple, temporal eigenvalue λ_0; see Proposition 2.5.6.

Consider a simple eigenvalue λ_0 in the extended point spectrum of $\mathcal{L}_*$, as defined in Section 2.5.3, with eigenfunction $u_0(r, \psi)$. Consider the roots $\nu_j^0 = \nu_j^0(\lambda_0)$ of the dispersion relation (2.196), associated with the asymptotic wavetrains, and the ordering (2.200)

$$\ldots \leq \operatorname{Re}\nu_{-j-1}^0 \leq \operatorname{Re}\nu_{-j}^0 \leq \ldots$$
$$\leq \operatorname{Re}\nu_{-1}^0 < \operatorname{Re}\nu_1^0 \leq \ldots \leq \operatorname{Re}\nu_{j-1}^0 \leq \operatorname{Re}\nu_j^0 \leq \ldots$$

Note that $\operatorname{Re}\nu_{-1}^0 < \operatorname{Re}\nu_1^0$ since λ_0 does not belong to the absolute spectrum (2.202). Assume in addition that ν_1^0 is a simple root of the dispersion relation (2.196).

Proposition 2.5.6 *[SaSc01a] In the setting of the preceding paragraph, the asymptotic shape of u_0 for $r \to \infty$ is given by*

$$|u_0(r, \psi) - u_\infty(r - (\psi/k_\infty))e^{-\nu_1 r}| = o(e^{-\nu_1 r}), \qquad (2.204)$$

uniformly in ψ. Here $u_\infty(\xi)$ is a spatially periodic solution to the Floquet problem (2.192) of the radially asymptotic wavetrain q_∞ with $\lambda = \lambda_0$.

Consider nonzero u_∞, in Proposition 2.5.6. Then exponential radial decay or growth of u_0 is determined by the real part of the first Floquet exponent ν_1 of the asymptotic wavetrain q_∞. Also the temporal eigenvalue λ_0 and the radial decay rate ν_1^0 of its eigenfunction u_0 are related by the dispersion relation (2.196) of the wavetrain.

We now turn to the particular case of a near-resonant, simple, imaginary eigenvalue

$$\lambda_0 = i\ell\omega_* + i\delta\omega,$$

where $\delta\omega$ indicates a small deviation from the $\ell : 1$-resonance with the spiral frequency ω_*.

At resonance, $\delta\omega = 0$, we obtain explicitly

$$u_\infty(r - \psi/k_\infty) = e^{-i\ell\psi} q'_\infty(r - \psi/k_\infty), \qquad (2.205)$$

simply because $q'_\infty(\xi)$ is the Floquet solution of the trivial Floquet exponent $\nu_1 = 0$ and due to the Floquet shift (2.197).

Slightly off resonance, that is, for small nonzero $\delta\omega$, we observe associated perturbations δu_∞, of the radially asymptotic Floquet eigenfunction $u_\infty(\xi)$, and $\delta\nu_1$, of the corresponding Floquet exponent. In view of Proposition 2.5.6, δu_∞ remains 2π-periodic, as u_∞ itself, and has little effect on the observed pattern. A variation of $\operatorname{Re}\delta\nu_1 \neq 0$ modifies the radial exponential decay or growth of the eigenfunction u_0. From the discussion in Section 2.5.3 on crossing of Floquet exponents near Fredholm borders, Proposition 2.5.4, we predict exponential growth in the Fredholm index -1 region and exponential decay in the Fredholm index 0 region; see Figure 2.20. However, $\operatorname{Re}\delta\nu_1 \neq 0$ does not influence the r-asymptotic shape of the pattern. Only $\operatorname{Im}\delta\nu_1 \neq 0$ interferes with the underlying periodicity of the primary spiral pattern, producing super-spiral patterns; see Figure 2.22. More specifically,

$$\begin{aligned} u_0(r,\psi) &\sim e^{-\nu_1^0 r} u_\infty(r - \psi/k_\infty) = e^{-\delta\nu_1 r} e^{-i\ell\psi} \tilde{q}'_\infty(\xi) \qquad (2.206) \\ &= e^{-(\operatorname{Re}\delta\nu_1)r} e^{-i\ell\psi} \tilde{q}'_\infty(\xi) e^{-i(\operatorname{Im}\delta\nu_1)r}. \end{aligned}$$

Here, $\tilde{q}'_\infty(\xi)$ is a small, but 2π-periodic perturbation of the wavetrain eigenfunction $q'_\infty(\xi)$, and $\sim$ indicates identity up to a term of order $o(e^{-\nu_1^0 r})$, for $r \to \infty$. The term $e^{-i\ell\psi}$ again accounts for the Floquet shift of u_∞. Since $\nu_1 = 0 + \delta\nu_1$ satisfies the dispersion relation (2.196), the definition of the group velocity c_g (2.198) implies

$$\operatorname{Im}\delta\nu_1 = \frac{\operatorname{Im}\delta\nu_1}{\operatorname{Im}\delta\lambda_1} \cdot \operatorname{Im}\delta\lambda_1 = \frac{\delta\omega_*}{c_g}, \qquad (2.207)$$

in the resonant limit $\delta\omega \to 0$. Substituting (2.207) into (2.206), and returning to the original coordinates $r, \varphi = \psi + \omega_* t$, and time t, the eigenfunction generates a perturbation

$$u_{\text{pert}}(t, r, \varphi) \sim e^{-(\operatorname{Re}\delta\nu_1)r} \cdot e^{i(\ell\varphi - (r - c_g t)\delta\omega/c_g)} \cdot \tilde{q}'_\infty(r - (\varphi - \omega_* t)/k_\infty), \qquad (2.208)$$

again for $r \to \infty$, and in the limit $\delta\omega \to 0$.

The eigenfunction shape u_{pert}, corresponding to a near-resonant temporal eigenvalue $\lambda_0 = i\ell\omega_* + i\delta\omega$, produces the pattern of a superspiral, for large r. Indeed, consider fixed t. Then u_{pert} is an ℓ-armed spiral, due to the term $\exp(i\ell\varphi - \delta\omega r/c_{\text{g}})$. The superspiral is Archimedean with large radial wavelength $2\pi c_{\text{g}}/\delta\omega$, for $\delta\omega \to 0$. The superspirals propagate in radial direction with speed c_{g} and rotate with the slow frequency $\delta\omega$.

In the experimentally observed pattern $q_* + \varepsilon u_{\text{pert}}$, the superspiral modulations manifest themselves in a phase shift of the emitted wavetrains

$$q_* + \varepsilon u_{\text{pert}} \sim q_\infty(r - (\varphi - \omega_* t)/k_\infty + e^{-(\operatorname{Re}\delta\nu_1)r}\cos(\ell\varphi - (r - c_{\text{g}}t)\delta\omega/c_{\text{g}})). \tag{2.209}$$

Here we exploited that $\tilde{q}'_\infty \sim q'_\infty$ in (2.208) amounts to a phase shift of the emitted wavetrains.

2.5.4 Comparison with Experiments

We explain the onset of spatio-temporal patterns and certain characteristics of incipient spiral wave instabilities, referring to the linear spectral analysis of the preceding section. In particular, we discuss the possibility of point spectrum crossing the imaginary axis, Section 2.5.4, and essential spectrum causing an instability, Section 2.5.4.

Meander instabilities Meander instabilities are caused by critical point spectrum, emerging from the essential spectrum; see Figure 2.21. In any disk of finite radius R, the essential spectrum disappears and five critical eigenvalues remain close to the imaginary axis. The influence of the boundary of the domain on the location of point spectrum is exponentially small for large disk radius R. Actually, also the nonlinear terms of the vector field on the center-manifold converge with exponential rate, as the size R of the disk tends to infinity [SaSc02b]. In the limit, we recover the $SE(2)$-equivariant skew-product description (2.152). The equations on the center manifold in a finite disk are therefore a small perturbation of the $SE(2)$-invariant skew-product system, discussed in Section 2.5.2 — as long as the tip stays far enough from the disk boundary.

Therefore, in any large disk, we recover the skew-product dynamics, even though a dynamical reduction in the unbounded domain is prohibited by the presence of essential spectrum.

Dynamics and tip motion are complemented in the experiments with striking super-patterns in the farfield, when sufficiently large spirals are observed. As a consequence of center manifold reduction, the spatial patterns are superpositions of the critical eigenfunctions, to leading order in the bifurcation parameter. For two purely imaginary eigenvalues close to resonance $i\omega_{\text{H}} \sim \ell i\omega_*$ we have computed the shape; see (2.209). Predictions coincide well with direct simulations of a specific reaction-diffusion system, based on Barkley's code EZSpiral [Do&al97]; see Figure 2.22. For marginally stable wavetrains,

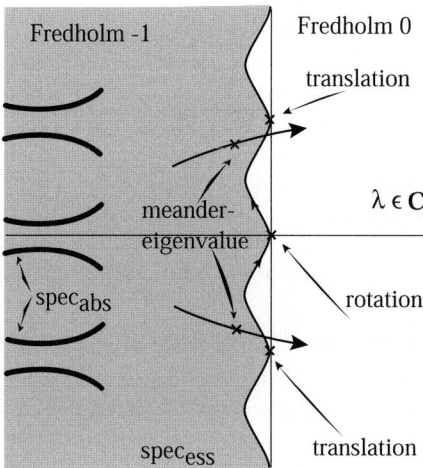

Fig. 2.21. Schematic plot of the meander eigenvalues popping out of the Fredholm index -1 region close to instability.

the essential spectrum touches the imaginary axis only at $i\ell\omega_*$, $\ell \in \mathbb{Z}$, with quadratic tangency. From the previous section, we conclude that, off resonance, super-patterns u_{pert} resulting from the instability will be strongly localized in the farfield (2.208). In particular, at the higher *drift*-resonances, $\omega_* = \ell\omega_H$, $\ell \geq 2$, the super-patterns will be strongly localized. At the inverse resonances, $\ell\omega_* = \omega_H$, $\ell \geq 2$ where the motion of the tip stays bounded, we do predict prominent, non-localized super-patterns in the form of weakly curved ℓ-armed super-spirals (2.208). The orientation of the super-spirals changes when the Hopf eigenvalue crosses through resonance. At transition, the super-pattern divides the domain in sectors. At the 1:1-resonance, a particularity occurs, since $i\omega_*$ is now algebraically double in the extended point spectrum. Drift resonance coincides with the emergence of strongly pronounced super-spirals. In [SaSc01a], it is shown that the principal eigenfunction is computed from integration of the eigenfunction in the radial direction. Its amplitude grows linearly with the radius r, such that the super-pattern exhibits an asymptotically r-independent shift of the wavenumber of the wavetrains; see the drift picture in Figure 2.22.

Experimentally, these striking superstructures, have first been observed in a biological system [Pe&al91]. An explanation based on a curvature description of the emitted pulse-trains, in the spirit of Section 2.4.1, has been attempted in [Pe&al93]. The simplest phenomenological explanation evokes the picture of a point source, alias the tip, emitting wavetrains. Variations of the local wavenumber are due to a Doppler-effect caused by the two-frequency motion of the spiral tip. Spatial decay as well as the correct speed of propagation are not predicted from these formal descriptions, which neglect the interaction between individual emitted pulses.

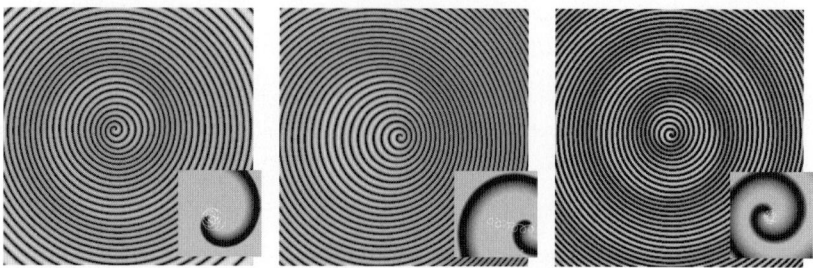

Fig. 2.22. Farfield patterns and tip motion in case of inward meander, drift, and outward meander. The large pictures show the superspiral-shaped deformations of the primary spiral. The small inlets show, on a smaller spatial scale, the tip motion. See Plate 4 in the Appendix for a version of this figure in colour.

The spectral assumptions of point spectrum crossing the axis were confirmed numerically by Barkley; see [Ba92]. In a disk, he found five critical eigenvalues at bifurcation. The eigenvalues from translation were extremely close to the axis, although the domain size was moderate, exhibiting about one turn of the spiral. This confirms the rapid (exponential) convergence of point spectrum in bounded domains of increasing size R; see Theorem 2.3.6. The numerically observed change of the growth behavior of the amplitude of the eigenfunction is in good agreement with the predictions from Section 2.5.3. Indeed, close to resonance, (2.208) predicts that the asymptotic behavior close to bifurcation changes: from exponential radial growth — when the eigenvalue is still located in the Fredholm index -1 region — to weak exponential radial decay at bifurcation.

Farfield and core breakup Farfield breakup is caused by essential spectrum crossing the imaginary axis [BäOr99, SaSc00d, ToKn98]. Mostly, a convective Eckhaus instability is observed, with unstable wavenumbers close to the wavenumber of the wavetrains, $\text{Im}\,\nu \sim k_\infty$. In the stable regime, the boundary curve Γ of the essential spectrum touches the imaginary axis at $i\omega_*\mathbb{Z}$ with quadratic tangency from the left. At criticality, the tangency is of fourth order. In the unstable regime, the tangency is from the right; see Figure 2.23. The absolute spectrum, however, remains confined in the open left half plane, close to this instability threshold. The group velocities (2.198) on the critical spectral curve Γ are always directed outwards close to $\lambda = 0$, by transversality of the spiral; see Definition 2.4.2. Therefore, the Eckhaus unstable modes inherit a positive group velocity, as well. In any large bounded disk, the Eckhaus instability remains transient, as long as the absolute spectrum stays confined to the left complex half plane. This was confirmed in direct simulations, see [SaSc00d]. We computed essential spectra of the wavetrains from the boundary-value problem (2.193),(2.194) following the dependence of a nontrivial solution $u(\xi; k), \lambda(ik)$ on the Bloch wavenumber k. The absolute spectrum was computed by a similar path-following procedure. Instability thresholds coincided well with the observations in direct simulations. In the convectively unstable regime, stability depends crucially on the size of the domain and the magnitude of the perturbation. The sensitivity is caused by the subcritical nature of the Eckhaus instability in the given parameter regime. Qualitatively, the sensitivity mimics the one-dimensional nonlinear advection-diffusion problem, addressed in Proposition 2.3.7: the minimal amount δ^u of a perturbation needed to cause breakup decays rapidly with the size R of the domain. In numerical simulations, again based on Barkley's code EZSpiral [Do&al97], the domain contained approximately 20 wavelengths; see Figure 2.25.

The absolute instability, if close enough to the convective instability, is produced by eigenfunctions which grow exponentially towards the boundary of the domain; see Figure 2.23. Their shape is again determined from the complex dispersion relation (2.196), which predicts prominent, exponentially growing super-spirals (2.208). Note that these super-spirals resemble the farfield pattern in the meander instability. The most striking difference is their growth towards the boundary. However, we emphasize that super-spirals in meandering and in farfield breakup occur for completely different reasons. In the farfield breakup, the motion of the tip is a regular one-frequency motion.

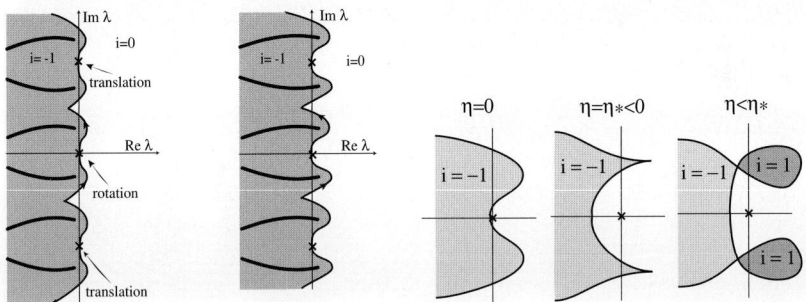

Fig. 2.23. Schematic plot of essential and absolute spectra crossing the axis successively at farfield breakup.

We discuss core breakup, next. Numerical computations of essential and absolute spectra indicate that parts of the critical curve Γ change slope to create group velocities directed towards the center of the spiral. The shape of the boundary curve of the essential spectrum becomes more complicated. In

Figure 2.24, two scenarios are suggested, that produce negative group velocities. Recall that the sign of the group velocity determined the vertical slope of the oriented curve Γ^ℓ. The onset of the second scenario in Figure 2.24 has been found in [SaSc00d] in a parameter regime close to core-breakup. In direct simulations of a nonlinear reaction-diffusion system, super-spirals appear before breakup, but are strongly localized. Much more detailed numerical investigation is necessary, however, to clarify the structure of this instability.

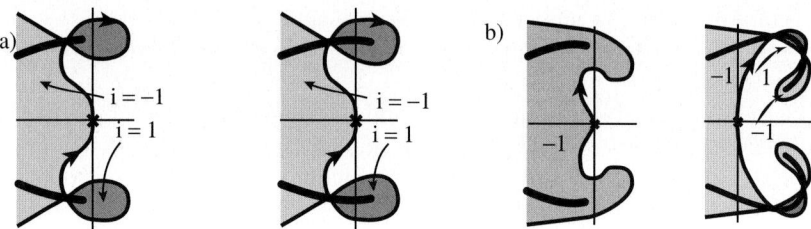

Fig. 2.24. Two scenarios (**a**), (**b**) of essential and absolute spectra crossing the axis successively at core-breakup.

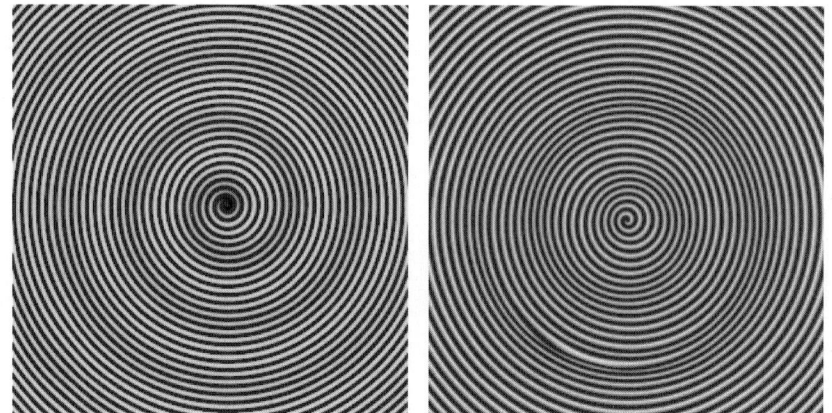

Fig. 2.25. Farfield patterns in case of farfield- and core-breakup. See Plate 5 in the Appendix for a version of this figure in colour.

The final states in farfield versus core breakup are quite different. Collision of wavetrains leads to the formation of new spirals. In the case of farfield breakup, the size of the various spiral domains remains large, close to the onset of instability, leading to a moderately complicated dynamical behavior. In the opposite case of core breakup, new spirals immediately break up, close to the core, and the final state is extremely complicated right after the instability has set in; see Figure 2.26.

Fig. 2.26. Final state after farfield and core breakup. See Plate 6 in the Appendix for a version of this figure in colour.

Meander instabilities in the presence of convective farfield instabilities enhance the effect of the instability. From the spectral analysis in Section 2.5.3 we predict the meander eigenfunctions to be exponentially growing. Indeed, at instability, they are embedded in the Fredholm index -1 region. Whereas the unstable essential spectrum disappears in bounded domains, the (extended) point spectrum still reproduces the effect of the essential spectrum. Indeed exponential radial growth of eigenfunctions is governed by the dispersion relation (2.196). In the simplistic phenomenological description by a Doppler effect, the meander movement of the spiral tip is amplified and transported by the wavetrains. Increasing the size of the domain, we find breakup in the farfield already for moderate amplitudes of the meander motion.

2.6 Three Space Dimensions: Scroll Waves

2.6.1 Filaments, Scrolls, and Twists

In this chapter we consider reaction-diffusion systems of $N = 2$ equations in $m = 3$ space dimensions,

$$u_t = D\Delta u + f(t, x, u). \tag{2.210}$$

Here $u = (u_1, u_2) \in \mathbb{R}^2$ and x is in a bounded domain $\Omega \subset \mathbb{R}^3$, say with Neumann boundary conditions. Let f be smooth. A motivating prototype is the singularly perturbed FitzHugh-Nagumo-like system with nonlinearity

$$f = \begin{pmatrix} \varepsilon^{-1} u_1 (1 - u_1)(u_1 - (u_2 + b)/a) \\ u_1 - u_2 \end{pmatrix} \quad (2.211)$$

of excitable media type, for small $\varepsilon > 0$. See Section 2.4.1, and (2.114). For simplicity of notation, we shift $u = (u_1, u_2)$ by suitable constants, in each component, so that the value $u = 0$ becomes a value inside the singular homoclinic pulse cycle of Figure 2.16(a).

Spirals, tips, and Brouwer degree In $m = 2$ space dimensions, $\Omega \subset \mathbb{R}^2$, the pre-images

$$z_*(t) = (u(t, \cdot))^{-1}(0) \quad (2.212)$$

may serve as a viable definition for a spiral tip in the sense of (2.155), (2.156). Both experimental evidence and numerical simulations suggest, in fact, that nondegenerate zeros z_* of $u(t, \cdot)$ are typically accompanied by local spiral wave patterns of (2.211) with rotation direction determined by the local Brouwer degree (see for example [De85, Du&al85, Ze93] for some background):

$$\deg_{loc} u(t, \cdot) := \text{sign det } u_x(t, z_*(t)) \quad (2.213)$$

at $x = z_*(t)$. So pronounced is this effect, indeed, that Winfree has proposed to model excitable media by only a phase variable $\tilde{u} = \tilde{u}(t, x) \in S^1$, with appropriate time evolution [Wi01, Wi87, Wi95]. Here the circle S^1 can model the homoclinic pulse cycle of Figure 2.16(a), or else just an abstract local "clock". Mathematically, the correspondence to our PDE approach can be achieved by simply defining $\tilde{u} := u/|u|$. The local Brouwer degree (2.213) coincides with the winding number of $\tilde{u}$, restricted to a small circle around the singularity of $\tilde{u}$ at $x = z_*$. Sometimes this local degree is called "topological charge" of the singularities of spirals. Conservation of "charge", as time t evolves, then corresponds to homotopy invariance of Brouwer degree: only anti-rotating spiral pairs can annihilate each other.

The restriction to $\tilde{u} \in S^1$ unfortunately produces a phase singularity of undefined $\tilde{u}$ at the spiral tips, where $u = 0$. Since we are moreover interested in collisions of such singularities, not only in two but also in three space dimensions, we therefore favor the reaction-diffusion description by smooth spatial profiles $u(t, \cdot) \in \mathbb{R}^2$ over the phase fields $\tilde{u}(t, \cdot) \in S^1$.

Scroll waves, filaments, and twists Returning to three-dimensional $x \in \Omega \subset \mathbb{R}^3$, the preimages

$$\tau \mapsto z = z(t, \tau) \in (u(t, \cdot))^{-1}(0) \subseteq \Omega \quad (2.214)$$

become differentiable curves $z(t, \cdot)$, at least as long as zero is a regular value of $u(t, \cdot)$. We call these curves *(tip) filaments*. Numerical observations in excitable media exhibit spiral wave patterns, when restricting the spatial profile

$u(t, \cdot)$ to a local, two-dimensional section $S^\perp \subset \Omega$ transverse to the filament curve $z(t, \cdot)$ at $z_* = z(t, \tau_0)$. Note, that we may always define an orientation of $S^\perp$ such that

$$\deg_{loc} u(t, \cdot)|_{S^\perp} = +1 \qquad (2.215)$$

at $x = z_*$. Requesting the pair $(S^\perp, z_\tau)$ of $S^\perp$ and the filament tangent z_τ to be oriented by the right-hand rule thus defines a unique *filament orientation* all along each filament curve $\tau \mapsto z(t, \tau)$.

A *scroll wave* is the wave pattern arising in a tubular neighborhood of a filament curve. By the excitable media paradigm, the scroll wave consists of a stack of spirals in the transverse sections S along the filament. See Figure 2.27 for a reproduction of Panfilov and Winfree's original drawings [PaWi85]. We emphasize that our mathematical discussion of scroll waves neither proves nor relies on the validity of the "stack of spirals" picture. Our results below will hold in a general framework of reaction-diffusion systems. In particular, they are not limited to a singular perturbation setting. It is nevertheless very useful to refer to Winfree's paradigm, for visual motivation and guidance.

For closed filaments $z(t, \tau), \tau \in S^1$, a *twist* of the oriented filament can be defined as follows. Choose $\delta > 0$ small enough and consider the closed curve

$$\tau \mapsto \tilde{z}(t, \tau) \subseteq (u(t, \cdot))^{-1}(\delta, 0) \qquad (2.216)$$

of the pre-image of $u_1 = \delta, u_2 = 0$ near $z(t, \tau)$, following the orientation of z itself. Geometrically, the curve $\tilde{z}(t, \cdot)$ indicates the local phases of the spiral motions in the stacked sections $S^\perp$. Then we call

$$\mathrm{twist}(z(t, \cdot)) := \mathrm{lk}(\tilde{z}(t, \cdot), z(t, \cdot)) \qquad (2.217)$$

the twist of the filament $z(t, \cdot)$. Here lk denotes the *linking number* of oriented curves, as explained below. A *twisted scroll* possesses a filament z with nonzero twist: the curves $z(t, \cdot)$ and $\tilde{z}(t, \cdot)$ are linked.

To define the *linking number* $\mathrm{lk}\,(\tilde{z}, z)$ of two closed oriented disjoint curves $\tilde{z} = \tilde{z}(\tau), z = z(\tau), \tau \in S^1$, we consider a suitable planar projection such that crossings of the two curves are transverse and distinct, in projection. See Figure 2.28, for examples of such projected crossings, indicating over- and under-crossings in the original three-dimensional view. Denoting the oriented unit tangent vectors of the over- and under-crossing curves by $\mathbf{t}_1, \mathbf{t}_2$, respectively, we associate a sign

$$\varepsilon_j := \mathrm{sign}\det(\mathbf{t}_1, \mathbf{t}_2) = \pm 1 \qquad (2.218)$$

to each crossing j. Then

$$\mathrm{lk}\,(\tilde{z}, z) := \frac{1}{2}\sum_j \varepsilon_j, \qquad (2.219)$$

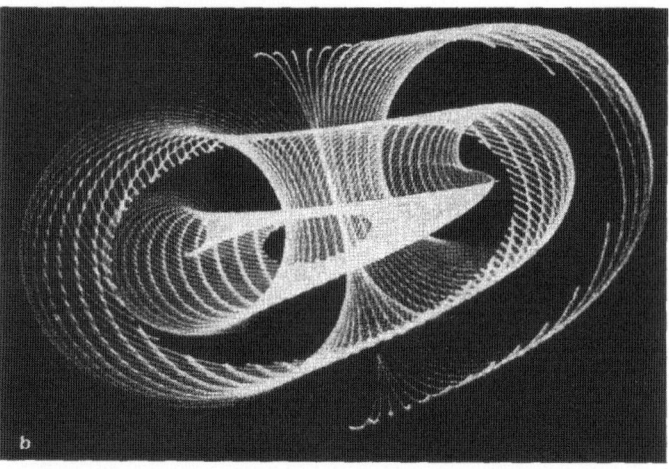

Fig. 2.27. Scroll waves are stacks of spirals (**a**). Twisted scroll wave (**b**). Courtesy of [PaWi85].

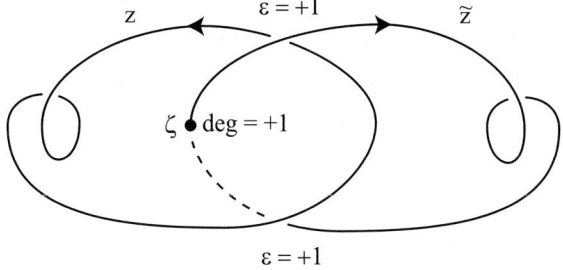

Fig. 2.28. Transverse crossings of planar projections of oriented curves, with associated signs $\varepsilon_j = \pm 1$. Note $\operatorname{lk}(z, \tilde{z}) = +1$, and $\operatorname{twist}(z) = \operatorname{twist}(\tilde{z}) = -1$.

where the sum extends over all mutual crossings, but not over self-crossings. Clearly, $\operatorname{lk}(\tilde{z}, z) = \operatorname{lk}(z, \tilde{z})$.

The linking number $\operatorname{lk}(\tilde{z}, z)$, and hence $\operatorname{twist}(z)$, is independent of the choice of small enough $\delta > 0$. The linking number does not depend on the choice of projection and is invariant under diffeotopies of $\mathbb{R}^3$; see for example [Ka87]. In particular, $\operatorname{lk} \neq 0$ implies $\tilde{z}, z$ cannot be unlinked, for example by a diffeotopy such that $\tilde{z}, z$ come to lie in the same plane.

The following proposition indicates how twisted scrolls have to be either knotted or else have to be linked to other twisted scrolls.

Proposition 2.6.1 *Let $S \subset \Omega$ be an embedded surface, diffeomorphic to a closed disk, such that ∂S is a closed filament curve $z(t, \cdot)$. Assume that S is transverse to all other filament curves $z_j(t, \cdot) \subset u(t, \cdot)^{-1}(0)$ intersecting S. Let ζ_k enumerate these intersection points. Choose an orientation for S such that the induced boundary orientation of ∂S coincides with the orientation of the filament $z(t, \cdot)$. Then*

$$\operatorname{twist}(z) + \sum_{x=\zeta_k} \deg_{loc} u(t, \cdot)|_S = 0 \qquad (2.220)$$

For the proof, see [Du&al85, §15.4].

Of course Proposition 2.6.1 holds only as long as $u = 0$ is a regular value of the spatial solution profile $u(t, \cdot)$, for fixed time $t > 0$. Neither linking nor knotting can change, in this situation. In the next section we will address the precise geometry of the failure of this assumption, which occurs at only discrete times $t > 0$, for generic initial conditions u_0 at $t = 0$.

2.6.2 Generic Changes of Scroll Filament Topology

How can two scroll wave filaments collide and "cross" each other as time t evolves? In Figure 2.29 we show two linked closed scroll wave filaments $z_1(t, \cdot), z_2(t, \cdot)$, each parameterized over $\tau \in S^1$. In scenario (a), called *"break through"*, the two circular filaments interpenetrate each other and restitute themselves as two unlinked circles $\tilde{z}_1, \tilde{z}_2$ after collision. The *"crossover"* scenario (b), in contrast, calls for a "kissing" of the two filaments $z_1(t,\cdot), z_2(t,\cdot)$ to occur, at time $t = t_0$, with a subsequent reconnection, such that a single closed filament z is generated. Now, which scenario is right?

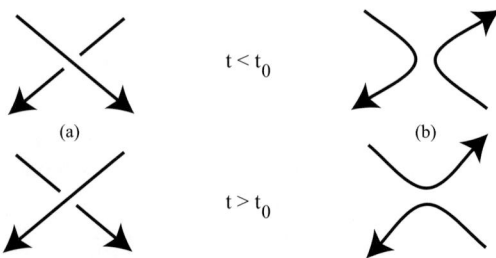

Fig. 2.29. Two scenarios for topology changes of linked scroll filaments: **(a)** break through; **(b)** crossover.

Note how filament orientation, as defined locally in (2.215), has to be respected by either scenario — and it is. Applying the scenario of Figure 2.29(a) to one of the mutual crossings in Figure 2.28 unlinks the two closed twisted filaments z and $\tilde{z}$. Scenario 2.29(b), in contrast, merges the two filaments to a single closed, untwisted filament, when applied between the top and bottom crossing.

Generic level sets To decide on the appropriate scenario we follow the approach in [FiMa00]. In Section 2.6.1 we have seen how the phenomenon of scroll wave filaments $x = z(t,\tau) \in \Omega$ mathematically corresponds to zeros $(t,x) \in (0,\infty) \times \Omega$ of solutions $u = u(t,x) \in \mathbb{R}^2$ of a reaction-diffusion system:

$$u(t,x) = 0. \quad (2.221)$$

If $u = u(t,x)$ was just "any" function, rather than a solution to a PDE, then singularity theory would provide ample information on the behavior of the zero set $\{(t,x); u(t,x) = 0\}$. Indeed, generic functions u only possess a finite number of singularities under our dimensional constraints. See for example [Ar&al85, GoSc85, Ar93, Ar94b] for more details. We give specific examples below. However, $u = u(t,x)$ is not just "any" function: it is a solution $u = u(t,x;u_0)$ of a reaction-diffusion system (2.210) subject to boundary conditions and, more importantly, with prescribed initial conditions

$$u(t=0,x) = u_0(x). \quad (2.222)$$

Theorem 2.6.2 *[FiMa00] Let $k \geq 1$. Consider initial conditions u_0 in a Sobolev space X which embeds into $C^k(\Omega, \mathbb{R}^N)$ and satisfies the (Neumann) boundary conditions. Then the solution $u(t,x)$ of (2.210) possesses only generic singularities, as determined by derivatives in (t,x) up to order k, for generic initial conditions $u_0 \in X$ and at any $t > 0$, $x \in \Omega \subset \mathbb{R}^m$.*

Note that we use the word singularity in the sense of singularity theory here. We do not discuss "singularities" in the PDE-sense of blow-up of solutions or their derivatives. Nor are "singularities" understood as phase singularities, where some $\tilde{u} = u/ \mid u \mid \in S^1$ remains undefined. Rather, a *singularity* is given by an algebraic variety S in the jet space J^k of Taylor polynomials up to order k in the variables (t, x). Consider the map

$$j^k : (0, \infty) \times \Omega \to J^k \\ (t, x) \mapsto j^k u(t, x) \qquad (2.223)$$

which maps (t, x) to the k-jet $j^k u(t, x)$, the Taylor polynomial of order k of the solution u, evaluated at (t, x). Since $t > 0$ and $x \in \Omega \subset \mathbb{R}^m$, together, are only $m+1$ independent variables, a generic function $u(t, x)$ will certainly miss a variety S unless

$$\text{codim } S \leq m + 1 \qquad (2.224)$$

in J^k. The theorem states that, likewise, solutions $u = u(t, x; u_0)$ of (2.210) will miss such singularities S, at least for a subset X' of initial conditions $u_0 \in X$ which is residual in the sense of Baire: X' contains at least a countable intersection of open dense subsets of X.

The proof of Theorem 2.6.2 is based on Thom transversality and backwards uniqueness for linear, nonautonomous parabolic systems. Essentially, Thom transversality reduces the problem to showing surjectivity of the linearization, with respect to u_0, of the k-jet of the solution u at any fixed (t, x). This finite-dimensional surjectivity property is then proved by a backwards uniqueness result which prevents solutions of linear equations to become zero in finite time. For a technically more complete account and slightly more general results we refer to [FiMa00]. For earlier results in a similar direction, of a purely local singularity theory flavor, see also [Dam97].

Sturm property, revisited We give some examples for our result, before returning to the problem of filament crossing. Since we are interested in zeros of solutions $u(t, x)$, the requirement $u = 0 \in \mathbb{R}^N$ will always contribute N to codim$S$ in (2.224). Consider scalar equations, $N = 1$, in one space dimension, $m = 1$ first. To exclude x-profiles $x \mapsto u(t, x)$ with zeros of multiplicity three or higher, define S by the requirement

$$u = u_x = u_{xx} = 0. \qquad (2.225)$$

Since this implies codim$S = 3 > 2 = m + 1$, the realization condition (2.224) is violated. Therefore (2.225) cannot occur, for generic initial conditions $u_0 \in X \hookrightarrow C^2$. Similarly, $u_t \neq 0$ at double zeros $u = u_x = 0$, because

$$u = u_x = u_t = 0 \qquad (2.226)$$

also implies codim$S = 3$. Therefore the only remaining local singularity basically takes the form

$$u(t,x) = \pm(t-t_0) \pm (x-x_0)^2; \qquad (2.227)$$

see Figure 2.30. In Section 2.2.1 we have seen how $f = f(x, u, u_x) = 0$ for $u = u_x = 0$ implies that the two signs in (2.227) have to coincide. This Sturm property excludes case (b) of creation of a new pair of zeros, admitting only annihilation (a). Of course the Sturm property of Proposition 2.2.1, unlike our present genericity argument, is not limited to generic initial conditions $u_0 \in X$.

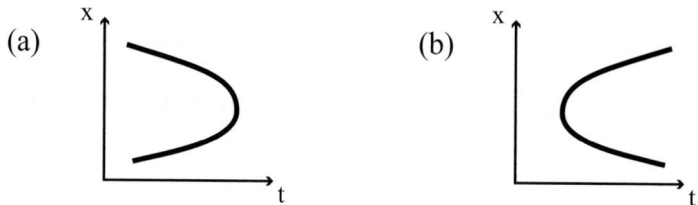

Fig. 2.30. Collision of zeros $u(t,x) = 0$ for scalar equations: **(a)** annihilation, **(b)** creation.

Comparison principle and nodal domains For scalar equations, $N = 1$, in arbitrary space dimension $m \geq 1$, Fig. 2.30 remains valid. Indeed the $m+1$ conditions

$$u = 0, \quad u_{x_1} = \cdots = u_{x_m} = 0 \qquad (2.228)$$

prevent further degeneracies to appear in generic singularities S, according to realization condition (2.224). In particular $u_t \neq 0$, and the Hessian matrix u_{xx} has to be nondegenerate.

If the Hessian matrix u_{xx} is strictly (positive or negative) definite, spherical bubbles of dimension $m-1$ are thus annihilated (a) or created (b). For $f = f(x, u, \nabla u)$ with $f(x, 0, 0) = 0$, however, creation (b) is again excluded by the partial differential equation which involves $\Delta u = \text{trace}(u_{xx})$. In the general, nongeneric case, this is due to the maximum (or comparison) principle; see Section 2.2.1, (2.12)–(2.14).

Suppose next that the Hessian u_{xx} is indefinite. For simplicity we consider $m = \dim x = 2$. In suitable local coordinates $x = (x_1, x_2)$, and normalizing to $x_0 = 0 \in \Omega$, we obtain the local form

$$u(t,x) = (t - t_0) + x_1^2 - x_2^2 \qquad (2.229)$$

See Figure 2.31 for the curves describing the local zero set. This "Morse scenario" of course coincides with the behavior of level curves near a nondegenerate saddle point, with level values parameterized by time. Indeed $N = \dim u = 1$, and $u_t \neq 0$ locally. The curves are oriented such that $u < 0$

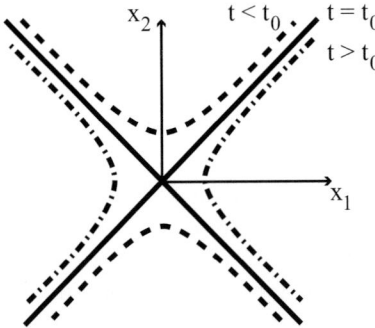

Fig. 2.31. Zero sets of $u(t,x)$ according to (2.229) as t passes through t_0.

to the left of each curve. This orientation is the analogue, for $N = 1$ and $m = 2$, of the orientation of filament curves which was defined for $N = 2$ and $m = 3$ in Section 2.6.1. Note how the connectivity of the nodal domains $\{u(t,\cdot) > 0\}$, $\{u(t,\cdot) < 0\}$ changes, as t increases through t_0. Because the sign of $u_t = \Delta u = \text{trace}(u_{xx})$ at (t_0, x_0) is indeterminate for indefinite Hessians u_{xx}, even when $f = f(x,0,0) = 0$, there is no analogue to the Sturm property in higher space dimensions $m \geq 2$ which could be based on counting nodal domains. See however [Uh76] for generic simplicity and sign change properties of eigenfunctions of the Laplacian in higher space dimensions.

Annihilation of spiral tips Finally we return to systems, $N = \dim u = 2$, for example of the singularly perturbed FitzHugh-Nagumo type (2.210), (2.211). Consider $S = \{u = 0, u_x = 0\}$ which describes total degeneracy of the $N \times m$ Jacobian matrix u_x. Since codim $S = (N+1)m > m+1$, this singularity does not occur, for generic initial conditions u_0. Lyapunov-Schmidt reduction, based on rank $u_x \geq 1$, then reduces the discussion of all singularities to the case $\dim u = N - 1 = 1$ and $\dim x = m - 1$ which was already discussed above.

For example Figure 2.30(a), now corresponding to $N = 2$, $m = 2$, describes the coalescence and subsequent annihilation of a pair of spiral tips. Note how the invariance of Brouwer degree forces the two annihilating spiral tips to have opposite degree (2.213). For the two annihilating spirals this implies opposite rotation direction, in the Winfree paradigm. See also Figure 2.33 for a numerical simulation. It is noteworthy that the opposite phenomenon of persistent spiral pair creation, Figure 2.30(b), has not been observed, to our knowledge, neither numerically nor experimentally. This unobserved effect cannot be excluded by comparison or genericity principles.

Collisions of scroll wave filaments Let us next apply the same reduction to the case $N = 2$, $m = 3$ of scroll filament collision. By the above arguments, Lyapunov-Schmidt reduces this case to $N = 1$, $m = 2$. The annihilation for $t \nearrow t_0$ according to Figure 2.30(a), now of a circular filament $z(t,\cdot)$, has been observed. It corresponds to the collapse and disappearance of a scroll ring. Note that the collapsing scroll filament has to be non-twisted, due to Proposition 2.6.1. Again, the antagonistic case of scroll ring creation analogous to Figure 2.30(b) has mysteriously escaped both numerical and experimental documentation.

The only remaining singularity for $N=2, m=3$ alias $N=1, m=2$ is the case of indefinite (reduced) Hessian as depicted in Figure 2.31. Note how the entire dynamic process of approach (a), kiss (b), and reconnection (c) occurs in a single tangent plane. Moreover, Figure 2.31 gives the definitive answer to the opening question of this section: for generic initial conditions $u_0 \in X \hookrightarrow C^2$, the topology changes of scroll wave filaments must occur by crossover collisions following the kissing scenario of Figure 2.29(b).

We summarize these results as follows.

Corollary 2.6.3 *[FiMa00] For generic initial conditions $u_0 \in X \hookrightarrow C^2$, only the following three collision scenarios of scroll wave filaments, as defined in (2.214) above, are possible for $t > 0$, $x \in \Omega$:*

(i) annihilation of untwisted filaments, according to Figure 2.30(a);
(ii) creation of untwisted filaments, according to Figure 2.30(b);
(iii) crossover collision of filaments, according to Figures 2.29(b) and 2.31

We emphasize that this result applies only in the interior of the domain Ω. We do not investigate, for example, the collision of filament points in the boundary $\partial\Omega$.

As another corollary, we note that

$$\sum_j \text{twist}(z_j) + 2 \sum_{j<k} \text{lk}(z_j, z_k) = const. \tag{2.230}$$

remains constant, for generic initial conditions $u_0 \in X$, independently of the time evolution of the closed filaments $z_j = z_j(t,\cdot)$ and of possible changes in their number. Mutual collision, or self-collision, of closed filaments may change the total number of closed filaments but not the total twist and linking according to (2.230). It is again assumed here — similarly to degree theory — that the closed filaments $z_j \subset \Omega$ neither collide with the boundary $\partial\Omega$ nor with filaments z which extend to the boundary.

For a proof we consider the accompanying filaments $\tilde{z}_j \subseteq u(t,\cdot)^{-1}(\delta, 0)$ which define the local phase along the scroll filament z_j. Then

$$\sum_j \text{twist}(z_j) + 2\sum_{j<k} \text{lk}(z_j, z_k) = \sum_{j,k} \text{lk}(z_j, \tilde{z}_k), \tag{2.231}$$

for small enough $\delta < 0$. None of the singularities (i)–(iii) of Corollary 2.6.3 changes the latter quantity.

If all filaments are closed, so that $u \neq 0$ on $\partial\Omega \cong S^2$, then we note in passing that

$$\sum_{j,k} \text{lk}\,(z_j, \tilde{z}_k) = 0, \qquad (2.232)$$

by direct homotopy $u(t,\cdot) \to const. \neq 0$.

For an example, we recall Figure 2.28. Another theoretical example well worth contemplating is the clover-shaped trefoil knot as shown in Figure 2.32. According to corollary (2.230) the initial torus knot must possess zero twist. One way to see this, directly, relies on the successive construction of *Seifert surfaces S*: these are compact, orientable, embedded surfaces of minimal genus, with the link, or knot, as boundary. We may interpret S as $u(t;\cdot)^{-1}([0,\infty) \times \{0\})$. We start with an embedded disk spanning the circle, in the lower right of Figure 2.32. Crossover collisions change the topology of S, but in the present example S will still qualify as a Seifert surface, both before and after collision. In the resulting Seifert surface of the original torus knot z, in the upper left of Figure 2.32, our interpretation produces $\tilde{z} = z(t;)^{-1}((\delta,0))$. Direct inspection then confirms zero twist of the torus knot, $\text{twist}(z) = \text{lk}\,(z,\tilde{z}) = 0$, as predicted by (2.230). Similarly (2.230) shows that any single knot is untwisted. Indeed any single knot can be decomposed into unlinked planar circles, under a suitable sequence of crossover collisions. Any twisted knot, in contrast, must be linked to another filament, again by (2.230). If two knots are mutually linked, but isolated otherwise, then at least one of the knots must be twisted. These remarks generalize our previous conclusions from proposition 6.1. For a numerical example see Section 2.6.3.

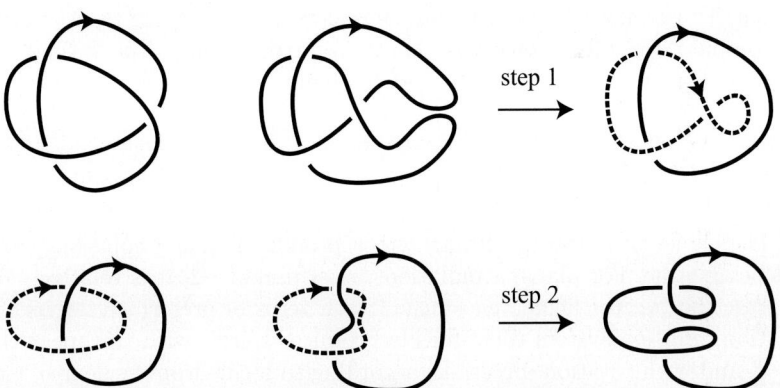

Fig. 2.32. Schematic two-step disintegration of a torus knot filament via linked filaments.

2.6.3 Numerical Simulations

In the previous two sections we have presented mathematically rigorous results concerning degrees, orientations, twists, linking, and collisions of zeros of solutions $u = u(t, x)$ of reaction-diffusion systems. The terminology of spiral tip, scroll waves, scroll wave filaments, etc. was largely motivated by experimental observations and numerical simulations. We present some such simulations in the present section. All results presented here are based on [FiMa00]. The online version contains downloadable animations, for the entertainment of our readers! As a model equation we consider the FitzHugh-Nagumo system (2.210), (2.211). For parameters we choose

$$
\begin{aligned}
&a = 0.8; \\
&b = 0.01 + A\cos\omega t \quad \text{with} \\
&\quad A = 0 \text{ or } 0.01; \ \omega = 3.21; \\
&D = \begin{pmatrix} 1 & 0 \\ 0 & 0.5 \end{pmatrix}; \\
&\varepsilon = 0.02; \\
&\Omega = [-15, 15]^3;
\end{aligned}
\tag{2.233}
$$

unless stated otherwise. To define tip and filament positions, we use the shifted reference values $u_1 = u_2 = 0.5$ rather than $u = 0$. The choices of A correspond to autonomous versus slightly periodically forced simulations.

For integration we use Barkley's code [Do&al97] with equidistant x-discretization by 125^3 points and a time step $\Delta t = 0.0172\cdots$. Filament positions are determined from the discrete data by a simplex method in the spirit of [AlGe90], compatible with the above degree and orientation approach. To visualize the local phase, as well as accompanying twist curves $\tilde{z}$, we compute and plot a piecewise linear approximation to the surface in Ω given by the preimage under $u(t, \cdot)$ of the band

$$u_2 = 0.5 \leq u_1 \leq 0.5 + \delta. \tag{2.234}$$

Following the Winfree paradigm, we call this surface with boundaries z and $\tilde{z}$ the *isochrone band*.

The choice of initial conditions which produce filament collisions can be a delicate issue. For planar simulations, $m = \dim x = 2$, it is relatively easy to prescribe "vector fields" $u_0 = u_0(x)$ with zeros of prescribed degree. The reaction-diffusion system then quickly produces spirals with tips, near these zeros, and with rotation directions according to local Brouwer degree. Collisions can then be produced, with a little bit of forcing. For three-dimensional simulations, [HeWi91] has described a method which can produce linked circular filaments and torus knots. This method is based on algebraic topology: the initial filament configurations in $x \in \Omega$ are realized by zeros z of appropriate complex polynomials $u(x) = p(z) \in \mathbb{C} \cong \mathbb{R}^2$, in two complex variables $z = (z_1, z_2)$. By inverse stereographic projection ι, we have $x \in \Omega \subset \mathbb{R}^3 \xhookrightarrow{\iota} S^3 \subset \mathbb{C}^2$ corresponding to the argument z of $p(z)$.

We now describe the individual runs. The planar simulation of Figure 2.33 shows color coded solution profiles $u_1(t, \cdot)$ corresponding to two meandering spirals which approach each other and annihilate. The domain $\Omega = [-15, 15]^2$ was discretized by 125^2 points. Slightly differently from Section 2.5.2, the meandering motion is produced by an external time-periodic, spatially homogeneous forcing $A = 0.01$. The frequency $\omega = 3.21$ and the phase were tuned to near drift resonance, so that the two spirals are moving towards each other, initially. Rather than mutual repulsion, annihilation occurs. Note the meandering, near drifting behavior of the spiral tips $u_1 = u_2 = 0.5$, traced in white.

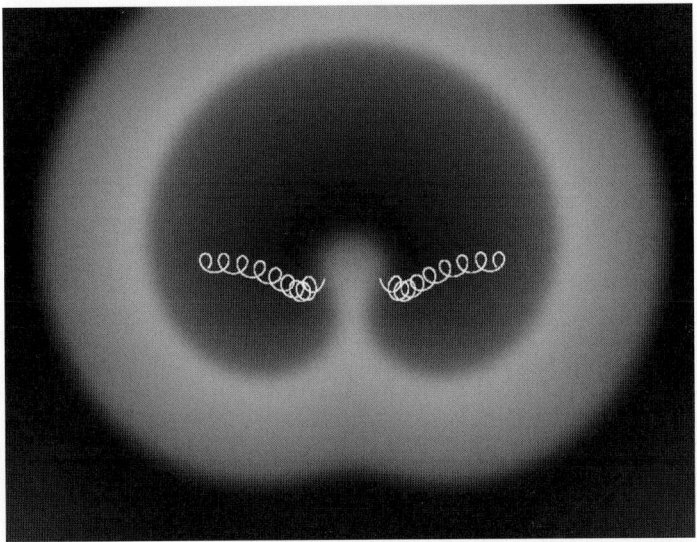

Fig. 2.33. [FiMa00, Fig. 7] Meandering interaction and collision of a pair of planar spiral waves. See Plate 7 in the Appendix for a version of this figure in colour.

The annihilation of a three-dimensional untwisted scroll ring in Figure 2.34 is an autonomous simulation, $A = 0$. Note the analogy to the planar simulation of spiral pair annihilation, by viewing adynsys/dynsysfig/ planar vertical slice through the toroidal scroll ring. The transparent full isochrone surfaces in $\Omega = [-15, 15]^3$ given by $u_2 = 0.5 \leq u_1$ and shown in this example would obscure the view on the scroll filaments $u_2 = 0.5 = u_1$ in more complicated simulations. Therefore they are replaced by the truncated isochrone bands $u_2 = 0.5 \leq u_1 \leq 0.5 + \delta$, as was explained above.

Crossover collision is documented in Figure 2.35. For finer resolution we choose a smaller domain $\Omega = [-10, 10]^3$, again with 125^3 discretization points. The appropriately adjusted time step is $\Delta t = 0.00765$. We choose two straight line horizontal filaments, orthogonal to each other and at a ver-

Fig. 2.34. [FiMa00, Fig. 8] Scroll ring annihilation of an untwisted filament. Surfaces $u_2 = 0 \leq u_1$ shown. See Plate 8 in the Appendix for a version of this figure in colour.

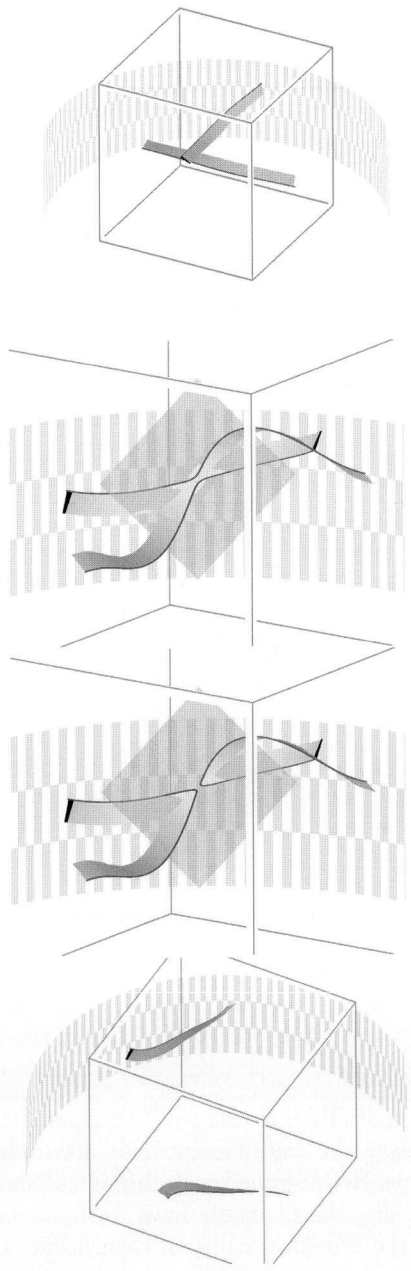

Fig. 2.35. [FiMa00, Fig. 9] Crossover collision of scroll waves including accompanying isochrone bands $u_2 = 0 \leq u_1 \leq \delta$. See Plate 9 in the Appendix for a version of this figure in colour.

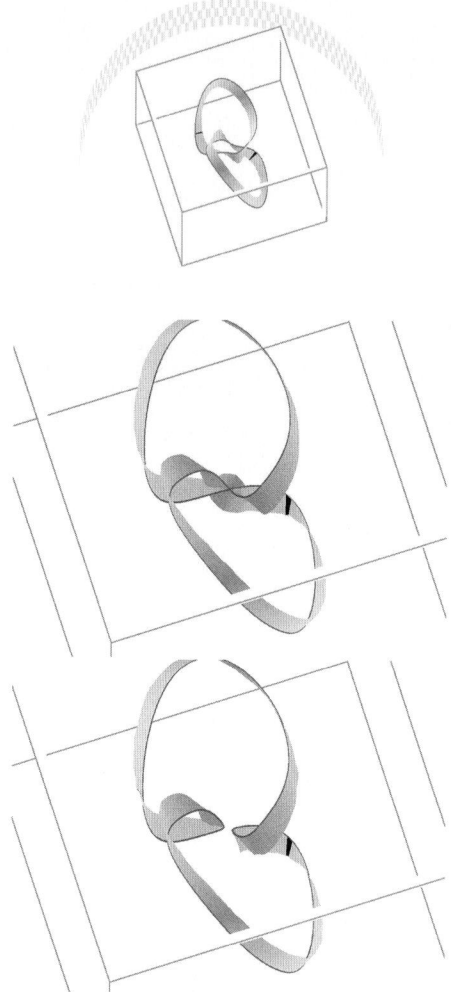

Fig. 2.36. [FiMa00, Fig. 11] Crossover collision of two linked, twisted, circular filaments. See Plate 10 in the Appendix for a version of this figure in colour.

tical distance. By a spatially homogeneous, time periodic forcing $A = 0.01$, as before, we can make the top filament drift downwards and the bottom filament drift upwards. Because the break through scenario of Figure 2.29(a) is forbidden, generically, the filaments have to first locate a tangent plane for kissing. Indeed, the crossover collision then occurs between $t = 35$ and $t = 40$.

Based on the polynomial $z(z) = z_1^2 - z_2^2$, an unlinking collision of two linked, twisted, circular scroll filaments is shown in Figure 2.36. The example is autonomous, $A = 0$, with data as in (2.233). Again a common tangent

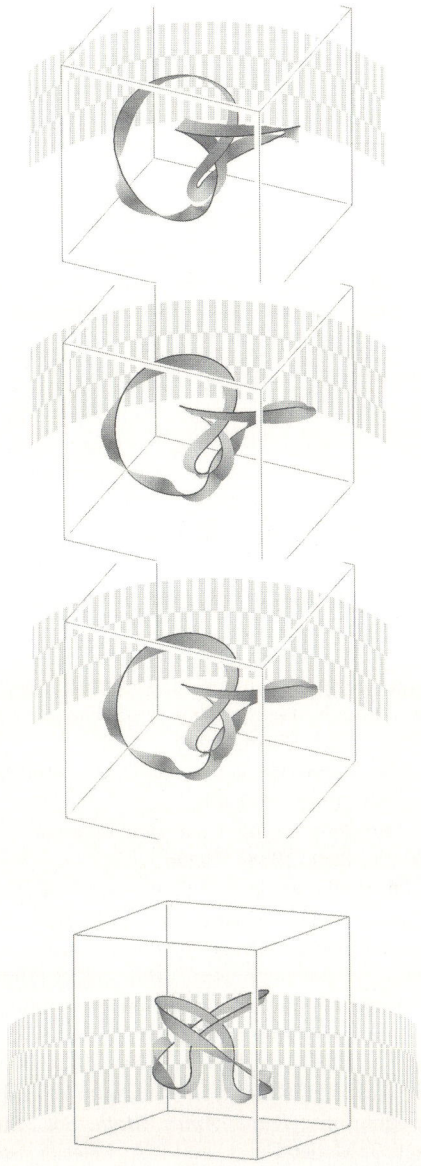

Fig. 2.37. [FiMa00, Fig. 13] Decomposition of the trefoil knot into two linked, twisted circular filaments. See Plate 11 in the Appendix for a version of this figure in colour.

plane is located. Kissing and reconnection according to the abstract scenario of Figures 2.28, 2.29(b), and 2.31 occurs between $t = 4.8$ and $t = 4.9$.

Our final simulation, the decomposition of a trefoil knot into two linked, twisted, circular scroll filaments, runs on a spatially coarser 125^3 grid on $\Omega = [-25, 25]^3$. The construction of the initial condition is based on the polynomial $p(z) = z_1^2 - z_2^3$. The autonomous run, $A = 0$, follows the theoretical scenario of Figures 2.29(b) and 2.32, fully compliant with our abstract theory.

Our theoretical results apply to any fixed value $u = c$ and the associated level curves of the spatial profiles. It is the prominence of the particular values associated to spiral tips, however, becoming visible only through experimental work and numerical simulation along the Winfree paradigm, which has motivated our mathematical analysis.

References

[AbRo67] R. Abraham and J. Robbin. *Transversal Mappings and Flows.* Benjamin Inc., Amsterdam, 1967.

[Al79] N. Alikakos. An application of the invariance principle to reaction diffusion equations. *J. Diff. Eqns.* **33** (1979), 201–225.

[Al&al89] N. Alikakos, P.W. Bates, and G. Fusco. Slow motion for the Cahn-Hilliard equation in one space dimension. *Preprint* (1989).

[AlGe90] E.L. Allgower and K. Georg. *Numerical Continuation Methods. An Introduction.* Springer-Verlag, Berlin, 1990.

[An86] S. Angenent. The Morse-Smale property for a semi-linear parabolic equation. *J. Diff. Eqns.* **62** (1986), 427–442.

[An88] S. Angenent. The zero set of a solution of a parabolic equation. *Crelle J. reine angew. Math.*, **390** (1988), 79–96.

[An90] S. Angenent. Parabolic equations for curves on surfaces. I: curves with p-integrable curvature. *Ann. Math.* **132** (1990), 451–483.

[An91] S. Angenent. Parabolic equations for curves on surfaces. II: Intersections, blow-up and generalized solutions. *Ann. Math.*, **133** (1991), 171–215.

[An93] S. Angenent. A variational interpretation of Melnikov's function and exponentially small separatrix splitting. *Lond. Math. Soc. Lect. Note Ser.*, **192** (1993), 5–35.

[AnFi88] S. Angenent and B. Fiedler. The dynamics of rotating waves in scalar reaction diffusion equations. *Trans. Amer. Math. Soc.*, **307** (1988), 545–568.

[An&al87] S. Angenent, J. Mallet-Paret, and L.A. Peletier. Stable transition layers in a semilinear boundary value problem. *J. Diff. Eqns.* **67** (1987), 212–242.

[An91] D.V. Anosov. *Dynamical Systems with Hyperbolic Behaviour.* Enc. Math. Sc. **66**, Dynamical Systems IX. Springer-Verlag, New York, 1991.

[AnAr88] D. V. Anosov and V. I. Arnol'd. *Ordinary differential equations and smooth dynamical systems.* Enc. Math. Sc. **1**, Dynamical Systems I. Springer-Verlag, Berlin, 1988.

[Ar&al92] I. S. Aranson, L. Aranson, L. Kramer, and A. Weber. Stability limits of spirals and travelling waves in nonequilibrium media. *Phys. Rev. A* **46** (1992), 2992–2995.

[Ar&al94] I.S. Aranson, L. Kramer, and A. Weber. Core instability and spatiotemporal intermittency of spiral waves in oscillatory media. *Phys. Rev. Lett.* **72**, 2316 (1994).
[Ar92] V.I. Arnol'd. *Theory of Bifurcations and Catastrophes.* Enc. Math. Sc. **5**, Dynamical Systems V. Springer-Verlag, Berlin, 1992.
[Ar93] V.I. Arnol'd. *Singularity Theory I.* Enc. Math. Sc. **6**, Dynamical Systems VI. Springer-Verlag, New York, 1993.
[Ar94a] V.I. Arnol'd. *Bifurcation Theory and Catastrophe Theory.* Enc. Math. Sc. **5**, Dynamical Systems V. Springer-Verlag, New York, 1994.
[Ar94b] V.I. Arnol'd. *Singularity theory II, Applications.* Enc. Math. Sc. **8**, Dynamical Systems VIII. Springer-Verlag, New York, 1993.
[Ar&al85] V.I. Arnol'd, S.M. Gusejn-Zade, and A.N. Varchenko. *Singularities of Differentiable Maps. Volume I: The Classification of Critical points, Caustics and Wave Fronts.* Birkhäuser, Boston, 1985.
[Ar&al88] V.I. Arnol'd, V.V. Kozlov, and A.I. Neishtadt. *Mathematical Aspects of Classical and Celestial Mechanics.* Enc. Math. Sc. **3**, Dynamical Systems III. Springer-Verlag, New York, 1988.
[ArNo90] V.I. Arnol'd and S.P. Novikov. *Symplectic Geometry and its Applications.* Enc. Math. Sc. **4**, Dynamical Systems IV. Springer-Verlag, New York, 1990.
[ArNo94] V.I. Arnol'd and S.P. Novikov. *Integrable Systems. Nonholonomic Dynamical Systems.* Enc. Math. Sc. **16**, Dynamical Systems VII. Springer-Verlag, New York, 1994.
[ArVi89] V.I. Arnol'd and M.I. Vishik et al. Some solved and unsolved problems in the theory of differential equations and mathematical physics. *Russian Math. Surveys*, **44** (1989), 157–171.
[AM97] P. Ashwin and I. Melbourne. Noncompact drift for relative equilibria and relative periodic orbits. *Nonlinearity*, **10** (1997), 595–616.
[As&al99] P. Ashwin, I. Melbourne, and M. Nicol. Drift bifurcations of relative equilibria and transitions of spiral waves. *Nonlinearity* **12** (1999), 741–755.
[As&al01] P. Ashwin, I. Melbourne, and M. Nicol. Hypermeander of spirals: local bifurcations and statistical properties. *Phys. D* **156** (2001), 364–382.
[BaVi92] A.V. Babin and M.I. Vishik. *Attractors of Evolution Equations.* North Holland, Amsterdam, 1992.
[BäEi93] M. Bär and M. Eiswirth. Turbulence due to spiral breakup in a continuous excitable medium. *Phys. Rev. E* **48** (1993), 1635–1637.
[BäOr99] M. Bär and M. Or-Guil. Alternative scenarios of spiral breakup in a reaction-diffusion model with excitable and oscillatory dynamics. *Phys. Rev. Lett.* **82** (1999), 1160–1163.
[Ba92] D. Barkley. Linear stability analysis of rotating spiral waves in excitable media. *Phys. Rev. Lett.* **68** (1992), 2090–2093.
[Ba93] D. Barkley. Euclidean symmetry and the dynamics of rotating spiral waves. *Phys. Rev. Lett.* **72** (1994), 164–167.
[Bar95] D. Barkley. Spiral meandering. In R. Kapral and K. Showalter (eds.), *Chemical Waves and Patterns, p.163–190*, Kluwer, 1995.
[Ba&al93] G. Barles, H.M. Soner, and P.E. Souganidis. Front propagation and phase field theory. *SIAM J. Contr. Optim.* **31** (1993), 439–469.
[Be&al97] A. Belmonte, J.-M. Flesselles, and Q. Ouyang. Experimental Survey of Spiral Dynamics in the Belousov-Zhabotinsky Reaction. *J. Physique II* **7** (1997), 1425–1468.

[BeNi90] H. Berestycki and L. Nirenberg. Some qualitative properties of solutions of semilinear elliptic equations in cylindrical domains. *Coll. Analysis, et cetera*, 115–164, Academic Press Boston, 1990.

[BePh96] I. Berkes and W. Philipp. Trigonometric series and uniform distribution mod 1. Stud. Sci. Math. Hung. **31** (1996), 15–25.

[Be87] W.J. Beyn. The effect of discretization on homoclinic orbits. In *Bifurcation: Analysis, Algorithms, Applications 1–8,* T. Küpper et al., (eds.). Birkhäuser Verlag, Basel, 1987.

[Be90] W.-J. Beyn. The numerical computation of connecting orbits in dynamical systems. *IMA Z. Numer. Anal,* **9** (1990), 379–405.

[Bi&al96] V. A. Biktashev, A. V. Holden, and E. V. Nikolaev. Spiral wave meander and symmetry of the plane. Preprint, University of Leeds, 1996.

[BiRo62] G. Birkhoff and G.-C. Rota. *Ordinary differential equations.* Ginn and Company, Boston, 1962.

[Bl&al00] P. Blancheau, J. Boissonade, and P. De Kepper. Theoretical and experimental studies of bistability in the chloride-dioxide-iodide reaction. *Physica D* **147** (2000), 283–299.

[Bo81a] R. Bogdanov. Bifurcation of the limit cycle of a family of plane vector fields. *Sel. Mat. Sov.* **1** (1981), 373–387.

[Bo81b] R. Bogdanov. Versal deformations of a singularity of a vector field on the plane in the case of zero eigenvalues. *Sel. Mat. Sov.*, **1** (1981), 389–421.

[BrEn93] M. Braune and H. Engel. Compound rotation of spiral waves in a light-sensitive Belousov-Zhabotinsky medium. *Chem. Phys. Lett.* **204** (1993), 257–264.

[Br64] R. J. Briggs. *Electron-Steam Interaction With Plasmas.* MIT press, Cambridge, 1964.

[BrtD85] T. Bröcker and T. tom Dieck. *Representations of Compact Lie Groups.* Springer-Verlag, Berlin, 1985.

[Br&al01] H.W. Broer, B. Krauskopf, and G. Vegter. *Global Analysis of Dynamical Systems.* IOP Publishing, Bristol, 2001.

[BrTa02] H. Broer and T. Takens (eds.). *Handbook of Dynamical Systems* **3**. Elsevier, Amsterdam, in preparation 2002.

[Br90] P. Brunovský. The attracor of the scalar reaction diffusion equation is a smooth graph. *J. Dynamics and Differential Equations,* **2** (1990), 293–323.

[BrCh84] P. Brunovský and S-N Chow. Generic properties of stationary state solutions of reaction-diffusion equations. *J. Diff. Eqns.* **53** (1984), 1–23.

[BrFi86] P. Brunovský and B. Fiedler. Numbers of zeros on invariant manifolds in reaction-diffusion equations. *Nonlin. Analysis, TMA*, **10** (1986), 179–194.

[BrFi88] P. Brunovský and B. Fiedler. Connecting orbits in scalar reaction diffusion equations. *Dynamics Reported* **1** (1988), 57–89.

[BrFi89] P. Brunovský and B. Fiedler. Connecting orbits in scalar reaction diffusion equations II: The complete solution. *J. Diff. Eqns.* **81** (1989), 106–135.

[Br&al92] P. Brunovský, P. Poláčik, and B. Sandstede. Convergence in general parabolic equations in one space dimension. *Nonl. Analysis TMA* **18** (1992), 209–215.

[Ca&al93] A. Calsina, X. Mora and J. Solà-Morales. The dynamical approach to clliptic problems in cylindrical domains and a study of their parabolic singular limit. *J. Diff. Eqns.* **102** (1993), 244–304.

[CaPe90] J. Carr and R. Pego.Invariant manifolds for metastable patterns in $u\bar{s}t = \epsilon^2 u\bar{s}xx - f(u)$. *Proc. Roy. Soc. Edinburgh A* **116** (1990), 133–160.

[Ca&al90] V. Castets, E. Dulos, J. Boissonade, and P. De Kepper. Experimental evidence of a sustained standing Turing-type nonequilibrium chemical pattern. *Phys. Rev. Lett.* **64** (1990), 2953–2956.

[ChIn74] N. Chafee and E. Infante. A bifurcation problem for a nonlinear parabolic equation. *J. Applicable Analysis* **4** (1974). 17–37.

[Ch98] X.-Y. Chen. A strong unique continuation theorem for parabolic equations. *Math. Ann.* **311** (1998), 603–630.

[ChVi02] V.V. Chepyzhov and M.I. Vishik. *Attractors for Equations of Mathematical Physics.* Colloq. AMS, Providence, 2002.

[ChLa00] P. Chossat and R. Lauterbach. *Methods in Equivariant Bifurcations and Dynamical Systems.* World Scientific, Singapore, 2000.

[ChHa82] S.-N. Chow and J. K. Hale. *Methods of Bifurcation Theory.* Springer-Verlag, New York, 1982.

[CoEc00] P. Collet and J.-P. Eckmann. Proof of the marginal stability bound for the Swift-Hohenberg equation and related equations. *Preprint*, 2000.

[Co68] W.A. Coppel. Dichotomies and reducibility II. *J. Diff. Eqns.* **4** (1968), 386–398.

[Co78] W.A. Coppel. *Dichotomies in Stability Theory.* Lect. Notes Math. **629**, Springer, Berlin, 1978.

[CrHo93] M.C. Cross and P.C. Hohenberg. Pattern formation outside equilibrium. *Rev. Modern Phys.* **65** (1993), 851–1112.

[Dam97] J. Damon. Generic properties of solutions to partial differential equations. *Arch. Rat. Mech. Analysis*, **140** (1997), 353–403.

[DaPo02] E.N. Dancer and P. Poláčik. *Realization of vector fields and dynamics of spatially homogeneous parabolic equations.* Mem. AMS, Providence, 2002, to appear.

[Da&al96] G. Dangelmayr, B. Fiedler, K. Kirchgässner, and A. Mielke. *Dynamics of Nonlinear Waves in Dissipative Systems: Reduction, Bifurcation and Stability.* Pitman **352**, Boston, 1996.

[De&al95] M. Dellnitz, M. Golubitsky, A. Hohmann,and I. Stewart. Spirals in scalar reaction-diffusion equations. *Internat. J. Bifur. Chaos Appl. Sci. Engrg.* **5** (1995), 1487–1501.

[De85] K. Deimling. *Nonlinear Functional Analysis.* Springer-Verlag, Berlin, 1985.

[dK&al94] P. de Kepper, J.-J. Perraud, B. Rudovics,and E. Dulos. Experimental study of stationary Turing patterns and their interaction with traveling waves in a chemical system. *Int. J. Bifurcation Chaos Appl. Sci. Eng.* **4** (1994), 1215-1231.

[Di&al95] O. Diekmann, S.A. v. Gils, S.M. Verduyn Lunel, and H.-O. Walther. *Delay Equations. Functional-, Complex-, and Nonlinear Analysis.* Springer-Verlag, New York, 1995.

[DoFr89] E. J. Doedel and M. J. Friedman. Numerical computation of heteroclinic orbits. *J. Comp. Appl. Math.* **26** (1989), 155–170.

[DoFr91] E. J. Doedel and M. J. Friedman. Numerical computation and continuation of invariant manifolds connecting fixed points. *SIAM J. Numer. Anal.* **28** (1991), 789–808.

[Do&al97] M. Dowle, M. Mantel, and D. Barkley. Fast simulations of waves in three-dimensional excitable media. *Int. J. Bifur. Chaos*, **7** (1997), 2529–2546.

[Du&al85] B.A. Dubrovin, A.T. Fomenko, and S.P. Novikov. *Modern Geometry - Methods and Applications. Part 2: The Geometry and Topology of Manifolds.* Springer-Verlag, New York, 1985.

[EiYa98] S.-I. Ei and E. Yanagida. Slow dynamics of interfaces in the allen-cahn equation on a strip-like domain. *SIAM J. Math. Anal.*, **29** (1998), 555–595.
[El&al87] C. Elphick, E. Tirapegui, M.E. Brachet, P. Coullet, and G. Iooss. A simple global characerization for normal forms of singular vector fields. *Physica* **29D** (1987), 95–127.
[Fe71] N. Fenichel. Persistence and smoothness of invariant manifolds for flows. *Indiana Univ. Math. J.* **21** (1971), 193–226.
[Fe74] N. Fenichel. Asymptotic stability with rate conditions. *Indiana Univ. Math. J.* **23** (1974), 1109–1137.
[Fe77] N. Fenichel. Asymptotic stability with rate conditions, II. *Indiana Univ. Math. J.* **26** (1977), 81–93.
[Fe79] N. Fenichel. Geometric singular perturbation theory for ordinary differential equations. *J. Diff. Eqns.*, **31** (1979), 53–98.
[Fi88] B. Fiedler. *Global Bifurcation of Periodic Solutions with Symmetry.* Springer-Verlag, Berlin, 1988.
[Fi89] B. Fiedler. Discrete Ljapunov functionals and ω-limit sets. *Math. Mod. Num. Analysis*, **23** (1989), 415–431.
[Fi94] B. Fiedler. Global attractors of one-dimensional parabolic equations: sixteen examples. *Tatra Mountains Math. Publ.*, **4** (1994), 67–92.
[Fi96] B. Fiedler. Do global attractors depend on boundary conditions? *Doc. Math.* **1** (1996), 215–228.
[Fi02] B. Fiedler (ed.) *Handbook of Dynamical Systems* **2**, Elsevier, Amsterdam. In press.
[FiGe98] B. Fiedler and T. Gedeon. A class of convergent neural network dynamics. *Physica D*, **111** (1998), 288–294,.
[FiGe99] B. Fiedler and T. Gedeon. A Lyapunov function for tridiagonal competitive-cooperative systems. *SIAM J. Math Analysis* **30** (1999), 469–478.
[Fi&al00] B. Fiedler, K. Gröger, and J. Sprekels (eds.). *Equadiff 99. International Conference on Differential Equations, Berlin 1999.* Vol.1,2. World Scientific, Singapore, 2000.
[FiMP89a] B. Fiedler and J. Mallet-Paret. Connections between Morse sets for delay-differential equations. *J. reine angew. Math.*, **397**:23–41, (1989).
[FiMP89b] B. Fiedler and J. Mallet-Paret. A Poincaré-Bendixson theorem for scalar reaction diffusion equations. *Arch. Rat. Mech. Analysis* **107** (1989), 325–345.
[FiMa00] B. Fiedler and R.-M. Mantel. Crossover collision of core filaments in three-dimensional scroll wave patterns. *Doc. Math.* **5** (2000), 695–731.
[FiPo90] B. Fiedler and P. Poláčik. Complicated dynamics of scalar reaction diffusion equations with a nonlocal term. *Proc. Royal Soc. Edinburgh* **115A** (1990), 167–192.
[FiRo96] B. Fiedler and C. Rocha. Heteroclinic orbits of semilinear parabolic equations. *J. Diff. Eq.* **125** (1996), 239–281.
[FiRo99] B. Fiedler and C. Rocha. Realization of meander permutations by boundary value problems. *J. Diff. Eqns.* **156** (1999), 282–308.
[FiRo00] B. Fiedler and C. Rocha. Orbit equivalence of global attractors of semilinear parabolic differential equations. *Trans. Amer. Math. Soc.*, **352** (2000), 257–284.
[Fi&al02a] B. Fiedler, C. Rocha, D. Salazar, and J. Sol-Morales. A note on the dynamics of piecewise-autonomous bistable parabolic equations. *Comm. Fields Inst.* (2002), in press.

[Fi&al02b] B. Fiedler, C. Rocha, and M. Wolfrum. Heteroclinic connections of S^1-equivariant parabolic equations on the circle. In preparation, 2002.

[Fi&al96] B. Fiedler, B. Sandstede, A. Scheel, and C. Wulff. Bifurcation from relative equilibria of noncompact group actions: skew products, meanders and drifts. *Doc. Math. J. DMV* **1** (1996), 479–505. See also http://www.mathematik.uni-bielefeld.de/documenta/vol-01/20.ps.gz

[Fi&al98] B. Fiedler, A. Scheel, and M. Vishik. Large patterns of elliptic systems in infinite cylinders. *J. Math. Pures Appl.* **77** (1998), 879–907.

[FiTu98] B. Fiedler and D. Turaev. Normal forms, resonances, and meandering tip motions near relative equilibria of Euclidean group actions. *Arch. Rat. Mech. Anal.* **145** (1998), 129–159.

[FiVi01a] B. Fiedler and M. Vishik. Quantitative homogenization of analytic semigroups and reaction diffusion equations with diophantine spatial frequencies. *Adv. in Diff. Eqns.* **6** (2001), 1377–1408.

[FiVi01b] B. Fiedler and M. Vishik. Quantitative homogenization of global attractors for reaction-diffusion systems with rapidly oscillating terms. *Preprint*, 2001.

[Fi88] P.C. Fife. *Dynamics of internal layers and diffusive interfaces, CBMS-NSF Reg. Conf. Ser. Appl. Math.* **53**, 1988.

[Fi84] G. Fischer. Zentrumsmannigfaltigkeiten bei elliptischen Differentialgleichungen. *Math. Nachr.* **115** (1984), 137–157.

[Fr64] A. Friedman. *Partial Differential Equations of Parabolic Type*. Prentice Hall, Englewood Cliffs, New Jersey, 1964.

[FuHa89] G. Fusco and J.K. Hale. Slow-motion manifolds, dormant instability, and singular perturbations. *J. Dyn. Diff. Eqns.* **1** (1989), 75–94.

[FuOl88] G. Fusco and W.M. Oliva. Jacobi matrices and transversality. *Proc. Royal Soc. Edinburgh* **A 109** (1988), 231–243.

[FuRo91] G. Fusco and C. Rocha. A permutation related to the dynamics of a scalar parabolic PDE. *J. Diff. Eqns.* **91** (1991), 75–94.

[GaHa86] M. Gage and R.S. Hamilton. The heat equation shrinking convex plane curves. *J. Diff. Geom.* **23** (1986), 69–96.

[GaSl02] T. Gallay and S. Slijepčevic. Personal communication, (2002).

[GiHi96a] M. Giaquinta and S. Hildebrandt. *Calculus of Variations 1. The Lagrangian Formalism*. Springer-Verlag, Berlin, 1996.

[GiHi96b] M. Giaquinta and S. Hildebrandt. *Calculus of Variations 2. The Hamiltonian Formalism*. Springer-Verlag, Berlin, 1996.

[Go&al00] M. Golubitsky, E. Knobloch, and I. Stewart. Target patterns and spirals in planar reaction-diffusion systems. *J. Nonlinear Sci.* **10** (2000), 333–354.

[Go&al97] M. Golubitsky, V. LeBlanc, and I. Melbourne. Meandering of the spiral tip: an alternative approach. *J. Nonl. Sci.* **7** (1997), 557–586.

[GoSc85] M. Golubitsky and D.G. Schaeffer. *Singularities and Groups in Bifurcation Theory I*. Springer-Verlag, 1985.

[Go&al88] M. Golubitsky, I. Stewart, and D.G. Schaeffer. *Singularities and Groups in Bifurcation Theory II*. Springer-Verlag, New York, 1988.

[Go&al98] A. Goryachev, H. Chaté, and R. Kapral. Synchronization defects and broken symmetry in spiral waves. *Phys. Rev. Lett.* **80** (1998), 873–876.

[Gr89] M. A. Grayson. Shortening embedded curves. *Ann. Math.* **129** (1989), 71–111.

[GuHo83] J. Guckenheimer and P. Holmes. *Nonlinear Oscillations, Dynamical Systems, and Bifurcations of Vector Fields.* Springer-Verlag, New York, 1983.

[Ha&al] G. Haas, M. Bär, and I.G. Kevrekidis et al. Observation of front bifurcations in controlled geometries: From one to two dimensions. *Phys. Rev. Lett.* **75** (1995), 3560–3563.

[Ha82] P.S. Hagan. Spiral waves in reaction-diffusion equations. *SIAM J. Appl. Math.* **42** (1982), 762–786.

[Ha69] J.K. Hale. *Ordinary Differential Equations.* John Wiley & Sons, New York, 1969.

[Ha85] J.K. Hale. Flows on centre manifolds for scalar functional differential equations. *Proc. R. Soc. Edinb., Sect. A* **101** (1985), 193–201.

[Ha88] J.K. Hale. *Asymptotic Behavior of Dissipative Systems.* Math. Surv. **25**. AMS Publications, Providence, 1988.

[HaRa92] J.K. Hale and G. Raugel. Reaction-diffusion equation on thin domains. *J. Math. Pures Appl.* **71** (1992), 33–95.

[Hä97] J. Härterich. Attractors of Viscous Balance Laws. Dissertation, Freie Universität Berlin, 1997.

[Hä98] J. Härterich. Attractors of viscous balance laws: Uniform estimates for the dimension. *J. Diff. Eqns.* **142** (1998), 188–211.

[Hä99] J. Härterich. Equilibrium solutions of viscous scalar balance laws with a convex flux. *Nonlin. Diff. Eqns. Appl.* **6** (1999), 413–436.

[HaMi91] H. Hattori and K. Mischaikow. A dynamical system approach to a phase transition problem. *J. Diff. Eqns.* **94** (1991), 340–378.

[He89] S. Heinze. *Travelling waves for semilinear parabolic partial differential equations in cylindrical domains.* Dissertation, Heidelberg, 1989.

[He81] D. Henry. *Geometric Theory of Semilinear Parabolic Equations.* Lect. Notes Math. **804**, Springer-Verlag, New York, Berlin, Heidelberg, 1981.

[He85] D. Henry. Some infinite dimensional Morse-Smale systems defined by parabolic differential equations. *J. Diff. Eqns.* **59** (1985), 165–205.

[HeWi91] C. Henze and A. T. Winfree. A stable knotted singularity in an excitable medium. *Int. J. Bif. Chaos* **1** (1991), 891–922.

[Hi83] M. W. Hirsch. Differential equations and convergence almost everywhere in strongly monotone semiflows. J. Smoller, (ed.). In *Nonlinear Partial Differential Equations.* p. 267–285, AMS Publications, Providence, 1983.

[Hi85] M. W. Hirsch. Systems of differential equations that are competitive or cooperative II. Convergence almost everywhere. *SIAM J. Math. Analysis* **16** (1985), 423–439.

[Hi88] M. W. Hirsch. Stability and convergence in strongly monotone dynamical systems. *Crelle J. reine angew. Math.* **383** (1988), 1–58.

[Hi&al77] M. W. Hirsch, C.C. Pugh, and M. Shub. *Invariant Manifolds.* Springer-Verlag, Berlin, 1977.

[IoMi91] G. Iooss and A. Mielke. Bifurcating time–periodic solutions of Navier–Stokes equations in infinite cylinders. *J. Nonlinear Science* **1** (1991), 107–146.

[JäLu92] W. Jäger and S. Luckhaus. On explosions of solutions to a system of partial differential equations modelling chemotaxis. *Trans. Am. Math. Soc.*, **329** (1992), 819–824.

[Ja&al88] W. Jahnke, C. Henze, and A.T. Winfree. Chemical vortex dynamics in the 3-dimensional excitable media. *Nature* **336** (1988), 662–665.

[Ja&al89] W. Jahnke, W.E. Skaggs, and A.T. Winfree. Chemical vortex dynamics in the Belousov-Zhabotinskii reaction and in the two-variable Oregonator model. *J. Chem. Phys.* **93** (1989), 740–749.

[Ka66] T. Kato. *Perturbation Theory for Linear Operators.* Springer, Berlin, Heidelberg, New York, 1966.

[KaHa95] A. Katok and B. Hasselblatt. *Introduction to the modern theory of dynamical systems.* With a supplementary chapter by Katok and Leonardo Mendoza. Encyclopedia of Mathematics and its Applications **54**, Cambridge University Press, Cambridge, 1995.

[KaHa02] A. Katok and B. Hasselblatt (eds.) *Handbook of Dynamical Systems* **1**, Elsevier, Amsterdam. to appear 2002.

[Ka87] L.H. Kauffman. *On Knots.* Princeton University Press, New Jersey, 1987.

[Ke92] J.P. Keener. The core of the spiral. *SIAM J. Appl. Math.* **52** (1992), 1370–1390.

[KeTy92] J.P. Keener and J.J. Tyson. The dynamics of scroll waves in excitable media. *SIAM Rev.*, **34** (1992), 1–39.

[Ki82] K. Kirchgässner. Wave-solutions of reversible systems and applications. *J. Differential Equations* **45** (1982), 113–127.

[Ki00] S.V. Kiyashko. The generation of stable waves in faraday experiment. *2000 Int. Symp. Nonlinear Theory and its Applications*, 2000.

[KoHo73] N. Kopell and L.N. Howard. Plane wave solutions to reaction-diffusion equations. *Studies in Appl. Math.* **52** (1973), 291–328.

[KoHo81] N. Kopell and L.N. Howard. Target patterns and spiral solutions to reaction-diffusion equations with more than one space dimension. *Adv. Appl. Math.* **2** (1981), 417–449.

[Ko02] V.V. Kozlov. *General Theory of Vortices.* Enc. Math. Sc. **67**, Dynamical Systems X. Springer-Verlag, New York, 2002.

[Kr90] M. Krupa. Bifurcations of relative equilibria. *SIAM J. Math. Analysis* **21** (1990), 1453–1486.

[KuMa83] M. Kubiček and M. Marek. *Computational Methods in Bifurcation Theory and Dissipative Structures.* Springer-Verlag, New York, 1983.

[Ku95] Y.A. Kuznetsov. *Elements of Applied Bifurcation Theory.* Springer-Verlag, Berlin, 1995.

[La91] O.A. Ladyzhenskaya. *Attractors for Semigroups and Evolution Equations.* Cambridge University Press, 1991.

[LaLi59] L.D. Landau and E.M. Lifschitz. *Fluid Mechanics.* Pergamon Press, London, 1959.

[Li90] X.-B. Lin. Using Melnikov's method to solve Shilnikov's problems. *Proc. Roy. Soc. Edinburgh*, **116A** (1990), 295–325.

[Ou&al] G. Li, Q. Ouyang, V. Petrov, and H. L. Swinney. Transition from simple rotating chemical spirals to meandering and traveling spirals. *Phys. Rev. Lett.* **77** (1996), 2105–2108.

[Lo04] A.E.H. Love. *A Treatise on the Mathematical Theory of Elasticity.* Dover Publications, New-York, 1904.

[MP88] J. Mallet-Paret. Morse decompositions for delay-differential equations. *J. Diff. Eqns.* **72** (1988), 270–315.

[MPSm90] J. Mallet-Paret and H. Smith. The Poincaré-Bendixson theorem for monotone cyclic feedback systems. *J. Diff. Eqns.* **4** (1990), 367–421.

[MPSe96a] J. Mallet-Paret and G.R. Sell. The Poincaré-Bendixson theorem for monotone cyclic feedback systems with delay. *J. Diff. Eqns.* **125** (1996), 441–489.

[MPSe96b] J. Mallet-Paret and G.R. Sell. Systems of differential delay equations: Floquet multipliers and discrete Lyapunov functions. *J. Diff. Eqns.* **125** (1996), 385–440.

[MaPa97] A.F.M. Maree and A.V. Panfilov. Spiral breakup in excitable tissue due to lateral instability. *Phys. Rev. Lett.* **78** (1997), 1819–1822.

[Ma78] H. Matano. Convergence of solutions of one-dimensional semilinear parabolic equations. *J. Math. Kyoto Univ.*, **18** (1978), 221–227.

[Ma79] H. Matano. Asymptotic behavior and stability of solutions of semilinear diffusion equations. *Publ. Res. Inst. Math. Sc. Kyoto Univ.* **15** (1979), 401–454.

[Ma82] H. Matano. Nonincrease of the lap-number of a solution for a one-dimensional semi-linear parabolic equation. *J. Fac. Sci. Univ. Tokyo Sec. IA*, **29**(1982), 401–441.

[Ma86] H. Matano. Strongly order-preserving local semi-dynamical systems — theory and applications. In *Semigroups, Theory and Applications*. H. Brezis, M.G. Crandall, F. Kappel (eds.), 178–189. John Wiley & Sons, New York, 1986.

[Ma87] H. Matano. Strong comparison principle in nonlinear parabolic equations. In *Nonlinear Parabolic Equations: Qualitative Properties of Solutions*, L. Bocardo, A. Tesei (eds.), 148–155. Pitman Res. Notes Math. Ser. **149** (1987).

[Ma88] H. Matano. Asymptotic behavior of solutions of semilinear heat equations on S^1. In *Nonlinear Diffusion Equations and their Equilibrium States II*. W.-M. Ni, L.A. Peletier, J. Serrin (eds.). 139–162. Springer-Verlag, New York, 1988.

[MaNa97] H. Matano and K.-I. Nakamura. The global attractor of semilinear parabolic equations on S^1. *Discr. Contin. Dyn. Syst.* **3** (1997), 1–24.

[MaWi89] J. Mawhin and M. Willem. *Critical Point Theory and Hamiltonian Systems*. Springer-Verlag, New York, 1989.

[Mi91] A. Mielke. *Hamiltonian and Lagrangian Flows on Center Manifolds*. Springer-Verlag, Berlin, 1991.

[Mi94] A. Mielke. Essential manifolds for elliptic problems in infinite cylinders. *J. Diff. Eqns.*, **110** (1994), 322–355.

[Mi97] A. Mielke. Instability and stability of rolls in the Swift-Hohenberg equation. *Comm. Math. Phys.* **189** (1997), 829–853.

[MiSc96] A. Mielke and G. Schneider. Derivation and justification of the complex Ginzburg-Landau equation as a modulation equation. Dynamical systems and probabilistic methods in partial differential equations (Berkeley, CA, 1994), 191–216, *Lectures in Appl. Math.* **31**, Amer. Math. Soc., Providence, RI, 1996.

[MiZy91] A.S. Mikhailov and V.S. Zykov. Kinematical theory of spiral waves in excitable media: comparison with numerical simulations. *Physica D* **52** (1991), 379–397.

[Mo73] J. Moser. *Stable and Random Motions in Dynamical Systems*. Princeton University Press, New York, 1973.

[MüZy94] S.C. Müller and V.S. Zykov. Simple and complex spiral wave dynamics. *Phil. Trans. Roy. Soc. Lond. A* **347** (1994), 677–685.

[Na90] N.S. Nadirashvili. On the dynamics of nonlinear parabolic equations. *Soviet Math. Dokl.* **40** (1990), 636–639.

[Ne&al] S. Nettesheim, A. von Oertzen, H.H. Rotermund, and G. Ertl. Reaction diffusion patterns in the catalytic CO-oxidation on Pt(110) — front propagation and spiral waves. *J. Chem. Phys.* **98** (1993), 9977–9985.

[OgNa00] T. Ogiwara and K.-I. Nakamura. Spiral traveling wave solutions of some parabolic equations on annulis. In *Nonlinear Analysis, Josai Math. Monogr., Nishizawa, Kiyoko (ed.)* **2** (2000), 15–34.

[Ol02] W.M. Oliva. Morse-Smale semiflows. Openess and A-stability. *Comm. Fields Inst.* (2002), in press.

[Pa61] R. S. Palais. On the existence of slices for actions of non-compact Lie groups. *Ann. of Math.* **73** (1961), 295–323.

[Pa69] J. Palis. On Morse-Smale dynamical systems. *Topology* **8** (1969), 385–404.

[PaSm70] J. Palis and S. Smale. Structural stability theorems. In *Global Analysis*. Proc. Symp. in Pure Math. vol. XIV. AMS, Providence, 1970. S. Chern, S. Smale (eds.).

[Pa88] K.J. Palmer. Exponential dichotomies and Fredholm operators. *Proc. Amer. Math. Soc.* **104** (1988), 149–156.

[PaWi85] A.V. Panfilov and A. T. Winfree. Dynamical simulations of twisted scroll rings in 3-dimensional excitable media. *Physica D* **17** (1985), 323–330.

[Pa83] A. Pazy. *Semigroups of Linear Operators and Applications to Partial Differential Equations.* Springer-Verlag, New York, 1983.

[Pe&al91] V. Perez-Munuzuri, R. Aliev, B. Vasiev, V. Perez-Villar, and V. I. Krinsky. Super-spiral structure in an excitable medium. *Nature* **353** (1991) 740–742.

[Pe&al93] V. Perez-Munuzuri, M. Gomez-Gesteira, and V. Perez-Villar. A geometrical-kinematical approach to spiral wave formation: Super-spiral waves. *Physica D* **64** (1993), 420–430.

[Pe&al97] D. Peterhof, A. Scheel, and B. Sandstede. Exponential dichotomies for solitary wave solutions of semilinear elliptic equations on infinite cylinders. *J. Diff. Eqns.* **140** (1997), 266–308.

[PlBo96] B.B. Plapp and E. Bodenschatz. Core dynamics of multiarmed spirals in Rayleigh-Bénard convection. *Physica Scripta* **67** (1996), 111–117.

[Po89] P. Poláčik. Convergence in strongly monotone flows defined by semilinear parabolic equations. *J. Diff. Eqs.* **79** (1989), 89–110.

[Po95] P. Poláčik. High-dimensional ω-limit sets and chaos in scalar parabolic equations. *J. Diff. Eqns.*, **119** (1995), 24–53.

[Po02] P. Poláčik. Parabolic equations: Asymptotic behavior and dynamics on invariant manifolds. In *Handbook of Dynamical Systems, Vol.* **2**. B. Fiedler (ed.), Elsevier, Amsterdam, 2002. In press.

[Po33] G. Polya. Qualitatives über Wärmeaustausch. *Z. Angew. Math. Mech.* **13** (1933), 125–128,.

[Po92] G. Pospiech. *Eigenschaften, Existenz und Stabilität von travelling wave Lösungen zu einem System von Reaktions-Diffusions-Gleichungen.* Dissertation, Universität Heidelberg, 1992.

[PrRy98a] M. Prizzi and K.P. Rybakowski. Complicated dynamics of parabolic equations with simple gradient dependence. *Trans. Am. Math. Soc.* **350** (1998), 3119–3130.

[PrRy98b] M. Prizzi and K.P. Rybakowski. Inverse problems and chaotic dynamics of parabolic equations on arbitrary spatial domains. *J. Diff. Eqns.* **142** (1998), 17–53.

[PrWe67] M.H. Protter and H.F. Weinberger. *Maximum Principles in Differential Equations.* Prentice Hall, Englewood Cliffs, New Jersey, 1967.

[Ra02] G. Raugel. Global attractors. In *Handbook of Dynamical Systems, Vol. 2*. B. Fiedler (ed.), Elsevier, Amsterdam, 2002. In press.

[ReSi78] M. Reed and B. Simon. *Methods of Modern Mathematical Physics IV*. Academic Press, 1978.

[RoSa95] J. Robbin and D. Salamon. The spectral flow and the Maslov index. *Bull. London Math. Soc.* **27** (1995), 1–33.

[Ro85] C. Rocha. Generic properties of equilibria of reaction-diffusion equations with variable diffusion.*Proc. R. Soc. Edinb.* A **101** (1985), 45–55.

[Ro91] C. Rocha. Properties of the attractor of a scalar parabolic PDE. *J. Dyn. Differ. Equations* **3** (1991), 575–591.

[Sa93a] B. Sandstede. *Verzweigungstheorie homokliner Verdopplungen*. Dissertation, Universität Stuttgart, 1993.

[Sa93b] B. Sandstede. Asymptotic behavior of solutions of non-autonomous scalar reaction-diffusion equations. *In Conf. Proceeding International Conference on Differential Equations, Barcelona 1991, C. Perello, C. Simo, and J. Sola-Morales (eds.)*, 888–892, World Scientific, Singapore, 1993.

[SaFi92] B. Sandstede and B. Fiedler. Dynamics of periodically forced parabolic equations on the circle. *Ergod. Theor. Dynam. Sys.* **12** (1992), 559–571.

[SaSc99] B. Sandstede and A. Scheel. Essential instability of pulses and bifurcations to modulated travelling waves. *Proc. Roy. Soc. Edinburgh.* A **129** (1999), 1263–1290.

[SaSc00a] B. Sandstede and A. Scheel.Gluing unstable fronts and backs together can produce stable pulses. *Nonlinearity* **13** (2000), 1465–1482.

[SaSc00b] B. Sandstede and A. Scheel. Spectral stability of modulated travelling waves bifurcating near essential instabilities. *Proc. R. Soc. Edinburgh A* **130** (2000), 419–448.

[SaSc00c] B. Sandstede and A. Scheel. Absolute and convective instabilities of waves on unbounded and large bounded domains. *Physica D* **145** (2000), 233–277.

[SaSc00d] B. Sandstede and A. Scheel. Absolute versus convective instability of spiral waves. *Phys. Rev. E.* **62** (2000), 7708–7714.

[SaSc01a] B. Sandstede and A. Scheel. Super-spiral structures of meandering and drifting spiral waves. *Phys. Rev. Lett.* **86** (2001), 171–174.

[SaSc01b] B. Sandstede and A. Scheel. Essential instabilities of fronts: bifurcation and bifurcation failure. *Dynamical Systems: An International Journal* **16** (2001), 1–28.

[SaSc01c] B. Sandstede and A. Scheel. On the stability of periodic travelling waves with large spatial period. *J. Diff. Eqns.* **172** (2001), 134–188.

[SaSc01d] B. Sandstede and A. Scheel. On the structure of spectra of modulated travelling waves.*Math. Nachr.* **232** (2001), 39–93.

[SaSc02a] B. Sandstede and A. Scheel. Nonlinear convective stability and instability — the role of absolute spectra and nonlinearities.*In preparation* (2002).

[SaSc02b] B. Sandstede and A. Scheel. Instabilities of spiral waves in large disks.*In preparation* (2002).

[Sa&al97a] B. Sandstede, A. Scheel, and C. Wulff. Center manifold reduction for spiral wave dynamics. *C. R. Acad. Sci. Paris, Série I* **324** (1997), 153–158.

[Sa&al97b] B. Sandstede, A. Scheel, and C. Wulff. Dynamics of spiral waves on unbounded domains using center-manifold reduction. *J. Diff. Eqns.* **141** (1997), 122–149.

[Sa&al99] B. Sandstede, A. Scheel, and C. Wulff. Bifurcations and dynamics of spiral waves. *J. Nonlinear Science* **9** (1999), 439–478.

[Sc90] R. Schaaf. *Global Solution Branches of Two Point Boundary Value Problems.* Springer-Verlag, New York, 1990.

[Sc96] A. Scheel. Existence of fast travelling waves for some parabolic equations — a dynamical systems approach. *J. Dyn. Diff. Eqns.* **8** (1996), 469–548.

[Sc97] A. Scheel. Subcritical bifurcation to infinitely many rotating waves. J. Math. Anal. Appl. **215** (1997), 252–261.

[Sc98] A. Scheel. Bifurcation to spiral waves in reaction-diffusion systems. *SIAM J. Math. Anal.* **29** (1998), 1399–1418.

[Sc01] A. Scheel.Radially symmetric patterns of reaction-diffusion systems. *Preprint* 2001.

[Sc98b] G. Schneider. Hopf bifurcation in spatially extended reaction-diffusion systems. *J. Nonlinear Sci.* **8** (1998), 17–41.

[Sc98c] G. Schneider. Nonlinear diffusive stability of spatially periodic solutions — abstract theorem and higher space dimensions. *Tohoku Math. Publ.* **8** (1998), 159–167.

[Si89] Ya.G. Sinai. *Ergodic theory with applications to dynamical systems and statistical mechanics.* Enc. Math. Sc. **2**, Dynamical Systems II. Springer-Verlag, Berlin, 1989.

[SkSw91] G. S. Skinner and H. L. Swinney. Periodic to quasiperiodic transition of chemical spiral rotation. *Physica D* **48** (1991), 1–16.

[Sm95] H. Smith. *Monotone Dynamical Systems: An Introduction to the Theory of Competitive and Cooperative Systems.* AMS, Providence, 1995.

[Sm83] J. Smoller. *Shock Waves and Reaction-Diffusion Equations.* Springer-Verlag, New York, 1983.

[St00] A. Steven. The derivation of chemotaxis equations as limit dynamics of moderately interacting stochastic many-particle systems. *SIAM J. Appl. Math.*, **61** (2000), 183–212.

[St90] M. Struwe. *Variational Methods.* Springer-Verlag, Berlin, 1990.

[St36] C. Sturm. Sur une classe d'équations à différences partielles. *J. Math. Pure Appl.* **1** (1836), 373–444,.

[Ta74] F. Takens. Singularities of vector fields. *Publ. IHES*, **43** (1974), 47–100.

[Ta79] H. Tanabe. *Equations of Evolution.* Pitman, Boston, 1979.

[Te88] R. Temam. *Infinite-Dimensional Dynamical Systems in Mechanics and Physics.* Springer-Verlag, New York, 1988.

[ToKn98] S.M. Tobias and E. Knobloch. Breakup of spiral waves into chemical turbulence. *Phys. Rev. Lett.* **80** (1998), 4811–4814.

[Tu52] A. Turing. The chemical basis of morphogenesis. *Phil. Trans. Roy. Soc. B* **237** (1952), 37–72.

[TySt91] J.J. Tyson and S.H. Strogatz. The differential geometry of scroll waves. *Int. J. Bif. Chaos*, **1** (1991), 723–744.

[Uh76] K. Uhlenbeck. Generic properties of eigenfunctions. *Amer. J. Math.*, **98** (1976), 1059–1078.

[Un&al93] Zs. Ungvarai-Nagy, J. Ungvarai, and S.C. Müller. Complexity in spiral wave dynamics. *Chaos* **3** (1993), 15–19.

[Va82] A. Vanderbauwhede. *Local Bifurcation and Symmetry.* Pitman, Boston, 1982.

[Va89] A. Vanderbauwhede. Center manifolds, normal forms and elementary bifurcations. *Dynamics Reported* **2** (1989), 89–169.

[Wa70] W. Walter. *Differential and Integral Inequalities*. Springer-Verlag, New York, 1970.

[WiRo46] N. Wiener and A. Rosenblueth. The mathematical formulation of the problem of conduction of impulses in a network of connected excitable elements, specifically in cardiac muscle. *Arch. Inst. Cardiol. Mexico* **16** (1946), 205–265.

[Wi72] A. T. Winfree. Spiral waves of chemical activity. *Science*, **175** (1972), 634–636.

[Wi73] A. T. Winfree. Scroll-shaped waves of chemical activity in three dimensions. *Science* **181** (1973), 937–939.

[Wi87] A. T. Winfree. *When Time Breaks Down*. Princeton University Press, Princeton, NJ, 1987.

[Wi91] A. T. Winfree. Varieties of spiral wave behavior: An experimentalist's approach to the theory of excitable media. *Chaos* **1** (1991), 303–334.

[Wi95] A. T. Winfree. Persistent tangles of vortex rings in excitable media. *Physica D* **84** (1995), 126–147.

[Wi01] A. T. Winfree. The geometry of biological time. *Biomathematics* **8**, Springer-Verlag, Berlin-New York, 2001.

[Wi&al95] A. T. Winfree, E.M. Winfree, and M. Seifert. Organizing centers in a cellular excitable medium. *Physica D*, **17** (1995), 109–115.

[Wo98] Matthias Wolfrum. *Geometry of Heteroclinic Cascades in Scalar Semilinear Parabolic Equations*. Dissertation, Freie Universität Berlin, 1998.

[Wo02a] M. Wolfrum. Personal communication, (2002).

[Wo02b] M. Wolfrum. A sequence of order relations, encoding heteroclinic connections in scalar parabolic PDEs. *J. Diff. Eqns.*, to appear (2002).

[Wu96] C. Wulff. *Theory of Meandering and Drifting Spiral Waves in Reaction-Diffusion Systems*. Dissertation, Berlin, 1996.

[Wu&al01] C. Wulff, J. Lamb, and I. Melbourne. Bifurcation from relative periodic solutions. *Ergodic Theory Dynam. Systems* **21** (2001), 605–635.

[Ya&al98] H. Yagisita, M. Mimura, and M. Yamada. Spiral wave behaviors in an excitable reaction-diffusion system on a sphere. *Physica D* **124** (1998), 126–136.

[Ze85] E. Zeidler. *Nonlinear functional analysis and its applications. III: Variational methods and optimization*. Springer-Verlag, New York, 1985.

[Ze93] E. Zeidler. *Nonlinear functional analysis and its applications. Volume I: Fixed-point theorems*. Springer-Verlag, New York, 1993.

[Ze68] T.I. Zelenyak. Stabilization of solutions of boundary value problems for a second order parabolic equation with one space variable. *Diff. Eqns.* **4** (1968), 17–22.

[ZhOu00] L. Q. Zhou and Q. Ouyang. Experimental studies on long-wavelength instability and spiral breakup in a reaction-diffusion system. *Phys. Rev. Lett.* **85** (2000), 1650–1653.

3 Some Nonclassical Trends in Parabolic and Parabolic-like Evolutions

Paul Fife

Department of Mathematics, University of Utah,
155 South 1400 East, Salt Lake City, UT 84112-0090, USA – fife@math.utah.edu

Abstract: An overview will be given of some nonlinear parabolic-like evolution problems which are off the classical beaten track, but have increased in importance during the past decade. The emphasis is on problems which are nonlocal, pattern-forming (including exhibiting propagative phenomena), and/or lead in some singular limit to free boundary problems. In all cases they have been proposed as models for phenomena in the natural sciences. Also emphasized are the relationships among these various trends.

3.1 Introduction

The particular nonclassical trends I want to emphasize are (spatially) nonlocal evolutions, pattern-forming processes, and reduction to free boundary problems. All of these trends rely on the problem at hand being nonlinear. "Classical" parabolic evolution problems are, by contrast, local and do not produce patterns. The concept of "pattern" I am using is not well defined, but includes solutions which exhibit geometrical structures which may be evolving in time.

Generally our context will be dissipative systems and gradient flows, so that in most cases, there is the concept of energy or free energy which decreases on solution paths.

Another recurring theme in this chapter is that of model reduction achieved through singular limits. Beginning with a system of parabolic PDE's, for example, it has become quite commonplace to let a parameter pass to some limit, or at least assume very large or very small values, thus obtaining a new type of problem: either one with nonlocal effects (an example being a "shadow" system) or a free boundary problem. Much of the important modern rigorous work in nonlinear PDE's has focussed on making such transitions from one kind of evolution to another rigorous.

Lifschitz-Slyosov-Wagner (LSW) theory, particularly its more recent extensions (described in Section 3.4.2), represents an excellent example of compound model reductions. The beginning free boundary problem on which the theory is based can be viewed as a model reduction from the Cahn-Hilliard PDE. The former may be further reduced, within the ripening scenario, to a system of evolution equations for the positions, shapes, and radii of a large number of particles. Under certain parameter assumptions, this system is reduced to ODE's for the radii alone. Finally, this system gives rise to an evolution equation for the size distribution function of the particles in the system.

As I mentioned before, another common thread here is that of gradient flows. In the context of flows on manifolds with general metrics, see e.g. the explanation in [93]. The gradient flow framework is often considered to be a natural one when modeling processes in materials science (see the discussion below in Section 3.2.4), and much of what we say will be in that context.

By necessity, this chapter can only be a partial survey of the topics I have set out to discuss, and many important results, and even large areas, will be omitted. For example I will not cover the huge field of pattern-forming reaction-diffusion systems, except to refer the reader to survey works [97, 90, 42, 101].

3.2 The Simplest Nonlocal Parabolic-like Evolution and its Relatives

Since our topic is parabolic-like nonlinear evolution processes, let us begin by looking for generalizations of the classical semilinear heat equation

$$u_t = \Delta u - f(u). \tag{3.1}$$

Focussing on nonlocal extensions, we first ask what is the most "natural" nonlocal counterpart of the "diffusion" term Δu? One way to motivate our answer is to characterize its negative $-\Delta u$ as the (L_2) functional derivative of the Dirichlet integral

$$\mathcal{E}_0[u] = \int_\Omega \frac{1}{2}|\nabla u|^2 dx. \tag{3.2}$$

(In this introductory section, I don't want to dwell on the spatial domain Ω of the function u or boundary conditions; suffice it to say that $\Omega = \mathbf{R}^n$ or else appropriate boundary conditions are to be imposed as part of the domain of definition of $\mathcal{E}_0$, $\mathcal{E}_{10}$, etc.) In fact, for any "nice enough" function $u(x,t)$,

$$\frac{d}{dt}\mathcal{E}_0[u(\cdot,t)] = \langle -\Delta u, u_t \rangle, \tag{3.3}$$

the scalar product on the right being in $L_2(\Omega)$. Thus

$$-\Delta u = \frac{\delta \mathcal{E}_0}{\delta u}. \tag{3.4}$$

3 Some Nonclassical Trends in Parabolic and Parabolic-like Evolutions

The energy $\mathcal{E}_0$ is a certain measure of how much u deviates from being constant. We have $\mathcal{E}_0[u] \geq 0$ and $\mathcal{E}_0[u] = 0$ only when $u = \text{const}$.

A very natural nonlocal energy functional which can act as an alternative measure of the same thing is

$$\mathcal{E}_{10}[u] = \int \int \frac{1}{4} J(x-y)(u(x) - u(y))^2 dx\, dy, \qquad (3.5)$$

for some nonnegative influence function $J(x)$. We take $J(x)$ to be even and to have unit integral: $\int J(x) dx = 1$. Again, $\mathcal{E}_{10}[u] \geq 0$ and $\mathcal{E}_{10}[u] = 0$ only when $u = \text{const}$. The L_2 functional derivative in this case is

$$\frac{\delta \mathcal{E}_{10}[u]}{\delta u}(x) = u(x) - \int J(x-y) u(y) dy = u(x) - J * u(x). \qquad (3.6)$$

In the sense that $\mathcal{E}_{10}$ is analogous to $\mathcal{E}_0$, the expression $J*u - u$ is analogous to Δu in (3.1). Going a bit further, we incorporate the nonlinear term $f(u)$ by defining

$$\mathcal{E}[u] = \mathcal{E}_0[u] + \int_\Omega W(u) dx, \qquad (3.7)$$

where $W'(u) = f(u)$, and

$$\mathcal{E}_1[u] = \mathcal{E}_{10}[u] + \int_\Omega W(u) dx. \qquad (3.8)$$

Then (3.1) becomes

$$u_t = -\frac{\delta \mathcal{E}}{\delta u}, \qquad (3.9)$$

and the corresponding nonlocal version $u_t = -\frac{\delta \mathcal{E}_1}{\delta u}$ is

$$u_t = J * u - u - f(u). \qquad (3.10)$$

The expression $\frac{\delta \mathcal{E}}{\delta u}$ is the L_2 gradient of the functional $\mathcal{E}[u]$, and the evolution laws (3.1), (3.10) and in general (3.9) are steepest descent laws for the respective energies. The "rate of descent" for functions satisfying (3.9) is given by

$$\frac{d}{dt} \mathcal{E}[u] = \left\langle \frac{\delta \mathcal{E}}{\delta u}, u_t \right\rangle = -\left\| \frac{\delta \mathcal{E}}{\delta u} \right\|^2. \qquad (3.11)$$

Another interpretation of (3.9) and its analog for $\mathcal{E}_1$ suggests itself: on the right side, the quantity $-\frac{\delta \mathcal{E}}{\delta u}$, being the negative of the rate at which $\mathcal{E}$ changes when subjected to the variation of u, can be considered a "force", in analogy with potential force fields. The left side of (3.9) is then the resulting "response" or "flux", in the sense of a movement or rate of change responding to a force. Thus (3.9) could represent some postulated linear force/response relation.

Finally, in modeling the dispersal of organisms in space when u is their density, the expression $J * u - u$ can represent transport due to long-range dispersal mechanisms. In fact let $J(r)$ (an even function with $\int J(r)dr = 1$) be the probability distribution of rates of dispersal over distance r. Then $J * u(x)$ is the rate at which organisms are arriving at position x from all other locations, and $-u(x) = -\int J(x-y)u(x)dy$ is the rate at which they are leaving location x to travel to all other sites. Therefore the expression $J * u - u$ on the right of (3.10) is the net rate of increase due to dispersal. Note that there exists a conservation principle, since $\int (J*u(x) - u(x))dx = 0$.

If $W(u)$ has exactly two local minima, then $\mathcal{E}$ is a particular case in the class of Ginzburg-Landau functionals (also used by van der Waals [124]), and (3.1) is a Ginzburg-Landau equation. It is also called the Allen-Cahn equation, especially when W takes the same value at its two minima.

A quite different nonlocal generalization of the Allen-Cahn equation is

$$u_t = \Delta u - f(u) + \overline{f(u)}, \qquad (3.12)$$

where the expression $\overline{f(u)}$ is the spatial average of $f(u)$; it may depend on time. As opposed to (3.1) and (3.10), this equation ensures that the mass $\int u\, dx$ is conserved during the evolution. It has been studied quite a bit; see e.g. [24].

The equation (3.1) has been researched intensively for a long time. Of special interest are propagative phenomena (traveling waves, interfacial motion) in the Ginzburg-Landau case, and that is what we will focus on. Some of the results will be cited in the rest of this chapter. Propagative phenomena for (3.10) have also been investigated.

It is worth summarizing the common features and differences between the evolution laws (3.1) and (3.10), and that will now be done.

3.2.1 Comparison Between the Local and Nonlocal Equations

The initial-value problem associated with (3.10) has some important properties to share with that for (3.1). Under certain natural assumptions on the functions J and W (such as $uW'(u) > 0$ for large $|u|$), it has a global solution for $t \geq 0$ for L_2 initial data.

Moreover, a maximum principle holds. If u_1 and u_2 are solutions for $t \geq 0$ with $u_1(x,0) \geq u_2(x,0)$ and $u_2(x,t) - u_1(x,t) \to 0$ as $|x| \to \infty$, then $u_1(x,t) \geq u_2(x,t)$ for all $t \geq 0$. When $W \equiv 0$, global existence for negative time also holds, but of course a priori estimates which hold globally in negative time are not to be expected.

On the other hand, the solution operator for (3.10) corresponding to this initial value problem is not a smoothing operator. For example if the initial function $u(x,0)$ is continuous except for a jump discontinuity at $x - 0$, the same will be true of the solution at all later times.

We here assume that the nonlinear term $f(u)$ is of bistable (Ginzburg-Landau) type. This means that $f(u) = W'(u)$ for some function W which

3 Some Nonclassical Trends in Parabolic and Parabolic-like Evolutions   157

has exactly two local minima, which are nondegenerate. For convenience, we place them at $u = \pm 1$. In this case, propagative phenomena represent the most important features of the two evolution problems (3.1) and (3.10). Those problems have the following common features:

- Both equations have $L_\infty-$ stable constant solutions equal to either of the two values $u = \pm 1$.
- There are unique traveling front solutions $u(x,t) = U(x - ct) = U(z)$ connecting a given one of the two values $U = \pm 1$ at $z = -\infty$ to the other one at $z = \infty$. They enjoy some stability properties [59, 18, 38].
- Motion by curvature holds approximately when the two minima of W are equal: $W(-1) = W(1)$. More specifically, when space is rescaled with a small parameter ϵ by $x = \tilde{x}/\epsilon$ and time is also rescaled to accelerate motion, solutions exist with thin (in $\tilde{x}$) interfacial regions whose thickness is of the order ϵ, and which migrate with a normal velocity approximately proportional to their mean curvature. [9, 91, 61, 118, 126]

The following are some features which differ between the two evolutions.

- Discontinuities in the initial data are retained in one case, but smoothed out in the other.
- The spatial decay rates of the traveling waves as $z \to \pm \infty$ differ in the two cases [126].

3.2.2 Models from Statistical Mechanics

Statistical mechanics has provided a good source of nonlocal evolution models. First, I point out that the relationship between the simple equation (3.10) and lattice models has been explored in [15]. We pass now to some relatives of the simplest nonlocal equation (3.10).

Dynamical spin models have been developed, whose starting points were the classical Ising model with (possibly) an imposed magnetic field h, based on accounting for the spin at points of a regular lattice. Rules for the rate of change of the spins have been proposed. An important one constitutes the Glauber dynamics model, a natural one when the total number of locations with given spin is not conserved.

Mesoscopic formations representing traveling fronts have been investigated in such dynamic lattice models in [79, 80, 81].

Passing to a certain limit [46] in a discrete Glauber system produces the analogous continuous system governed by the nonlinear nonlocal scalar evolution equation

$$\frac{\partial m}{\partial t} = -m + \tanh\{\beta J * m + h\}, \tag{3.13}$$

where m is a coarse-grained magnetization, h (as mentioned before) is the imposed magnetic field strength, β is the inverse temperature, and the kernel

J is like that in (3.5). It turns out that an energy similar to (3.8) is a Lyapunov functional for this evolution.

The theory of traveling waves for (3.13) was given in [44] and [105], and the motion by curvature law when $h = 0$ under rescaling of time was proved in [45].

3.2.3 Related Nonlocal Evolutions

For a more general context see [118] and [14]. In these papers the focus is on the global existence and definition of weak solutions of moving interface problems; they show that existence follows, in this class, from the local existence of classical solutions. Moreover this provides the possibility of global validation of singular limit reductions.

There are nonlocal parabolic-like evolution equations in the theory of neural networks; for a mathematical treatment, especially of travelling waves, we merely mention the paper [48] and the references found in it.

The theory in [18], as well as previous theories of traveling waves for the bistable equation (3.1), the most recent being that of Shen [115], was generalized by F. Chen [36] to a wide class of nonlocal nonlinear evolution problems

$$u_t = Du_{xx} + g(u, J_1 * s^1(u), \ldots, J_n * s^n(u), t) = 0. \tag{3.14}$$

Here the nonnegative kernels J_i are like the one in (3.10). The coefficient $D \geq 0$, so that the diffusion term Du_{xx} may or may not be present. The functions $s^i(u)$ are monotone increasing. The nonlinear function g has a special bistable feature: there exist exactly 3 almost periodic solutions $u^\pm(t), u^0(t)$ of the ODE

$$\frac{du}{dt} = G(u,t) \equiv g(u, s^1(u), \ldots, s^n(u), t) \tag{3.15}$$

whose ranges are bounded away from one another, the outer two ($u = u^\pm$) being stable. It is also assumed that $G_u < 0$ at $u = u^\pm$ (this condition is not needed if $D > 0$).

Consider first the case when g does not depend explicitly on t. Then the existence, uniqueness, and stability of traveling waves was proved in [36]. When almost periodic dependence of g on t does occur, then quite analogous results also hold. Since strict traveling waves generally do not exist in that case, the analog has to be defined. The proofs are an adaptation to this equation of proofs of similar results for (3.1) with almost periodic t-dependence in the function f by W. Shen [115].

3.2.4 Digression on the Role of Gradient Flows in Modeling

Since the simple equation (3.10) was motivated as a gradient flow, and that concept has considerable bearing on the following material as well, I shall

3 Some Nonclassical Trends in Parabolic and Parabolic-like Evolutions

comment now on the relevance of modeling with gradient flows in materials science and some other areas of applied mathematics.

First, I refer the reader to the lucid discussions of this concept in the context of flows on a manifold of function space with general metric in [104, 93]. It turns out that the Wasserstein metric on spaces of probability distributions is a convenient one to use in many cases [103, 104, 96, 66]. There is also the deep thermodynamic theory of phase-field models in [10, 11, 12].

Besides the relevance to modeling, it is clear from many examples that the gradient nature of evolutions may introduce considerable mathematical advantage in the analysis of those models; for a recent example, see [93].

Let me begin by giving a brief background perspective on phase-field models, about which there is now an enormous literature.

These models, as applied to phase transitions in materials, are characterized by one or more "phase functions" which take on a continuum of values. These functions are intended to be rough microscopic descriptors of the state of the material relevant to its transition between phases. The models usually involve other descriptor variables as well.

The dynamics of the phase functions ϕ is often given in terms of a free energy functional $\mathcal{F}[\phi]$ (I am suppressing dependence on the other variables). A linear force/response relation

$$\phi_t = -C \frac{\delta \mathcal{F}}{\delta \phi} \qquad (3.16)$$

with some positive coefficient C (possibly a function) is postulated (again, we disregard the effects on this of the other variables). In addition to choosing the functional $\mathcal{F}$, the type of functional derivative $\frac{\delta \mathcal{F}}{\delta \phi}$ must be selected.

In any case, the latter is a measure of the rate at which $\mathcal{F}$ can be changed by changing ϕ. As such, its negative can be considered a "force", in analogy to the relation between force and potential in classical mechanics (as was brought out in Section 3.2). The response ϕ_t may be considered a "flux", in the generalized sense of a motion or change which occurs in response to a force. The relation (3.16) is called a gradient flow law associated with the functional $\mathcal{F}$.

A precise definition of the phase function ϕ and a detailed knowledge of the physical microscopic change mechanisms are typically lacking. Without this specific knowledge, general models within this framework have been studied. The reasonableness of any particular model would depend, among other things, on whether it predicts known results and whether it leads to sensible mathematics.

Some of the recent papers using this approach to modeling in physics and materials science are [108, 75, 98, 102, 30, 56, 127, 128, 34, 35].

3.2.5 The Issue of Discontinuous Profiles in the Nonlocal Problem

It was mentioned before, and is quite clear, that solutions of (3.10) preserve the locations of spatial jump discontinuities. Moreover, it was shown in [18] for one dimension and [16] for higher dimensions that stationary solutions with discontinuities can exist. In two dimensions, for example, stationary solutions with discontinuities on rather arbitrary curves in space are possible. Nevertheless, if a mechanism were introduced allowing those locations to migrate (this cannot be done by the evolution law (3.10)), then the free energy (3.8) would be thereby decreased. In this sense that law, despite its being of gradient type, is not an altogether efficient way to decrease $\mathcal{E}_1$.

One way to make it more efficient is to alter the choice of gradient flow for $\mathcal{E}_1$ [54]. In fact this entails a slight alteration also of the domain of $\mathcal{E}_1$. It is most clearly explained in the case when the solution has but one point of discontinuity. We label that point $\xi(t)$ and allow it now to change in time.

The new functional $\mathcal{E}_1$, while retaining the exact formal appearance in (3.8), is now to be defined for all pairs (u, ξ), where $\xi \in \mathbf{R}$ and $u(x)$ is a function which is continuous except at $x = \xi$, uniformly continuous on $(-\infty, \xi)$ as well as on (ξ, ∞).

The new gradient flow takes the form $\frac{\partial}{\partial t}(u, \xi) = -C \frac{\delta \mathcal{E}_1}{\delta(u, \xi)}$. The initial value problem for this evolution turns out to be ill-posed, in that it has many solutions. The nonuniqueness can be cured by altering the definition of $\mathcal{E}_1$. If an extra term is adjoined representing free energy concentrated at the discontinuity ξ, it was shown in [54] that the initial value problem becomes well posed.

3.3 The Simplest Pattern-Forming Parabolic Equation

We go now to our second major theme: the pattern-forming capabilities of nonlinear parabolic equations.

3.3.1 Overview

The Cahn-Hilliard equation

$$u_t = -\Delta(\Delta u - f(u)), \tag{3.17}$$

is the simplest conserved gradient flow [53] for the Ginzburg-Landau energy function given already in (3.7):

$$\mathcal{E} = \int \left[\frac{1}{2}|\nabla u|^2 + W(u)\right] dx. \tag{3.18}$$

Specifically, in some sense it (with proper boundary conditions) is the simplest local gradient flow on which the mass $\int u \, dx$ is conserved.

3 Some Nonclassical Trends in Parabolic and Parabolic-like Evolutions

A popular nonlocal alternative constrained gradient flow for (3.18) is the "nonlocal Allen-Cahn equation" (3.12). This alternative does not have the pattern-forming capabilities of (3.17).

The equation (3.17) was proposed in [29, 31] as a model for some important features of solidifying molten binary alloys, such as spinodal decomposition. One interprets u as the relative concentration of one component of the alloy. The equation (3.17) has undergone extensive mathematical investigation. For the initial value problem for (3.17), global existence theories are available; for example this equation is included in a class treated by Henry [71].

When looking for the typical features of solutions, one natural beginning has been to examine the global minimizers of (3.17) constrained by a given mass, say $\bar{u} = b$, where $\bar{u}$ is the constant spatial average of u. However, those global minimizers in one space dimension do not always provide interesting patterns. Nor are these patterns the most relevant in typical applications. The reason is that other temporary but long-lasting (metastable) patterned solutions may result, due to a competition between two important physical events.

To examine this competition, we fix f to be any bistable function such as $u^3 - u$ with stable zeros at $u = \pm 1$. In this section consider the problem in one space dimension with the spatial domain of u being $[0, L]$, where L is large. Then the minimization of the last term of $\mathcal{E}$, i.e. $\int_0^1 W(u)dx$, is achieved when u is ± 1. To achieve this minimum, one must therefore require the function u to take the value 1 on one set in $[0, L]$ and -1 in the remainder of that domain. This is often called "phase separation". Since the total mass $\int u\, dx$ is prescribed, the measures of these two sets are chosen to satisfy this constraint.

However, we must also consider the additional term $\frac{1}{2}|\nabla u|^2 dx$ in the energy (3.18). It should be as small as possible; this dictates that the transitions between ± 1 should be smooth rather than abrupt, and that there should be as few of them as possible. In other words, the trend to minimize the energy drives the configuration toward longer wavelengths (but they must all fit into the given interval (0,L)). This is the first of our new pair of competing influences.

The second of these influences involves the kinetics of diffusion. It arises because to effect this separation with mass conserved, a redistribution of the material particles whose concentration is u will be necessary. This must be accomplished by diffusion (diffusion is provided for by the fourth order term of (3.17) except when u is near ± 1; then the diffusion is governed by the other two terms on the right). Diffusion takes time, and the time required is longest when the sizes of the regions where $u \approx 1$ or $u \approx -1$ are largest. The competition is therefore between (1) the drive to minimize energy, coupled with the fact that the energy is least when the length scale of the separation is largest, and (2) the fact that the time it takes to do that increases with this same length scale.

This competition results in a compromise: there is a preferred range of characteristic lengths for the patterning mechanism which is neither too small nor too large. Here the terminology "preferred" takes the following dynamical meaning. Suppose that the initial state $u(x, 0)$ is a small random perturbation of the constant solution $u \equiv \mu$ and that constant state is unstable. Then a whole range of patterned modes, whose amplitudes start small, will grow. Each mode has its own characteristic spacing. Those modes which have the fastest rates of growth will dictate the characteristic length scale of the emerging pattern. This is the length scale which results from the competition described above. As will be explained below, this emerging pattern does not persist for all time; new structures with larger space scales develop afterwards on longer time scales.

The modes of growth from those almost constant initial data can be determined from a linearized stability analysis around the constant solution $u \equiv \mu$ (see e.g. [69, 17]). This linear analysis provides a dispersion relation between the growth rate of each mode and the wavenumber k characterizing that mode.

The first rigorous analysis of this part of the separation process was given by Grant [69] for the 1D case. In this stage, we may take "separation" to be synonymous with "patterning". He proved roughly that if the initial state is a sufficiently small random perturbation of the constant state, then with high probability the evolution drives the system to a neighborhood of a stationary periodic exact solution with wavelength predicted by the linear dispersion relation. Such periodic solutions are unstable for λ large enough, and the solution will not remain in that neighborhood forever.

This temporary patterning process is the mathematical analog of spinodal decomposition (e.g. [29]), an event common in the processing of alloys. The subsequent evolution of the system away from the unstable separated solution takes place more slowly. It is also the mathematical analog, in this model, of an important physical process, namely the coarsening of alloys. An extensive study involving both the modeling and numerical simulation of the statistics of coarsening in 1D was undertaken by Eyre [50].

Modeling more complicated alloys leads to systems of Cahn-Hilliard equations, which have even more interesting separation dynamics. Among the notable mathematical studies of these models in the initial phase separation stage, I mention the work of Eyre in [51].

Next, we review some mathematical results on separation in higher dimensions.

3.3.2 Spinodal Decomposition in Higher Dimensions

The process by which patterns are formed spontaneously under Cahn-Hilliard dynamics is considerably more involved in dimensions greater than 1. One

3 Some Nonclassical Trends in Parabolic and Parabolic-like Evolutions

reason is that although in 1D the patterns which are observed have maxima and minima and zeros which are approximately evenly spaced (so that the spacing is approximately equal to a number which we may call the characteristic length of the pattern), no such property takes place in 2 or 3 (say) dimensions. The zero set of a solution does not consist of approximately evenly spaced lines, which might be the analog of the 1D situation. Instead, they typically are irregular and curved. Yet they do appear to the eye to possess some characteristic length.

To put the phenomenon of spinodal decomposition on a more satisfactory mathematical footing in higher dimensions, therefore, one needs to provide a notion of characteristic length for a given function and somehow to quantify the assertion that most solutions starting near an unstable constant evolve temporarily to a configuration with an identifiable characteristic length.

We shall consider the Cahn-Hilliard equation (3.17) with a small parameter ϵ inserted as follows:

$$u_t = -\Delta(\epsilon^2 \Delta u - f(u)), \tag{3.19}$$

Where f is some bistable cubic function as before, or more generally a function with the same qualitative properties. Then the characteristic length referred to turns out to be $O(\epsilon)$.

The papers by Maier-Paape, Sanders and Wanner; also Blömker when some stochastic effects are included [88, 89, 112, 113, 21] serve to quantify the intuitive picture we have been describing and to prove some form of it. They consider solutions of the Cahn-Hilliard equation which have evolved from a small neighborhood of an unstable constant solution μ, but which have not evolved too far from it. Thus these solutions start in a neighborhood N_ϵ^0 of it in H^2, and are considered before they leave a larger neighborhood N_ϵ. In 3-dimensional space, both of these neighborhoods have size of order ϵ^3. In other dimensions, this exponent 3 here and below should be replaced by the dimension of the space.

To develop a reasonable assertion, one first considers the eigenvalues of the Cahn-Hilliard operator linearized about μ, and collects in a group those which are close to the maximal one, which is positive. The number of these eigenvalues is strongly ϵ-dependent and will be very large when ϵ is small. They generate a strongly unstable finite dimensional invariant manifold M_ϵ for the flow, which is imbedded within another finite dimensional inertial manifold M'_ϵ with much larger dimension. The crucial statements are that solutions starting in N_ϵ^0 are exponentially attracted to M'_ϵ, and that most of them exit N_ϵ being close to M_ϵ. The meaning of "close" and "most" is this: one can choose an arbitrarily small number δ_ϵ proportional to ϵ^3 and a probability p arbitrarily close to 1 and conclude that if the diameter of N_ϵ^0 is taken small enough relative to ϵ^3 and initial conditions are taken with projections onto the tangent space of M'_ϵ in a random manner, then the probability of the orbit exiting N_ϵ at a point closer than δ_ϵ to M_ϵ is at least p.

The second conceptual result is that those orbits which are close to M_ϵ have a characteristic length which is $O(\epsilon)$. This idea needs to be clarified, and the authors do so. They succeed in proving that with high probability, solutions exhibit a mathematical phenomenon like spinodal decomposition, a type of phase separation, by the time they leave a ball of radius $O(\epsilon^3)$ in H^2 centered at μ, provided that they begin randomly in a smaller neighborhood with radius of the same order of magnitude. Numerical simulations (of which there are many; see for example [47]) show that separation with characteristic length persists outside of that small neighborhood, when in fact the amplitude of the separated solution is $O(1)$.

Extensive Monte-Carlo simulations in 1D were performed by Sander and Wanner [112] in order to gain insight into the question of how long the decomposition persists. They made a systematic study of the distribution of exit points on the boundaries of balls V_ϵ of various radii centered at μ. Their observations indeed confirm the persistence of the separation mechanism beyond the confines of the small neighborhood figuring in the proofs in [88, 89]. In fact they indicate that the solutions of the Cahn-Hilliard equation remain close to the solutions of the equation linearized about the constant $u = \mu$ much longer than one would expect—far into the nonlinear regime. This is important since the solutions of the linearized problem are built up from the various linear modes growing according to their known growth rates, and it is easily seen that the fastest growing ones will eventually dominate. As mentioned above, this domination is another way of expressing the spinodal decomposition event.

The unexpected validity of the linear approximation for solutions growing to larger amplitudes can be given a mathematical justification. This was done in [112, 113].

3.4 Layer Phenomena Related to the Cahn-Hilliard Equation

3.4.1 The Slowness of Some Motions

Phenomena in 1D As in the preceding section, we use $f(u) = -u + u^3$, or any similar function $f = W'$ with the two minima of W at equal levels. It was shown by Fusco, Hale, Carr and Pego [65, 32] that the Allen-Cahn-Ginzburg-Landau equation (3.1) in one dimension with a small parameter ϵ^2 appearing as coefficient of the second derivative and Neumann boundary conditions has solutions with internal layers effecting transitions between the two stable zeros of f. The locations of these layers move at an exponentially slow rate $O(\exp(-\gamma/\epsilon))$, $\gamma > 0$ depending on the location.

Slow motion results in higher dimensions were proven in [26, 70, 3] and other places.

3 Some Nonclassical Trends in Parabolic and Parabolic-like Evolutions

Similar results hold for the Cahn-Hilliard equation (3.19) in one space dimension:

$$u_t = -(\epsilon^2 u_{xx} - f(u))_{xx}, \quad x \in (0,1); \quad u_x = u_{xxx} = 0 \text{ for } x = 0, 1. \quad (3.20)$$

Because this equation conserves "mass" $\int_0^1 u(x,t)\,dx$, the dynamics of the layers is different, and in fact when there is only one of them, it does not move.

The extension of the results of [65, 32] to analogous results for this equation in the case of two layers was done in [2]. Later, this was done for more layers in [19, 20]. The careful construction of approximate unstable manifolds was a key feature of the proof. Some of the spectral estimates used the results of [17].

It should be mentioned that analogous slow motion phenomena appear also in very different contexts; see for example [82].

Bubbles and such Consider now solutions of (3.19) in higher dimensions. The analog of the layers discussed above are curves (in 2D) or surfaces which are either closed or intersect the boundary at a right angle. Again, these curves separate domains in which u takes values near one of the two stable zeros of f. When there is only one such curve and it is very close to being circular, the motion is again exponentially slow.

The study of solutions structured in this manner relies, as before, on careful estimates of the spectrum of the operator linearized about a layered function [4, 37].

In [6], the exponentially slow motion of layered solutions which approximate a single bubble was proved in higher dimensions, with the aid of results in [5]. (If more than one bubble are present with different diameters, the motion speeds up; the larger one will grow while the other decreases, at a rate faster than exponentially slow. Thus motion with more than one bubble is inherently unstable.) One of the new features of this paper is the development of a method for determining the equilibrium positions of the bubbles. The authors use what they term the "manifold approach".

3.4.2 Reduction to the Mullins-Sekerka Problem

More generally, there are solutions of (3.19) which exhibit internal mobile transition layers of arbitrary shape. (Finding such layer phenomena has been a great industry also in phase-field models; see e.g. [52, 27, 28, 60].) This possibility was first explored for (3.19) by Pego [106], who considered singular limit problems taking the form of free boundary problems obtained formally for small ϵ. One of these, obtained when time is scaled so that (3.19) becomes

$$u_t = -\Delta(\epsilon \Delta u - \frac{1}{\epsilon} f(u)), \quad (3.21)$$

is called the Mullins-Sekerka problem. The free boundary $\Gamma(t)$ is the locus of the internal layer. We give that free boundary problem in the case that $\Gamma(t)$ is a bounded connected closed surface enclosing a domain $\Omega_+ \subset \mathbf{R}^3$. We call the exterior domain Ω_-. We suppose that $u = \pm 1$ are the two stable zeros of f. If the spatial setting is the entire space $\mathbf{R}^3$, say, then the problem is the following: Find an evolving closed surface $\Gamma(t)$ and a function (chemical potential) $w(x,t)$ such that

$$\begin{aligned} \Delta w &= 0 \text{ in } \mathbf{R}^3\backslash\Gamma, \\ w &= \tfrac{1}{2}\alpha\kappa \text{ on } \Gamma, \\ 2v &= [\partial_n w] \text{ on } \Gamma, \\ w &\to 0 \text{ as } x \to \infty. \end{aligned} \qquad (3.22)$$

Here v is the normal velocity of Γ, κ is twice its mean curvature, $\alpha = \int_0^1 \sqrt{2F(u)}du$, $F'(u) = f(u)$, $F(0) = F(1) = 0$, and the brackets on the right represent the jump in the normal derivative of w across Γ. The sign convention is that velocity is considered positive if motion is toward Ω_+, and $\kappa > 0$ if the centers of curvature are in Ω_+. Also ∂_n is the directional derivative in the direction toward Ω_+, and the jump means the value on the Ω_+ side minus that on the Ω_- side.

This free boundary problem (3.22) arises in a number of other connections. But in the context of a singular limit problem for (3.19), we may identify $w = \tfrac{1}{\epsilon}f(u) - \epsilon\Delta u$. Also in that context, we are to suppose that $u = \pm 1$ in $\Omega_\pm$. We shall call the domain Ω_+ a "particle".

The relation between these two problems, (3.19) and (3.22), was investigated rigorously by Alikakos, Bates, and Chen [1, 39]. The gradient nature of both of them has been discussed in e.g. [53, 93].

A fascinating extension of the reduction described in this section has been applied to phase separation by layers in the context of materials with at least three phases. The intersection of layers, forming "triple junctions" can occur. See e.g. [25].

3.4.3 Further Reductions: Ripening

The form (3.22) of the Mullins-Sekerka problem given above represents a single "particle" Ω_+ of one phase where $u = 1$, surrounded by a sea (Ω_-) of the other phase, $u = -1$. But there could be more particles; then Γ would consist of the union of many such closed surfaces, pairwise disjoint. This is a case of special interest and practical importance. When the particle density is small, it models Ostwald ripening [122, 123]. This latter is the process by which a large system of particles in one phase, embedded in a matrix of the other phase, evolves in such a way that the mean size distribution of the particles increases in time, the total volume of the particles being approximately constant.

The modeling of this process was initially done by Lifschitz and Slyozov [86] and independently by Wagner [125]. Their results are commonly called the LSW theory. The assumption was that each particle is spherical, that the particles are far apart, and that their centers do not move. The restriction of (3.22), or rather its appropriate analog, to spherical particles then yielded a system of ODE's in time for the radii. There is a nonlocal element in this reduction, since the system of equations involves a quantity $\overline{w}$, the mean value of the chemical potential w in the region outside the particles.

This system was further reduced to an evolution equation for the size distribution of the particle system. The authors then exhibited a similarity solution of this evolution equation, which they stated would be the limiting distribution for reasonable initial data. That similarity solution predicts that the mean size of the particles in the system grows like $Ct^{1/3}$ for large time t, C being a constant.

The LSW reduction has recently been justified mathematically in different senses by Niethammer [92] and by Alikakos and Fusco [7, 8]. In this scenario there are actually 5 characteristic lengths which are of interest [93], at least for an infinite system: the typical particle diameter r, the typical deviation δ of a particle from being exactly spherical, the typical distance d between nearest particles, the correlation distance ξ_{cor} (characterized as the minimal size of a box big enough that the particle distribution inside that box is typical of the distribution in all larger boxes), and the screening distance $\xi_{scr} \approx \left(\frac{d^3}{r}\right)^{1/2}$ (characterized as the maximal size of a box small enough that the particle distribution inside that box is dominated by the boundary conditions generated by particles outside the box). The meaning of r and δ is that the boundary of say the i-th particle can be expressed in parametric form in a canonical manner for some $\boldsymbol{x}_{i0}$ as $\boldsymbol{x}_i(\boldsymbol{\sigma}) = \boldsymbol{x}_{i0} + r_i \boldsymbol{\sigma}(1+\rho_i(\boldsymbol{\sigma})), \boldsymbol{\sigma} \in S^2$. Then r is a typical value of the r_i and δ a typical oscillation $r_i[\max \rho_i(\boldsymbol{\sigma}) - \min \rho_i(\boldsymbol{\sigma})]$.

As explained in [93], the classical derivations of the LSW ripening theory and the rigorous derivation in [92] assumed $\delta = 0$ (or rather that $(3.22)_2$, $(3.22)_3$ are understood in an averaged sense), $r \ll d$ and $\xi_{cor} \ll \xi_{scr}$. In [93], this last relation was curtailed, it being assumed only that $\xi_{cor} \approx \xi_{scr}$. The classical evolution problem for the size distribution is then no longer necessarily valid. Rather, it is replaced by an evolution for the joint distribution of $(r, \boldsymbol{x})$, the radii r and locations $\boldsymbol{x}$ of the particles. Finally, in [7, 8], if $\epsilon = \frac{r}{d}$, it is assumed that $\epsilon \ll 1$, $\frac{\delta}{r} = O(\epsilon)$ (initially), and that the particles are finite in number, lying in a bounded domain, so that the concepts ξ_{cor} and ξ_{scr} do not apply, although the size of the domain does. A system of ODE's for the time evolution of all of the functions $r_i(t)$, $\boldsymbol{x}_i(t)$, $\rho_i(\boldsymbol{\sigma},t)$ is derived, and it is shown to be the classical system for the $r_i(t)$ plus error terms which can be estimated rigorously. Global estimates show that the solutions are close to the classical solutions.

About the similarity solutions, whose supposed roles are limiting states attained after a long time, it was shown in [94] that the set of possible limiting states is not unique (unlike the implication in [86]), but rather depends sensitively on initial conditions; specifically on the nature of those conditions near the boundary of their support. In [33], similar conclusions were obtained for a simplified model. In [96], the authors provided a rigorous treatment of the Lifschitz-Slyosov initial-value problem, and in [95], results similar to [92] were proved in a modified model: the constraint was that the total "mass" was constant, rather then just the total volume of the particles.

3.5 Patterning Due to Competition in General Gradient Systems

In Section 3.3.1 we discussed how the initial spinodal decomposition patterns exhibited by solutions of the Cahn-Hilliard equation may result from a competition between the driving force of the energy functional and the kinetics of diffusion. In other evolution processes, patterns may result from a different kind of competition: that between stabilizing and destabilizing effects (see Section 3.8 for still another kind).

The way this latter phenomenon works in many cases is perhaps best understood by viewing it in an abstract setting [58, 55]. So doing will bring us back once again to nonlocal models.

3.5.1 An Abstract Setting

So consider the parabolic-like evolution

$$u_t = Au - \rho Bu - f(u), \qquad (3.23)$$

in the Hilbert space $L_2(\lambda)$ consisting of λ-periodic functions of the space variable $x \in \mathbf{R}$, where A and B are negative (not necessarily differential) self-adjoint operators, and f is a function such that $uf(u)$ grows rapidly enough as $|u| \to \infty$.

For our purposes, we shall define a "patterned solution of (3.23)" as a stationary stable nonconstant solution.

At this initial stage, we shall understand u to be complex-valued, and take f to be of the form

$$f(u) = 2uF'(|u|^2). \qquad (3.24)$$

The function $F(w)$ is a real function of a real variable $w \geq 0$ with a minimum of 0 attained at some point $w_0 = u_0^2$. It is assumed to grow faster than quadratically as $w \to \infty$. Since $A, B < 0$, the first term on the right of (3.23) represents a stabilizing influence in the evolution process, and the second a destabilizing influence. The role of the last (nonlinear) term will be to limit the growth of patterned solutions, whose amplitude might otherwise increase without bound.

The real number ρ will be our control parameter; stable patterns typically arise when it surpasses a critical value. We make certain other assumptions on

A and B. Most importantly, A dominates B in a sense to be given below (**A2**), and the nullspaces of the two operators consist of the constant functions.

The existence of **stable** patterned solutions can be derived from the fact that this evolution (3.23) is the L_2–gradient flow for the energy functional

$$\mathcal{E}[u] = -\frac{1}{2}\langle Au, u\rangle + \frac{\rho}{2}\langle Bu, u\rangle + \frac{1}{\lambda}\int_0^\lambda F(|u|^2)dx. \tag{3.25}$$

Further assumptions on A and B are:

A1. Smooth symbols $\hat{A}(k)$ and $\hat{B}(k)$, defined for all real k, independent of λ, exist such that when e^{ikx} is λ-periodic, $\hat{A}(k) = e^{-ikx}A[e^{ikx}]$, and the same for $\hat{B}$. These functions are real and even.

A2.
$$\lim_{k\to\infty}\frac{\hat{A}(k)}{\hat{B}(k)} = \infty, \quad \limsup_{k\to\infty}\hat{B}(k) < 0. \tag{3.26}$$

A3. The nullspaces of A and B are the set of constant functions. It follows that $\hat{A}(k)$ and $\hat{B}(k)$ are strictly negative for $k > 0$.

Examples

Example 1
$$u_t = -u_{xxxx} - \rho u_{xx} + f(u). \tag{3.27}$$

Here $Au = -u_{xxxx}$ and $Bu = u_{xx}$.

The Swift-Hohenberg equation can be put into this form (usually with a different control parameter). The original derivation [120] of this equation was in connection with the onset of convective instabilities in fluid dynamics. It was produced by a weakly nonlinear analysis of the Boussinesq model near the critical value of the Rayleigh number at which instabilities appear.

Authors dealing with this equation are too numerous to present here with any hope of completeness; however I refer the reader to an excellent thorough treatment by Peletier and Troy [107] of the time-independent solutions. The comprehensive review article by Cross and Hohenberg [42] of pattern formation covers this and many other contexts.

Example 2
$$u_t = u_{xx} - \rho(J * u - u) + f(u). \tag{3.28}$$

Here $Au = u_{xx}$ and $Bu = J * u - u$. In this convolution operator, the kernel J is like the one in (3.5):

$$J(x) \geq 0, \quad \int J(x)\,dx = 1. \tag{3.29}$$

We have

$$\hat{A} = -k^2, \ \hat{B} = \sqrt{2\pi} \text{ times the Fourier Transform of } J. \tag{3.30}$$

We shall have a great deal to say about this example later in Section 3.6.1.

Threshold results Here are some results from [55, 58]. There is a critical number $\rho^* > 0$, which can be calculated explicitly, such that for $\rho < \rho^*$, there is no patterned global minimizer, and for $\rho > \rho^*$ there is one. These patterns are λ-periodic sinusoidal functions of the form $u = a(\rho)e^{ik(\rho)x}$.

Properties of the minimizers The following are some of the qualitative properties of the patterned solutions found this way.

(i) The functions a and k satisfy
$$\lim_{\rho \to \infty} a(\rho, \lambda) = \infty, \quad \lim_{\rho \downarrow \rho^*(\lambda)} a(\rho, \lambda) = u_0. \tag{3.31}$$

(ii) If
$$\inf_{\kappa} \hat{B}(\kappa) < \hat{B}(k) \text{ for each } k \tag{3.32}$$

(the former could be $-\infty$), then $\lim_{\rho \to \infty} k(\rho, \lambda) = \infty$.

(iii) Suppose $F'(w) < cw^r$ for some $r > 0$, $c > 0$, and all large enough w. Then for some $C > 0$,
$$a(\rho, \lambda) \geq C\rho^{1/2r} \tag{3.33}$$

for sufficiently large ρ.

Example 3
$$u_t = (J_1 * u - u) - \rho(J_2 * u - u) + f(u).$$

We assume that the kernels J_i satisfy (3.29). Depending on J_1, J_2, and ρ, we may find the existence of such a threshold value ρ^* of ρ, or a minimizing sequence of patterned states with finer and finer wavelengths, and amplitudes which approach a limit.

Restriction to real-valued functions All of the above was for minimizers in the field of complex-valued functions. Consider now the analogous situation for real-valued functions. More or less similar results can be obtained; however there are fewer details available. For example, there is a critical value ρ^* of ρ, but the minimizers are no longer sinusoidal. The value of ρ^* is no longer given precisely.

An important question for the real case is the λ-dependence of the minimizers. As $\lambda \to \infty$, do they approach a periodic solution with fixed finite period? There are some important results about this question in the case of Example 1 above; see the references given in [58, 55]. But in general this is a very difficult issue.

3.5.2 Conserved Evolutions

Recall that the Ginzburg-Landau (3.1) and Cahn-Hilliard (3.17) equations are both gradient flows for the same Ginzburg-Landau energy (3.18), but using different metrics. The form of (3.17) is a simple way to ensure that the flow conserves mass $\int u(x,t)\,dx$. In the same way, the evolution equation

$$u_t = -\Delta[Au - \rho Bu - f(u)] \tag{3.34}$$

can be proposed as a general possible pattern-forming evolution which conserved mass $m = \frac{1}{\lambda}\int_0^\lambda u\,dx$. In the case of (3.23), stable patterned solutions were found among the minimizers of $\mathcal{E}$ (3.25), and in (3.34), they will be minimizers of $\mathcal{E}$ under the constraint that m be a given value. Call them $U(x)$. They will satisfy the equation

$$AU(x) - \rho BU(x) - f(U) = \mu, \tag{3.35}$$

μ being a constant.

3.5.3 A Paradigm

Although there is no general theory, it can be expected that problems of the form (3.23) or (3.34), but in other settings, may have pattern-forming capabilities when ρ surpasses a critical value. For example, the function u may be generalized to have m components and be a function of $\boldsymbol{x}$ in a some domain $\Omega \subset \boldsymbol{R}^n$, the basic Hilbert space then being $L_2(\Omega)^m$. The operators A and B would be negative self-adjoint operators with A dominating B in an appropriate sense, and f the u-gradient of a scalar function $H(u) \geq 0$.

3.6 Ginzburg-Landau Energies with Nonlocal Additions

We consider here the Allen-Cahn or Cahn-Hilliard equation with an extra term corresponding to the term $-\rho Bu$ in (3.23) or (3.34):

$$u_t = \Delta u - f(u) - \rho Bu \tag{3.36}$$

or

$$u_t = -\Delta[\Delta u - f(u) - \rho Bu]. \tag{3.37}$$

We will be putting a small parameter ϵ at strategic places in these two equations, to see what reductions to free boundary problems they offer.

3.6.1 A Prototypical Inverse Elliptic Reduction

An example of how (3.28) might arise from a system of parabolic PDE's by a reduction process is the following system of activator-inhibitor type, for a pair of real-valued functions $u(x,t)$ and $v(x,t)$:

$$\begin{aligned} u_t &= D\Delta u - f(u) - \rho(v-u), \\ \epsilon v_t &= \Delta v + u - v. \end{aligned} \quad (3.38)$$

The parameter $\epsilon \ll 1$. If D is an $O(1)$ quantity as $\epsilon \to 0$, the parameter $\frac{1}{\epsilon}$ is a measure of the relative magnitude of the diffusivity of v, vs that of u, as well as of the strength of the v-reaction. The parameter ρ represents a measure of the strength of v's inhibition and u's activation of the first reaction.

Consider bounded solutions of (3.38) in the entire space $\mathbf{R}^3$, in the formal limit $\epsilon \to 0$. The second equation can then be solved for v in terms of u:

$$v = (-\Delta + I)^{-1} u = G * u, \quad (3.39)$$

where $G > 0$ is Green's function for the operator $(-\Delta + I)$. With this, (3.38) becomes

$$u_t = D\Delta u - \rho(G * u - u) - f(u), \quad (3.40)$$

which is an example of the equation (3.36).

Models of this sort were presented in [104, 114, 67]. Models bearing some similarity were studied in [40, 98, 87, 22].

If we wish for this to fit into the category (3.23), we should stay in 1D. The convolution then involves an integral over the whole line with the Green's function kernel $G(y) = \frac{1}{2}e^{-|y|}$, and the function u is periodic with period λ. It is easily checked that $G * u$ is then also periodic with the same period. The operator B in fact satisfies all the requirements of the theory in Section 3.5.1. We have

$$\hat{A}(k) = -Dk^2, \quad \hat{B}(k) = \hat{G}(k) - 1 = \frac{-k^2}{1+k^2} \leq 0. \quad (3.41)$$

Another example is the pair of equations

$$u_t = -\Delta(\Delta u - f(u) - \rho(v-u)) \quad (3.42)$$

and

$$0 = \Delta v + \overline{u} - u \quad (3.43)$$

in a bounded smooth domain Ω, homogeneous Neumann conditions being applied to both u and v. Here $\overline{u} = Mu = \frac{1}{|\Omega|}\int_\Omega u\,dx$ is the average of u, and M is the averaging operator. Again, (3.43) can be solved uniquely for v satisfying $\overline{v} = 0$. We call the solution $v = Bu$, where $B = (-\Delta)^{-1}(M - I)$. It can be checked that B is indeed negative, self-adjoint, and has exactly the constants as its nullspace. We therefore obtain (3.37) with this definition of B.

This is one of many possible generalizations of the Cahn-Hilliard equation. Another one with many applications is the convective CH equation [85, 68].

3.7 Free Boundary Reductions

Already in the chapter we have encountered examples of the reduction of a nonlinear evolution problem to a "singular limit" free boundary problem, the principal one being in Section 3.4.2.

We give now an example, which was shown in [67] to be significant in a number of physical contexts, of a nonlinear evolution process which can be reduced to a nonlocal free boundary problem. We shall be content with a cursory look at the results, which as yet lack proofs in the general context.

Consider the nonlocal parameter-dependent parabolic equation

$$\epsilon^2 u_t^\epsilon = \epsilon^2 \Delta u^\epsilon - f^\epsilon(u^\epsilon) - \epsilon\rho B u^\epsilon. \tag{3.44}$$

Here the operator B is a negative self-adjoint operator as in (3.36) above. The function f^ϵ has the form

$$f^\epsilon(u) = f^0(u) + \epsilon\sigma, \tag{3.45}$$

where σ is a constant, f^0 is bistable with the two stable zeros being 0 and 1, $\int_0^1 f^0(u)du = 0$, and σ is a constant. We could take σ instead to be a function of u. But this way, σ can be interpreted as a measure of the deviation of the bulk energy $W^\epsilon(u) = \int f^\epsilon(u)du$ from being "balanced", this term referring to the case when its two local minima are equal in depth. The superscript "ϵ" applied to the solution indicates that we will be considering a family of solutions as $\epsilon \to 0$. For the same reason, we denote the interface (to be constructed–the set where $u^\epsilon = \frac{1}{2}$) by $\Gamma^\epsilon(t)$. A family endowed with internal layers can be constructed formally by standard methods (e.g. [52, 60]).

The method of doing so consists in positing an "outer" expansion in formal powers of ϵ, supposedly valid in regions of Ω away from the interface $\Gamma^\epsilon(t)$, and also an "inner" expansion valid near $\Gamma^\epsilon(t)$. If $r(x, t; \epsilon)$ is the distance from x to $\Gamma^\epsilon(t)$, defined for points near the latter, the inner expansion is constructed by rewriting (3.44) in terms of the coordinate system (z, t) instead of (x, t), where $z = r/\epsilon$ is a "stretched" variable; then expanding the solution formally again in powers of ϵ.

The outer expansion yields, to lowest order in ϵ, the approximation $f^0(u^0) = 0$. This means that u^0 is equal to one of the zeros of f^0 everywhere in the outer region. We suppose that $\Gamma^\epsilon(t)$ separates Ω into two open regions $\Omega_\pm^\epsilon$. We select the outer solution so that

$$u^0(x,t) = \begin{cases} 1, & x \in \Omega_+^0(t), \\ 0, & x \in \Omega_-^0(t); \end{cases} \tag{3.46}$$

In other words,

$$u^0(x,t) = \chi_{\Omega_+^0(t)}(x,t), \tag{3.47}$$

χ denoting the characteristic function.

At this point, we make an assumption about the nonlocal operator B. We assume that it smooths out discontinuities, in the sense that when applied to a function u^0 with jump discontinuities, it yields a continuous function. In particular, $B(\chi_{\Omega_+^0})$ is a continuous function of x. This value can then be evaluated at points $x \in \Gamma^0$. Since dependence on Ω_+ is really dependence on its boundary Γ, we can write to lowest order (omitting the superscript 0)

$$B\chi_{\Omega_+} = B(\Gamma). \tag{3.48}$$

The lowest order result of the inner expansion, taking into account (3.47), is the following expression for the normal velocity V of the interface Γ. At each point $x \in \Gamma(t)$,

$$V(x,t) = \kappa(x,t) + \frac{\sigma}{\alpha} + \frac{\rho}{\alpha} B(\Gamma(t))(x,t), \tag{3.49}$$

where $\alpha = \int_0^1 \sqrt{2F(u)} du$, $F'(u) = f^0(u)$, $F(0) = F(1) = 0$. If we are in two dimensional space, κ is the curvature of the curve $\Gamma(t)$ at the point x, and in 3D it is twice the mean curvature of the surface $\Gamma(t)$. The velocity is considered positive if motion is toward Ω_+, and $\kappa > 0$ if the centers of curvature are in Ω_+.

This equation (3.49), then, is the singular limit free boundary problem corresponding to (3.44). What one expects is: (a) Given a smooth solution of (3.49) for some time interval, there is a solution u^ϵ of (3.44) for every small enough ϵ in that same time interval which approximates it in the sense that there is an abrupt spatial transition layer located near $\Gamma(t)$ causing u^ϵ to change from 0 to 1 in that layer. (b) Given a wide class of ϵ-dependent initial data, the solutions of (3.44) with those initial data play the role indicated in (a) for some solutions of (3.49).

Let me comment on a possible physical interpretation of the terms in (3.49). The first term on the right, κ, represents the effect of line tension or surface tension of Γ. It opposes the lengthening of Γ and is a stabilizing effect. It comes from the corresponding stabilizing term $\epsilon^2 \Delta u^\epsilon$ in (3.44); more precisely from the pair $\epsilon^2 \Delta u^\epsilon - f^0(u^\epsilon)$.

The second term, $\frac{\sigma}{\alpha}$, is neither stabilizing nor destabilizing. It is a systematic forcing term causing Γ to move independently of anything else. It comes from the unbalanced nature of the bistable function f^ϵ, when $\epsilon \neq 0$. In some cases it plays the role of a pressure [67].

The last term $\rho B(\Gamma)$ comes from the destabilizing term $\rho B u$ in (3.44), which is expected to be the cause of patterns. In the present context, by analogy we also expect it to be destabilizing and to cause "irregularities" to form in the solution Γ of the free boundary problem. Indeed there are arguments to this effect in the citations given below.

The example in [67] is such that $B(\Gamma)$ is an integral operator over Γ.

Examples of models in which nonlocal free boundary problems with pattern-forming capabilities of this type are found in [67, 75, 83, 82, 77].

3.8 Another Kind of Competition

We now return to the conserved evolution equation (3.34), but replace $-\rho$ by $\rho > 0$:
$$u_t = -\Delta[Au + \rho Bu - f(u)]. \tag{3.50}$$
So doing immediately removes this model from the considerations given in Section 3.5.2, if we continue to assume that both A and B are negative operators (which we do). However, there is another pattern-forming mechanism in operation. It was explained in connection with the copolymer model (Section 3.8.1) in [98], and we now sketch the idea.

We shall now operate in the Hilbert space $L_2(\Omega)$, where Ω is bounded and smooth, and take A and B to be negative with respect to that space. The new pattern-forming mechanism is most easily understood in the case when we restrict attention to real valued functions u and f. As in the previous section, we assume that $f(u) = W'(u)$, and that W has equal minima at $u = 0, 1$. Since mass is conserved with (3.50), stable patterns may be sought as mass-constrained global minimizers of the energy analogous to (3.25). The appropriate energy is

$$\mathcal{E}[u] = -\frac{1}{2}\langle Au, u \rangle - \frac{\rho}{2}\langle Bu, u \rangle + \int_\Omega W(u)dx \equiv \mathcal{E}_A[u] + \mathcal{E}_B[u] + \mathcal{E}_W[u]. \tag{3.51}$$

The constraint is
$$\int_\Omega u(x,t)dx = |\Omega|\bar{u}, \text{ where } 0 < \bar{u} < 1, \tag{3.52}$$

$\bar{u}$, the average of u, being known from the value of this integral at $t = 0$. The three parts of the energy may interact in the following manner.

We assume that $\mathcal{E}_A[u]$ penalizes fine spatial structure. In the scenario of Section 3.5.1, that simply means that $\lim_{k\to\infty} \hat{A}(k) = -\infty$. Or suppose that A has a complete set of eigenfunctions ϕ_n whose spatial structure is finer and finer as $n\to\infty$; then we assume its eigenvalues λ_n satisfy $\lim_{n\to\infty} \lambda_n = -\infty$. We then have $\mathcal{E}_A[\phi_n] = -\frac{1}{2}\lambda_n \|\phi_n\|^2$, so that if $\|\phi_n\| = 1$, $\mathcal{E}_A[\phi_n] \to \infty$ as $n\to\infty$. Since any function u in the domain of A can be expanded in these eigenfunctions, this gives some rough meaning to the claim that fine structure is penalized by $\mathcal{E}_A$.

We make a different pair of assumptions on B, namely (a) that for each function satisfying the mass constraint, $\mathcal{E}_B[u] > 0$ and (b) there is a minimizing sequence u_n of functions with finer and finer spatial structure: $\lim_{n\to\infty} \mathcal{E}_B[u_n] = 0$. Again in the context of Section 3.5.1, this would imply that $0 > \hat{B}(k) \to 0$ as $k \to \infty$.

The presence of this part $\mathcal{E}_B$ of the energy is what distinguishes this model from the Cahn-Hilliard equation. And indeed, as we have seen, constrained minimizers associated with the latter equation have "maximal spatial scale", which of course depends on the spatial domain. That is contrary to what we expect in the present case.

In all, we have one part, namely $\mathcal{E}_W[u]$, which drives u to attain values as close as possible to 0 and 1, which together with the mass constraint penalizes solutions which are nearly constant; another part ($\mathcal{E}_A$) which penalizes functions with small characteristic lengths; and a third part ($\mathcal{E}_B$) which prefers small characteristic lengths. Under these competitive conditions, it is natural to expect that the constrained minimizers will have spatial structure whose characteristic length assumes, by compromise, some value between the two extremes. In other words, we expect that stable spatial structure (patterning) will appear with some preferred characteristic length.

3.8.1 Models for Copolymers

An important example of the foregoing is found in the modeling of block copolymer melts, where $Bu = -(-\Delta)^{-1}(u-\overline{u})$, $Au = \epsilon \Delta u$, and f is replaced by $\frac{1}{\epsilon} f$, this $f(u)$ being as described above. We use the symbol σ in place of ρ to conform to tradition. If we set $v = (-\Delta)^{-1}(u - \overline{u})$, the PDE (3.50) can be written

$$u_t = \Delta w, \qquad (3.53)$$

$$w = -\left(\epsilon \Delta u - \frac{1}{\epsilon} f(u) - \sigma v\right), \qquad (3.54)$$

$$-\Delta v = u - \overline{u}, \qquad (3.55)$$

$$\overline{u} = \frac{1}{|\Omega|} \int_\Omega u(x,0) dx, \qquad (3.56)$$

under zero Neumann conditions on a bounded domain Ω for both u and v. Since mass is conserved, $\overline{u}$ is known beforehand from (3.56). If we solve (3.55) under the given boundary conditions to obtain $v = (-\Delta)^{-1}(u - \overline{u})$ and substitute into (3.53), (3.54), then we obtain the equation (3.50) with the identification of A and B as given above. The corresponding free energy functional is [100]

$$\mathcal{E}^\epsilon[u] = \int_\Omega \left[\frac{\epsilon}{2}|\nabla u|^2 + \frac{1}{\epsilon} W(u) + \frac{\sigma}{2}\left|(-\Delta)^{-1/2}(u-\overline{u})\right|^2\right] dx. \qquad (3.57)$$

Numerical simulations in 2D spatial domains have verified the existence of labyrinthine patterns with finite characteristic spatial scales.

Let me give a brief background description of this model. Block copolymer melts have been the subject of a number of field-theoretic models and analyses. Equilibrium models based on a free energy functional were given by [84] and Ohta et al [102, 78]. Bahiana and Oono [13] suggested a phenomenological cell dynamical system leading to an evolutionary partial differential equation for the composition $u(x,t)$ of the melt, as a function of space and time. In [98], a similar PDE was proposed by Nishiura and Ohnishi as a gradient flow for the free energy functional (3.57). This PDE was investigated in [98, 99].

In [98] the authors also derived formally a limit free boundary problem (FBP) as a parameter ϵ in the equation approaches zero. This limiting process was investigated rigorously by Henry [72] for radial solutions in 3D.

In 1D, which is of course the easiest case, the local and global minimizers of $\mathcal{E}^\epsilon$ and of related functionals were considered by Ohnishi et al [100] and by Ren and Wei [109, 110]. In [109], the local minimizers for small ϵ were found to be close, in the L^2 sense, to piecewise constant functions whose intervals of constancy alternated in length. Among them, the global minimizers were specified. In [100, 110], the global minimizers of a rescaled free energy functional were characterized. Finally the 1D dynamic free boundary problem was completely solved in [57]. That paper illustrates the way a general distribution of 1D intervals between interfaces evolve so as to achieve uniformity.

Choksi in [41] was able to show, in higher dimensions, that the minimal energy satisfies the same asymptotic relation as $\epsilon \to 0$ for a certain range of the parameters ϵ and σ, as was found in 1 dimension in [109].

3.9 Conclusion

We have explored a range of nonclassical trends among nonlinear parabolic-like evolutions:

- Nonlocal problems (NL)
- Pattern-forming problems (PAT)
- Free boundary problems (FBP)

And we have also considered their interconnections. Throughout, the ideas of gradient flows and model reduction provided threads of continuity. The interconnections can be described by the diagram below, where the arrows represent transitions from one kind of model to another, through model reduction by singular limits, by an extension (adding extra terms), or by the inverse elliptic reduction. Classical problems are denoted by CL.

CL $\longrightarrow$ NL $\longrightarrow$ PAT $\longrightarrow$ FBP

NL $\longrightarrow$ FBP

CL $\longrightarrow$ PAT

CL $\longrightarrow$ FBP

But I have left out reference to a great deal of other research on nonlocal, pattern-forming, and free boundary phenomena. Much additional information can be found in the surveys [97, 90, 42, 101].

For example, many types of exotic patterns and interfacial dynamics have been elucidated in other papers. I mention here only the recent paper [119]. There has also been considerable interest in the phenomenon of spatial segregation of species in population dynamics models [43, 76, 74], the boundary between habitats forming a moving interface akin to those discussed above.

There is a very different class of nonlocal elliptic and parabolic problems reviewed in the monograph by Skubachevskii [116], to which I direct the reader's attention. Still another direction in nonlocal problems is exemplified by [73].

Finally in materials science, there have been a number of nonlocal generalizations of existing models; see for example [63, 64, 111, 23, 62, 35].

References

1. N. Alikakos, P. Bates, and X. Chen, Convergence of the Cahn-Hilliard equation to the Hele-Shaw model, Arch. Rat. Mech. Anal. *128*, 165-205 (1994). [The object of study is (3.17) in the scaled form $u_t + \Delta[\epsilon\Delta u - \epsilon^{-1}f(u)] = 0$ in a bounded domain. Assume the Mullins-Sekerka problem (3.22) has a classical solution in $[0, T]$. It divides the domain into two subdomains. Then with appropriate initial and boundary conditions, there is a family of solutions (depending on ϵ) converging as $\epsilon \to 0$ to ± 1 in the two subdomains. From *Math. Reviews 97b:35174*: In this paper, the authors carry out a program of rigorous justification of the connection between the Cahn-Hilliard equation and the Hele-Shaw problem in an appropriate asymptotic limit which corresponds to a sharpening of the transition layer between phases... The Cahn-Hilliard equation, a fourth-order nonlinear partial differential equation, serves as a well-regarded model for the process of phase separation and coarsening in a melted alloy. It has been the subject of much work over the past ten to fifteen years. In particular, formal asymptotic analysis [106] established the connection between the level sets of the solution to the Cahn-Hilliard equation and interfacial motion in the Hele-Shaw (or Mullins-Sekerka) problem. For the time interval in which the Hele-Shaw problem possesses a classical solution, the present paper now makes this formal connection rigorous. To set this result in context, one can compare it to the large body of work devoted to an asymptotic analysis of the (second-order) Allen-Cahn equation—which tends to dissipate the same energy as the Cahn-Hilliard equation but which, unlike the Cahn-Hilliard equation, does not respect conservation of mass. The connection between the level sets of solutions to the Allen-Cahn equation and the problem of motion by mean curvature was made rigorous by Evans, Soner and Souganidis [49] , using a weak formulation of the curvature flow via viscosity theory. At roughly the same time, de Mottoni and Schatzman [91] used a detailed spectral analysis to justify the formal asymptotics for the Allen-Cahn equation for as long as the mean curvature flow retained a classi-

cal solution. The paper under review is in spirit the Cahn-Hilliard cousin of this last reference. (We should note that more recently Chen [39] developed a weak formulation of the Hele-Shaw problem using varifolds and carried out the asymptotic connection in this setting as well.) The procedure combines a delicate and nonstandard application of the method of matched asympotic expansions with detailed spectral analysis for the Cahn-Hilliard equation, previously worked out in [4, 37].]

2. N. Alikakos, P. Bates, and G. Fusco, Slow motion for the Cahn-Hilliard equation in one space dimension, J. Diff. Equations *90*, 81–135 (1991). [See Section 3.4.1.]
3. N. Alikakos, L. Bronsard, and G. Fusco, Slow motion in the gradient theory of phase transitions via energy and spectrum, Calculus of Variations *6*, 39–66 (1998).
4. N. Alikakos, G. Fusco, The spectrum of the Cahn-Hilliard operator for generic interface in higher space dimensions, Indiana Mathematics Journal *42*, 637–674 (1993). [See Section 3.4.1.]
5. N. Alikakos, G. Fusco, Slow dynamics for the Cahn-Hilliard equation in higher space dimensions. Part I: Spectral estimates, Comm. in P.D.E. *19*, 1397–1447 (1994).
6. N. Alikakos, G. Fusco, Slow dynamics for the Cahn-Hilliard equation in higher space dimensions: the motion of bubbles, Arch. Rat. Mech. Anal. *141*, 1–61 (1998). [See Section 3.4.1.]
7. N. Alikakos, G. Fusco, The equations of Ostwald ripening for dilute systems, J. Stat. Phys. *95*, 851–866 (1999). [See Section 3.4.3.]
8. N. Alikakos, G. Fusco, Ostwald ripening for dilute systems under quasistationary dynamics, preprint. [This is an expanded version, with proofs, of [7]. See Section 3.4.3.]
9. S. M. Allen, J. W. Cahn, A microscopic theory for antiphase boundary motion and its application to antiphase domain coarsening, Acta. Mater. *27*, 1084–1095 (1979).
10. H. W. Alt and I. Pawlow, Existence of solutions for non-isothermal phase separation. Adv. Math. Sci. Appl. *1*, 319–409 (1992). [From *Math. Reviews 95h:80003*: A model describing the dynamics of nonisothermal phase separation of a two-component system is studied. The coupled concentration and temperature fields are governed by thermodynamically unstable mechanisms. The model has the form of a nonlinear parabolic system consisting of a fourth-order equation for the concentration and a second-order equation for the temperature. Existence of solutions is established. The proof is based on implicit time discretization of the problem and appropriate energy and time compactness estimates for the time-discrete solutions.]
11. H. W. Alt and I. Pawlow, A mathematical model of dynamics of nonisothermal phase separation, Physica D *59*, 389–416, (1992). [From *Math. Reviews 92f:35080*: A Landau-Ginzburg free energy functional approach to the kinetics of a phase transition in a binary system is presented. Thermodynamic compatibility of the model is demonstrated by showing that the Clausius-Duhem inequality is satisfied, and the linear stability of stationary solutions of the kinetic equations, giving the rates of change of mass and energy, is investigated.]

12. H. W. Alt and I. Pawlow, A mathematical model and an existence theory for nonisothermal phase separation, Numerical methods for free boundary problems Internat. Schriftenreihe Numer. Math. **99**, 1-32, Birkhäuser, Basel, 1991. [Summary: "We present a mathematical model for nonisothermal phase separation in binary systems. The model is constructed within the Landau-Ginzburg theory of phase transitions and nonequilibrium thermodynamics. It consists of a system of nonlinear parabolic differential equations for the order parameter and the energy as conserved quantities. The model is conformable with the first principles of thermodynamics. Besides a description of the model we study the stability of stationary solutions, and give a survey of the existence theory for the system of governing differential equations."]
13. M. Bahiana and Y. Oono, Cell dynamical system approach to block copolymers, Phys. Rev. A **41**, 6763–6771 (1990). [See Sec. 3.8.1.]
14. G. Barles, P. E. Souganidis, A new approach to front propagation problems: theory and applications. Arch. Rational Mech. Anal. 141 (1998), no. 3, 237–296. [The authors' summary: "In this paper we present a new definition for the global-in-time propagation (motion) of fronts (hypersurfaces, boundaries) with a prescribed normal velocity, past the first time they develop singularities. We show that if this propagation satisfies a geometric maximum principle (inclusion-avoidance-type property), then the normal velocity must depend only on the position of the front, its normal direction and principal curvatures. This new approach, which is more geometric and, as it turns out, equivalent to the level-set method, is then used to develop a very general and simple method to rigorously validate the appearance of moving interfaces at the asymptotic limit of general evolving systems like interacting particles and reaction-diffusion equations. We finally present a number of new asymptotic results. Among them are the asymptotics of (i) reaction-diffusion equations with rapidly oscillating coefficients, (ii) fully nonlinear nonlocal (integral differential) equations, and (iii) stochastic Ising models with long-range anisotropic interactions and general spin-flip dynamics."]
15. P. Bates and A. Chmaj, A discrete convolution model for phase transitions, Arch. Rat. Mech. Anal. *150* 281–305 (1999). [From *Math. Reviews 2001c:82026*: In this interesting paper, the authors consider an infinite system of coupled ODEs $\dot{u}\bar{s}n = (J*u)\bar{s}n - u\bar{s}n - \lambda f(u\bar{s}n)$, where $n \in \mathbf{Z}$ or, in some cases, $n \in \mathbf{Z}^d$ with $d = 1, 2, 3, \cdots$; $\{J(i)\}_{i=11,12,\cdots}$ is a given sequence such that $\sum_{i \neq 0} J(i) = 1$ and $(J*u)_n = \sum_{i \neq 0} J(i) u_{n-i}$; $\lambda > 0$ is a parameter, and f, as they indicate, is bistable. First, they demonstrate a connection of this system with a model of statistical mechanics and with a semilinear heat equation. Next, under certain hypotheses, they show the existence of travelling waves and investigate their properties, including the stability and the uniqueness of a travelling wave with a nonzero speed. In the last part of the paper, the authors consider the existence of stationary solutions, present a characterization of all such solutions, give a criterion for their stability and show the admissibility, for sufficiently large $\lambda > 0$, of a property that they call pinning.]
16. P. Bates and A. Chmaj, An integrodifferential model for phase transitions: Stationary solutions in higher space dimensions, J. Stat. Phys. **95**, 1119–1139 (1999). [From *Math. Reviews 2000j:82020* "A large variety of integrodifferential equations describing various physically interesting phenomena are

considered intensively in the current literature. The non-local version of the Cahn-Hilliard equation for the phase segregation dynamics in certain particle systems was studied very recently in [G. Giacomin and J. L. Lebowitz, Phys. Rev. Lett. 76 (1996), no. 7, 1094–1097]; and in M. A. Katsoulakis and P. E. Souganidis [80]. A non-local, natural generalization of the familiar Landau-Ginzburg free energy functional is provided by $\mathcal{E}_1[u]$ (3.8). In particular, the L^2 gradient of this functional leads to the so-called non-local version of the Allen-Cahn equation, which is the main topic of the present paper. Different behaviours of the kernel j model a variety of interesting physical phenomena, which is the motivation for considering particular versions of the non-local equations arising from minizations of the free energy functional..." The main results in the paper relate to the possibility, under some conditions, of the existence of stationary solutions in higher space dimensions whose set of discontinuity is rather arbitrary.]

17. P. Bates and P. Fife, Spectral comparison principles for the Cahn-Hilliard and phase-field equations and time scales for coarsening, Physica D *43*, 335-348 (1990). [The Allen-Cahn, Cahn-Hilliard, and phase field models share certain features; for example they have similar stationary solutions and the last two exhibit phenomena like "spinodal decomposition" and coarsening. Insight into these phenomena can be gained by spectral analysis. This paper compares corresponding spectra for these models.]

18. P. Bates, P. Fife, X. Ren, and X. Wang, Traveling waves in a convolution model for phase transitions, Arch. Rat. Mech. Anal. *138* (1997), 105–136. [This gives the basic theory for traveling waves of the "nonlocal Allen-Cahn" equation (3.10).]

19. P. Bates and J. Xun, Metastable patterns for the Cahn-Hilliard equation: Part I, J. Differential Equations *111*, 421–457 (1995). [See Section 3.4.1.]

20. P. Bates and J. Xun, Metastable patterns for the Cahn-Hilliard equation: Part II, J. Differential Equations *117*, 165–216 (1995).

21. D. Blömker, S. Maier-Paape, and T. Wanner, Spinodal decomposition for the Cahn-Hilliard-Cook equation, preprint.

22. A. Bonami, D. Hilhorst and E. Logak, Modified motion by mean curvature: local existence and uniqueness and qualitative properties, Differential Integral Equations *13*, 1371–1392 (2000).

23. D. Brandon and R. Rogers, Nonlocal superconductivity, Z. angew. Math. Phys. *45*, 135–152 (1994).

24. L. Bronsard and B. Stoth, Volume-preserving mean curvature flow as a limit of a nonlocal Ginzburg-Landau equation, SIAM J. Math. Anal. *28*, 769–807 (1997). [Summary: "We study the asymptotic behavior of radially symmetric solutions of the nonlocal equation $\epsilon \phi \bar{s} t - \epsilon \Delta \phi + \epsilon^{-1} W'(\phi) - \lambda \bar{s} \epsilon(t) = 0$ in a bounded spherically symmetric domain $\Omega \subset R^n$, where $\lambda \bar{s} \epsilon(t) = \epsilon^{-1} \int \bar{s} \Omega W'(\phi) dx$, with a Neumann boundary condition. The analysis is based on 'energy methods' combined with some a priori estimates, the latter being used to approximate the solution by the first two terms of an asymptotic expansion. We only need to assume that the initial data as well as their energy are bounded. We show that, in the limit as $\epsilon \to 0$, the interfaces move by a nonlocal mean curvature flow, which preserves mass. As a by-product of our analysis, we obtain an L^2 estimate on the 'Lagrange multiplier' $\lambda \bar{s} \epsilon(t)$, which holds in the nonradial case as well. In addition, we show rigorously (in

general geometry) that the nonlocal Ginzburg-Landau equation and the Cahn-Hilliard equation occur as special degenerate limits of a viscous Cahn-Hilliard equation."]

25. L. Bronsard and F. Reitich, On three-phase boundary motion and the singular limit of a vector-valued Ginzburg-Landau equation, Arch. Rational Mech. Anal. *124* 355–379 (1993). [From *Math. Reviews 94h:35122:* The authors study the vector-valued Ginzburg-Landau equation $u\bar{s}t = 2\epsilon\Delta u - \nabla_{\bar{s}u}W(u)$ as a model for three-phase boundary motion, i.e. W has three minima. Formal asymptotic analysis is used to derive equations for the evolution of the phase boundaries, with one time scale used to analyze the evolution near the triple junction and the boundary of the domain of u, and another time scale used to analyze the motion on the rest of the boundary. Short term existence and uniqueness of solutions are shown by first proving existence for the linearized system, then constructing a contraction mapping whose fixed point is the desired solution. This method applies to a class of parabolic problems, as the authors illustrate by applying the method to an equation modeling eutectic solidification.]

26. L. Bronsard and R. Kohn, On the slowness of the phase boundary motion in one space dimension, Comm. Pure Appl. Math. *43*, 983–997 (1990).

27. G. Caginalp, Stefan and Hele-Shaw type models as asymptotic limits of the phase field equations, Phys. Rev. A *39*, 5887–5896 (1989).

28. G. Caginalp and P. Fife, Dynamics of layered interfaces arising from phase boundaries, SIAM J. Appl. Math. *48*, 506–518 (1988). [A systematic formal layer reduction analysis for a simple phase-field model. It includes the reduction of (3.1) (with parameter ϵ) to the motion-by-curvature law as a special case.]

29. J. W. Cahn, On spinodal decomposition, Acta Metall. *9*, 795–801 (1961). [The linearized Cahn-Hilliard evolution equation is proposed, based on the flux being proportional to the gradient of the chemical potential. The implications for stability are explored.]

30. J. Cahn, P. Fife and O. Penrose, A phase-field model for diffusion-induced grain boundary motion, Acta Materialia *45*, 4397–4413 (1997). [A gradient flow with Ginzburg-Landau energy plus elastic energy additions is proposed as a model for diffusion induced grain boundary motion.]

31. J. Cahn and J. Hilliard, Free energy of a nonuniform system. I. Interfacial free energy. J. Chem. Physics *28*, 258–267 (1958). [The free energy of a system with nonuniform concentration or density is proposed to be an integral with bulk free energy and quadratic gradient terms. On this basis, the character of a flat interface between phases, including the interfacial free energy and interface thickness, is investigated.]

32. J. Carr and R. Pego, Metastable patterns in solutions of $u_t = \epsilon^2 u_{xx} - f(u)$, Comm. Pure Appl. Math. *42*, 523–576 (1989). [This and [65] are the first papers to prove super slow motion.]

33. J. Carr and O. Penrose, Asymptotic behaviour in a simplified Lifschitz-Slyosov equation, Physica D *124*, 166–176 (1998).

34. C. Charach and P. Fife, Solidification fronts and solute trapping in a binary alloy, SIAM J. Appl. Math. *58*, 1826–1851 (1998). [Elaborate asymptotics applied to a phase field model for a solidifying alloy.]

35. C-K Chen and P. Fife, Nonlocal models of phase transitions in solids, Adv. Math. Sciences and Applications, to appear.
36. F. Chen, Almost periodic traveling waves of nonlocal evolution equations, preprint. [Existence, uniqueness, and stability of almost periodic traveling wave solutions of (3.14).]
37. X. Chen, Spectrum for the Allen-Cahn, Cahn-Hilliard, and phase-field equations for generic interfaces, Commun. Partial Differential Equations *19*, 1371–1395 (1994).
38. X. Chen, Existence, uniqueness, and asymptotic stability of travelling waves in nonlocal evolution equations, Adv. Differential Equations *2*, 125–160 (1997). [Uniqueness and global asymptotic stability are addressed.]
39. X. Chen, Global asymptotic limit of solutions of the Cahn-Hilliard equation, J. Differential Geometry *44*, 262–311 (1996). [Limit of solutions with bounded energy as $\epsilon \to 0$. There is a subsequence converging to a certain type of weak solution. This is a weak formulation of the Mullins-Sekerka problem using varifolds. See review of [1].]
40. X. Chen, D. Hilhorst and E. Logak, Asymptotic behavior of solutions of an Allen-Cahn equation with a nonlocal term, Nonlinear Analysis, TMA *28*, 1283–1298 (1997). [The limiting FBP for $u_t = \Delta u + \frac{1}{\epsilon^2} f(u, \epsilon \int u)$ with Neumann BC's and IC. This in turn is the shadow limit of the system $u_t = \Delta u + \frac{1}{\epsilon^2} f\left(u, \frac{\epsilon |\Omega|}{\gamma} v\right)$, $\tau v_t = \frac{1}{\sigma} \Delta v + u - \frac{1}{\gamma} v$ as σ and $\tau \to 0$. (The FitzHugh-Nagumo system with certain parameters fits in here.) The FBP is $V = -K + c_0 \left(|\Omega^+| - |\Omega^-|\right)$. The free boundary separates Ω^+ from Ω^-, where in the limit $u = \pm 1$. Here c_0 depends on f. An existence theory for the FBP is included.]
41. R. Choksi, Scaling laws in microphase separation of diblock copolymers, J. Nonlinear Science *11–3*, 223–236 (2001). [See Section 3.8.1.]
42. M. C. Cross and P. C. Hohenberg, Pattern formation outside of equilibrium, Rev. Modern Phys. *65*(3), 851–1112 (1993). [Comprehensive review of patterning in physical models.]
43. E. N. Dancer, D. Hilhorst, M. Mimura, and L. A. Peletier, Spatial segregation limit of a competition-diffusion system, European J. Appl. Math. *10*, 97–115 (1999). [A system of reaction-diffusion equations for competing species. When the competition strength becomes large, the habitats of (regions dominated by) the two species become disjoint, with a single curve separating them. The limiting problem is similar to the Stefan problem of phase transformation, except that there is zero latent heat of transformation. The behavior of the limiting problem is discussed, particularly the large time evolution of the habitats.]
44. A. De Masi, T. Gobron, and E. Presutti, Travelling fronts in nonlocal evolution equations, Arch. Rat. Mech. Anal. *132*, 143–205 (1995). [This is the model from [46]. Existence of traveling waves for small h is proved, and well as their nonlinear global stability.]
45. A. De Masi, E. Orlandi, E. Presutti, and L. Triolo, Motion by curvature by scaling nonlocal evolution equations, J. Stat. Physics *73*, 543–570 (1993). [Model introduced in [46] is studied: $\frac{\partial m}{\partial t} = -m + \tanh\{\beta J * m\}$. Convergence to a motion by mean curvature is proved under proper scaling.]
46. A. De Masi, E. Orlandi, E. Presutti, and L. Triolo, Glauber evolution with Kac potentials: I. Mesoscopic and macroscopic limits, interface dynamics, Nonlin-

earity 7, 633–696 (1994). [Evolution of an Ising spin system with Kac potential J in the limit when the inverse range of the potential γ goes to 0.]

47. K. Elder, T. Rogers, and R. Desai, Early stages of spinodal decomposition for the Cahn-Hilliard-Cook model of phase separation. Phys. Rev. B *38*, 4725–4739 (1988).

48. G. B. Ermentrout and J. B. McLeod, Existence and uniqueness of traveling waves for a neural network, Proc. Roy. Soc. Edin. *123A*, 461–478 (1993). [Summary: "A one-dimensional scalar neural network with two stable steady states is analysed. It is shown that there exists a unique monotone travelling wave front which joins the two stable states. Some additional properties of the wave such as the direction of its velocity are discussed."]

49. L. C. Evans, H. M. Soner and P. E. Souganidis, Comm. Pure Appl. Math. *45*, 1097–1123 (1992).

50. D. Eyre, Coarsening dynamics for solutions of the Cahn-Hilliard equation in one dimension, preprint.

51. D. Eyre, Systems of Cahn-Hilliard equations, SIAM J. Appl. Math. *53*, 1686–1712 (1993). [Summary: "The phase separation of alloys with two or more components is studied, with emphasis on more than two components. Particular attention is given to differences between multicomponent and binary alloys. Specific topics of the paper include equilibrium theory, aspects of the dynamics, and numerical simulations. In the equilibrium theory, it is found that there is an enriched equilibrium structure that allows for multiple coexisting phases and the presence of triple points in the solution. Dynamic results include the characterization of the spinodal region and of the concentration variations that lead to spinodal decomposition. Unlike the binary theory, not all composition variations lead to separation, and the compositions are not restricted to the convex hull of the equilibrium concentrations. Linear analysis is used to predict that a pseudo-binary will initially result from spinodal decomposition. Numerical simulations of the dynamics for a ternary alloy verify this initially, but more than two phases often separate. A sequential application of the dominant growth mode in linearly independent directions of composition variations appears to explain these additional phases. Finally, intermediate products are found that have both separated and metastable phases. This is not seen in binary materials."]

52. P. Fife, Dynamics of Internal Layers and Diffusive Interfaces, CBMS-NSF Regional Conference Series in Applied Mathematics 53, Soc. Ind. Appl. Math, Philadelphia (1988).

53. P. Fife, Models for phase separation and their mathematics, Electron. J. Diff. Eqns. Vol. *2000*, No. 48, 1 – 26 (2000). (Part of the proceedings of a workshop held in Katata, Japan in August, 1990 [Nonlinear Partial Differential Equations with Applications to Patterns, Waves, and Interfaces, M. Mimura and T. Nishida, eds.]).

54. P. Fife, Well-posedness issues in models for phase transition with weak interaction, Nonlinearity *14*, 221– 238 (2001). [See Section 3.2.5.]

55. P. Fife, Pattern formation in gradient systems, to appear as a chapter in Handbook for Dynamical Systems, Vol. 3, Applications, B. Fiedler and N. Kopell, eds.

56. P. Fife, J. Cahn, and C. Elliott, A free boundary model for diffusion induced grain boundary motion, Interfaces and Free Boundaries *3*, 291– 336 (2001).

[Extensive further development of the model in [30]. The difficulties of degenerate mobility are addressed.]
57. P. Fife and D. Hilhorst, The Nishiura-Ohnishi free boundary problem in the 1D case, SIAM J. Math. Anal. *33*, 589– 606 (2001). [See Section 3.8.1.]
58. P. Fife and M. Kowalczyk, A class of pattern-forming models, J. Nonlinear Science *9*, 641– 669 (1999). [Development of ideas in Section 3.5.1. See also [55].]
59. P. Fife and J. B. McLeod, The approach of solutions of nonlinear diffusion equations to travelling front solutions, Arch. Rat. Mech. Anal. *65*, 335– 361 (1977).
60. P. Fife and O. Penrose, Interfacial dynamics for thermodynamically consistent phase-field models with non-conserved order parameter, Elect. J. Differential Equations *1995*, 1– 49 (1995). [Full asymptotics for a class of phase field models.]
61. P. Fife and X. Wang, A convolution model for interfacial motion: the generation and propagation of internal layers in higher space dimensions, Advances in Differential Equations *5*, 85–110 (1998). [Reduction of (3.10) (with parameter ϵ) to motion by curvature.]
62. N. A. Fleck and J. W. Hutchinson, Strain gradient plasticity, Adv. Appl. Mech. *3* 296– 381 (1997). [Survey of phenomenological theories of plastic materials in which stress depends not only on the strain but also on the spatial gradient of the strain. In this sense it is slightly "nonlocal".]
63. R. Fosdick and D. Mason, Single phase energy minimizers for materials with nonlocal spatial dependence, Quart. Appl. Math. *54*, 161– 195 (1996).
64. R. Fosdick and D. Mason, On a model of nonlocal continuum mechanics, Part I: Existence and regularity, SIAM J. Appl. Math. *58*, 1278– 1306 (1998). [The energy field for a body is assumed to depend not only on the local strain field but also on the field's average with a weighting kernel. Minimizers are considered. They must be periodic and piecewise smooth. An existence theorem is given.]
65. G. Fusco and J. Hale, Slow motion manifolds, dormant instability and singular perturbation, in Dynamics and Differential Equations *1*, 1989. [See comment under [32].]
66. K. Glasner, A diffuse interface approach to Hele-Shaw flow, preprint. [Gradient nature of the problem is exploited.]
67. R. Goldstein, D. J. Muraki, and D. M. Petrich, Interface proliferation and the growth of labyrinths in a reaction-diffusion system, Phys. Rev. E *53*, 3933– 3957 (1996). [The paper starts with the FitzHugh-Nagumo model and goes to a shadow system. This is the opposite limit from the usual excitable structures. Thus, we get a nonlocal gradient flow. A formal expansion can lead to a 4th order problem. If one assumes the fronts are narrow in 1D, one gets a 1D FBP. Formal asymptotics are provided. Heuristics and asymptotics are given for the 2D front case. Linear stability for fronts, stripes, and discs is investigated. Patterns are revealed by simulations. Similarities are brought out with models from superconductors, magnetic fluids, and monolayers.]
68. A. A. Golovin, S. H. Davis and A. A. Nepomnyashchy, A convective Cahn-Hilliard model for the formation of facts and corners in crystal growth, Physics D *122*, 202– 230 (1998). [In 1D, the derivative u_t is replaced by the convective derivative $u_t - Duu_x$.]

69. C. P. Grant, Spinodal decomposition for the Cahn-Hilliard equation, Commun. in Partial Differential Equations *18*, 453–490 (1993). [See Section 3.3.1.]
70. C. P. Grant, Slow motion in one-dimensional Cahn-Morral systems, SIAM J. Math. Anal. *26*, 21–34 (1995).
71. D. Henry, Geometric Theory of Semilinear Parabolic Equations, Vol. 840 of Lecture Notes in Mathematics, Springer-Verlag, Berlin, 1981.
72. M. Henry, Singular limit of a fourth order problem arising in the micro-phase separation of diblock copolymers, Advances in Differential Equations, to appear.
73. D. Hilhorst and J.F. Rodrigues, On a nonlocal diffusion equation with discontinuous reaction, Advances in Differential Equations *5*, 657– 680 (2000). [$\partial_t u = a\left(\int u\right)\Delta u + f\left(u, \int u\right)$. Start with $\partial_t u = a(v)\Delta u + f(u,v)$ + an eqn for v, and make the shadow system. Here f could be discontinuous; we get nonuniqueness then.]
74. T. Ikeda and M. Mimura, An interfacial approach to regional segregation of two competing species mediated by a predator, J. Math. Biol. *31*, 215–240 (1993). [The possibility of coexistence (with spatial segregation) is considered when the diffusion rate of the prey tends to zero. The dynamics of spatial segregation is discussed by means of an interfacial dynamics approach.]
75. D. Jackson, R. Goldstein and A. Cebers, Hydrodynamics of fingering instabilities in dipolar fluids, Phys. Rev. E *50*, 298–307 (1994). [The subject is ferrofluids under magnetic field. In the introduction, there is a discussion of general patterning from competition between surface tension and body forces. The present context could be generalized. Reference is made to [83] as the simplest gradient-flow model for this physics. In that one, viscous dissipation by Darcy's law was neglected except at the boundary. Here it is not. Law of motion for the boundary: $v = -\nabla \Pi$ (generalized Darcy's law), $\nabla^2 \Pi = 0$, $\Pi = \kappa + I$, where I an integral operator along the boundary C. This is a gradient flow for $\mathcal{E} = L - \int_C ds \int_C ds' \hat{t}(s)\cdot\hat{t}(s')\Phi(R(s,s'))$. Here $\Phi 0$ is a nonlinear function, the contour C is given parametrically by $r(s)$, $R = |r(s) - r(s')|$, $\hat{t}$ is the unit tangent vector, and L is the length of C. Time-dependence of the fastest growing modes is studied.]
76. Y. Kan-on and M. Mimura, Singular perturbation approach to a 3-component reaction-diffusion system arising in population dynamics, SIAM J. Math. Anal. 29, 1519–1536 (1998) (electronic). [The effect of a predator on the coexistence question for competing species is considered in this model. Singular perturbation analysis may reveal coexisting spatially segregated species.]
77. S. Kawaguchi, and M. Mimura, Collision of travelling waves in a reaction-diffusion system with global coupling effect, SIAM J. Appl. Math. *59*, 920–941 (1999). [The system is that proposed by Krischer and Mikhailov. The destabilization of radially symmetric equilibria, the formation of traveling waves, and the collision of the latter are investigated.]
78. K. Kawasaki, T. Ohta, and M. Kohrogui, Equilibrium morphology of block copolymer melts. 2, Macromolecules *21*, 2972– 2980 (1988).
79. M. Katsoulakis and P. E. Souganidis, Interacting particle systems and generalized mean curvature evolution, Arch. Rat. Mech. Anal. *127*, 133– 157 (1994). [Authors' summary: "We study the limiting behavior (the macroscopic limit) of an appropriately scaled spin system with Glauber-Kawasaki dynamics. We

rigorously establish the existence in the limit of an interface evolving according to motion by mean curvature. This limit is valid for all positive times, past possible geometric singularities of the motion, which is interpreted in the viscosity sense."]

80. M. Katsoulakis and P. E. Souganidis, Generalized motion by mean curvature as a macroscopic limit of stochastic Ising models with long range interactions and Glauber dynamics, Comm. Math. Phys. *169*, 61– 97 (1995). [*Math. Reviews 96m:60238:* The mesoscopic limit of a ferromagnetic stochastic Ising model under Glauber dynamics and Kac potential leads to the non-local mean-field equation $\partial m/\partial t = \tanh \beta (J*m + h) - m$ in the limit $\gamma \to 0$. The above solution $m(t,x), x \in \mathbf{R}^d$, is the limit of the expected variable located at lattice point $i = x/\gamma$ in $\mathbf{Z}^d$, evolving under an interaction potential with couplings $\gamma^d J(\gamma|i-j|)$ between particles at $i,j \in \mathbf{Z}^d$ and external field h. The macroscopic limit considered here is the large time–large space asymptotics of the model. The authors prove that the rescaled solution $m(t/\lambda^2, x/\lambda)$ tends as $\lambda \to 0$ to an interface motion which propagates with normal velocity $\theta 7$(mean curvature)+const, when the external field brings some speed-up: $h = \text{const}\cdot\lambda$. For $\lambda = \lambda(\gamma) \to 0$ slowly enough, they prove that the stochastic density field itself, rescaled as before, converges to the mean curvature flow. The flow is viewed as a viscosity solution, and convergence holds past the occurrence of singularities. The transport coefficient θ has an explicit form, coming from the averaging effect of small λ's.]

81. M. Katsoulakis and P. E. Souganidis, Stochastic Ising models and anisotropic front propagation, J. Statist. Phys. *87*, 63–89 (1997). [From *Math. Reviews 98k:82118:* The Ising models studied here have general spin-flip dynamics in $\mathbf{Z}^{d+1}$ obeying the detailed balance law with respect to Gibbs measures with long range interactions (Kac potentials [46]). Rescaling time, space and the interactions together, the authors obtain interfaces moving with normal velocity depending anisotropically on their principal curvatures and direction. The mobility depends on the direction and is computed explicitly in terms of the microscopic dynamics.]

82. M. Kuwamura, S.-I. Ei, and M. Mimura, Very slow dynamics for some reaction-diffusion systems of the activator-inhibitor type, Japan J. Indust. Appl. Math. *9*, 35–77 (1992). [Slow dynamics is considered for the shadow system consisting of the following two equations: (1) $u\bar{s}t = \epsilon^2 u\bar{s}xx + \sigma^{-2}[u(1-u^2) - \zeta]$, (2) $\zeta \bar{s}t = -\gamma\zeta + \int^1 \bar{s}0 u(x,t)dx$, where $x \in (0,1)$ and $t > 0$.]

83. S. Langer, R. Goldstein, and D. Jackson, Dynamics of labyrinthine pattern formation in magnetic fluids, Phys. Rev. A *46*, 4894–4904 (1992). [See commentary under [75].]

84. L. Leibler, Theory of microphase separation in block copolymers, Macromolecules *13*, 1602– 1617 (1980).

85. K. Leung, Theory of morphological instability in driven systems, J. Stat. Phys. *61*, 345 (1990).

86. I. M. Lifshitz and V. V. Slyozov, The kinetics of precipitation from supersaturated solid solutions, J. Phys. Chem. Solids *19*, 35–50 (1961). [This and [125] are the original papers presenting LSW theory.]

87. E. Logak, Singular limit of reaction-diffusion systems and modified motion by mean curvature, preprint.

88. S. Maier-Paape and T. Wanner, Spinodal decomposition for the Cahn-Hilliard equation in higher dimensions. Part I: Probability and wavelength estimate, Comm. Math. Physics *195*, 435–464 (1998). [See Section 3.3.2.]
89. S. Maier-Paape and T. Wanner, Spinodal decomposition for the Cahn-Hilliard equation in higher dimensions. Part II: Nonlinear dynamics, Arch. Rat. Mech. Anal. *151*, 187– 219 (2000).
90. E. Meron, Pattern formation in excitable media, Physics Reports 218, No. *1*, 1-66 (1992). [Survey.]
91. P. de Mottoni and M. Schatzman, Geometrical evolution of developed interfaces, Trans. Amer. Math. Soc. it 347, 1533–1589 (1995). [The first rigorous justification of the motion-by-curvature property of Allen-Cahn layers.]
92. B. Niethammer, Derivation of the LSW-theory for Ostwald ripening by homogenization methods, Arch. Rat. Mech. Anal. *147*, 119– 178 (1999). [Rigorous derivation of LSW theory in the regime $\theta \ll d$ and $\xi_{cor} \ll \xi_{scr}$. See [93] and Sec. 3.4.3.]
93. B. Niethammer and F. Otto, Ostwald ripening: The screening length revisited, Calc. Var. Partial Differential Equations *13*, 33–68 (2001). [See Section 3.4.3.]
94. B. Niethammer and R. Pego, Non-self-similar behavior in the LSW theory of Ostwald ripening, J. Stat. Phys. *95*, 867– 902 (1999). [The supposedly long time behavior of the LSW model (convergence to a self-similar solution) is not universal. If the particle distribution has compact support, the asymptotics depend sensitively on the behavior of the initial distribution near its support boundary.]
95. B. Niethammer and R. Pego, The LSW model for domain coarsening: Asymptotic behavior for conserved total mass, preprint. [The variation this time is that a constraint is made that the total mass, including the mean-field mass, is conserved, rather than just the total volume fraction of one phase. Some results are similar to the classical case.]
96. B. Niethammer and R. Pego, On the initial-value problem in the Lifshitz-Slyozov-Wagner theory of Ostwald ripening, SIAM J. Math. Anal. *31*, 467– 485 (2000). [Global existence, uniqueness and continuous dependence on initial data for measurue-valued solutions with compact support in particle size. The topology of solutions used is the Wasserstein metric, a natural one for the space of size distributions.]
97. Y. Nishiura, Nonlinear Problems, Part I, Amer. Math. Soc. (2001).
98. Y. Nishiura and I. Ohnishi, Some mathematical aspects of the micro-phase separation in diblock copolymers, Physica D *84*, 31–39 (1995). [Functional introduced in [102, 13]. It involves parameters ϵ and σ. The H^{-1} gradient dynamics introduced. Space and time rescaled according to powers of ϵ and σ. Powers chosen so that the rescaled problem involves only one parameter δ. Then letting $\delta \to 0$, one gets a meaningful free boundary problem with no parameter in it. This is formal. Stability considerations discussed, a la [17].]
99. I. Ohnishi and Y. Nishiura, Spectral comparison between the second and fourth order equations of conservative type with non-local terms, Japan Jour. Industrial and Applied Mathematics *15*, 253– 262 (1998). [A comparison result for the spectra of two nonlocal operators is given. The corresponding evolutions are (3.50) and a similar but second order equation. The two equations have the same steady states, and their stabilities coincide.]

100. I. Ohnishi, Y. Nishiura, M. Imai, and Y. Matsushita, Analytical solutions describing the phase separation driven by a free energy functional containing a long-range interaction term, Chaos *9*, 329– 341 (1999). [The free energy is given by (3.57). A rigorous asymptotic expansion in ϵ of the global minimizer in 1D. Numerics on the dynamical problem reveal local minima.]
101. T. Ohta and M. Mimura, Pattern dynamics in excitable reaction-diffusion media. Formation, dynamics and statistics of patterns, Vol. 1, 55–112, World Sci. Publishing, River Edge, NJ, 1990. [Recent investigations of pattern formation and pattern evolution in excitable reaction-diffusion media by means of the interface dynamics are reviewed. This involves interfacial equations of motion and the associated stability considerations. An example is a phase field model of crystal growth.]
102. T. Ohta and K. Kawasaki, Equilibrium morphology of block copolymer melts, Macromolecules *19*, 2621–2632 (1986). [Stat. mech. theory developed by Helfand and Wasserman; mean field in the weak segregation limit was given by Leibler. Here, such a theory is given in the strong segregation limit. They use a local order parameter ψ and develop a free energy functional.]
103. F. Otto, Evolution of microstructure in unstable porous media flow: A relaxational approach, Comm. Pure Appl. Math. *LII*, 0873–0915 (1999). [Uses the notion of mass transference plan in study of microstructure evolution in unstable porous medium flow.]
104. F. Otto, Dynamics of labyrinthine pattern formation in magnetic fluids: A mean-field theory, Arch. Rat. Mech. Anal. *141*, 63–103 (1998). [Hele-Shaw with magnetic fluid under magnetic field. Patterns have been observed in the laboratory. Reference is made to [75, 83]. The formulation of the dynamics is that given in [75]. This is shown to be a gradient flow. The author uses the Wasserstein metric. He gets an evolution law for the envelop of the labyrinth. It is a porous medium equation.]
105. E. Orlandi and L. Triolo, Travelling fronts in nonlocal models for phase separation in an external field, Proc. Roy. Soc. Edinburgh Sect. A *127*, 823–835 (1997). [Model nonlocal equation in [46]. Existence, uniqueness, and linear stability of traveling waves. The existence in [44] was for small h; that is removed. Other results are similar; some different methods.]
106. R. Pego, Front migration in the nonlinear Cahn-Hilliard equation, Proc. R. Soc. Lond. A *422*, 261– 278 (1989). [Various kinds of singular reductions of the Cahn-Hilliard equation are revealed when a parameter is small. They lead to free boundary problems. One of them corresponds to what would later be called the Mullins-Sekerka problem.]
107. L. A. Peletier and W. C. Troy, Spatial Patterns, Higher Order Models in Physics and Mechanics, Birkhäuser, Boston, 2001. [An extensive study of the existence and properties of solutions of $\frac{d^4u}{dx^4} + q\frac{d^2u}{dx^2} + f(u) = 0$, which includes the Swift-Hohenberg equation. Special attention is given to the case when f is bistable.]
108. G. Peng, F. Qiu, V. Ginzburg, D. Jasnow and A. Balazs, Forming supramolecular networks from nanoscale rods in binary, phase-separating mixtures, Science *288*, 1802– 1804 (2000). [Patterning in a mixture with two fluids and rods. Modeling is done with a phase-field system.]
109. X. Ren and J. Wei, On the multiplicity of solutions of two nonlocal variational problems, SIAM J. Math. Anal. *31*, 909–924 (2000). [Two mass-constrained

variational problems are considered, associated with the energy functional (3.57) and a variation on it arising from a model by Truskinovsky and Ren. The Γ-limits of these problems as $\epsilon \to 0$ are found. In 1D, all local minima are characterized, using perturbation from the Γ-limit.]
110. X. Ren and J. Wei, Analysis of global minima of a nonlocal variational problem, in preparation. [A different parameter range from that in [109].]
111. R. Rogers, Nonlocal variational problems in nonlinear electromagnetoelastostatics, SIAM J. Math. Anal. *19*, 1329–1347 (1988). [The setting is a nonlinear deformable magnetizable nonconducting body. It is unshielded, which makes the problem nonlocal. Reference is made to a previous nonlocal problem from astrophysics.]
112. E. Sander and T. Wanner, Monte Carlo simulations for spinodal decomposition, J. Stat. Phys. *95*, 925– 948 (1999). [Simulations on the Cahn-Hilliard equation showing phase separation.]
113. E. Sander and T. Wanner, Unexpectedly linear behavior for the Cahn-Hilliard equation, SIAM J. Appl. Math. *60*, 2182–2202 (2000). [See Section 3.3.2. Some results already conjectured in [112].]
114. M. Seul and D. Andelman, Domain shapes and patterns: the phenomenology of modulated phases, Science *267*, 476–483 (1995).
115. W. Shen, Traveling waves in time almost periodic structures governed by bistable nonlinearities I. Stability and uniqueness, J. Differential Equations *159*, 1–54 (1999) II. Existence, op. cit. *159*, 55-101 (1999). [See Sec. 3.2.3.]
116. A. L. Skubachevskii, Elliptic Functional Differential Equations and Applications, Birkhäuser, Basel-Boston-Berlin, 1997.
117. H. M. Soner, Convergence of the phase-field equations to the Mullins-Sekerka problem with kinetic undercooling, Arch. Rat. Mech. Anal. *131*, 139–197 (1995).
118. P. E. Souganidis, Recent developments in the theory of interface dynamics. Differential equations and applications (Hangzhou, 1996), 286–300, Internat. Press, Cambridge, MA, 1997. [From *Math. Reviews 2000d:35238:* The author presents a survey of recent results in the theory of interface dynamics, and some applications. The related proofs are contained in [14]. More precisely, a new definition of motion is described, which holds for any time and coincides with the regular solution up to the first time of singularity. This evolution has a simple geometric definition based on a maximum principle type avoidance-inclusion property, and turns out to be equivalent to the viscosity-type solution obtained through the level-set method [see Y. G. Chen, Y. Giga and S. Goto, J. Differential Geom. 33 (1991), no. 3, 749–786; MR 93a:35093]. A similar definition of weak evolution has been proposed by E. De Giorgi ["Barriers, boundaries, motion of manifolds", in Proceedings of Capri Workshop (1990)] and further developed by G. Bellettini and Novaga [Ann. Scuola Norm. Sup. Pisa Cl. Sci. (4) 26 (1998), no. 1, 97–131; MR 99i:35081]. The author uses similar techniques to introduce a general method in order to study the large time asymptotics of a class of fully nonlinear differential equations including, among others, reaction-diffusion equations and limits of stochastic Ising models with long-range interactions. The advantage of the method is that the rigorous justification of the appearance of an interface is reduced to proving a similar result only in the case where everything is smooth and for small time intervals.]

119. M. Suzuki, T. Ohta, M. Mimura, and H. Sakaguchi, Breathing and wiggling motions in three-species laterally inhibitory systems, Phys. Rev. E *52*, 3645–3655 (1995). [Authors' summary: "We study the layer dynamics in a coupled set of reaction-diffusion systems describing the interaction of one activator and two inhibitors. In some special limits, this set of equations is reduced to a Bonhoeffer-van der Pol-type excitable system or a model for electric glow discharge. Dynamics of layers is investigated by an interfacial approach and complementarily by computer simulations. By changing the parameters, these layers undergo several sustained oscillations such as breathing, wiggling, and quasiperiodic motions, which are due to the competition of two inhibitors and one activator."]
120. J. Swift and P. C. Hohenberg, Hydrodynamic fluctuations at the convective instability, Phys. Rev. A *15*, 319–328 (1977). [Derivation of the Swift-Hohenberg equation in the context indicated.]
121. J. J. L. Velázquez, The Becker-Döring equations and the Lifshitz-Slyozov theory of coarsening, J. Stat. Phys. *92*, 195–236 (1998). [The LSW evolution law is approximated by a Fokker-Planck equation.]
122. P. W. Voorhees, The theory of Ostwald ripening, J. Stat. Phys. *38*, 231–252 (1985). [Review]
123. P. W. Voorhees, Ostwald ripening of two-phase mixtures, A. Rev. Mater. Sci. *22*, 197–215 (1992). [Review]
124. J. D. van der Waals, The thermodynamic theory of capillarity under the hypothesis of a continuous variation of density, Verh. Konink. Acad. Wetensch. Amsterdam *1* (1893). [First use of Ginzburg-Landau style energy.]
125. C. Wagner, Theorie der Alterung von Neiderschlägen durch Umlösen, Z. Elektrochem. *65*, 581–594 (1961). [This and [86] are the original papers presenting LSW theory.]
126. X. Wang, Metastability and stability of patterns in a convolution model for phase transitions, preprint. [Metastability for (3.10). The results depend on how fast J decays. Results analogous to local case. Persistence, annihilation, stability of patterns.]
127. J. A. Warren and W. J. Boettinger, Prediction of dendritic growth and microsegregation patterns in a binary alloy using the phase-field method, Acta Metal. Mater. *43*, 689–703 (1995).
128. A. A. Wheeler, G. B. McFadden and W. J. Boettinger, Phase field model for solidification of a eutectic alloy, Proc. Royal Soc. London A *495*, 525 (1996).

4 Mathematical Aspects of Design of Beam Shaping Surfaces in Geometrical Optics

Vladimir Oliker

Emory University, Department of Mathematics and Computer Science, Mathematics and Sciences Center, Suite W401, 400 Dowman Drive, Atlanta, Georgia 30322, USA – *oliker@mathcs.emory.edu*

To Professor Willi Jäger on his 60-th Birthday

Abstract: Numerous optical and electromagnetic applications require synthesis of reflecting and refracting surfaces capable of reshaping the energy radiation intensity of a given source into a prescribed output irradiance distribution. Determination of such surfaces requires investigation of nonlinear, second order partial differential equations of Monge-Ampère type and development of computational algorithms for constructing their numerical solutions. These equations are very far from being standard and it is quite remarkable that geometric ideas not only provide natural means for their analysis but also means for computing solutions numerically. In this paper we survey some of these problems and describe the current progress in their study.

4.1 Introduction

Synthesis of a system for shaping a beam of rays from a given source usually involves determination of reflecting surfaces with capabilities to control the energy intensity distribution and/or phase on the far-field or on the near-field output apertures. Numerous optics applications, especially in laser optics, require such shaped reflectors [49, 31, 10, 9, 36]. Similar synthesis problems arise also in design of reflector antennas with pre-specified signal directivity properties [25, 26, 18, 53]. The traditional design techniques typically utilize rotationally symmetric solutions with the source on the axis of revolution. However, it has been known for a long time that a significant gain in efficiency of the system can be achieved by using non-rotationally symmetric surfaces, since the latter provide more freedom in controlling the output phase, intensity distribution, and polarization [17, 25, 26, 23, 24, 47, 53, 55]. This class of design problems is not limited to reflectors. The same type of problems have to be solved in design of various refractive systems, for example, in design of lenses [46].

Usually, the derivation of the synthesis equations is based on the geometrical optics approximation and amounts to systematic application of the ray tracing and energy conservation laws [25, 26, 48, 47, 53, 31]. However, because the intensity of the source and the intensity distribution required on the output aperture are related by the Jacobian of the ray tracing map defined by the reflecting surface(s), the resulting equations are quite complex. Typically, it is a highly nonlinear system of partial differential equations (PDE's) for the position vector (vectors) of the required reflector surface (surfaces).

In some cases it is possible to find a scalar quasi-potential such that the corresponding problem involves a second order partial differential equation of Monge-Ampère type. In these cases the problems are more manageable though, still, the resulting equations are usually complicated and, in spite of important recent progress in the theory of Monge-Ampère equations, the investigation of the synthesis equations requires new and significant efforts.

Furthermore, solving numerically the beam shaping equations is a particularly challenging problem. The engineering literature contains numerous papers in which attempts are made to solve numerically the equations of beam shaping. However, the authors always rely on specific numerical examples to justify their approaches [17, 23, 24, 47, 53, 55]. The difficulties here are connected with the fact that because of strong nonlinearities the traditional discretization schemes based on finite differences fail to converge [42] and, for the same reason, the discretization schemes based on projection methods are not applicable. This is true for beam shaping problems formulated as systems of PDE's and as Monge-Ampère equations. As far as the equations of Monge-Ampère type are concerned, it should be noted that apart from the papers [42, 28, 14] and the preprint [4], we are not aware of any other work where convergent computational algorithms for solving such equations have been studied.

In recent years a very general geometric method for solving several classes of equations arising in beam shaping problems has been developed and applied in [38, 35, 36, 15, 34, 14, 27, 28, 30] The developed method is based on new geometric ideas for constructing weak (viscosity) solutions of Monge-Ampère equations and it provides a rigorous and unified framework for establishing existence of solutions and calculating them numerically.

Still, many theoretical and computational issues important for applications remain open. In particular, the problem of solving the beam shaping equations of a two-reflector system with a point source is open (see section 4.5) in spite of its importance in design of two-reflector antennas. The numerous papers in the engineering literature on this problem do not provide a rigorous treatment and, in fact, contain conflicting statements [25, 17, 53].

Another important issue is the development of appropriate computational methods for solving the resulting equations numerically. For problems with one reflecting surface computational algorithms were proposed in [28, 14]. The algorithms described in these papers are motivated by the physical and geometric properties of special solutions to these problems (see section 4.2.5 for more details) and are shown to converge. However, these algorithms become inefficient when the distribution density of discretization nodes becomes high. This phenomena is quite typical for problems with strong nonlinearities, including equations of Monge-Ampère type. Analysis and enhancements of computational algorithms for solving Monge-Ampère equations numerically is an important direction of research.

This survey is written from the mathematical point of view and the selection of problems is based primarily on the work that the author has been involved in during the last 15 years. Consequently, many related problems important for applications are not discussed here. We refer the reader to a very stimulating survey by Kinber [25] written from an engineering point of view. Reading that survey in conjuction with the present paper should be useful.

In this paper we do not limit ourselves to two-dimensional problems which actually arise in applications. The geometric nature of these problems is so prominent that it makes perfect sense to formulate and study these problems for any number of independent variables.

Overall, the reader should look at this survey as a sample of problems from an area very rich with interesting, difficult, and genuinely nonlinear problems.

4.2 Creating a Prescribed Intensity Distribution in the Far-Field

4.2.1 Statement of the Problem

Consider a non-isotropic point source of light positioned at the origin of a Cartesian coordinate system O in R^{n+2}, $n \geq 2$ and emitting rays in a set of directions defined by the aperture $\bar{D}$ given as a closed set on a unit sphere S^n centered at O. Usually, we will assume that D is a domain on S^n; the case $D = S^n$ is not excluded. Denote by $I(m)$ the intensity of this source in direction $m \in D$. Suppose that a light ray emitted by the source O in direction $m \in \bar{D}$ is incident on some smooth perfectly reflecting hypersurface R at point $r(m)$ and reflects off it in direction $y(m)$, $|y(m)| = 1$.

If we denote by u the unit normal field on the hypersurface R, then the reflected direction y is given by Snell's law as

$$y = m - 2\langle m, u \rangle u. \tag{4.1}$$

Thus, the hypersurface R defines a map $\gamma : m \longrightarrow y$ which maps the input aperture $\bar{D} \subset S^n$ onto some set $\bar{T} \subset S^n$; see Fig. 1, where for convenience the

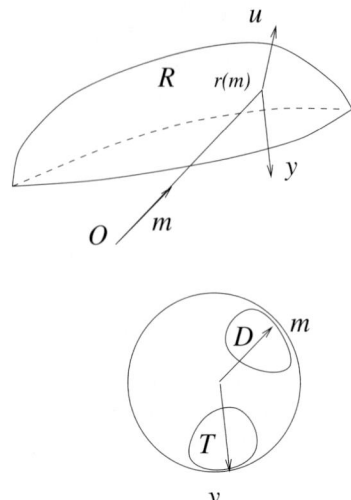

Fig. 4.1. Formulation of the far-field beam shaping problem.

input aperture D and the set of reflected directions T (the "output aperture" in the "far-field") are shown on the same unit sphere S^n.

If γ is differentiable then the intensity of the light reflected in direction $y = \gamma(m)$ is given by
$$\frac{I(m)}{|J(\gamma(m))|},$$
where $J(\gamma)$ is the Jacobian of the map γ.

Suppose now that the sets D and T on S^n are given as well as positive functions $I(m)$ on D and $L(y)$ on T. We will assume that T is also a domain on S^n and the case when $T = S^n$ is not excluded. Consider the problem of finding a reflector (hypersurface) R such that for the map γ defined by R the following conditions are satisfied:

$$\gamma(\bar{D}) \supset \bar{T}, \tag{4.2}$$

$$L(y) = I(\gamma^{-1}(y))|J(\gamma^{-1}(y))|, \quad y \in T. \tag{4.3}$$

Note that the last equation is formulated on the output aperture T rather than on the input aperture D. One could also set up this problem on D. For our purposes the formulation on T is more convenient because the geometric constructions in such setting are more transparent.

4.2.2 Weak Formulation of the Problem

For each $y \in \bar{T}$ denote by $P(y)$ a paraboloid of revolution (about y as its axis) with focus at the point O and polar radius

$$\rho(m) = \frac{p(y)}{1 - \langle m, y \rangle}, \quad m \in S \setminus \{y\},$$

where $p(y) > 0$ is the focal parameter of $P(y)$. Denote by $B(y)$ the convex body bounded by $P(y)$ and consider a closed convex hypersurface R defined by

$$R = \partial B \quad \text{where} \quad B = \bigcap_{y \in \bar{T}} B(y). \tag{4.4}$$

Definition 1. *The hypersurface R is called a (closed) reflector defined by the set of paraboloids $\{P(y)\}_{y \in \bar{T}}$ (with the light source at O).*

In the special case when the set $\bar{T}$ consists of a finite number of points $y_1, y_2, ..., y_k, k \geq 2$, we say that the reflector R is defined by a finite number of paraboloids of revolution $P(y_i)$, $i = 1, 2, ..., k$.

The set B is a closed convex body with interior points and, since $p' > 0$ on Ω, the origin O is strictly inside B. The closed convex hypersurface R is star-shaped relative to O.

Obviously, for any positive and continuous function $p'(y), y \in \bar{T}$, one can define a reflector which will be a convex hypersurface star-shaped relative to O. Because some of the paraboloids in the family $\{P(y)\}_{y \in \bar{T}}$ may have no common points with R we need the following

Definition 2. *Let $P(y)$ be a paraboloid of revolution with focus at O and axis of direction y, $B(y)$ the convex body bounded by $P(y)$, and $R \subset B(y)$ a reflector. If there exists a point $z \in P(y) \bigcap R$ then we say that $P(y)$ is* **supporting to R at point** z.

Observe that since B is compact, for every $y \in S^n$ there exists a paraboloid of revolution with focus O and axis y supporting to R. Therefore, with each R defined as above one can always associate some function $p(y), y \in S^n$ such that each paraboloid $P(y)$ defined by $p(y)$ is supporting to R. Such function will be called the *focal function of the reflector R*. Note that $p(y) > 0 \; \forall y \in S^n$, since the origin O is strictly inside the convex body B bounded by R.

Thus, the focal function of a closed convex reflector is defined on the entire S^n regardless of the domain of definition of the function p'. For that reason it is convenient to consider the set of all closed convex reflectors with the source O strictly inside the convex body bounded by R and irrespective of the originally given set $\bar{T}$. We denote this set of reflectors by $\mathcal{R}$.

For reflectors in $\mathcal{R}$ we can define a generalized reflector map $S^n \to S^n$ as follows. For any unit vector $m \in S^n$ we consider a ray originating at O in direction m. Let

$$\rho(m) = max\{\lambda \geq 0 \mid \lambda m \in B\} \tag{4.5}$$

be the radial function of the reflector R. Consider the point $r(m) = \rho(m)m$ on reflector R. By construction, at every point of R there exists at least one supporting paraboloid. Let $P(y)$ be a supporting paraboloid at $r(m)$. It follows from the well known reflection property of parabolas that the ray of direction m reflects off $P(y)$ at $r(m)$ in direction y. If u is the exterior unit normal to the paraboloid $P(y)$ at $r(m)$ then, according to Snell's law,

$$y = m - 2\langle m, u\rangle u. \tag{4.6}$$

The generalized reflector map (denoted again by γ) is defined by the reflector R as the map $\gamma : S^n \to S^n$ by setting

$$\gamma(m) = \bigcup \{y\}, \tag{4.7}$$

where y is the axis of a supporting paraboloid at $r(m)$ and the union is taken over all such supporting paraboloids. The map γ is, in general multivalued. Also, it maps S^n onto itself, since, as it was mentioned earlier, for any $y \in S^n$ there exists a supporting paraboloid with axis y.

It can be shown (see, [40]) that the focal function of a reflector in R can be extended from S^n to the entire R^{n+1} as a positively homogeneous of order one function and the extended function is convex. The converse, in general, is not true. If a given function $p \in C^2(S^n)$ is such that $p > 0$ on S^n and

$$p_{\alpha\alpha} + p - \frac{|\nabla p|^2 + p^2}{2p} > 0, \tag{4.8}$$

where $p_{\alpha\alpha}$ denotes differentiation along the arc length of any large circle of S^n and ∇ denotes the gradient in the metric of S^n. Then the map

$$-r(y) = \nabla p(y) + p(y)y - \rho(y)y, \tag{4.9}$$

where $\rho = \frac{|\nabla p|^2 + p^2}{2p}$, defines a closed convex reflector in $\mathcal{R}$ of class C^1. Furthermore, $|r(y)| = \rho(y)$, that is, the radial function in direction $m(y) = |r(y)|/\rho(y)$ is $\rho(y)$ [40].

The question when a positive continuous function $p(y)$, $y \in S^n$, generates a reflector in $\mathcal{R}$ such that for each $y \in S^n$ the paraboloid

$$P(y) : \rho(m) = \frac{p(y)}{1 - \langle m, y\rangle}$$

is supporting is open.

From now on it will be convenient to assume that the input energy distribution is a non-negative integrable function $I(m)$ on the sphere S^n. If it is defined only on a subset D of S^n we will assume that it is extended to the entire S^n by setting it equal to zero on $S^n \setminus D$.

The amount of energy transferred from the source O in a given set of directions $\omega \subset S^n$ is best described with the use of the following notion of "visibility" sets.

Definition 3. Let $R \in \mathcal{R}$, ω a subset of S^n, and

$$M(\omega) = \{z \in R \mid \exists\, y \in \omega \text{ such that } P(y) \text{ is supporting to } R \text{ at } z.\}$$

Let $V(\omega)$ be the radial projection of $M(\omega)$ on S^n by rays from O. This set is called the **visibility** set of ω.

Obviously, $V(\omega) = \gamma^{-1}(\omega)$, where g is the generalized reflector map. It can be shown [15, 40] that for a Borel set $\omega \subset S^n$ the set $V(\omega)$ is Lebesgue measurable on S^n, and, therefore, the function

$$G(R, \omega) = \int_{V(\omega)} I(m) d\sigma(m),$$

where $d\sigma$ is the standard measure on S^n, is a Borel measure on S^n.

For a given nonnegative and integrable function L on S^n we say that a closed convex reflector R is a *weak solution* of the reflector problem if

$$G(R, \omega) = \int_{\omega} L(y) d\sigma(y) \quad \text{for any Borel set} \quad \omega \subset S^n. \tag{4.10}$$

The equation (4.10) is the weak form of the equation (4.3) as we now explain by showing that for a smooth reflector the function $G(R, \omega)$ and the relation (4.3) can be written explicitly in terms of the focal function. More precisely, let $p \in C^2(S^n)$, $p > 0$, satisfy (4.8), and $R \in \mathcal{R}$ is the corresponding reflector given by (4.9). The outward unit normal vector field on R is given by

$$u = -\frac{\nabla p + py}{\sqrt{|\nabla p|^2 + p^2}}. \tag{4.11}$$

Also, $|r(y)| = \rho(y) > 0$ and $m(y) = r(y)/\rho(y)$ satisfies the reflection law (4.1). Thus, in this case $m(y) = \gamma^{-1}(y)$, where γ is the C^1 reflector map associated with R. Its Jacobian can be computed as

$$[J(\gamma^{-1})]^2 = \frac{\det[\langle \partial_i(r/\rho), \partial_j(r/\rho)\rangle]}{\det[e_{ij}]}, \quad i,j = 1,2,$$

where $e = [e_{ij}]$ is the matrix of coefficients of the standard metric on S^n. Taking the positive value of the square root, we get

$$M(p) \equiv |J(\gamma^{-1})| = \frac{\det[Hess(p) + (p-s)e]}{s^2 \det(e)}, \tag{4.12}$$

where $Hess(p)$ is the matrix of second covariant derivatives computed in the metric e. Then, for a Borel set $\omega \subset S^n$

$$G(R, \omega) = \int_{V(\omega)} I(m) d\sigma(m) = \int_{\omega} I(r(y)/\rho(y)) M(p(y)) d\sigma(y). \tag{4.13}$$

Consequently, the point-wise version of (4.10) is

$$I(r(y)/\rho(y)) M(p) = L(y) \quad \text{on} \quad S^n. \tag{4.14}$$

4.2.3 Strong Solutions of the Reflector Problem

The considerations leading to (4.14) are essentially local and we could have restated the relation (4.3) as

$$I(r/\rho)M(p) = L \quad \text{on} \quad T \subset S^n. \tag{4.15}$$

For a more detailed derivation of this equation and some related results see [54], [43]. The corresponding reflector problem in this case is formulated as the problem of finding a function $p \in C^2(T) \cap C^1(\bar{T})$ such that the normalized map (4.9)

$$r/\rho : \bar{T} \to \bar{D} \tag{4.16}$$

is one-to-one between $\bar{D}$ and $\bar{T}$ and satisfies (4.15) for given D, positive function I on D, T, and positive function L on T. Such function p will be called a *strong* solution of the reflector problem.

4.2.4 Existence, Uniqueness and Regularity

Existence of weak solutions to the reflector problem was established in [15, 19]; see also [14].

Theorem 1. *Let I and L be nonnegative integrable functions on S^n. The reflector problem in weak formulation admits a solution if and only if*

$$\int_{S^n} I(m) d\sigma(m) = \int_{S^n} L(y) d\sigma(y). \tag{4.17}$$

In [15] it was shown that the weak solution to the *reflector problem* can be obtained as a limit of a sequence of solutions to "discrete" problems. This approach, in addition to providing a proof of existence, is also suitable for computing solutions numerically (see section 4.2.5 below).

The main idea of the proof is as follows. Approximate the measure $\int_{S^n} L(y) d\sigma(y)$ by a sequence of Dirac measures F_k for each of which one can find a reflector in $\mathcal{R}$ defined by k supporting paraboloids and such that the equation (4.10) with the right hand replaced by F_k is satisfied. It is shown that the family of such special solutions is compact in $C(S^n)$. Then, selecting a convergent sequence of reflectors and using weak continuity of $G(R_k, \omega)$ we obtain a weak solution of (4.10).

We describe now this procedure in some more detail. Fix some integer $k \geq 2$ and let

$$F_k = \sum_{i=1}^{k} f_i \delta(y_i), \quad f_i > 0,$$

be a discrete measure concentrated at some distinct points $y_1, \cdots, y_k \in S^n$. The "discrete" version of the problem (4.10) consists of constructing a hypersurface $R^k \in \mathcal{R}$ defined by k paraboloids of revolution with foci at O

and axes $y_1, \cdots, y_k$ in such a way that for each $i = 1, ..., k$ the input energy $\int_{S^n} I(m) d\sigma(m)$ is redistributed so that the amount of it transferred in direction y_i is equal to f_i.

Note that when points $y_1, \cdots, y_k$ are fixed, the reflector R^k is completely determined by the k-tuple $(p_1, \cdots, p_k)$ of their focal parameters.

On the other hand, for each $i = 1, ..., k$ the visibility set $V(y_i)$ is the set of all rays reflected by the supporting paraboloid P_i in the direction y_i. Then, the amount of energy sent by the reflector R^k in direction y_i is given by the integral

$$G(R^k, y_i) \equiv G_i(R^k) = \int_{V(y_i)} I(m) d\sigma(m).$$

Thus to solve the discrete reflector problem we need to construct reflector R^k defined by paraboloids $P_1, \cdots, P_k$ such that

$$G_i(R^k) = f_i, \quad i = 1, \cdots, k. \tag{4.18}$$

It is shown in [15] that if the energy balance condition

$$\int_{S^n} I(m) d\sigma(m) = \sum_{i=1}^{k} f_i \tag{4.19}$$

is satisfied then the problem (4.18) has a solution and if $inf_{S^n} I > 0$ then this solution is unique up to a homothety with respect to O. The key ingredient of the proof is a C^0 a priori estimate of the radial function ρ^k of R^k

$$0 < c_1 \leq \rho^k(m) \leq c_2 < \infty, \tag{4.20}$$

valid whenever the total output energy $\sum_{i=1}^{k} f_i$ is not concentrated at one point; one of the constants, for example, c_2, can always be selected equal to 1 (due to invariance of solutions with respect to homotheties), while the constant c_1 then depends only on the $max_{1 \leq i \leq k} f_i$.

The procedure for finding a solution of (4.18) is constructive and will be outlined in section 4.2.5 where we discuss a scheme for solving (4.18) numerically.

The weak solution to the general *reflector problem* (4.10) is constructed as a limit of solutions to a sequence of discrete reflector problems. This is done in two stages. On the first stage S^n is partitioned into k subsets $\omega_1, \cdots, \omega_k$, $k \geq 2$, such that

$$\bigcup_{i=1}^{k} \omega_i = S^n \quad \text{and the numbers} \quad f_i = \int_{\omega_i} L(y) d\sigma(y), \quad i = 1, \cdots, k,$$

are introduced. Let $y_i \in \omega_i$. If the balance equation (4.17) is satisfied then the equation (4.19) is also satisfied, and, consequently, for the data I, $y_1, ..., y_k$

and $f_1, ..., f_k$ there exists a reflector $R^k \in \mathcal{R}$ defined by k paraboloids solving the problem (4.18).

By repeating this process for a sequence of partitions of S^n into subsets with decreasing diameters, we obtain an infinite sequence of reflectors $R^k, k = 2, ...$ The estimate (4.20) implies uniform bounds away from zero and from $+\infty$ on the diameters of R^k. Applying Blashke's selection principle, one can conclude that there exists a subsequence of reflectors converging in $C^0(S^n)$ to a convex reflector $R \in \mathcal{R}$. It is shown that the corresponding measures $G(R^k, \cdot)$ converge weakly to $G(R, \cdot)$ satisfying (4.10) and, therefore, R is the required weak solution.

If $inf_{S^n} I > 0$ then the solution R is unique up to a homothetic transformation relative to O. In [15] this was shown in case of reflectors defined by a finite number of confocal paraboloids; the general case was treated in [19] under a slightly more general condition on I. The proof of uniqueness is based on the "touching" principle which is essentially a special geometric form of the maximum principle for elliptic equations geometric form of the "touching" principle; see [1], ch. IX, [2].

The following regularity result was established in [19].

Theorem 2. *If I, $L \in C^{1,1}(S^n)$, I, $L > 0$, and the condition (4.17) holds then the radial function ρ of the reflector R is of class $C^{3,\alpha}(S^n)$ for any $\alpha \in (0, 1)$.*

Further regularity follows from standard elliptic theory. In particular, if $I, L \in C^\infty(S^n)$ then $\rho \in C^\infty(S^n)$.

In practical applications the reflector problem must be solved when the input and output apertures $\bar{D}$ and $\bar{T}$ do not coincide with S^n. The construction of the weak solutions described earlier can be easily adapted to this formulation as well. Namely, when $\bar{D} \subset S^n$ and $\bar{T} \subset S^n$ we extend the functions I and L to the entire S^n by setting them equal to zero on $S^n \setminus \bar{D}$ and $S^n \setminus \bar{T}$. Assuming that the balance equation is satisfied with D and T, it will also be satisfied with the functions I and L extended to the entire S^n. Under such circumstances the Theorem 1 can be applied. Next, we delete from the resulting closed reflector all of the supporting paraboloids with axes of directions in $S^n \setminus \bar{T}$. This determines a reflector R' with a boundary and whose radial projection on S^n is some closed set $D' \subset S^n$. Since

$$G(R', \omega) = \int_{V(\omega)} I(m) d\sigma(m) = \int_\omega L(y) d\sigma(y) \quad \text{for any Borel set} \quad \omega \subset T,$$

$suppI \subset D'$. The uniqueness (up to a homothety with respect to O) holds also in this case.

A different approach to construction of solutions in the when $n = 2$ and $\bar{D}, \bar{T} \neq S^n$) was considered in [52] where the uniqueness and regularity was also studied. The uniqueness of smooth solutions to (4.2), (4.3) when $n = 2$ was also considered in [32].

We also note that the special case of the rotationally symmetric solutions of the reflector problem was investigated in detail in [43]. In this connection see also [37, 44]. Various geometric properties of reflectors have been studied in [20, 34].

4.2.5 Computational Methods

Because of important practical applications, this problem has been discusses very extensively in engineering literature. During the last four decades essentially three different approaches have been used for formulating the general problem analytically and solving it numerically. In the first approach the problem is formulated as a system of first order partial differential equations and a method resembling the method of characteristics is applied to solve the system numerically; see [17, 23], and other references there. In the second approach the same problem is formulated as a boundary value problem for a second order partial differential equation of Monge-Ampère type for a certain complex-valued function [53]. Then a linearization procedure is used to construct a solution close to some a priori selected solution. In both approaches a rigorous mathematical analysis of the resulting equations is lacking and consequently the validity of the numerics is never fully established. This led to a controversy regarding existence and uniqueness of solutions that until now has not been resolved [26, 17, 23]. The only computational algorithm for this problem for which convergence has been rigorously established is given in [14]. We outline in the following subsection.

The method of supporting paraboloids (SP method) The method described in section 4.2.4 for proving existence of weak solutions of (4.10) can also be applied to construct numerical solutions. The main idea is as follows. Since a weak solution is obtained as a limit in C^0 of a sequence of closed convex reflectors solving (4.18) and defined by a finite number of supporting paraboloids, one needs a method for constructing numerically a solution to the problem (4.18). The latter requires finding k parameters $p_1, ..., p_k$ satisfying (4.18) and therefore can be viewed as a finite dimensional version of the problem (4.10).

Such a method (SP method) and its numerical implementation were given in [14] where it was shown that if the energy balance condition (4.19) holds then for any given $\epsilon > 0$ the corresponding algorithm generates in a finite number of steps a reflector R^k defined by k confocal paraboloids of revolution for which[1]

$$E(R^k) \equiv \frac{\sqrt{\sum_{i=1}^{k}(G_i(R^k) - f_i)^2}}{\int_D I} \leq \epsilon. \tag{4.21}$$

---

[1] We have to use here the expression E as the measure of error because the error defined in a usual way, for example, as $\sqrt{\sum(G_i(R_k) - f_i)^2}$, changes linearly when the total energy $\int_D I$ is rescaled. To neutralize this effect, we use the definition of error as in (4.21).

When $\epsilon \to 0$ the corresponding sequence of reflectors converges to the exact solution of (4.18).

The main steps of the SP method are as follows. Fix some $\epsilon > 0$ and k confocal paraboloids $P_1, ..., P_k$ with foci at O and respective focal parameters $p_1 = 1, p_2 > 0, ..., p_k > 0$, such that for the reflector defined by $P_1, ..., P_k$ the following inequalities hold

$$G_i(R^k) \equiv G_i(1, p_2, \cdots, p_k) \leq f_i, \quad i = 2, \cdots, k. \quad (4.22)$$

hold. This is always possible, since for each $i = 2, \cdots, k$

$$G_i(1, p_2, \cdots, p_k) \longrightarrow 0 \text{ as all } p_2, \cdots, p_k \longrightarrow +\infty; \quad (4.23)$$

see [14]. Denote by $\mathcal{R}^{\|}$ the class of reflectors which satisfy (4.22). Obviously, each reflector in $\mathcal{R}^{\|}$ corresponds to a point $(p_2, \cdots, p_k)$ in the positive part of the Euclidean space R^{k-1}. Because of the estimate (4.20) the set of points corresponding to reflectors in R^{k-1} is compact and at a positive distance from coordinate hyperplanes.

The important property of the functions $G_i(1, p_2, \cdots, p_k)$ is that when p_i is decreasing the function $G_i(1, p_2, \cdots, p_k)$ is strictly increasing, while $G_j(1, d_2, \cdots, d_k)$ are non-increasing for all $j \neq i$. Consequently, we can decrease sequentially and monotonically each of the $p_i, i = 2, \cdots, k$ in a manner that allows to remain in the class $\mathcal{R}^{\|}$. Since the set R^{k-1} is compact this process can be continued without any of the parameters p_i vanishing until (4.21) is satisfied. Note that because the balance equation is satisfied on each iteration only the paraboloids $P_2, ..., P_k$ must be modified.

The SP algorithm can be initialized with any reflector in $\mathcal{R}^{\|}$. In our implementation, using the property (4.23), the algorithm automatically determines from the data some values of $p_2, \cdots, p_k$ for which it is guaranteed that $R = (1, p_2, \cdots, p_k)$ is in $\mathcal{R}^{\|}$. Once the algorithm is initialized it produces iteratively a sequence of reflectors $R_s^k, s = 0, 1, 2, ...$, such that a) each member of this sequence belongs to $\mathcal{R}^{\|}$, b) the sequence monotonically converges to a solution of (4.18) and c) there exists some finite N for which R_N^k satisfies (4.21). This N can be estimated in terms of the given data. More details on the proof of convergence, including an estimate of the convergence rate, can be found in [14].

4.2.6 Open Problems

For convenience of the reader we collect here some of the open questions related to the problem discussed in this section. Some of these questions have already been mentioned.

1. The solutions to the reflector problem described above are convex hypersurfaces with the direction of convexity away from the source. In applications, concave reflectors (that is, with direction of convexity towards the

origin) are also very important [21]. The problems of existence, uniqueness and regularity of solutions in this case have not been investigated (see, however, a short remark in Wang [52]).

2. Is there a "natural" variational problem for which the equation (4.10) or (4.14) is the Euler-Lagrange equation? Some results in [38] may be useful here.

3. Our numerical implementation of the SP method works quite well in problems in which the number of paraboloids k is not too large. This is an obvious drawback of the SP method. Some improvements have been proposed in [29], but many important issues still remain open. The principal one is concerned with finding a way to increase significantly the efficiency of the SP method for large k. Also, accurate error estimates between solutions of (4.10) and approximate solutions of (4.18) are not available. It is not difficult to see that the SP method has features which are very useful for its implementation in a parallel architecture on a multi-processor computer. However, such an implementation has not been carried out yet.

It may be noted that the importance of investigating numerical schemes for the reflector problem goes far beyond this specific application. This is connected with the fact that elliptic equations of Monge–Ampère type are representative of fully nonlinear elliptic equations, and any results obtained for Monge–Ampère equations will be useful for understanding the fully nonlinear case.

4.3 Creating a Prescribed Intensity Distribution in the Near-Field

The configuration of the system is shown schematically on Fig. 4.2. In contrast to the problem described in section 4.2 in this problem the output aperture T is a specified region in space, usually, a two-dimensional surface, for example, a bounded domain on a plane. The objective is to determine a surface R which reflects the rays emitted by the source O through the aperture $\bar{D}$ in such a manner that the reflected rays reach every point of $\bar{T}$ and the intensity I of the source is transformed into a pre-specified intensity distribution on T.

This is a practical design problem from applied optics. It was formulated by J. Schruben [48] in 1972 as a problem of designing a reflector that "spreads" the light of a small high-intensity discharge source over a large plane aperture. In [48] Schruben derived an implicit integro-differential equation for the polar radius of the reflector but did not suggest any ways for solving it. Later some results in the special case of rotationally symmetric reflectors were obtained in [49]. In fact, the case of rotationally symmetric reflectors has been considered quite extensively in optical research; see, [7], [22], [33], [43], and other references there. General formulas for energy density produced by reflected fronts can be found in [6] and [8].

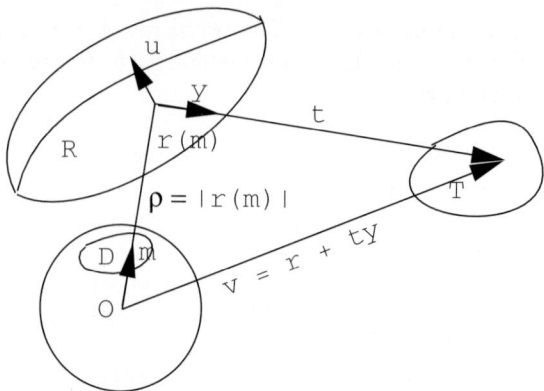

Fig. 4.2. Formulation of the near-field problem

4.3.1 Statement of the Near-Field (NF) Reflector Problem

Let S^n be a unit sphere centered at the coordinate origin O of R^{n+1} and D a domain on S^n. The case $D = S^n$ is not excluded. The point O is considered as a source of rays passing through $\bar{D}$. Let $I(m)$, $m \in D$, be a nonnegative integrable function on D. In agreement with the geometrical optics origin of the problem we will refer to D as the "input aperture" and to I as the "input intensity".

Let h be some hyperplane in R^{n+1} and T a bounded set on h. Consider a hypersurface R given by the vector function $r(m) = \rho(m)m$, $m \in \bar{D}$, where m is treated as a point on S^n and a unit vector in space, and $\rho : \bar{D} \to (0, \infty)$, $\rho \in C^2(D) \bigcap C^1(\bar{D})$. Assume that the hypersurface R is such that any ray of direction $m \in \bar{D}$ from the source O reflects off R in direction $y(m)$ according to (4.1) and hits $\bar{T}$. Then we obtain a map $\Gamma(m) : \bar{D} \to \bar{T}$

$$\Gamma(m) = \rho(m)m + t(m)y(m). \tag{4.24}$$

Here, $t(m)$ is the distance from the point $r(m)$ on R to a point in $\bar{T}$ along the reflected ray.

The intensity I of the source O and the resulting intensity distribution L on T are related by

$$L(\Gamma(m))|J(\Gamma(m))| = I(m), \quad m \in D, \tag{4.25}$$

where $J(\Gamma)$ is the Jacobian of the map Γ [5]. The explicit expression for $J(\Gamma)$ in terms of the polar radius ρ is very complicated [41]. A somewhat simpler expression was derived in [38] by considering the inverse map Γ^{-1} and expressing it in terms of the function $l = \rho + t$ defined as a function on $\bar{T}$. Thus, assume that Γ is a diffeomorphism and let $m \in \bar{D}$ be such that $m = \Gamma^{-1}(x)$, where $x \in \bar{T}$. We assume here that $\bar{T}$ is given in parametric form as $x = x(\xi^1, \xi^2, ..., \xi^n)$ for some Cartesian coordinates $\xi^1, \xi^2, ..., \xi^n$. Thus, x

denotes a point in $\bar{T}$ as well as the position vector in R^{n+1} defining this point. Using Γ^{-1} we introduce the local coordinates $\xi^1, \xi^2..., \xi^n$ on the reflector R and it can be shown that

$$y = \sum_{i=1}^{i=n} \frac{\partial l}{\partial \xi^i} \frac{\partial x}{\partial \xi^i} - \psi \eta \qquad \rho = l - \frac{l^2 - x^2}{2(l - \langle x, u \rangle)}, \qquad (4.26)$$

where $\psi = (1 - |grad\ l|^2)^{1/2}$. Also,

$$Q(l) \equiv |J(\Gamma^{-1})| = \frac{\det[(l - \rho)Hess(l) + grad\ l \otimes grad\ l - Id]}{\rho^2 \psi}, \qquad (4.27)$$

where $grad\ l \otimes grad\ l$ is the matrix $[\frac{\partial l}{\partial \xi^i} \frac{\partial l}{\partial \xi^j}]$ and Id is the identity matrix.

Now, suppose that D, T and the nonnegative, integrable functions I on D and L on T are given. Analytically, the NF reflector problem consists in determining a smooth hypersurface R such that the corresponding map

$$\Gamma \in Diff(\bar{D}, \bar{T}) \qquad (4.28)$$

and

$$I(\Gamma^{-1}(x))Q(l) = L(x), \quad x \in T. \qquad (4.29)$$

The rotationally symmetric case of this problem was studied in [38]. The general case in weak formulation was investigated in [27]. The main idea here is to use families of confocal ellipsoids with one common focus at the source O and the second foci positioned at points of $\bar{T}$. While this construction of weak solutions is similar to that in section 4.2, there are some important differences. In [28] a convergent computational scheme for calculating weak solutions numerically was proposed and analyzed.

Finally, we note that the assumption that T is a set on a hyperplane was made only in order to simplify the expression for $Q(l)$. A similar expression can be derived in the more general case when $\bar{T}$ is a smooth hypersurface in R^{n+1} [38] (where the exposition is for the case $n = 2$ but is essentially the same for any n).

4.3.2 Weak Formulation and Solution of the NF Reflector Problem

Here we follow closely our paper [27]. Let d be a continuous function defined on $\bar{T}$ and $d : \bar{T} \longrightarrow (0, +\infty)$. For each $x \in \bar{T}$ denote by $E(x)$ an ellipsoid of revolution, about Ox as the axis, with foci at points O and x and polar radius

$$\rho(m) = \frac{d(x)}{1 - \epsilon \langle m, k \rangle}, \quad m \in S^n, \qquad (4.30)$$

where ϵ is the eccentricity, $k = \frac{x}{|x|}$, and $d(x)$ is the focal parameter of $E(x)$. It is assumed that $dist(O, \bar{T}) > 0$. Note that for any ellipsoid $E(x)$ the eccentricity $\epsilon \in (0,1)$ and $\epsilon = 1$ if only if the ellipsoid degenerates into the linear segment $[O, x]$.

The solid closed convex body bounded by $E(x)$ we denote by $B(x)$. Let

$$B = \cap_{x \in \bar{T}} B(x), \quad R = \partial B. \tag{4.31}$$

It follows from our assumptions that $d \geq const > 0$ on $\bar{T}$ and therefore the closed convex hypersurface R is star-shaped relative to O.

Definition 4. *Assume that the function $d(x)$ is large enough so that for any $x \in \bar{T}$ $B(x) \supset \bar{T}$. In this case we call R a closed convex **reflector** with the light source at O. The class of closed convex reflectors with the source at O defined by the set $\bar{T}$ is denoted by $\mathcal{R}_e$. We will refer to $\bar{T}$ as the "output aperture".*

In the special case when the set $\bar{T}$ consists of a finite number of points $x_1, x_2, ..., x_k$ we assume that the function d is replaced by a k-tuple of positive numbers $d_1, d_2, ..., d_k$ and we say that the reflector R is defined by a finite number of ellipsoids $E(x_i), i = 1, 2, ..., k$.

Definition 5. *Let $R \in \mathcal{R}_e$ and let $E(x)$ be n ellipsoid with foci at O and $x \in R^{n+1}$. Assume that R is contained inside the convex body bounded by $E(x)$. If $E(x) \cap R \neq \emptyset$, and $z \in E(x) \cap R$ then we call $E(x)$ **an ellipsoid supporting to R at the point** z.*

Note that at every point of R there exists a supporting ellipsoid from the family $E_d(x), x \in \bar{T}$. However, this family may include ellipsoids which are not supporting to R. Also, note that for any $x \in R^{n+1}$, $x \neq O$, there exists an ellipsoid with focus at O and axis of direction $x/|x|$ supporting to R. Such an ellipsoid can be constructed by first taking an ellipsoid with one focus at O, the other focus on the ray of direction $x/|x|$ and then choosing focal parameter large enough so that it contains R. Then, by shrinking this ellipsoid homothetically to O until it touches R we obtain the required ellipsoid.

Definition 6. *Let $R \in \mathcal{R}_e$ be a reflector with the output aperture $\bar{T}$. Let ω be a subset of $\bar{T}$ and let*

$$M(\omega) = \{z \in R \mid \exists\, x \in \omega \text{ such that } E(x) \text{ is supporting to } R \text{ at } z\}.$$

*Since R is star-shaped with respect to the point O, we may consider the radial projection of $M(\omega)$ by rays from O on S^n. The image of the set $M(\omega)$ under this projection is called the **visibility** set of ω. The corresponding map from $\bar{T}$ to S^n we denote by V.*

It was shown in [27] that for any Borel set $\omega \in T$ the set $V(\omega)$ is Lebesgue measurable. Clearly, the map V is the analogue of the map Γ^{-1} described earlier.

Let the domain D and function I be as in the beginning of section 4.3.1. Extend the function I from D to the entire S^n by setting $I = 0$ on $S^n \setminus D$.

For each reflector $R \in \mathcal{R}_e$ we define a function on Borel subsets of T by the formula
$$G(R,\omega) = \int_{V(\omega)} I(m)d\sigma(m). \tag{4.32}$$

It was shown in [27] that $G(R,\omega)$ is a finite, completely additive measure on T. In addition, the measure $G(R,\omega)$ is weakly continuous, that is, if a sequence of convex reflectors R_s (with the same source O and output aperture $\bar{T}$) converges to a convex reflector R then the measures $G(R_s,\omega)$ converge weakly to the measure $G(R,\omega)$. Convergence of R_s to R can be understood here as convergence in $C^0(S^n)$ of respective radial functions.

Let L be a nonnegative integrable function on T. A reflector $R \in \mathcal{R}_e$ is called a *weak solution* of the NF reflector problem if
$$G(R,\omega) = \int_\omega L(x)d\mu \quad \text{for any Borel set} \quad \omega \subset T, \tag{4.33}$$

where $d\mu$ is the standard Lebesgue measure on the hyperplane h.

Theorem 3. *[27] Let $\bar{T}$ be a compact set on a hyperplane in R^{n+1}, $dist(O,\bar{T}) > 0$, I a nonnegative integrable function on S^n and L a nonnegative integrable function on T. Suppose*
$$\int_{S^n} I(m)d\sigma(m) = \int_T L(x)d\mu(x). \tag{4.34}$$

Then there exists a reflector $R \in \mathcal{R}_e$ which is a weak solution of the NF reflector problem. If $I(m) \geq const > 0$ on S^n then the reflector R is unique provided one of the supporting ellipsoids is fixed.

The additional condition for uniqueness requiring fixing one of the supporting ellipsoids is not superfluous. This can be seen from the following simple example. Suppose $\bar{T}$ consists of one point $\{x\}$. Then any ellipsoid of revolution with foci at O and $\{x\}$ is a weak solution of the NF reflector problem. Note that these solutions are not homothetic to each other relative to O and, using the value of the focal parameter, they can be parametrized by the positive half-axis. The uniqueness part in Theorem 3 has been established in the case of reflectors defined by a finite number of ellipsoids, but the result is true in general, though it has not been published.

The proof of Theorem 3 is based on an a priori C^0 estimates from below (away from zero) and from above of the focal parameters of supporting ellipsoids to reflectors satisfying (4.33). The proof of this estimate is more delicate than that of the similar estimate (4.20). However, the general idea of constructing a weak solution as a limit of reflectors defined by a finite number of supporting ellipsoids is the same as in the proof of Theorem 1 (see section 4.2.4). Uniqueness is proved with the use of the touching principle.

A numerical scheme for finding weak solutions when the right hand side of (4.33) is a Dirac measure concentrated at a finite number of points was investigated in [28]. It is based on a monotonicity property of $G(R,\vec{\ })$ similar to the one described in section 4.2.5. Several important improvements to this scheme were described in [45].

4.3.3 Some Open Problems

The problem of regularity of weak solutions to the NF problem is open. Also, existence of regular solutions to the problem (4.28), (4.29) is open. Note, however, that for an input aperture $D \subset S^n, \bar{D} \neq S^n$, and the input intensity I extended by zero from D to the entire S^n the above weak solution gives a weak solution of the problem (4.28), (4.29). In this case, the reflector with the boundary is defined as a portion of the closed convex reflector which projects radially from O onto onto $\bar{D}$.

The reflectors in class $\mathcal{R}_e$ are closed convex reflectors. Can one construct reflectors with boundaries, star-shaped relative to O and convex towards O? Perhaps, it is possible to use for this purpose families of confocal hyperboloids of revolution; some results in this direction can be found in [30]. If the answer is "yes", what are the regularity properties of such reflectors?

The issues regarding efficiency of the numerical method described in section 4.2.6 apply also in case of the problem (4.33).

4.4 Two-Reflector System for Transforming a Beam of Parallel Rays

In contrast with the problems described in previous sections in which the input wave is spherical, here we consider a problem for the case when the incoming beam is formed by parallel rays. The output beam is required to have a plane wave front propagating in the same direction as the incoming one. In addition, it is required that the input intensity distribution be transformed into a prescribed in advance output intensity distribution. Because of these two requirements a system with one reflector surface will not have sufficient degrees of freedom to control both characteristics. However, this can be done with two reflector surfaces.

The known designs of optical systems with two reflectors are usually limited to rotationally symmetric reflectors and/or rotationally symmetric input and output intensities [31]. Designs in which rotational symmetry either of reflectors or of intensities is not imposed a priori are generally expected to be much more energy efficient because there is more control over the shape, position and intensity of the output beam. Malyak [31] studied the case when the input and output intensities are assumed a priori rotationally symmetric but the reflectors, in general, are not required to be rotationally symmetric. However, the design described in [31] leads to non-convex reflectors and manufacturing of such reflectors presents significant difficulties. In [31] substantial efforts were applied to determine parameters under which the "wells"

4 Beam Shaping Surfaces in Geometrical Optics   211

in the reflectors would have the lowest possible depth. The results described below show that with the same data and requirements as in [31] it is possible to design a two-reflector system with convex/concave reflectors and without requiring the input and output intensities to be rotationally symmetric. Thus, these particular difficulties are completely resolved with the proposed approach.

4.4.1 Statement of the Problem

Denote by z the horizontal axis and by $x_1, ..., x_n$ the Cartesian coordinates on the hyperplane $\alpha : z = 0$. A point in space is denoted by (x, z), where $x = (x_1, ..., x_n)$. The configuration of the system is shown schematically on Fig. 4.3. A beam of rays parallel to the z−axis and defined by its cross section $\bar{D}$ by the hyperplane α propagates in the positive direction k of the z−axis. On the figure the incoming beam is denoted by I. The set $\bar{D}$ is assumed to be a bounded convex set on α. The intensity of the incoming rays is denoted by $I(x)$, $x \in D$, where I is a nonnegative integrable function. The beam

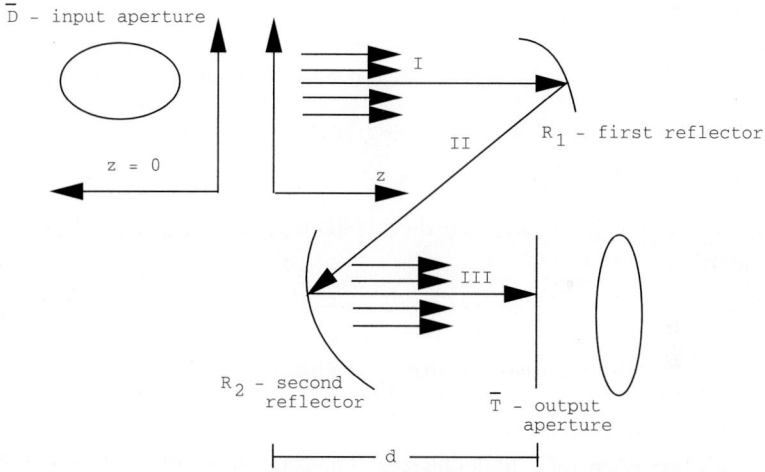

Fig. 4.3. Two-reflector system

shaping system consists of two reflecting hypersurfaces R_1 and R_2, both of which are assumed to be in nonparametric form and such that the incoming beam is transformed into the output beam III. It is required that the output beam III be formed by parallel rays propagating in the same direction k as the incoming beam. It is important that the reflectors be separated so that no blockage can occur. Consequently, the position and, in addition, the shape of the output beam of rays are specified by prescribing in advance its cross section $\bar{T}$ by a given hyperplane $z = d$. It is assumed that $\bar{T}$ can be any bounded convex domain.

Furthermore, it is required that the intensity of the input beam I be transformed so that intensity of the output beam III is a prescribed in advance function $L(p)$, where $(p, d) \in T$.

The geometrical optics approximation is used to describe the propagation and the energy conservation. In particular, it is assumed that the optical path length must be the same for all rays in the system and no energy is lost in the process.

The objective is, of course, to determine the reflectors R_1 and R_2 so that all of the above requirements are satisfied; see [31] and other references there.

We review now the steps involved in deriving the required relations. More details can be found in [39] where a more general situation is considered. We begin by tracing a typical ray through the system. Such a ray is "marked" by a point $x \in \bar{D}$. Thus, assuming that we already have two hypersurfaces R_1 and R_2 which satisfy the stated requirements, we derive the equations of the problem.

A ray marked by a point $x \in \bar{D}$ propagates in direction k, strikes the first reflector R_1, reflects off it in the direction $y(x)$, strikes the second reflector R_2, and reflects off it, again, in direction k. Since the hypersurfaces R_1 and R_2 are assumed to be in nonparametric form, they project univalently onto $\bar{D}$ and $\bar{T}$, respectively. Let R_1 be given by the position vector $r_1(x) = (x, z(x))$, $x \in \bar{D}$, $z \in C^2(\bar{D})$. Denote by u the unit normal vector on R_1 given by

$$u = \frac{(-z_x, 1)}{\sqrt{1 + z_x^2}},$$

where we put, for brevity, $z_x \equiv \mathrm{grad}\ z$. This together with the reflection law (4.1) gives

$$y = k - 2\frac{(-z_x, 1)}{1 + z_x^2}. \tag{4.35}$$

Denote by $t(x)$ the distance from reflector R_1 to reflector R_2 along the ray reflected in the direction $y(x)$ and let $s(x)$ be the distance from R_2 to $\bar{T}$ along the corresponding ray reflected off R_2. Assume for the time being that $t \in C^1(\bar{D})$ and R_2 is a C^1 hypersurface. The total optical path length (OPL) corresponding to the ray associated with the point $x \in \bar{D}$ is denoted by $l(x)$ and it is given by

$$l(x) = z(x) + t(x) + s(x). \tag{4.36}$$

Because $l(x, y)$ is the optical distance between input and output fronts, we have (see [5]):

$$l(x) = const \equiv l. \tag{4.37}$$

Let

$$R_2: \quad r_2(x) = r_1(x) + t(x)y(x), \quad x \in \bar{D}, \tag{4.38}$$

$$\bar{T}: \quad P(x) = r_2(x) + s(x)k, \quad x \in \bar{D}. \tag{4.39}$$

Thus, a point $(p,d) \in \bar{T}$ is the image of some $x \in \bar{D}$ under the ray tracing map P, that is, $(p,d) = P(x)$, $x \in \bar{D}$.

Let $P_\alpha(x)$ denote the projection of $P(x)$ on hyperplane α and T_α the projection of T on α. Then

$$P(x) = P_\alpha(x) + dk, \tag{4.40}$$

$$P_\alpha : \bar{D} \to \bar{T}_\alpha, \quad P_\alpha(x) = x + t(x)\frac{2z_x}{1+z_x^2}. \tag{4.41}$$

For the projection of $P(x)$ onto direction k we have

$$z + t(1 - \frac{2}{1+z_x^2}) + s = d.$$

Taking into account (4.35), (4.36), (4.37), and (4.40), we find

$$t = \frac{\beta}{2}(1+z_x^2), \tag{4.42}$$

where we put $\beta = l - d$. Then, using (4.41), we obtain

$$p = P_\alpha(x) = x + \beta z_x, \quad x \in \bar{D}. \tag{4.43}$$

Next, we relate the input intensity $I(x)$, $x \in D$, to the output intensity $L(P(x))$ on T. According to the differential form of the energy conservation law (see [5], p. 115),

$$L(P(x))|J(P(x))| = I(x), \quad x \in D, \tag{4.44}$$

where $J(P)$ is the Jacobian of the map P. It follows from (4.44) that D, T, I, and L must satisfy the necessary condition

$$\int_T L(p)dp = \int_D I(x)dx. \tag{4.45}$$

Using (4.40) and (4.43), we see that

$$J(P) = J(P_\alpha) = \det\left[Id + \beta Hess(z)\right],$$

where Id is the identity matrix. Hence, the equation (4.44) assumes the form:

$$L(x + \beta z_x)|\det\left[Id + \beta Hess(z)\right]| = I \text{ in } D. \tag{4.46}$$

Thus, to determine the reflectors R_1 and R_2 that satisfy the requirements stated in the beginning of this section one may attempt to find a function $z \in C^2(\bar{D})$ such that the map

$$P_\alpha = x + \beta z_x : \bar{D} \to \bar{T}_\alpha \tag{4.47}$$

is onto and satisfies (4.46). Once such a function z is found, it is necessary to check that the reflection off R_1 onto R_2 and off R_2 onto $\bar{T}$ are diffeomorphisms, so that the procedure leading to the map P_α is valid.

4.4.2 Properties of Reflectors R_1 and R_2

First note that for the function $\tilde{z} = z + \langle a, x \rangle + c$, where z as above, a a constant vector in the hyperplane α, and $c = const$, the image of $\bar{D}$ under $\tilde{P}_\alpha$ (generated with $\tilde{z}$) is the set $\bar{T}_\alpha$ translated by the vector βa. Also, $J(\tilde{P}_\alpha) = J(P_\alpha)$.

Next, we describe an interesting relation between the hypersurfaces R_1 and R_2 and the classical Legendre transform. To establish this relation we first express the position vector of R_2 in terms of z. Using (4.38), (4.42), and (4.35), we get

$$r_2 = (x + \beta z_x, z + \frac{\beta}{2} z_x^2 - \frac{\beta}{2}), \quad x \in \bar{D}. \tag{4.48}$$

We rearrange the z-component of r_2 as follows:

$$\frac{1}{\beta}\left[\beta z + \frac{\beta^2}{2} z_x^2 - \frac{\beta^2}{2}\right] = \frac{1}{\beta}\left[v - \langle x, v_x \rangle + \frac{1}{2} v_x^2\right],$$

where we put

$$v = \frac{x^2}{2} + \beta z - \frac{\beta^2}{2}, \quad v_x = grad\ v. \tag{4.49}$$

The function $v - \langle x, v_x \rangle$ is the negative of the usual Legendre transform of v.

Consider the transformation

$$p = v_x, \quad w = \frac{1}{\beta}\left[v - \langle x, v_x \rangle + \frac{1}{2} v_x^2\right]. \tag{4.50}$$

If the map $P_\alpha : \bar{D} \to \bar{T}$, $p = v_x$, is a diffeomorphism then we can reparametrize the position vector r_2 of R_2 as

$$r_2(p) = (p, w(p)), \quad p \in \bar{T}. \tag{4.51}$$

Put $w_p = grad\ w$ with respect to p and $p_x = Hess(v)$. Then, since

$$w_p p_x = \frac{1}{\beta}(-x + v_x) Hess(v),$$

we get

$$w_p = \frac{1}{\beta}(-x + p), \tag{4.52}$$

and

$$x(p) = p - \beta w_p(p), \quad v(p) = \beta[w(p) - \langle p, w_p(p) \rangle] + \frac{p^2}{2}. \tag{4.53}$$

Consequently, using (4.49) and (4.53), we get an explicit expression for the hypersurface R_1 in terms of p:

$$r_1(p) = (x(p) = p - \beta w_p(p), \quad z(p) = w(p) - \frac{\beta}{2} w_p^2(p) + \frac{\beta}{2}), \quad p \in \bar{T}_\alpha. \tag{4.54}$$

A comparison of this with (4.48) shows a curious almost "symmetry" between the two reflectors, which is physically very natural, since one could start the ray tracing from the output front and "mark" the rays by points in $\bar{T}$.

The relations (4.50), (4.53) can be used to establish some useful properties of the reflectors R_1 and R_2. For example, let the function v defined by (4.49) be strictly concave, that is, $Hess(v) < 0$ in D. Differentiating the first of the equations in (4.50) and using (4.49), we obtain

$$Id = Hess(v)x_p = [Id + \beta Hess(z)]x_p,$$

where x_p is the matrix of the Jacobian of the map P_α^{-1}. It follows that x_p is negative. This and differentiation of (4.52) with respect to p give

$$Hess(w) = \frac{1}{\beta}(-x_p + Id) > 0.$$

Hence, the function w is strictly convex. Note, that in this case the function z is concave because it is a sum of two concave functions. Note that this property remains true if it is only assumed that $Hess(v) \leq 0$. In this case it follows by considering $v - \epsilon x^2/2$ and letting ϵ tend to zero in the inequality for $Hess(w)$.

This observation is very important for applications. Indeed, a concave v will produce a concave reflector R_1 while a convex v may lead to R_1 which is neither concave nor convex. As it was mentioned earlier, manufacturing a reflector which is neither concave nor convex is significantly more difficult than a concave (or convex) one [31]. Therefore, solutions produced with concave v are more useful in practical applications.

Our final remark in this section is related to the fact that if $z \in C^2(\bar{D})$ and $J(P_\alpha) \neq 0$ then the hypersurface R_2 defined by (4.38) with t defined by (4.42) is an embedding with no tangent spaces parallel to the z−axis. To establish this property we note first that a direct calculation using (4.48) (we omit it here) shows that the tangent spaces at corresponding points of R_1 and R_2 are parallel. Let τ be the map of R_1 to R_2 such that the corresponding points have the same x. Then a straight forward computation shows that $J(\tau) = J(P_\alpha)$, where $J(\tau)$ is the Jacobian of τ. Hence, $d\tau$ is nondegenerate. Since R_1 has no tangent spaces parallel to the z−axis, the same property holds also for R_2. It follows now from (4.1) applied to R_2 that the rays reflect off R_2 in direction k.

4.4.3 Weak Formulation and Weak Solutions

It is convenient to rewrite (4.46) in terms of the function v. Then (4.46) and the condition (4.47) assume the form

$$L(v_x)|det[Hess(v)]| = I \quad \text{in} \quad D, \quad v_x : \bar{D} \to \bar{T}_\alpha. \tag{4.55}$$

It is amazing that in this form the problem can be interpreted as the so-called "second boundary value problem" for Monge-Ampère equation arising in the geometry of convex hypersurfaces [3], Ch. 5, and in the theory of mappings with convex potentials [12]. The theory of such equations is available only when v is either convex or concave and most results are usually stated for convex solutions. Existence and uniqueness of weak solutions among convex (and concave) functions under the necessary condition (4.45) can be established by geometric methods and will be discussed below. The geometric approach is important because it is also used for solving the problem numerically.

We now describe the construction of weak solutions to (4.55). It suffices to consider the case of convex v because one can pass from concave solutions to convex with second condition in (4.55) replaced by $v_x : \bar{D} \to -\bar{T}_\alpha$, where $p' \in -\bar{T}_\alpha$ when $p' = -p$, $p \in \bar{T}_\alpha$. Let $W^+(\equiv W^+(\alpha))$ be the class of convex functions (not necessarily smooth) defined over the hyperplane α. Fix some $x_0 \in D$ and for $p \in \bar{T}_\alpha$ consider the hyperplanes $Q(p) : z = \langle p, x - x_0 \rangle$ and the half-spaces $Q^+(p) : z - \langle p, x - x_0 \rangle \geq 0$. The intersection of these half-spaces defines an infinite convex body whose boundary is a convex cone $K_T : z = k(x)$, with $k \in W^+$ and vertex at $(x_0, 0)$. Each of $Q(p)$, $p \in \bar{T}_\alpha$, is supporting to K_T and $grad\ Q \in \bar{T}_\alpha$. Note that a hyperplane supporting to K_T at any $(x, k(x))$, $x \in \alpha$, is also supporting at $(x_0, 0)$. Hence, the image of α under the map $\nu : x \to grad\ Q$, where Q is a supporting hyperplane at $(x, k(x))$, is $\bar{T}_\alpha$. Note that at $(x_0, 0)$ the map ν is multivalued. Thus, for any bounded convex set in α one can construct a convex cone whose image under the map ν is the closure of that set. Obviously, such cone is defined uniquely up to a parallel translation.

For any function $v \in W^+$ the map ν is defined similarly by putting $\nu(x) = \cup grad\ Q$, where the union is taken over all hyperplanes supporting to the graph of v at $(x, v(x))$. It can be shown that for any such v and Borel set $e \subset \alpha$ the image $\nu(e)$ is Lebesgue measurable [3], p. 116.

We need the following notion of an *asymptotic cone* [3]. Let $v \in W^+$, $S = (x, v(x))$, $x \in \alpha$, and let B_v be the infinite convex body bounded by the hypersurface S. Let A be an arbitrary point in the interior of B_v. Consider the collection of all infinite rays emanating at A and which do not intersect S. This collection is a cone which may be of dimension $1, ..., n+1$. The boundary of this cone is called the *asymptotic cone* of S. The asymptotic cone does not depend on the choice of the point $A \in B_v$.

The following result is adapted from [3], ch. 5.

Theorem 4. *Let D and T_α be bounded convex domains on α, $I \geq 0$ and $L > 0$ integrable functions in D and T_α, respectively, and assume that condition (4.45) is satisfied. Let K_T be a convex cone constructed as above from T_α. Then there exists a function $v \in W^+$ such that the asymptotic cone of the graph $(x, v(x))$, $x \in \alpha$, is K_T and for each Borel set $e \subset \alpha$*

4 Beam Shaping Surfaces in Geometrical Optics   217

$$\int_{\nu(e)} L(p)dp = \int_e I(x)dx, \qquad (4.56)$$

where it is assumed that $I \equiv 0$ for $x \notin \bar{D}$. The function v is unique up to an additive constant.

Theorem 4 is proved in two steps. On the first step the measure $\int_e I(x)dx$ is approximated by a sequence of Dirac measures μ_k, $k = 1, 2, ...$, concentrated at a finite number of points $x_1, x_2, ..., x_k$ in D and satisfying (4.45). For each μ_k the equation

$$\int_{\nu(e)} L(p)dp = \mu_k(e) \qquad (4.57)$$

is considered in the class of convex piece-wise linear functions from W^+ whose graphs may only have vertices projecting into the points $x_1, x_2, ..., x_k$. It is shown by a variational argument that this problem has a solution v_k which is unique up to an additive constant. In addition, this solution satisfies Lipschitz condition with a constant depending only on $\bar{T}_\alpha$. Consequently, one obtains a family of convex piece-wise linear functions $\{v_k\}_{k=1}^\infty$ compact in $W^+(\bar{D})$ (with the usual maximum-norm). For a sequence of v_k converging to some $v \in W^+$ the measures $\int_{\nu_k(e)} L(p)dp$ converge weakly to $\int_{\nu(e)} L(p)dp$ defined by v. The corresponding measures $\mu_k(e)$ converge to $\int_e I(x)dx$. Since the weak limit is unique, we conclude that v is the required weak solution. The proof of uniqueness is based on a weak form of the maximum principle for convex solutions of (4.56). Detailed proofs of existence and uniqueness can be found in [3].

This theorem contains everything we need except for the assertion that $\bar{T}_\alpha = \nu(\bar{D})$. The inclusion $\nu(\bar{D}) \subset \bar{T}_\alpha$ follows from the known fact that $\nu(\alpha) = \nu_{K_T}(\alpha)$ and $\nu(\bar{D}) \subset \nu(\alpha)$; here ν_{K_T} denotes the map ν applied to the asymptotic cone of the graph of v.

Let us show that the inclusion $\bar{T}_\alpha \subset \nu(\bar{D})$ is true. Let $S = (x, v(x))$, $x \in \alpha$. First we note that at any $x \notin \bar{D}$ there are no strictly supporting hyperplanes to S at $(x, v(x))$. Indeed, if such hyperplane were to exist then for some neighborhood U of x, $U \cap \bar{D} = \emptyset$, we should have $|\nu(U)| \neq 0$, where $|\ |$ denotes the Lebesgue measure. Since $\nu(U) \subseteq \bar{T}_\alpha$ we get a contradiction with (4.56). Let $X = (x, v(x))$ be a point on S with $x \notin \bar{D}$ and Q a supporting hyperplane containing X. The projection of the contact set of $Q \cap S$ on α can not have an empty intersection with $\bar{D}$. Otherwise, at some boundary point of the contact set there would exist a strictly supporting hyperplane. Hence, for any $x \notin \bar{D}$ the supporting hyperplane at $(x, v(x))$ is also supporting at some point $x' \in \bar{D}$ and we conclude that $\bar{T}_\alpha \subset \nu(\bar{D})$. Thus, $\nu(\bar{D}) = \bar{T}_\alpha$.

Corollary 1. *Let D, T_α, I and L be as in Theorem 4. Then the function v from that theorem is such that the map $\nu : \bar{D} \to \bar{T}_\alpha$ is onto and for each Borel set $e \subset D$ the equation (4.56) is satisfied.*

Using this corollary we can now construct a concave solution (in weak sense) to the original problem of recovering the hypersurfaces R_1 and R_2. We summarize the corresponding procedure. First we replace the given T_α by $-T_\alpha$. According to the corollary, there exists a convex v such that $\nu(\bar{D}) \to -\bar{T}_\alpha$ is onto and (4.56) is satisfied. Put $z(x) = -(v(x) + x^2/2)/\beta$, $x \in \bar{D}$. Then z is concave and the map $x + \beta\nu_z(x) = \nu_{-v}(x) : \bar{D} \to \bar{T}_a$, where ν_z and ν_{-v} denote the map ν evaluated for z and $-v$, respectively. Clearly, for any Borel set $e \subset D$ its image under the map $x + \beta\nu_z(x)$ is the same as under ν_{-v} and the equation (4.56) holds.

The reflector R_2 is constructed as follows. For a point $x \in \bar{D}$ we consider the point $X = (x, z(x))$. If at X there is a unique supporting hyperplane with angular coefficients $q = (q_1, ..., q_n)$ then $x + \beta\nu_z(x) = x + q = p$ and the corresponding point on R_2 has coordinates $(x + q, z + (\beta/2)q^2 - \beta/2)$. If at X there is more than one supporting hyperplane then the image $x + \beta\nu_z(x)$ of x is the $\cup\{x + q\} = \cup p$, where the union is taken over all hyperplanes supporting to the graph of z at X, and the corresponding set of points on R_2 is described by $\cup(x+q, z+(\beta/2)q^2 - \beta/2)$. Thus, a point X on the graph of z at which there is more than one supporting hyperplane can be interpreted as a point at which the light ray hitting this point splits and reflects in directions corresponding to each of the supporting hyperplanes at X.

As it has already been mentioned if instead of a concave v one uses a convex solution then z may not be convex. It is this non-convex solution for rotationally symmetric input and output intensities that was constructed in [31].

4.4.4 Regularity and Numerics

Regularity of weak convex solutions to problem (4.55) was investigated by L. Caffarelli in the context of studying gradient mappings with convex potentials [12, 11, 13]. The following results have been established. Let D and T be convex, I and L are both bounded away from zero and infinity, and both are Holder continuous with exponent $s > 0$. Then the weak solution of (4.55) $v \in C^{2,s}(D) \cap C^{1,s'}(\bar{D})$ for some $s' > 0$. Further, if D and T are strictly convex with C^2 boundaries and I and L are in $C^{0,s}$ up to the boundary then $v \in C^{2,s}(\bar{D})$. Similar results were proved by P. Delanoe [16] for $n = 2$ and J. Urbas [51] for any n.

Numerical solutions of the problem (4.55) and actual construction of reflectors of R_1 and R_2 were studied by V. Oliker in an unpublished manuscript. The results of applying the numerical scheme for solving a concrete problem arising in applications were described in [35].

Essentially, the approach to solving (4.55) numerically is based on the procedure used for proving existence of weak solutions. In that approach the weak solutions are approximated uniformly by convex piece-wise linear functions solving the special case of (4.55) with data constructed from the data in the original problem (see the outline of the proof of Theorem 5.1).

The convex piece-wise linear solutions are constructed by an iterative procedure producing a converging monotone sequence of approximations. However, dealing with an asymptotic cone as the boundary condition is inconvenient and the problem (4.55) is reformulated and solved as a sequence of Dirichlet problems approximating the solution of (4.57). A numerical scheme for solving such Dirichlet problems was investigated in [42]. It would be interesting to find other approaches to numerical solution of (4.55).

4.5 Two-Reflector System with a Point Source

In this section we formulate a beam shaping problem involving two reflectors and a point source O generating a spherical wavefront over a given input aperture $\bar{D}$. The latter is required to be transformed into a beam of parallel rays propagating in a given direction. A cross section of the output beam is usually specified as a set on a plane perpendicular to the direction of propagation. We will refer to this cross section as the output aperture. In addition, the intensity of the source must be transformed into a prescribed intensity distribution on the output aperture.

This problem arises in synthesis of reflector antennas and an extensive engineering literature has been devoted to its various aspects; see [25, 26, 18, 23, 53, 56, 47, 50] and other references there. The approaches described in the engineering literature are usually justified only by specific numerical examples which are not easily reproducible. Consequently, the applicability of these approaches and the level of confidence in them are hard to determine. As far as we know, currently, no rigorous mathematical analysis of this problem is available.

Similar to the case considered in section 4.4 the laws of geometrical optics are used for deriving the equations of the problem. The Fig. 4.4 shows schematically the configuration of the system. Analytically, this problem can be formulated in several different forms all of which are quite involved [17, 53, 39]. In engineering literature this problem was formulated (for n=2) either as a system of first order partial differential equations [25, 26, 17] or as a Monge-Ampère for a complex-valued function [53]. In [39] this problem was reformulated in a form involving a real-valued equation of Monge-Ampère type for the polar radius of the sub-reflector R_1. Below, following [39], we briefly outline the main steps of this formulation.

The input aperture $\bar{D}$ is given as a closed set on a unit sphere S^n with the center at O where a point energy source is located. The input intensity I is a given nonnegative integrable function on D. The position vector of the sub-reflector R_1 which needs to be determined is denoted by vector function $r_1(m) = \rho(m)m$, $m \in \bar{D}$, where ρ is the polar radius of R_1. The unit normal vector field u on R_1 can be described in terms of ρ as

$$u = \frac{1}{\sqrt{\rho^2 + |\nabla \rho|^2}}(\rho m - \nabla \rho),$$

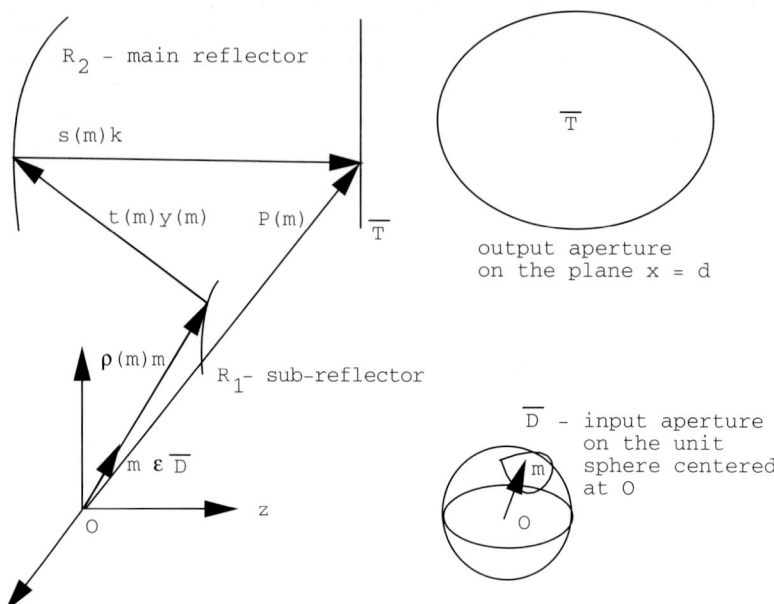

Fig. 4.4. Schematic diagram of a two-reflector system with a point source.

where $\nabla \rho$ is the gradient in the standard metric of S^n. The unit vector $y(m)$ in the direction of the ray reflected off R_1 for the incidence direction m is given by the equation (4.1) applied to sub-reflector R_1. It can also be expressed in terms of ρ and its derivatives.

Denote by $t(m)$ the distance from R_1 to the main reflector R_2 along the direction $y(m)$ (see Fig. 4.4).

The output beam is required to be a set of parallel rays reflected off R_2 and propagating in the positive direction k of the z axis. Its cross section is specified as the output aperture $\bar{T}$ on a some fixed hyperplane $z = d > 0$. The points on $\bar{T}$ are denoted by (x, d), where $x = (x_1, ..., x_n)$ is a point on the hyperplane $z = 0$. Denote by $L(x, d)$, $(x, d) \in T$, the required output intensity distribution on T.

The phase, that is, the total optical path length for an input direction $m \in \bar{D}$ is denoted by $l(m) = \rho(m) + t(m) + s(m)$, where $s(m)$ is the distance from the point $\rho(m)m + t(m)y$ on R_2 to the hyperplane $z = d$. The fact that $\bar{T}$ is a cross section of the output beam implies that the phase is constant $l(m) = const \equiv l$. Thus, it can be assumed that $l - d$ is some fixed positive constant.

Finally, the vector function

$$P(m) = \rho(m)m + t(m)y(m) + (l - \rho(m) - t(m))k, \quad m \in \bar{D},$$

gives a parametrization of $\bar{T}$ connected with the two reflections.

Projecting $P(m)$ onto the z-axis and using the fact that the output aperture $\bar{T}$ and the quantity $l - d$ are specified, it is possible to express the function t in terms of ρ and $l - d$. Namely,

$$t(m) = \frac{\beta + \rho(m)(\langle m, k \rangle - 1)}{1 - \langle y, k \rangle}.$$

The input and output intensities I and L, respectively, are assumed to satisfy the total energy conservation condition

$$\int_D I(m) d\sigma = \int_T L(x, d) dx. \tag{4.58}$$

The differential form of the energy conservation law [5] is used to relate the input and output intensity distributions:

$$L(P)|J(P)| = I,$$

where $J(P)$ is the Jacobian of the map $P: \bar{D} \to \bar{T}$. In more explicit form,

$$L(P) \frac{\det[M_{ij}]}{\det[e_{ij}]} = I, \quad m \in D, \tag{4.59}$$

where

$$M_{ij} = \frac{2t}{\rho} b_{ij} \langle m, u \rangle + (\rho + t) e_{ij},$$

$[b_{ij}]$, $i, j = 1, ..., n$, is the matrix of coefficients of the second fundamental form of the sub-reflector R_1 and $[e_{ij}]$, $i, j = 1, 2, ..., n$, is the matrix of coefficients of the standard metric of S^n. The coefficients b_{ij} can be expressed completely in terms of ρ:

$$b_{ij} = \frac{\rho \nabla_{ij} \rho - 2 \rho_i \rho_j - \rho^2 e_{ij}}{\sqrt{\rho^2 + |\nabla \rho|^2}},$$

where

$$\rho_i = \frac{\partial \rho}{\partial u^i}, \quad \nabla_{ij} \rho = \frac{\partial^2 \rho}{\partial u^i \partial u^j} - \sum_{k=1}^n \Gamma_{ij}^k \rho_k,$$

with $u^1, u^2, ..., u^n$ being some local coordinates on S^n and Γ_{ij}^k the Christoffel symbols of the second kind of the standard metric on S^n.

The problem consists in constructing the reflectors R_1 and R_2 so that $\bar{D}$ is mapped onto $\bar{T}$ by the map P and the equation (4.59) is satisfied. In addition, the reflector R_1 should be in a position in which it does not obstruct propagation of rays reflected off R_2.

Acknowledgement: Research was partially supported by Emory University Research Committee.

References

1. A.D. Aleksandrov. *Convex Polyhedra.* GITTL, Moskow, USSR (In Russian), German transl.: Konvexe Polyeder, Akademie-Verlag, Berlin, 1958, 1950.
2. A.D. Aleksandrov. Uniqueness theorems for hypersurfaces in the large, i. *Vestnik LGU*, 19:5–17, 1956.
3. I.J. Bakelman. *Convex Analysis and Nonlinear Geometric Elliptic Equations.* Springer-Verlag, Berlin, 1994.
4. A. Baldes and O. Wohlrab. Convex discrete solutions of generalized Monge-Ampère equations. *Preprint, SFB 256, Universitat Bonn, Germany*, 1989.
5. M. Born and E. Wolf. *Principles of Optics.* Pergamon Press, Elmsford, NY, 6 edition, 1989.
6. D.G. Burkhard and D. L. Shealy. View function in generalized curvilinear coordinates for specular reflection of radiation from a curved surface. *Int. J. Heat Mass Transfer*, 16:1492–1496, 1973.
7. D.G. Burkhard and D.L. Shealy. Specular aspheric surface to obtain a specified irradiance from discrete or continuous line source radiation: design. *Applied Optics*, 14(6):1279–1284, 1975.
8. D.G. Burkhard and D.L. Shealy. Simplified formula for the illuminance in an optical system. *Applied Optics*, 20(5):897–909, 1981.
9. D.G. Burkhard and D.L. Shealy. A different approach to lighting and imaging: formulas for flux density, exact lens and mirror equations and caustic surfaces in terms of the differential geometry of surfaces. In *SPIE Vol. 692 Materials and Optics for Solar Energy Conversion and Advanced Lighting Technology*, pages 248–272, 1987.
10. D.G. Burkhard and G.L. Strobel. Reflective optics system for multiple beam laser fusion. *Applied Optics*, 22(9):1313–1317, 1983.
11. L.A. Caffarelli. Boundary regularity of maps with convex potentials. *Comm. in Pure and Applied Math.*, 45(9):1141–1151, 1992.
12. L.A. Caffarelli. The regularity of mappings with convex potentials. *J. of AMS*, 5:99–104, 1992.
13. L.A. Caffarelli. Boundary regularity of maps with convex potentials. ii. *Ann. of Math.*, 144(3):453–496, 1996.
14. L.A. Caffarelli, S. Kochengin, and V.I. Oliker. On the numerical solution of the problem of reflector design with given far-field scattering data. *Contemporary Mathematics*, 226:13–32, 1999.
15. L.A. Caffarelli and V.I. Oliker. Weak solutions of one inverse problem in geometric optics. *Preprint*, 1994.
16. P. Delanoe. Classical solvability in dimension two of the second boundary-value problem associated with the monge-ampre operator. *Ann. Inst. H. Poincar Anal. Non Linaire*, 8(5):443–457, 1991.
17. V. Galindo-Israel, W.A. Imbriale, and R. Mittra. On the theory of the synthesis of single and dual offset shaped reflector antennas. *IEEE Transactions on Antennas and Propagation*, AP-35(8):887–896, 1987.
18. V. Galindo-Israel, R. Mittra, and A.G. Cha. Aperture amplitude and phase control on offset dual reflectors. *IEEE Transactions on Antennas and Propagation*, AP-27:154–164, 1979.
19. Pengfei Guan and Xu-Jia Wang. On a Monge-Ampère equation arising in geometric optics. *J. Differential Geometry*, 48:205–223, 1998.

20. T. Hasanis and D. Koutroufiotis. The characteristic mapping of a reflector. *J. of Geometry*, 24:131–167, 1985.
21. M. Herzberger. *Modern Geometrical Optics*. Interscience Publishers, New York, 1959.
22. T.E. Horton and J.H. McDermit. Design of a specular aspheric surface to uniformly radiate a flat surface using a nonuniform collimated radiation source. *J. Heat Transfer*, pages 453–458, November 1972.
23. V. Galindo Israel, W.A. Imbriale, R. Mittra, and K. Shogen. On the theory of the synthesis of offset dual-shpaed reflectors – case examples. *IEEE Transactions on Antennas and Propagation*, 39(5):620–626, May 1991.
24. V. Galindo Israel, S. Rengarajan, W.A. Imbriale, and R. Mittra. Offset dual-shaped reflectors for dual chamber compact ranges. *IEEE Transactions on Antennas and Propagation*, 39(7):1007–1013, July 1991.
25. B.E. Kinber. On two reflector antennas. *Radio Eng. Electron. Phys.*, 7(6):973–979, 1962.
26. B.E. Kinber. *Inverse problems of the reflector antennas theory - geometric optics approximation.* preprint No. 38, Academy of Sc., USSR, 1984. in Russian.
27. S. Kochengin and V.I. Oliker. Determination of reflector surfaces from near-field scattering data. *Inverse Problems*, 13(2):363–373, 1997.
28. S. Kochengin and V.I. Oliker. Determination of reflector surfaces from near-field scattering data II. Numerical solution. *Numerishe Mathematik*, 79(4):553–568, 1998.
29. S. Kochengin and V.I. Oliker. Computational algorithms for constructing reflectors. *Computing and Visualization in Science*, to appear, 2001.
30. S. Kochengin, V.I. Oliker, and O. von Tempski. On the design of reflectors with prespecified distribution of virtual sources and intensities. *Inverse Problems*, 14:661–678, 1998.
31. P.H. Malyak. Two-mirror unobscured optical system for reshaping irradiance distribution of a laser beam. *Applied Optics*, 31(22):4377–4383, August 1992.
32. L. Marder. Uniqueness in reflector mappings and the Monge-Ampere equation. *Proc. R. Soc. Lond.*, A(378):529–537, 1981.
33. J.H. McDermit and T.E. Horton. Reflective optics for obtaining prescribed irradiative distributions from collimated sources. *Applied Optics*, 13(6):1444–1450, 1974.
34. E. Newman and V. Oliker. Differential-geometric methods in design of reflector antennas. In *Symposia Mathematica*, volume 35, pages 205–223, 1994.
35. V. Oliker and L.D. Prussner. A new technique for synthesis of offset dual reflector systems. In *10-th Annual Review of Progress in Applied Computational Electromagnetics*, pages 45–52, 1994.
36. V. Oliker, L.D. Prussner, D. Shealy, and S. Mirov. Optical design of a two-mirror asymmetrical reshaping system and its application in superbroadband color center lasers. In *SPIE Proceedings*, volume 2263, pages 10–18, San Diego, CA, 1994.
37. V.I. Oliker. Near radially symmetric solutions of an inverse problem in geometric optics. *Inverse Problems*, 3:743–756, 1987.
38. V.I. Oliker. On reconstructing a reflecting surface from the scattering data in the geometric optics approximation. *Inverse Problems*, 5:51–65, 1989.
39. V.I. Oliker. The ray tracing and energy conservation equations for mirror systems with two reflecting surfaces. *Computers and Mathematics with Applications*, 26(7):9–18, 1993.

40. V.I. Oliker. On the geometry of convex reflectors. *Banach Center Publications*, vol. 57, 2001.
41. V.I. Oliker, E.J. Newman, and L. Prussner. A formula for computing illumination intensity in a mirror optical system. *Journal of the Optical Society of America, A,*, 10(9):1895–1901, 1993.
42. V.I. Oliker and L.D. Prussner. On the numerical solution of the equation $\frac{\partial^2 z}{\partial x^2}\frac{\partial^2 z}{\partial y^2} - (\frac{\partial^2 z}{\partial x \partial y})^2 = f$ and its discretizations, I. *Numerische Math.*, 54:271–293, 1988.
43. V.I. Oliker and P. Waltman. Radially symmetric solutions of a Monge-Ampere equation arising in a reflector mapping problem. In I. Knowles and Y. Saito, editors, *Proc. UAB Int. Conf. on Diff. Equations and Math. Physics*, pages 361–374. Lecture Notes in Math. 1285, 1987.
44. V.I. Oliker. The reflector problem for closed surfaces. *Partial Differential Equations and Applications*, 177:265–270, 1996.
45. V.I. Oliker. A rigorous method of synthesis of offset shaped reflector antennas. *preprint*, 2000.
46. P.W. Rhodes and D.L. Shealy. Refractive optical systems for irradiance redistribution of collimated radiation: their design and analysis. *Applied Optics*, 19(20):3545–3553, 1980.
47. W.V.T. Rusch. Quasioptical antenna design (section 3.4). In A.W. Rudge, K. Milne, A.D. Olver, and P. Knight, editors, *The handbook of antenna design, Volumes 1 and 2*. Peter Peregrims Ltd., London, UK, 1986.
48. J.S. Schruben. Formulation of relector-design problem for a lighting fixture. *J. of the Optical Society of America*, 62(12):1498–1501, 1972.
49. J.S. Schruben. Analysis of rotationally symmetric reflectors for illuminating systems. *J. of the Optical Society of America*, 64(1):55–58, 1974.
50. C.J. Sletten. Reflector antennas and surface shaping techniques. In C.J. Sletten, editor, *Reflector and Lens Antennas*. Artech House, Norwood, MA, 1988.
51. J. Urbas. On the second boundary value problem for equations of monge-ampre type. *J. Reine Angew. Math.*, 487:115–124, 1997.
52. Xu-Jia Wang. On design of reflector antenna. *Inverse Problems*, 12(2):351–375, 1996.
53. B.S. Westcott. *Shaped Reflector Antenna Design*. Research Studies Press, Letchworth, UK, 1983.
54. B.S. Westcott and A.P. Norris. Reflector synthesis for generalised far fields. *J. Phys. A: Math. Gen.*, 8:521–532, 1975.
55. B.S. Westcott, F. Brickell, and I.C. Wolton. Crosspolar control in far-field synthesis of dual offset reflectors. *IEE Proceedings:H*, 137(1):31–38, February 1990.
56. B.S. Westcott, R.K. Graham, and F. Brickell. Systematic design of a dual offset reflector antenna with an elliptical aperture. *IEE Proceedings:H*, 131(6):365–370, December 1984.

5 Recent Developments in Multiscale Problems Coming from Fluid Mechanics

Andro Mikelić

Analyse Numérique, UFR Mathématiques, Université Claude Bernard Lyon 1, Bât. 101, 43, bd. du 11 novembre 1918, 69622 Villeurbanne Cedex, France – andro@iris.univ-lyon1.fr

Abstract: Two topics will be discussed:

(i) Homogenization of Flow Problems in the Presence of Rough Boundaries and Interfaces

   We consider tangential viscous flows over rough surfaces and obtain the effective boundary condition. It is the Navier's slip condition, used in the computations of viscous flows in complex geometries. The effective coefficients, the Navier's matrix, is determined by upscaling. It is given by solving an appropriate boundary layer problem. Then we address application to the drag reduction. Finally, we consider viscous flows through domains containing two or more sub-domains, separated by interfaces. Sub-domains are supposed to be geometrically different. A typical example is a viscous flow over a porous bed. It will be shown that in the upscaled problem, it is enough to consider the free fluid part and impose the law of Beavers and Joseph at the interface. Another class of problems is a flow through 2 different porous media. The homogenization, coupled with the boundary layers, gives the effective law at the interface. In this chapter we'll explain how those results are obtained, give precise references for technical details and present open problems.

(ii) Interactions Flow-Structures

   We consider viscous and inviscid fluid flows through a deformable porous medium. The solid skeleton is supposed to be elastic. We present the modeling of the problem and a derivation of Biot's type equations by homogenization.

5.1 Homogenization of Flow Problems in the Presence of Rough Boundaries and Interfaces

5.1.1 Wall Laws at Rough Boundaries

Introduction From a physical point of view, the no-slip condition $v = 0$, at an immobile solid boundary, is only justified where the molecular viscosity is concerned. Since the fluid cannot penetrate the solid, its normal velocity is equal to zero. This is the condition of non-penetration. To the contrary, the absence of slip is not very intuitive. For the Newtonian fluids, it was established experimentally and contested even by Navier himself (see [49]). He claimed that the slip velocity should be proportional to the shear stress. The kinetic-theory calculations have confirmed Navier's boundary condition, but they give the slip length proportional to the mean free path divided by the continuum length (see [53]). For practical purposes it means a zero slip length, justifying the use of the no-slip condition.

In many cases of practical significance the boundary is rough. An example are complex boundaries in the geophysical fluid dynamics. Compared with the characteristic size of a computational domain, such boundaries could be considered as rough. Other examples involve sea bottoms of random roughness and artificial bodies with periodic distribution of small bumps. A numerical simulation of the flow problems in the presence of a rough boundary is very difficult since it requires many mesh nodes and handling of many data. For computational purposes, an artificial smooth boundary, close to the original one, is taken and the equations are solved in the new domain. This way the rough boundary is avoided, but the boundary conditions at the artificial boundary aren't given. It is clear that the non-penetration condition $v \cdot n = 0$ should be kept, but there are no reasons to keep the no-slip. Usually it is supposed that the shear stress is a non-linear function F of the tangential velocity. F is determined empirically and its form varies for different problems. Such relations are called the *wall laws* and classical Navier's condition is one example. Another well-known example is modeling of the turbulent boundary layer close to the rough surface by a *logarithmic velocity profile*

$$v_\tau = \sqrt{\frac{\tau_w}{\rho}} \left(\frac{1}{\kappa} \ln \left(\frac{y}{\mu} \sqrt{\frac{\tau_w}{\rho}} \right) + C^+(k_s^+) \right) \tag{5.1}$$

where v_τ is the tangential velocity, y is the vertical coordinate and τ_w the shear stress. ρ denotes the density and μ the viscosity. $\kappa \approx 0.41$ is the von Kármán's constant and C^+ is a function of the ratio k_s^+ of the roughness height k_s and the thin wall sublayer thickness $\delta_v = \frac{\mu}{v_\tau}$. For more details we refer to the book of Schlichting [56].

Justifying the logarithmic velocity profile in the overlap layer is mathematically out of reach for the moment. Nevertheless, after recent results we are able to justify the Navier's condition for the laminar incompressible viscous flows over periodic rough boundaries. In the text which follows, we are going to give a review of recent rigorous results on Navier's condition.

Somewhat related problem is the homogenization of the Poisson equation in a domain with a periodic oscillating boundary. The amplitude and the period of the oscillations are of order ε and the homogeneous Dirichlet condition is imposed on the solution. This problem was considered in the paper by Achdou and Pironneau [2] and a homogenized problem was obtained. The rough boundary was replaced by a smooth artificial one and the corresponding wall law was the Robin boundary condition, saying that the effective solution u was proportional to the characteristic roughness ε times its normal derivative. The proportionality constant was calculated using an auxiliary problem for Laplace's operator in a finite cell. Such approach has similarities with problems of reinforcement by thin layers with oscillating thickness (see e.g. Buttazzo and Kohn [22] and references therein). Reinforcement problems lead to a Robin type boundary condition with coefficients of order 1, calculated using finite cell auxiliary problems. Nevertheless, in [2] the conductivity of the thin layer close to the boundary is not small and, contrary to [22], the homogenized boundary condition contains an ε. Consequently, it is not clear that the H^1-error estimate in [2] is of order ε. The detailed estimates for a simpler geometry are in the paper by Allaire and Amar [7]. They considered a rectangular domain having one face which was a periodic repetition of $\varepsilon \Gamma_g$ and the same boundary value problem as in [2] except periodic lateral conditions. Then they introduced the following auxiliary boundary layer problem in the infinite strip $\Gamma_g \times]0, +\infty[$: Find a harmonic function ψ, $\nabla \psi \in L^2$, periodic in $y' = (y_1, \ldots, y_{n-1})$ and having a value on Γ_g equal to its parametric form. The classical theory (see e.g. [52] or [41]) gives existence of a unique solution which decays exponentially to a constant d. The conclusion of [7] was that the homogenized solution $\bar{u}^\varepsilon$ obeyed the wall law $\bar{u}^\varepsilon = \varepsilon d \dfrac{\partial \bar{u}^\varepsilon}{\partial x_n}$ on the artificial boundary and gave an interior H^1-approximation of order $\varepsilon^{3/2}$. We note the difference in determination of the proportionality constant in the wall law between papers [2] and [7].

In the text which follows we concentrate on the incompressible Stokes and Navier-Stokes equations and we are going to present a different construction of the boundary layer corrections, following the approach from [33].

Flow problems over rough surfaces were considered by O. Pironneau and collaborators in [48], [3] and [4]. The paper [48] considers the flow over a rough surface and the flow over a wavy sea surface. It discusses a number of problems and announces a rigorous result for an approximation of the Stokes flow. Similarly, in the paper [3] numerical calculations are presented and rigorous results in [4] are announced. Finally, in the paper [4] the stationary incompressible flow at high Reynolds number $\mathbf{Re} \sim \frac{1}{\varepsilon}$ over a periodic rough boundary, with the roughness period ε, is considered. An asymptotic expansion is constructed and, with the help of boundary layer correctors defined in a semi-infinite cell, effective wall laws are obtained. A numerical validation is presented, but there are no mathematically rigorous convergence results. The error estimate for the approximation, announced in [3], was not proved in [4].

We mention also the results of Y. Amirat, J. Simon and collaborators on the Couette flow over a rough plate (see [8], [9] and [10]). They concern principally the drag reduction for the Couette flow and we compare their results with our approach below.

In this section we are going to justify the Navier's slip law by the technique developed in [32] for Laplace's operator and then in [33] for the Stokes system. The result for a 2D laminar stationary incompressible viscous flow over a rough boundary is in [37]. In the subsections which follow we consider a 3D Couette flow over a rough boundary. In this section we introduce the corresponding boundary layer problem and we obtain the Navier slip condition.

Navier's boundary layer As observed in hydrodynamics, the phenomena relevant to the boundary occur in a thin layer surrounding it. We are not interested in the boundary layers corresponding to the inviscid limit of the Navier-Stokes equations, but we are going to construct the viscous boundary layer describing effects of the roughness. There is a similarity with boundary layers describing effects of interfaces between a perforated and a non-perforated domain. Such boundary layers were first developed in [32] for Laplace's operator with homogeneous Dirichlet boundary conditions at the perforations. The corresponding theory for the Stokes system is in [33] and, in a more pedagogical way, in [47]. We will be back to the construction of the effective laws at interfaces between a free flow and a porous medium in the last subsection. In this subsection we are going to present a self-contained construction of the main boundary layer, used for determining the coefficient in Navier's condition. It is natural to call it the *Navier's boundary layer*. In [37] the 2D boundary layer was constructed. We generalize here the results from [37] to a 3D situation.

Let b_j, $j = 1, 2, 3$ be 3 positive constants. Let $Z = (0, b_1) \times (0, b_2) \times (0, b_3)$ and let Υ be a Lipschitzian surface $y_3 = \Upsilon(y_1, y_2)$, taking values between 0 and b_3. We suppose that the rough surface $\cup_{k \in \mathbb{Z}^2} (\Upsilon + k)$ is also a Lipschitz surface. We introduce the canonical cell of roughness (the canonical hump) by $Y = \{y \in Z \mid b_3 > y_3 > \max\{0, \Upsilon(y_1, y_2)\}\}$.

Following the construction from [37], the crucial role is played by an auxiliary problem. It reads as follows:

For a given constant vector $\lambda \in \mathbb{R}^2$, find $\{\beta^\lambda, \omega^\lambda\}$ that solve

$$-\triangle_y \beta^\lambda + \nabla_y \omega^\lambda = 0 \quad \text{in } Z^+ \cup (Y - b_3 \mathbf{e}_3) \tag{5.2}$$

$$\text{div}_y \beta^\lambda = 0 \quad \text{in } Z_{bl} \tag{5.3}$$

$$[\beta^\lambda]_S(\cdot, 0) = 0 \quad \text{on } S \tag{5.4}$$

$$\left[\{\nabla_y \beta^\lambda - \omega^\lambda I\} \mathbf{e}_3\right]_S(\cdot, 0) = \lambda \quad \text{on } S \tag{5.5}$$

$$\beta^\lambda = 0 \quad \text{on } (\Upsilon - b_3 \mathbf{e}_3), \quad \{\beta^\lambda, \omega^\lambda\} \text{ is } y' = (y_1, y_2) - \text{periodic}, \tag{5.6}$$

where $S = (0, b_1) \times (0, b_2) \times \{0\}$, $Z^+ = (0, b_1) \times (0, b_2) \times (0, +\infty)$, and $Z_{bl} = Z^+ \cup S \cup (Y - b_3 \mathbf{e}_3)$.

Let $V = \{z \in L^2_{loc}(Z_{bl})^3 : \nabla_y z \in L^2(Z_{bl})^9$; $z = 0$ on $(\Upsilon - b_3 \mathbf{e}_3)$; $\text{div}_y z = 0$ in Z_{bl} and z is $y' = (y_1, y_2)$-periodic $\}$. Then, by Lax-Milgram lemma, there is a unique $\beta^\lambda \in V$ satisfying

$$\int_{Z_{bl}} \nabla \beta^\lambda \nabla \varphi \, dy = -\int_S \varphi \lambda \, dy_1 dy_2, \quad \forall \varphi \in V. \tag{5.7}$$

Using De Rham's theorem we obtain a function $\omega^\lambda \in L^2_{loc}(Z_{bl})$, unique up to a constant and satisfying (5.2). By the elliptic theory, $\{\beta^\lambda, \omega^\lambda\} \in V \cap C^\infty(Z^+ \cup (Y - b_3 \mathbf{e}_3))^3 \times C^\infty(Z^+ \cup (Y - b_3 \mathbf{e}_3))$ to (5.2) – (5.6).

In the neighborhood of S we have $\beta^\lambda - (\lambda_1, \lambda_2, 0)(y_3 - y_3^2/2)e^{-y_3} H(y_3) \in W^{2,q}$ and $\omega^\lambda \in W^{1,q}$, $\forall q \in [1, \infty)$.

Then we have

Lemma 1. ([33], [34], [47]). *For any positive a, a_1 and a_2, $a_1 > a_2$, the solution $\{\beta^\lambda, \omega^\lambda\}$ satisfies*

$$\begin{cases}
\int_0^{b_1} \int_0^{b_2} \beta_2^\lambda(y_1, y_2, a) \, dy_1 dy_2 = 0, \\
\int_0^{b_1} \int_0^{b_2} \omega^\lambda(y_1, y_2, a_1) \, dy_1 dy_2 = \int_0^{b_1} \int_0^{b_2} \omega^\lambda(y_1, y_2, a_2) \, dy_1 dy_2, \\
\int_0^{b_1} \int_0^{b_2} \beta_j^\lambda(y_1, y_2, a_1) dy_1 dy_2 = \int_0^{b_1} \int_0^{b_2} \beta_j^\lambda(y_1, y_2, a_2) dy_1 dy_2, j = 1, 2 \\
C_\lambda^{bl} = \sum_{j=1}^{2} C_\lambda^{j,bl} \lambda_j = \int_S \beta^\lambda \lambda \, dy_1 dy_2 = -\int_{Z_{bl}} |\nabla \beta^\lambda(y)|^2 \, dy < 0
\end{cases} \tag{5.8}$$

Lemma 2. *Let $\lambda \in \mathbb{R}^2$ and let $\{\beta^\lambda, \omega^\lambda\}$ be the solution for (5.2) – (5.6) satisfying $\int_S \omega^\lambda \, dy_1 dy_2 = 0$. Then $\beta^\lambda = \sum_{j=1}^{2} \beta^j \lambda_j$ and $\omega^\lambda = \sum_{j=1}^{2} \omega^j \lambda_j$, where $\{\beta^j, \omega^j\} \in V \times L^2_{loc}(Z_{bl})$, $\int_S \omega^j \, dy_1 dy_2 = 0$, is the solution for (5.2) – (5.6) for $\lambda = \mathbf{e}_j$, $j = 1, 2$.*

Lemma 3. *Let $a > 0$ and let $\beta^{a,\lambda}$ be the solution for (5.2) – (5.6) with S replaced by $S_a = (0, b_1) \times (0, b_2) \times \{a\}$ and Z^+ by $Z_a^+ = (0, b_1) \times (0, b_2) \times (a, +\infty)$. Then we have*

$$C_\lambda^{a,bl} = \int_0^{b_1} \int_0^{b_2} \beta^{a,\lambda}(y_1, y_2, a)\lambda \, dy_1 = C_\lambda^{bl} - a \mid \lambda \mid^2 b_1 b_2 \qquad (5.9)$$

Proof: It goes along the same lines as Lemma 2 from [37]. ■

This simple result will imply the invariance of the obtained law on the position of the artificial boundary.

Corollary 1. $\mid C_\lambda^{a,bl} \mid$ is smallest for $a = 0$.

Remark 1. If the boundary is flat, i.e. $\Upsilon = $ cte., then the smallest constant $\mid C_\lambda^{a,bl} \mid$ equals zero. We will see that this means that the no-slip condition remains. If b_3 is the height of the biggest peak of Υ, then we can't extend Lemma 3 for $a < 0$.

Lemma 4. Let $\{\beta^j, \omega^j\}$ be as in Lemma 2 and let $M_{ij} = \frac{1}{b_1 b_2} \int_S \beta_i^j \, dy_1 dy_2$ be the Navier's matrix. Then the matrix M is symmetric negatively definite.

Lemma 5. Let $\{\beta^j, \omega^j\}, j = 1$ and $j = 3$, be as in Lemma 2. Then we have

$$\begin{cases} \mid D^\alpha \, curl_y \beta^j(y) \mid \leq Ce^{-2\pi y_3 \min\{1/b_1, 1/b_2\}}, \; y_3 > 0, \; \alpha \in \mathbb{N}^2 \cup (0,0) \\ \mid \beta^j(y) - (M_{1j}, M_{2j}, 0) \mid \leq C(\delta)e^{-\delta y_3}, \; y_3 > 0, \; \forall \delta < \dfrac{2\pi}{\max\{b_1, b_2\}} \\ \mid D^\alpha \beta^j(y) \mid \leq C(\delta)e^{-\delta y_3}, \; y_3 > 0, \; \alpha \in \mathbb{N}^2, \; \forall \delta < \dfrac{2\pi}{\max\{b_1, b_2\}} \\ \mid \omega^j(y) \mid \leq Ce^{-2\pi y_3 \min\{1/b_1, 1/b_2\}}, \qquad y_3 > 0. \end{cases} \qquad (5.10)$$

Corollary 2. The system $(5.2) - (5.6)$ defines a boundary layer.

Justification of the Navier's slip condition for the laminar 3D Couette flow A mathematically rigorous justification of the Navier's slip condition for the 2D Poiseuille flow over a rough boundary is in [37]. Rough boundary was the periodic repetition of a basic cell of roughness, with characteristic heights and lengths of the impurities equal to a small parameter ε. Then the flow domain was decomposed on a rough layer and its complement.

The no-slip condition was imposed on the rough boundary and there were inflow and outflow boundaries, not interacting with the humps. The flow was governed by a given constant pressure drop. The mathematical model were the stationary Navier-Stokes equations. In [37] the flow under moderate Reynolds numbers was considered and the following results were proved: a) A non-linear stability result with respect to small perturbations of the smooth boundary with a rough one; b) An approximation result of order $\varepsilon^{3/2}$; c) Navier's slip condition was justified.

In this review we are going to establish analogous results for a 3D Couette flow.

We consider a viscous incompressible fluid flow in a domain Ω^ε consisting of the parallelepiped $P = (0, L_1) \times (0, L_2) \times (0, L_3)$, the interface $\Sigma = (0, L_1) \times (0, L_2) \times \{0\}$ and the layer of roughness $R^\varepsilon = \left(\cup_{\{k \in \mathbb{Z}^2\}} \varepsilon \big(Y + (k_1, k_2, -b_3)\big) \right) \cap \left((0, L_1) \times (0, L_2) \times (-\varepsilon b_3, 0) \right)$. The canonical cell of roughness $Y \subset (0, b_1) \times (0, b_2) \times (0, b_3)$ is defined in section 5.1.2. For simplicity we suppose that $L_1/(\varepsilon b_1)$ and $L_2/(\varepsilon b_2)$ are integers. Let $\mathcal{I} = \{0 \leq k_1 \leq L_1/b_1;\ 0 \leq k_2 \leq L_2/b_2;\ k \in \mathbb{Z}^2\}$. Then, our rough boundary $\mathcal{B}^\varepsilon = \cup_{\{k \in \mathcal{I}\}} \varepsilon \big(\Upsilon + (k_1, k_2, -b_3)\big)$ is supposed to consist of a large number of periodically distributed humps of characteristic length and amplitude ε, small compared with a characteristic length of the macroscopic domain.

Then, for a fixed $\varepsilon > 0$ and a given constant velocity $\mathbf{U} = (U_1, U_2, 0)$, the Couette flow is described the following system

$$-\nu \Delta v^\varepsilon + (v^\varepsilon \nabla) v^\varepsilon + \nabla p^\varepsilon = 0 \quad \text{in } \Omega^\varepsilon, \tag{5.11}$$

$$\operatorname{div} v^\varepsilon = 0 \quad \text{in } \Omega^\varepsilon, \tag{5.12}$$

$$v^\varepsilon = 0 \quad \text{on } \mathcal{B}^\varepsilon, \tag{5.13}$$

$$v^\varepsilon = \mathbf{U} \quad \text{on } \Sigma_2 = (0, L_1) \times (0, L_2) \times \{L_3\} \tag{5.14}$$

$$\{v^\varepsilon, p^\varepsilon\} \quad \text{is periodic in } (x_1, x_2) \text{ with period } (L_1, L_2) \tag{5.15}$$

where $\nu > 0$ is the kinematic viscosity and $\int_{\Omega}^\varepsilon p^\varepsilon\, dx = 0$.

Let us note that a similar problem was considered in [10], but in an infinite strip with a rough boundary. In [10] the authors were looking for solutions periodic in (x_1, x_2), with the period $\varepsilon(b_1, b_2)$.

Since we need not only existence for a given ε, but also the a priori estimates independent of ε, we give a non-linear stability result with respect to rough perturbations of the boundary, leading to uniform a priori estimates. Our proof follows the corresponding one from [37].

First, we observe that the Couette flow in P, satisfying the no-slip conditions at Σ, is given by

$$v^0 = \frac{U_1 x_3}{L_3} \mathbf{e}_1 + \frac{U_2 x_3}{L_3} \mathbf{e}_2 = \mathbf{U} \frac{x_3}{L_3}, \quad p^0 = 0. \tag{5.16}$$

Let $|U| = \sqrt{U_1^2 + U_2^2}$. Then it is easy to see that it is unique if $|U| L_3 < 2\nu$ i.e. if the Reynolds number is moderate.

We extend the velocity field to $\Omega^\varepsilon \setminus P$ by zero.

The idea is to construct the solution to (5.11) – (5.15) as a small perturbation to the Couette flow (5.16). Before the existence result, we prove an auxiliary lemma:

Lemma 6. *([37]) Let $\varphi \in H^1(\Omega^\varepsilon \setminus P)$ be such that $\varphi = 0$ on $\mathcal{B}^\varepsilon$. Then we have*

$$\|\varphi\|_{L^2(\Omega^\varepsilon \setminus P)} \leq C\varepsilon \|\nabla\varphi\|_{L^2(\Omega^\varepsilon \setminus P)^3} \tag{5.17}$$

$$\|\varphi\|_{L^2(\Sigma)} \leq C\varepsilon^{1/2}\|\nabla\varphi\|_{L^2(\Omega^\varepsilon \setminus P)^3} \tag{5.18}$$

Now we are in position to prove the desired non-linear stability result:

Theorem 1. *([39]) Let $|U|L_3 \leq \nu$. Then there exists a constant $C_0 = C_0(b_1, b_2, b_3, L_1, L_2)$ such that for $\varepsilon \leq C_0(\frac{L_3}{|U|})^{3/4}\nu^{3/4}$ the problem (5.11) – (5.15) has a unique solution $\{v^\varepsilon, p^\varepsilon\} \in H^2(\Omega^\varepsilon)^3 \times H^1(\Omega^\varepsilon)$, $\int_\Omega^\varepsilon p^\varepsilon\, dx = 0$, satisfying*

$$\|\nabla(v^\varepsilon - v^0)\|_{L^2(\Omega^\varepsilon)^9} \leq C\sqrt{\varepsilon}\frac{|U|}{L_3}. \tag{5.19}$$

Moreover,

$$\|v^\varepsilon\|_{L^2(\Omega^\varepsilon \setminus P)^3} \leq C\varepsilon\sqrt{\varepsilon}\frac{|U|}{L_3} \tag{5.20}$$

$$\|v^\varepsilon\|_{L^2(\Sigma)^3} + \|v^\varepsilon - v^0\|_{L^2(P)^3} \leq C\varepsilon\frac{|U|}{L_3} \tag{5.21}$$

$$\|p^\varepsilon - p^0\|_{L^2(P)} \leq C\frac{|U|}{L_3}\sqrt{\varepsilon}, \tag{5.22}$$

where $C = C(b_1, b_2, b_3, L_1, L_2)$.

Therefore, we have obtained the uniform a priori estimates for $\{v^\varepsilon, p^\varepsilon\}$. Moreover, we have found that Couette's flow in P is an $O(\varepsilon)$ L^2-approximation for v^ε.

Following the approach from [37], the Navier slip condition should correspond to taking into the account the next order corrections for the velocity. Then formally we get

$$v^\varepsilon = v^0 - \frac{\varepsilon}{L_3}\sum_{j=1}^{2}U_j\big(\beta^j(\tfrac{x}{\varepsilon}) - (M_{j1}, M_{j2}, 0)H(x_3)\big) -$$
$$\frac{\varepsilon}{L_3}\sum_{j=1}^{2}U_j(1 - \tfrac{x_3}{L_3})(M_{j1}, M_{j2}, 0)H(x_3) + O(\varepsilon^2)$$

where v^0 is the Couette velocity in P and the last term corresponds to the counterflow generated by the motion of Σ. Then on the interface Σ

$$\frac{\partial v_j^\varepsilon}{\partial x_3} = \frac{U_j}{L_3} - \frac{1}{L_3}\sum_{i=1}^{2}U_i\frac{\partial \beta_j^i}{\partial y_3} + O(\varepsilon) \quad \text{and} \quad \frac{1}{\varepsilon}v_j^\varepsilon = -\frac{1}{L_3}\sum_{i=1}^{2}U_i\beta_j^i(\tfrac{x}{\varepsilon}) + O(\varepsilon).$$

After averaging we obtain the familiar form of the Navier's slip condition

$$u_j^{eff} = -\varepsilon \sum_{i=1}^{2} M_{ji} \frac{\partial u_i^{eff}}{\partial x_3} \quad \text{on} \quad \Sigma, \qquad (NFC)$$

where u^{eff} is the average over the impurities and the matrix M is defined in Lemma 4. The higher order terms are neglected.

Now let us make this formal asymptotic expansion rigorous.

It is clear that in P the flow continues to be governed by the Navier-Stokes system. The presence of the irregularities would only contribute to the effective boundary conditions at the lateral boundary. The leading contribution for the estimate (5.19) were the interface integral terms $\int_\Sigma \varphi_j$. Following the approach from [37], we eliminate it by using the boundary layer-type functions

$$\beta^{j,\varepsilon}(x) = \varepsilon \beta^j(\frac{x}{\varepsilon}) \quad \text{and} \quad \omega^{j,\varepsilon}(x) = \omega^j(\frac{x}{\varepsilon}), \quad x \in \Omega^\varepsilon, \ j = 1, 2, \quad (5.23)$$

where $\{\beta^j, \omega^j\}$ is defined in Lemma 2. We have, for all $q \geq 1$ and $j = 1, 2$,

$$\frac{1}{\varepsilon}\|\beta^{j,\varepsilon} - \varepsilon(M_{1j}, M_{2j}, 0)\|_{L^q(P)^3} + \|\omega^{j,\varepsilon}\|_{L^q(P)} + \|\nabla \beta^{j,\varepsilon}\|_{L^q(\Omega)^9} = C\varepsilon^{1/q} \quad (5.24)$$

and

$$-\Delta \beta^{j,\varepsilon} + \nabla \omega^{j,\varepsilon} = 0 \quad \text{in} \ \Omega^\varepsilon \setminus \Sigma \qquad (5.25)$$

$$\text{div} \, \beta^{j,\varepsilon} = 0 \quad \text{in} \ \Omega^\varepsilon \qquad (5.26)$$

$$[\beta^{j,\varepsilon}]_\Sigma(\cdot, 0) = 0 \quad \text{on} \ \Sigma \qquad (5.27)$$

$$[\{\nabla \beta^{j,\varepsilon} - \omega^{j,\varepsilon} I\} e_3]_\Sigma(\cdot, 0) = e_j \quad \text{on} \ \Sigma. \qquad (5.28)$$

As in [37] stabilization of $\beta^{j,\varepsilon}$ towards a nonzero constant velocity $\varepsilon(M_{1j}, M_{2j}, 0)$, at the upper boundary, generates a counterflow. It is given by the 3D Couette flow $d^i = (1 - \frac{x_3}{L_3})e_i$ and $g^i = 0$.

Now, we would like to prove that the following quantities are $o(\varepsilon)$ for the velocity and $O(\varepsilon)$ for the pressure:

$$\mathcal{U}^\varepsilon(x) = v^\varepsilon - \frac{1}{L_3}\left(x_3^+ \mathbf{U} - \varepsilon \sum_{j=1}^{2} U_j \beta^j(\frac{x}{\varepsilon}) + \varepsilon \frac{x_3^+}{L_3} M \mathbf{U}\right) \qquad (5.29)$$

$$\mathcal{P}^\varepsilon = p^\varepsilon + \frac{\nu}{L_3} \sum_{j=1}^{2} U_j \omega^{j,\varepsilon}. \qquad (5.30)$$

Then we have the following result:

Theorem 2. *([39]) Let $\mathcal{U}^\varepsilon$ be given by (5.29) and $\mathcal{P}^\varepsilon$ by (5.30). Then $\mathcal{U}^\varepsilon \in H^1(\Omega^\varepsilon)^3$, $\mathcal{U}^\varepsilon = 0$ on Σ, it is periodic in (x_1, x_2), exponentially small on Σ_2 and div $\mathcal{U}^\varepsilon = 0$ in Ω^ε. Furthermore, $\forall \varphi$ satisfying the same boundary conditions, we have the following estimate*

$$|\nu \int_{\Omega^\varepsilon} \nabla \mathcal{U}^\varepsilon \nabla \varphi - \int_{\Omega^\varepsilon} \mathcal{P}^\varepsilon \mathrm{div}\varphi + \int_{\Omega^\varepsilon} \frac{x_3^+}{L_3} \sum_{j=1}^{2} U_j \frac{\partial \mathcal{U}^\varepsilon}{\partial x_j} \varphi + \int_{\Omega^\varepsilon} \mathcal{U}_3^\varepsilon \frac{\mathbf{U}}{L_3} \varphi$$

$$+|\int_{\Omega^\varepsilon} ((v^\varepsilon - v^0)\nabla)(v^\varepsilon - v^0)\varphi| \leq C\varepsilon^{3/2} \|\nabla \varphi\|_{L^2(\Omega^\varepsilon)^9} \frac{|U|^2}{L_3}. \quad (5.31)$$

Corollary 3. *([39]) Let $\mathcal{U}^\varepsilon(x)$ and $\mathcal{P}^\varepsilon$ be defined by (5.29) – (5.30) and let*

$$\varepsilon \leq \frac{\nu^{6/7}}{|U|} \min \left\{ \frac{\nu^{1/7}}{4(|M| + \|\beta\|_{L^\infty})}, C(b_1, b_2, b_3, L_1, L_2) L_3^{3/7} |U|^{1/7} \right\}. \quad (5.32)$$

Then v^ε, constructed in Theorem 1, is a unique solution to (5.11) – (5.15) and

$$\|\nabla \mathcal{U}^\varepsilon\|_{L^2(\Omega^\varepsilon)^9} + \mathcal{P}^\varepsilon\|_{L^2(P)} \leq C\varepsilon^{3/2} \frac{|U|^2}{\nu L_3} \quad (5.33)$$

$$\|\mathcal{U}^\varepsilon\|_{L^2(P)^3} + \|\mathcal{U}^\varepsilon\|_{L^2(\Sigma)^3} \leq C\varepsilon^2 \frac{|U|^2}{\nu L_3} \quad (5.34)$$

The estimates (5.33) – (5.34) allow to justify Navier's slip condition.

Remark 2. It is possible to add further correctors and then our problem would contain an exponentially decreasing forcing term. This is in accordance with [10] for the Navier-Stokes system and with [8] and [9] for the Stokes system. For the case of rough boundaries with different characteristic heights and lengths we refer to the doctoral dissertation of I. Cotoi [24]. The estimate (5.32) is of the same order in ε as the H^1-estimate in [7], obtained for the Laplace operator. The advantage of our approach is that we are going to obtain the Navier's slip condition with a negatively definite matricial coefficient.

Now we introduce the effective Couette-Navier flow through the following boundary value problem:

Find a velocity field u^{eff} and a pressure field p^{eff} such that

$$-\nu \triangle u^{eff} + (u^{eff}\nabla)u^{eff} + \nabla p^{eff} = 0 \quad \text{in } P, \quad (5.35)$$

$$\mathrm{div}\, u^{eff} = 0 \quad \text{in } P, \quad u^{eff} = (U_1, U_2, 0) \text{ on } \Sigma_2, \quad u_3^{eff} = 0 \text{ on } \Sigma \quad (5.36)$$

$$u_j^{eff} = -\varepsilon \sum_{i=1}^{2} M_{ji} \frac{\partial u_i^{eff}}{\partial x_3}, \ j = 1, 2 \text{ on } \Sigma \quad (5.37)$$

$$\{u^{eff}, p^{eff}\} \text{ is periodic in } (x_1, x_2) \text{ with period } (L_1, L_2) \quad (5.38)$$

If $|U|L_3 \leq \nu$, the problem (5.35) – (5.38) has a unique solution

$$\begin{cases} u^{eff} = (\tilde{u}^{eff}, 0), \tilde{u}^{eff} = \mathbf{U} + \left(\dfrac{x_3}{L_3} - 1\right) & \text{for } x \in P \\ p^{eff} = 0 & \text{for } x \in P \end{cases} \quad (5.39)$$

Let us estimate the error made when replacing $\{v^\varepsilon, p^\varepsilon, \mathcal{M}^\varepsilon\}$ by $\{u^{eff}, p^{eff}, \mathcal{M}^{eff}\}$.

Theorem 3. *([39]) Under the assumptions of Theorem 1 we have*

$$\|\nabla(v^\varepsilon - u^{eff})\|_{L^1(P)^9} \leq C\varepsilon, \quad (5.40)$$

$$\sqrt{\varepsilon}\|v^\varepsilon - u^{eff}\|_{L^2(P)^3} + \|v^\varepsilon - u^{eff}\|_{L^1(P)^3} \leq C\varepsilon^2 \frac{|U|}{L_3}, \quad (5.41)$$

Our next step is to calculate the *tangential drag force* or the *skin friction*

$$\mathcal{F}^\varepsilon_{t,j} = \frac{1}{L_1 L_2} \int_\Sigma \nu \frac{\partial v^\varepsilon_j}{\partial x_3}(x_1, x_2, 0)\, dx_1 dx_2, \quad j = 1, 2. \quad (5.42)$$

Theorem 4. *([39]) Let the skin friction $\mathcal{F}^\varepsilon_t$ be defined by (5.41). Then we have*

$$\left|\mathcal{F}^\varepsilon_t - \nu \frac{1}{L_3}\left(\mathbf{U} + \frac{\varepsilon}{L_3}M\mathbf{U}\right)\right| \leq C\varepsilon^2 \frac{|U|^2}{\nu L_3}\left(1 + \frac{\nu}{L_3|U|}\right) \quad (5.43)$$

Corollary 4. *Let $\mathcal{F}^{eff}_t = \nu \dfrac{1}{L_3}\left(I - \dfrac{\varepsilon}{L_3}M\right)^{-1}\mathbf{U}$ be the tangential drag force corresponding to the effective velocity u^{eff}. Then we have*

$$|\mathcal{F}^{eff}_t - \mathcal{F}^\varepsilon_t| \leq C\varepsilon^2 \frac{|U|^2}{\nu L_3}\left(1 + \frac{\nu}{L_3|U|}\right) \quad (5.44)$$

Remark 3. We see that the presence of the periodic roughness diminishes the tangential drag. The contribution is linear in ε, and consequently rather small. It coincides with the conclusion from [9] that for laminar flows there are no palpable drag reduction. Nevertheless, we are going to see in the next subsection that the calculations from the laminar case could be useful for turbulent Couette flow.

Remark 4. As in [37] we prove that a perturbation of the interface position of order $O(\varepsilon)$ implies a perturbation in the solution of $O(\varepsilon^2)$. The result is a consequence of Lemma 3. In fact the matrix M would change but it is compensated by the change of the position of Σ. Consequently, there is a freedom in fixing position of Σ. It influences the result only at the next order of the asymptotic expansion.

5.1.2 Drag Reduction and Homogenization

Drag reduction for planes, ships and cars reduces significantly the spending of the energy, and consequently the cost for all type of land, sea and air transportation.

Drag-reduction adaptations were important for the survival of Avians and Nektons, since their efficiency or speed, or both, have improved.

Essentially, there are three forms of drag. The largest drag component is pressure or form drag. It is particularly troublesome when flow separation occurs. The two remaining drag components are skin-friction drag and drag due to lift. Skin-friction drag is the result of the no-slip condition on the surface. Those components are present for both laminar (low Reynolds number) or turbulent (high Reynolds number) flows.

There are several drag-reduction methods and here we discuss only the use of drag-reducing surfaces. For an overview of other techniques we refer to Bushnell, Moore [21].

The inspiration comes from morphological observations. It is known that the skin of fast sharks is covered with tiny scales having little longitudinal ribs on their surface (shark dermal denticles). These are tiny ridges, closely spaced (less than 100 μm apart and still less in height). We note that the considered sharks have a length of approximately 2 m and swim at Reynolds numbers **Re**$\approx 3 \cdot 10^7$ (see e.g. Vogel [58]). Such grooves are similar to ones used on the yacht "Stars and Strips" in America's Cup finals and seem to reduce the skin-friction for $\mathcal{O}(10\%)$ (see [21]).

In the applications, the main interest is in the turbulent case. Mathematical modeling of the turbulent flows in the presence of solid walls is still out of reach. However the turbulent boundary layers on surfaces with fine roughness contain a viscous sublayer. It was found that the viscous sublayer exhibits a streaky structure. Those " low-speed streaks " are believed to be produced by slowly rotating longitudinal vortices. For a streaky structure, with a preferred lateral wavelength, a turbulent shear stress reduction was observed.

The experimental facts were theoretically explained in the papers by Bechert and Bartenwerfer [13] and Luchini, Manzo and Pozzi [42] (see also [14] and references in mentioned articles).

Bechert and Bertenwerfer argued that the velocity profile, which is asymptotically linear Couette flow in the viscous sublayer, appears as it was originated from a smooth solid wall placed at some distance below the riblet tips, but still above the bottom of roughness. They called that distance the *protrusion height*. In [13] the protrusion heights were calculated for the mean longitudinal flows for groove shapes. Similar calculations, but for the cross flow problems are in [42] . If Δh is the difference between the origins of the longitudinal and cross flows, v the friction velocity and if s is the lateral rib spacing, than the quantity $\dfrac{\Delta h}{s}$ plays an important role. In [14] it is noted that, in particular for small Reynolds numbers $s^+ = \dfrac{sv}{\nu}$, the drag reduction is proportional to $\dfrac{\Delta h}{s}$.

Let us apply the theory developed in the laminar situation to the turbulent flow. It is known that the turbulent Couette flow has a 2-layer structure. There is a large core layer where the molecular momentum transfer can be neglected and a thin wall layer (or sublayer) were both turbulent and molecular momentum transfer are important. The flow in the wall layer is governed by the turbulent viscous shear stress τ_w, supposed to depend only on time. In connection with τ_w we introduce the friction velocity $v = \sqrt{\dfrac{\tau_w}{\rho}}$, where ρ is the density. Then the wall layer thickness is $\delta_v = \dfrac{\nu}{v}$ and the theory of the turbulent Couette flow (see e.g. [56]) gives

$$u^+ = f(y^+), \quad y^+ = \dfrac{x_3}{\delta_v} \quad \text{and} \quad u^+ = \dfrac{\bar{u}}{v}$$

where $\bar{u}$ is the mean velocity, as the universal law for the velocity distribution in the pure viscous layer. y^+ is the characteristic wall coordinate and $u^+ = \dfrac{\bar{u}}{v}$ is the rescaled velocity. The above universal law is valid for $0 \leq y^+ < 5$, then for $5 < y^+ < 70$ we have the buffer layer and for $y^+ > 70$ the overlap layer, where the logarithmic profile of the velocity $u^+ = \dfrac{1}{\kappa} \ln y^+ + C^+$ is used. We refer to [56] for more details. For a smooth wall $C^+ = 5$. The theory from [56] is also applied to rough boundaries. It is believed that the reduction of the friction drag by small regular roughness elements is due to a damping of the turbulent fluctuations. Then the riblets with depth and height of about $15\delta_v$ have been used to achieve up to 8% reduction of the friction drag. We note that this means riblets in the buffer layer. Effects of riblets entering the buffer layer where modeled in [56] by adjusting C^+. C^+ depends on $k_s^+ = \dfrac{k_s}{\delta_v}$, where k_s is the roughness height. Then $\lim_{k_s^+ \to 0} C^+(k_s^+) = 5$ (smooth boundary) and $\lim_{k_s^+ \to +\infty} \left(C^+(k_s^+) + \dfrac{1}{\kappa} \ln k_s^+\right) = 8$ (fully rough boundary).

Mathematical analysis of a flow in the buffer and overlapping layer is out of reach for the moment. Nevertheless we suppose that our riblets remain all the time in the pure viscous sublayer and try to apply the analysis from section 5.1.1.

The corresponding equations are (5.11)–(5.15) with $L_3 = \delta_v$ and velocity $v = \sqrt{\dfrac{\tau_w}{\rho}} = (v_1, v_2, 0)$ at $x_3 = \delta_v$. Since $\delta_v \sqrt{\dfrac{\tau_w}{\rho}} = \nu < 2\nu$, our results from section 5.1.1 are applicable and we get

$$|\mathcal{F}_t^\varepsilon - \dfrac{\nu}{\delta_v}(v + \dfrac{\varepsilon}{\delta_v} Mv)| \leq C\left(\dfrac{\varepsilon |U|}{\delta_v}\right)^2. \tag{5.45}$$

Since $\delta_v = \nu\sqrt{\dfrac{\rho}{\tau_w}}$, we see that the effects of roughness are significant.

For the shark skin $\varepsilon/\delta_v = 0.1, L_3 = \delta_v = 10^{-3} = \sqrt{\nu}$ and $|U| = \sqrt{\nu} = 10^{-3}$. The uniqueness condition from Corollary 3 applies if $\varepsilon \leq C\nu^{9/4}$. Since $\varepsilon \approx 10^{-4}$ and $\nu^{9/14} \approx 1.389 \cdot 10^{-4}$. We see that our theory is applicable to the swimming of Nektons. For more details we refer to [38].

Furthermore, let us suppose the geometry of the rough boundary from [13] and [42]. Then M is diagonal and the origins of the cross and longitudinal flows are at $y^+ = \dfrac{\varepsilon}{\delta_v}M_{11}$ and $y^+ = \dfrac{\varepsilon}{\delta_v}M_{22}$, respectively. Hence the proposition is to model the flow in the viscous sublayer in the presence of the rough boundary by the Couette-Navier profile (5.38) instead of the simple Couette profile in the smooth case.

5.1.3 Law of Beavers and Joseph

Introduction For creeping flows through a porous medium, the momentum equation, valid in the pores, is replaced by an effective law linking the filtration velocity and the effective pressure gradient. This law was found experimentally by H. Darcy and it is called Darcy's law. It was justified on mathematically rigorous way in a number of papers starting from L. Tartar's appendix in the book by Sanchez-Palencia [55]. For its derivation the periodicity of the porous medium was required and then the Stokes and the incompressible Navier-Stokes system with a small Reynolds number were homogenized. The periodicity condition can be relaxed to a kind of statistical homogeneity and ergodicity, but clearly such assumptions break down close to the boundaries. Deviations from Darcy's law are expected only in thin layers near the interfaces, however they can significantly change the structure of coefficients and even the effective constitutive law.

Homogeneity is broken if we have presence of boundaries or interfaces. The simplest possible problem to be considered is finding relationship between the seepage velocity and the pressure gradient for an incompressible viscous flow through a domain consisting of two different periodic porous media separated by an interface. Homogenization processes is local and we get the Darcy law in every porous piece. However, due to the different geometric structures, the permeability matrices are different. Coupling the flow requires conditions at the interface. Incompressibility of the flow implies continuity of the normal components of the seepage velocities. Another physically natural interface condition is the continuity of the effective pressure field. It is usually imposed without discussion. The rigorous derivation we undertaken in the paper [35] by Jäger and Mikelić, where the pressure continuity was justified using the corresponding boundary layers from [33]. It is interesting to note the deterioration of the order of approximation close to the interface. It is due to the jump of the normal derivatives of the effective pressure at the boundary. For details we refer to [35]. A physical discussion of these interface conditions, sometimes called the *refraction at a boundary between two porous bodies*, is in [25]. Clearly, a good physical intuition allowed direct modeling in this case.

The problem of finding effective boundary conditions at a naturally permeable wall is much more complicated. Namely, if we have a porous medium in the contact with a free viscous flow, direct computations of the velocity and pressure fields are complicated and costly. Furthermore, the exact geometry of the porous part isn't really known. As in the case of flows through a porous medium, an upscaled model is needed. Let us limit, as before, our considerations to a slow viscous and incompressible flow. Since the homogenization is local, we find immediately the upscaled laws in the homogeneous parts: the filtration velocity through a porous medium is given by the Darcy law and the free fluid flow remains governed by the Stokes system (or by the Navier-Stokes system if the inertia effects in the free fluid are important). This means that one should couple two systems of PDE's, one being a second order system for the velocity and a first order equation for the pressure, respectively, and the other being a scalar second order equation for the pressure and a first order system for the seepage velocity. The coupling conditions should be imposed at the interface. That interface is not a physical surface, but an artificial computational boundary. One coupling condition is very simple. It is a consequence of the fluid incompressibility and says that we have the continuity of the normal mass flux. Knowing that the permeability tensor in Darcy's law is always of order ε^2, where ε is the characteristic pore size, we obtain that the normal component of the effective velocity of the free fluid at the interface is equal to zero. This is the *non-penetration* condition. Since the free fluid is viscous, this condition doesn't suffice and one should specify more conditions. Classically, the tangential velocity of the free fluid velocity was set to zero at the interface. This condition corresponds to an impervious boundary and could not be justified, neither from mathematical nor modeling or experimental point of view.

In the fluid mechanics literature we find two different conditions which should complete the flow equations. The first condition was found experimentally by G.S. Beavers and D.D. Joseph and published in [12]. It says that the difference, between the slip velocity of the free fluid and the tangential component of the seepage velocity at the interface, is proportional to the shear stress from the free fluid. This law was justified at a physical level of rigor by P.G. Saffman in [54], where it was observed that the seepage velocity contribution could be neglected. He used the ensemble averaging to obtain the Stokes system for the mean velocity U and the mean pressure P. Averaging process introduced an unknown effective force F, corresponding to the ensemble average of the contact forces over interfaces. For this force the following representation was supposed

$$F_i(x) = \mu \sum_{j=1}^{3} \int K_{ij}(x,y) U_j(y)\, dy - P(x) \frac{\partial}{\partial x_i} \ln \eta(x) \qquad (5.46)$$

where μ was the fluid viscosity and $\eta(x)$ the porosity. K_{ij} is a kernel satisfying
 a) $K_{ij}(x,y) \to 0$ as x moves out of the porous medium and
 b) $K_{ij}(x,y) \to \frac{\eta}{k}\delta_{ij}\delta(x-y)$ as x moves into the porous medium.
Then he used the matching of asymptotic expansions at the interface in order to obtain a variant of Beaver's and Joseph's law. The Saffman's modification of the law by Beavers and Joseph is widely used and it reads

$$\sqrt{k}\frac{\partial v_\tau}{\partial \nu} = \alpha v_\tau. \qquad (5.47)$$

Here α is a non-dimensional parameter depending on the geometrical structure of the porous medium and k is the (scalar) permeability. We note k is of order ε^2 and in (5.47) the terms of order k are neglected. ν denotes the unit normal vector at the interface and v_τ is the slip velocity of the free fluid in the channel.

A similar argument is developed in [25], where a linear relationship between the ∇P and U was postulated. Then the Slattery's relationship

$$\frac{\partial P}{\partial x_i} = \mu\Big(-\sum_{j=1}^{3} r_{ij}^0 U_j + \sum_{j,l=1}^{3} r_{ijl}^1 \frac{\partial U_j}{\partial x_l} + \sum_{j,l,m=1}^{3} r_{ijlm}^2 \frac{\partial^2 U_j}{\partial x_l \partial x_m} + \ldots\Big)(5.48)$$

between the pressure gradient and a combination of derivatives of the seepage velocity was assumed. Here r_{ij}^0, r_{ijl}^1 and r_{ijlm}^2 are the macroscopic resistivity tensors. The asymptotic matching at the boundary lead once more to the law (5.47).

Clearly these arguments are not entirely satisfactory since (5.46) and (5.48) were supposed a priori with unknown kernels. Neither paper [54] nor [25] study boundary layers describing the flow behavior close to the interface.

The second condition is at first glance different and was introduced in the paper [27] by H. Ene and E. Sanchez-Palencia. After studying orders of magnitude of the pressures and the velocities in 2 media and introducing an intermediate layer, they concluded the equality of the free fluid pressure and the mean pressure of the porous medium at the interface. In the paper [27] there is a kind of discussion, arguing with the help of an intermediate boundary layer, but without a rigorous argument or an asymptotic expansion. This modeling could be viewed as a physical intuition. As in the simple situations we always had the pressure continuity, this interface law could considered as acceptable. It is as an alternative to (5.47), however it should be noted that the well-posedness of the coupled averaged problem is not clear. Numerically, the law (5.47) is much more simple to implement.

Finally, let us mention that the case of the flow in a cavity, lying inside a porous matrix, was considered by Th. Levy and E. Sanchez-Palencia in [43]. By comparing the orders of the magnitude of characteristic quantities, it was found out that the effective pressure should be constant at the interface. This conclusion was rigorously justified in [33], after constructing the appropriate boundary layers.

Our goal is to study the interface laws between a free viscous flow and a porous medium using a combination of the homogenization and the singular perturbation. This program was carried out in the papers [36] and [38]. Using essentially the tools developed in [33], the Saffman's variant (5.47) of the law by Beavers and Joseph was rigorously obtained from the first principles. Coupled with the non-penetration condition, it gives a well-posed boundary value problem for the Stokes or, respectively, the Navier-Stokes system describing the free fluid flow. We note that the effective free fluid flow is totally decoupled from the porous media flow. Then, one could turn to the porous medium. Even if the Darcy velocity is of order ε^2, where ε is the characteristic pore size, the effective pressure is of order 1. The effective pressure satisfies a 2nd order linear elliptic equation, obtained by introducing the Darcy law into the incompressibility condition. We miss only an interface condition. The question was discussed in [38] and [47] and it was found that the effective pressure in the porous medium equals the free fluid pressure plus a constant times the viscous shear stress at the boundary. This constant is zero for an isotropic porous medium, but in general it is different from zero. We refer to [38] for a numerical example contradicting pressure continuity at the interface.

Reference [47] contains a detailed derivation of the law (5.47) and we give here only the basic ideas. They are similar to the construction used to derived the Navier slip condition in the previous subsection.

Modeling of the experiment by Beavers and Joseph In their experiment described in [12], G.S. Beavers and D.D. Joseph have used a two-dimensional Poiseuille flow above a fluid saturated, permeable block to infer the value of the velocity at the interface. Consequently, we consider the laminar viscous two-dimensional incompressible flow through a domain Ω consisting of the porous medium $\Omega_2 = (0,b) \times (-L,0)$, the channel $\Omega_1 = (0,b) \times (0,h)$ and the permeable interface $\Sigma = (0,b) \times \{0\}$ between them. We assume that the structure of the porous medium is periodic and generated by translations of a cell $Y^\varepsilon = \varepsilon Y$, where Y is the standard cell, $Y = (0,1)^2$, containing an open Lipschitzian set Z^*, strictly included in Y. Let $Y_F = Y \setminus \overline{Z}^*$ and let χ be the characteristic function of Y_F, extended by periodicity to $\mathbb{R}^2$. We set $\chi^\varepsilon(x) = \chi(\frac{x}{\varepsilon})$, $x \in \mathbb{R}^2$, and define Ω_2^ε by $\Omega_2^\varepsilon = \{x \mid x \in \Omega_2, \chi^\varepsilon(x) = 1\}$. Furthermore, $\Omega^\varepsilon = \Omega_1 \cup \Sigma \cup \Omega_2^\varepsilon$ is the fluid part of Ω. It is supposed that $(b/\varepsilon, L/\varepsilon) \in \mathbb{N}^2$.

Therefore, we have a porous medium consisting of a large number of periodically distributed channels of characteristic length ε, being small compared with a characteristic length of the macroscopic domain.

The flow is supposed to be slow and governed by the following equations

$$-\nu\Delta v^\varepsilon + \nabla p^\varepsilon + (v^\varepsilon \nabla)v^\varepsilon = -\frac{p_b - p_0}{b}e_1 \quad \text{in } \Omega^\varepsilon, \quad (5.49)$$

$$\text{div } v^\varepsilon = 0 \quad \text{in } \Omega^\varepsilon, \quad (5.50)$$

$$v^\varepsilon = 0 \quad \text{on } \partial\Omega^\varepsilon \setminus \partial\Omega \quad \text{and on} \quad (0,b) \times (\{-L\} \cup \{h\}), \quad (5.51)$$

$$\{v^\varepsilon, p^\varepsilon\} \quad \text{is } b-\text{periodic in } x_1 \quad (5.52)$$

where $\nu > 0$ is the kinematic viscosity and p_0 and p_b are given constants. $\varepsilon > 0$ is the characteristic pore size, v^ε is the velocity and p^ε is the pressure field. Problem (5.49) – (5.52) has a unique solution $\{v^\varepsilon, p^\varepsilon\} \in H^1(\Omega^\varepsilon)^2 \times L_0^2(\Omega^\varepsilon)$.

Now one would like to study of the effective behavior of the velocities v^ε and pressures p^ε as $\varepsilon \to 0$. We follow the decomposition approach from [36]. Firstly we observe that the classic Poiseuille flow in Ω_1, satisfying the no-slip conditions at Σ, is given by

$$\begin{cases} v^0 = \left(\dfrac{p_b - p_0}{2b\nu}x_2(x_2 - h), 0\right) & \text{for } 0 \leq x_2 \leq h; \\ p^0 = \dfrac{p_b - p_0}{b}x_1 + p_0 & \text{for } 0 \leq x_1 \leq b \end{cases} \quad (5.53)$$

We extend this solution to Ω_2 by setting $v^0 = 0$ for $-L \leq x_2 \leq 0$ and keeping the same form of p^0. Now, the idea is to construct the solution to (5.49) – (5.52) as a small perturbation to the Poiseuille flow (5.53). This could be true only if the Reynolds number

$$\mathbf{Re} = \frac{|p_b - p_0|}{\nu^2}\frac{h^2}{8}$$

is not too large. First, there is a $C = C(b/h)$ such that for $\mathbf{Re} \leq C(b/h)$ (53) is the unique solution for the Poiseuille flow in Ω_1, between all these lying in the ball

$$B = \left\{ z \in H^1(\Omega_1)^2 \,|\, \|z\|_{L^4(\Omega_1)^2} \leq \nu C_0(b,h) \right\}.$$

Then we have the following nonlinear stability result:

Proposition 1. *Let us suppose* $\mathbf{Re} \leq C(b/h)$ *and* $\varepsilon \leq \varepsilon_0$. *Let* $\{v^0, \pi^0\}$ *be defined by (5.53). Then the problem* (5.49) – (5.52) *has a solution* $\{v^\varepsilon, p^\varepsilon\} \in H^2(\Omega^\varepsilon)^2 \times H^1(\Omega^\varepsilon)$ *satisfying*

$$\|\nabla(v^\varepsilon - v^0)\|_{L^2(\Omega^\varepsilon)^4} + \|p^\varepsilon - p^0\|_{L^2(\Omega_1)} \leq C\sqrt{\varepsilon} \quad (5.54)$$

$$\|v^\varepsilon\|_{L^2(\Omega_2^\varepsilon)^2} \leq C\varepsilon\sqrt{\varepsilon} \quad (5.55)$$

$$\|v^\varepsilon\|_{L^2(\Sigma)} + \|v^\varepsilon - v^0\|_{L^2(\Omega_1)^2} \leq C\varepsilon \quad (5.56)$$

Moreover, all solutions lying in the ball B *are equal to* $\{v^\varepsilon, p^\varepsilon\}$.

Proof: See [36]. ∎

Therefore, we have obtained the uniform a priori estimates for $\{v^\varepsilon, p^\varepsilon\}$. Moreover, we have found that Poiseuille's flow in Ω_1 is an $O(\varepsilon)$ L^2-approxi"-mation for v^ε. Beavers and Joseph's law should correspond to the next order velocity correction.

Navier's boundary layer As in section 5.1.1., the leading order in the estimate (5.54) was coming from the stresses at the interface. In order to correct them we introduce the corresponding Navier's boundary layer by

Find $\{\beta^{bl}, \omega^{bl}\}$ with square-integrable gradients satisfying

$$-\Delta_y \beta^{bl} + \nabla_y \omega^{bl} = 0 \quad \text{in } Z^+ \cup Z^- \quad (5.57)$$

$$\text{div}_y \beta^{bl} = 0 \quad \text{in } Z^+ \cup Z^- \quad (5.58)$$

$$[\beta^{bl}]_S(\cdot, 0) = 0 \quad \text{on } S \quad (5.59)$$

$$[\{\nabla_y \beta^{bl} - \omega^{bl} I\} e_2]_S(\cdot, 0) = e_1 \quad \text{on } S \quad (5.60)$$

$$\beta^{bl} = 0 \text{ on } \cup_{k=1}^{\infty} (\partial Z^* - \{0, k\}), \{\beta^{bl}, \omega^{bl}\} \text{ is } 1 - \text{periodic in } y_1 \ (5.61)$$

where $S = (0,1) \times \{0\}$, $Z^+ = (0,1) \times (0,+\infty)$, $Z^- = (0,1) \times (-\infty, 0) \setminus \cup_{k=1}^{\infty}(Z^* - \{0,k\})$ and $Z_{BL} = Z^+ \cup S \cup Z^-$.

Let $V = \{z \in L^2_{loc}(Z_{BL})^2 : \nabla_y z \in L^2(Z_{BL})^4; z \in L^2(Z^-)^2; z = 0 \text{ on } \cup_{k=1}^{\infty}(\partial Z^* - \{0,k\}); \text{div}_{yz} = 0 \text{ in } Z_{BL} \text{ and } z \text{ is 1-periodic in } y_1\}$. Then, by Lax-Milgram lemma, there is a unique $\beta^{bl} \in V$ satisfying

$$\int_{Z_{BL}} \nabla \beta^{bl} \nabla \varphi \, dy = -\int_S \varphi_1 \, dy, \quad \forall \varphi \in V.$$

The pressure field is easily reconstructed from the equations and the regularity results are immediate. We have to prove that the system $(5.57) - (5.61)$ describes a boundary layer, i.e. that β^{bl} and ω^{bl} stabilize exponentially towards constants, when $|y_2| \to \infty$.

Lemma 7. *(see [47])* a) *Any solution $\{\beta^{bl}, \omega^{bl}\}$ satisfies*

$$\int_0^1 \beta_2^{bl}(y_1, b) \, dy_1 = 0, \quad \forall b \in \mathbb{R} \quad (5.62)$$

$$\int_0^1 \omega^{bl}(y_1, b_1) \, dy_1 = \int_0^1 \omega^{bl}(y_1, b_2) \, dy_1, \quad \forall b_1 > b_2 \geq 0 \quad (5.63)$$

$$\int_0^1 \beta_1^{bl}(y_1, b_1) \, dy_1 = \int_0^1 \beta_1^{bl}(y_1, b_2) \, dy_1, \quad \forall b_1 > b_2 \geq 0 \quad (5.64)$$

$$\int_0^1 \beta_1^{bl}(y_1, 0) \, dy_1 = -\int_{Z_{BL}} |\nabla \beta^{bl}(y)|^2 \, dy \quad (5.65)$$

b) Let $\xi^{bl} = \text{curl } \beta^{bl} = \dfrac{\partial \beta_1^{bl}}{\partial y_2} - \dfrac{\partial \beta_2^{bl}}{\partial y_1}$. Then for every $a \in (0,1)$ and $y \in Z^+ \cap \{y_2 > a\}$ we have

$$\begin{cases} |\xi^{bl}(y_1, y_2)| \leq C(a) e^{-2\pi y_2} \quad ; \\ |D^\alpha \xi^{bl}(y_1, y_2)| \leq C(\alpha, a) e^{-2\pi y_2}. \end{cases} \quad (5.66)$$

c) Let

$$C_1^{bl} = \int_0^1 \beta_1^{bl}(y_1, 0) dy_1. \quad (5.67)$$

Then for every $y_2 \geq 0$ and $y_1 \in (0,1)$

$$|\beta^{bl}(y_1,y_2) - (C_1^{bl},0)| \leq Ce^{-\delta y_2}, \qquad \forall \delta < 2\pi.$$

d) Let $C_\omega^{bl} = \int_0^1 \omega^{bl}(y_1,0)\,dy_1$. Then for every $y_2 \geq 0$ and $y_1 \in (0,1)$

$$|\omega^{bl}(y_1,y_2) - C_\omega^{bl}| \leq e^{-2\pi y_2}.$$

We note that the above proposition is analogous to the corresponding results in the case of a rough boundary.

The presence of the porous medium necessitates some additional results. They concern the decay of β^{bl} and ω^{bl} in Z^-.

We start with an estimate on ω^{bl} over the cell $Z_k = Z^- \cap (]0,1[\times]k,k+1[)$.

Proposition 2. *(see [47]) Let β^{bl} and ω^{bl} be defined by (5.57) – (5.61) and let r_k be given by*

$$r_k = \frac{1}{|Y_F|} \int_{Z_k} \omega^{bl}(y)\,dy.$$

Then we have

$$\|\omega^{bl} - r_k\|_{L^2(Z_k)} \leq C\|\nabla \beta^{bl}\|_{L^2(Z_k)^4} \text{ and}$$

$$|r_{k+1} - r_k| \leq C\|\nabla \beta^{bl}\|_{L^2(Z_k \cup Z_{k+1} \cup (]0,1[\times\{k+1\}))^4}.$$

This results allows to eliminate the pressure and we are ready to prove the exponential decay for the Navier's system. Let $Z^-(k) = Z^- \cap (]0,1[\times]-\infty,k[)$. We have

Proposition 3. *(see [47]) Let β^{bl} and ω^{bl} be defined by (5.57) – (5.61). Then there exist positive constants C and γ_0, independents of k, such that*

$$\int_{Z^-(k)} |\nabla \beta^{bl}|^2\,dy_1 dy_2 \leq Ce^{\gamma_0 k} \text{ for every negative integer } k.$$

Corollary 5. *Let us suppose the assumptions of the previous proposition. Then there exist constants κ_∞, given by*

$$\kappa_\infty = \lim_{k \to -\infty} \frac{1}{|Y_F|} \int_{Z_k} \omega^{bl}(y)\,dy$$

and C_1, independent of k, such that $\forall k \in \mathbb{N}_-$ we have

$$\|\omega^{bl} - \kappa_\infty\|^2_{L^2(Z^-(k))} \leq C_1 e^{\gamma_0 k}.$$

Since ω^{bl} is unique up to a constant, we fix it by setting $\kappa_\infty = 0$. Hence in the text which follows C_ω^{bl} has the meaning of the pressure drop between $y_2 = +\infty$ and $y_2 = -\infty$.

Remark 5. If the geometry of Z^- is axisymmetric with respect to reflections around the axis $y_1 = 1/2$, then $C_\omega^{bl} = 0$. For the proof we refer to [38]. In [38] a detailed numerical analysis of the problem (5.57) – (5.61) is given. Through numerical experiments it is shown that for a general geometry of Z^-, $C_\omega^{bl} \neq 0$.

5 Multiscale Problems Coming from Fluid Mechanics

Justification of the law by Beavers and Joseph Following the approach from [36], we eliminate the leading contribution by using the boundary layer-type functions

$$\beta^{bl,\varepsilon}(x) = \varepsilon\beta^{bl}(\frac{x}{\varepsilon}) \quad \text{and} \quad \omega^{bl,\varepsilon}(x) = \omega^{bl}(\frac{x}{\varepsilon}), \quad x \in \Omega^\varepsilon, \quad (5.68)$$

We extend $\beta^{bl,\varepsilon}$ by zero to $\Omega \setminus \Omega^\varepsilon$.

Stabilization of $\beta^{bl,\varepsilon}$ towards a nonzero constant velocity $\varepsilon(C_1^{bl}, 0)$, at the upper boundary, generates a counterflow. It has form of the 2D Couette flow $d = (1 - \frac{x_2}{h})e_1$. Let us correct the zeroth order approximation by the new terms : we would like to prove that the following quantities are $o(\varepsilon)$ for the velocity and $O(1)$ for the pressure:

$$\mathcal{U}_0^\varepsilon(x) = u^\varepsilon - v^0 + \left(\beta^{bl,\varepsilon} - \varepsilon(C_1^{bl}, 0)\frac{x_2}{h}H(x_2)\right)\frac{\partial v_1^0}{\partial x_2}(0) \quad (5.69)$$

$$\mathcal{P}_0^\varepsilon = p^\varepsilon - p^0 H(x_2) - p^{1,\varepsilon} H(-x_2) + (\omega^{bl,\varepsilon} - H(x_2)C_\omega^{bl})\nu \cdot \frac{\partial v_1^0}{\partial x_2}(0) \quad (5.70)$$

where $p^{1,\varepsilon} \in H^1(\Omega_2)$.

Let

$$\mathcal{Z}^\varepsilon = \left\{z \in H^1(\Omega^\varepsilon)^2 \mid z = 0 \text{ on } \partial\Omega^\varepsilon \setminus \Omega; z_2 = 0 \text{ on } \partial\Omega; z_1 = 0 \right.$$
$$\left. \text{on } (0,b) \times (\{0\} \cup \{h\})\right\}.$$

Then we have the following result

Proposition 4. *(see [36]) Let $\mathcal{U}_0^\varepsilon$ be given by (5.69) and $\mathcal{P}_0^\varepsilon$ by (5.70). Then $\mathcal{U}_0^\varepsilon \in H^1(\Omega^\varepsilon)^2$, $\mathcal{U}_0^\varepsilon = 0$ on $\partial\Omega^\varepsilon \setminus \partial\Omega$ and div $\mathcal{U}_0^\varepsilon = 0$ in Ω^ε. Furthermore, $\forall \varphi \in \mathcal{Z}^\varepsilon$ we have the following estimate*

$$|\nu \int_{\Omega^\varepsilon} \nabla \mathcal{U}_0^\varepsilon \nabla \varphi - \int_{\Omega^\varepsilon} \mathcal{P}_0^\varepsilon \text{div}\varphi + \int_{\Omega^\varepsilon} v_1^0 \frac{\partial \mathcal{U}_0^\varepsilon}{\partial x_1}\varphi + \int_{\Omega^\varepsilon} (\mathcal{U}_0^\varepsilon)_2 \frac{\partial v_1^0}{\partial x_2}\varphi_1|$$

$$\leq C\varepsilon^{3/2}\|\nabla\varphi\|_{L^2(\Omega^\varepsilon)^4} + |-\nu C_\omega^{bl}\frac{\partial v_1^0}{\partial x_2}(0)\int_\Sigma \varphi_2 \quad (5.71)$$

$$+ \int_{\Omega_2^\varepsilon}(-\frac{p_b - p_0}{b}\varphi_1 + (p^{1,\varepsilon} - p^0) \text{ div } \varphi)|$$

Obviously, the estimate (5.71) is useful only if the pressure field $p^{1,\varepsilon}$ is chosen in an appropriate way. In analogy with the Darcy's law, one can try to take as $p^{1,\varepsilon}$ the solution p for the problem

$$\text{div}\,(K\nabla p) = 0 \quad \text{in } \Omega_2,$$
$$p(x_1, -0) = p^0(x_1) + \nu C_\omega^{bl} \frac{\partial v_1^0}{\partial x_2}(0) \quad \text{on } \Sigma,$$
$$K\nabla p\, e_2 = 0 \quad \text{on } (0,b) \times \{-L\},$$
$$p = p_0 \quad \text{on } \{0\} \times (-L, 0) \quad \text{and} \quad p = p_b \quad \text{on } \{b\} \times (-L, 0),$$

where the permeability tensor K is defined through the problem

$$\begin{cases} -\Delta_y w^j + \nabla_y \pi^j = e^j \quad \text{in} \quad Y_F \\ \text{div}_y w^j = 0 \quad \text{in } Y_F, \quad \int_{Y_F} \pi^j = 0 \\ w^j = 0 \quad \text{on } \partial Y_F, \quad \{w^j, \pi^j\} \text{ is } Z-\text{periodic}, \end{cases}$$

by $K_{ij} = \int_{Y_F} w_i^j\, dy$. The pressure field p is an element of $W^{1,q}(\Omega_2)$, $\forall q \in (1,2)$, and it is a C^∞-function outside the corners. However, due to the discontinuities of the traces at $(0,0)$ and at $(b,0)$, it is not an element of $H^1(\Omega_2)$. We have to regularize the values at the upper corners. Let δ^ε be a Lipschitzian function defined in a neighborhood of $(0,0)$ by

$$\frac{x_1}{\varepsilon} \text{ if } x_1^2 + x_2^2 \leq \varepsilon^2 \quad \text{and} \quad \frac{x_1}{\sqrt{x_1^2 + x_2^2}} \text{ if } x_1^2 + x_2^2 > \varepsilon^2,$$

analogously in a neighborhood of $(b,0)$ and by a smooth extension elsewhere in Ω_2. Then we introduce $p^{1,\varepsilon}$ as the solution to the problem

$$\text{div}\,(K\nabla p^{1,\varepsilon}) = 0 \quad \text{in } \Omega_2, \tag{5.72}$$
$$p^{1,\varepsilon}(x_1, -0) = p^0(x_1) + C_\omega^{bl} \frac{\partial v_1^0}{\partial x_2}(0)\delta^\varepsilon(x_1, -0) \quad \text{on } \Sigma, \tag{5.73}$$
$$K\nabla p^{1,\varepsilon}\, e_2 = 0 \quad \text{on } (0,b) \times \{-L\}, \tag{5.74}$$
$$p^{1,\varepsilon} = p_0 \text{ on } \{0\} \times (-L, 0) \quad \text{and} \quad p^{1,\varepsilon} = p_b \text{ on } \{b\} \times (-L, 0), \tag{5.75}$$

Then

$$\|p^{1,\varepsilon} - p^0 - C_\omega^{bl} \frac{\partial v_1^0}{\partial x_2}(0)\|_{L^2(\Sigma)} \leq C\sqrt{\varepsilon}$$
$$\|\nabla p^{1,\varepsilon}\|_{L^2(\Omega_2^\varepsilon)^2} \leq C|\log \varepsilon|$$

and, consequently, for every $\varphi \in \mathcal{Z}^\varepsilon$ we have

$$|-\nu C_\omega^{bl} \frac{\partial v_1^0}{\partial x_2}(0) \int_\Sigma \varphi_2 + \int_{\Omega_2^\varepsilon} (-\frac{p_b - p_0}{b}\varphi_1 + (p^{1,\varepsilon} - p^0)\,\text{div}\,\varphi)|$$
$$\leq C\varepsilon|\log \varepsilon|\|\nabla \varphi\|_{L^2(\Omega_2^\varepsilon)^4}.$$

Thus, we have obtained the following result

Corollary 6. *Let $p^{1,\varepsilon}$ be defined by (5.72) – (5.75). Then for $\{\mathcal{U}_0^\varepsilon, \mathcal{P}_0^\varepsilon\}$ given by (5.69) – (5.70), we have*

$$|\nu \int_{\Omega^\varepsilon} \nabla \mathcal{U}_0^\varepsilon \nabla \varphi - \int_{\Omega^\varepsilon} \mathcal{P}_0^\varepsilon \mathrm{div}\varphi + \int_{\Omega^\varepsilon} v_1^0 \frac{\partial \mathcal{U}_0^\varepsilon}{\partial x_1} \varphi + \int_{\Omega^\varepsilon} (\mathcal{U}_0^\varepsilon)_2 \frac{\partial v_1^0}{\partial x_2} \varphi_1|$$
$$\leq C\varepsilon |\log \varepsilon| \|\nabla \varphi\|_{L^2(\Omega^\varepsilon)^4}, \qquad \forall \varphi \in \mathcal{Z}^\varepsilon.$$

At this stage we would like to follow the ideas from [33], take $\varphi = \mathcal{U}_0^\varepsilon$ as the test function and get the required higher order a priori estimate. Nevertheless here we are in the presence of the physical outer boundaries and $\mathcal{U}_0^\varepsilon \notin \mathcal{Z}^\varepsilon$. At $(0,b) \times (\{-L\} \cup \{h\})$ the velocity field $\mathcal{U}_0^\varepsilon$ is exponentially small and we can suppose it being zero without loosing generality.

At the inflow/outflow boundaries $(\{0\} \cup \{b\}) \times (-L, h)$ the situation is different. One has to correct the values of $\mathcal{U}_0^\varepsilon$ there. For this purpose an outer boundary layer $\{s^\varepsilon, \vartheta^\varepsilon\}$ is necessary. It doesn't enter the effective law but deteriorates the final estimate. For the details we refer to [36]. After introducing all auxiliary functions we are in position to state the main approximation result

Theorem 5. *(see [36]) Let*

$$\mathcal{U}^\varepsilon(x) = u^\varepsilon - v^0 + \left(\beta^{bl,\varepsilon} - s^\varepsilon\right) \frac{\partial v_1^0}{\partial x_2}(0) - \varepsilon C_1^{bl} \frac{\partial v_1^0}{\partial x_2}(0) H(x_2) \frac{x_2}{h} e_1$$

$$\mathcal{P}^\varepsilon = p^\varepsilon - p^0 H(x_2) - p^{1,\varepsilon} H(-x_2) + \left(\omega^{bl,\varepsilon} - \vartheta^\varepsilon - H(x_2) C_\omega^{bl}\right) \nu \frac{\partial v_1^0}{\partial x_2}(0),$$

where $\{v^0, p^0\}$ is defined by (5.53), $p^{1,\varepsilon}$ by (5.72) – (5.75) and $\{\beta^{bl,\varepsilon}, \omega^{bl,\varepsilon}\}$ by (5.68).

Then we have the following estimates

$$\|\nabla \mathcal{U}^\varepsilon\|_{L^2(\Omega^\varepsilon)^4} + \|\mathcal{P}^\varepsilon\|_{L^2(\Omega_1)} \leq C\varepsilon |\log \varepsilon| \qquad (5.76)$$
$$\|\mathcal{U}^\varepsilon\|_{L^2(\Omega_2^\varepsilon)^2} \leq C\varepsilon^2 |\log \varepsilon| \qquad (5.77)$$
$$\|\mathcal{U}^\varepsilon\|_{L^2(\Sigma)^2} + \|\mathcal{U}^\varepsilon\|_{L^2(\Omega_1)^2} \leq C\varepsilon^{3/2} |\log \varepsilon| \qquad (5.78)$$

The estimates (5.76) – (5.78) allow us to justify Saffman's modification of the Beavers and Joseph law. First, we have a result analogous to the theorem 3 from section 5.1.1. It is precisely stated in [36], we give here only the comparison between the starting problem and the effective upscaled equations.

The effective flow equations in Ω_1 are given through the following boundary value problem:

Find a velocity field u^{eff} and a pressure field p^{eff} such that

$$-\nu\Delta u^{eff} + (u^{eff}\nabla)u^{eff} + \nabla p^{eff} = 0 \quad \text{in } \Omega_1, \tag{5.79}$$
$$\text{div } u^{eff} = 0 \quad \text{in } \Omega_1, \tag{5.80}$$
$$u^{eff} = 0 \quad \text{on } (0,b) \times \{h\}, \tag{5.81}$$
$$u_2^{eff} = 0 \quad \text{on } (\{0\} \cup \{b\}) \times (0,h), \tag{5.82}$$
$$p^{eff} = p_0 \quad \text{on } \{0\} \times (0,h) \quad \text{and} \quad p^{eff} = p_b \quad \text{on } \{b\} \times (0,h), \tag{5.83}$$
$$u_2^{eff} = 0 \quad \text{and} \quad u_1^{eff} + \varepsilon C_1^{bl}\frac{\partial u_1^{eff}}{\partial x_2} = 0 \quad \text{on } \Sigma. \tag{5.84}$$

Under the assumptions of Proposition 1, the problem (5.79) – (5.84) has a unique solution

$$\begin{cases} u^{eff} = \left(\dfrac{p_b - p_0}{2b\nu}\left(x_2 - \dfrac{\varepsilon C_1^{bl} h}{h - \varepsilon C_1^{bl}}\right)(x_2 - h), 0\right) & \text{for } 0 \leq x_2 \leq h; \\ p^{eff} = p^0 = \dfrac{p_b - p_0}{b}x_1 + p_0 & \text{for } 0 \leq x_1 \leq b \end{cases}$$

The effective mass flow rate through the channel is then

$$M^{eff} = b\int_0^h u_1^{eff}(x_2)\,dx_2 = -\frac{p_b - p_0}{12\nu}h^3\frac{h - 4\varepsilon C_1^{bl}}{h - \varepsilon C_1^{bl}},$$

where $C_1^{bl} < 0$.

Let us estimate the error made when replacing $\{v^\varepsilon, p^\varepsilon, M^\varepsilon\}$ by $\{u^{eff}, p^{eff}, M^{eff}\}$. We have

Proposition 5. *(see [36]) Under the assumptions of Proposition 1 we have*

$$\|\nabla(v^\varepsilon - u^{eff})\|_{L^1(\Omega_1)^4} \leq C\varepsilon|\log\varepsilon|, \tag{5.85}$$
$$\|v^\varepsilon - u^{eff}\|_{H^{1/2-\gamma}(\Omega_1)^2} \leq C\varepsilon^{3/2}|\log\varepsilon|, \quad 1/2 > \gamma > 0, \tag{5.86}$$
$$|M^\varepsilon - M^{eff}| \leq C\varepsilon^{3/2}|\log\varepsilon|. \tag{5.87}$$

Our interface is a mathematical one and it doesn't exist as a physical boundary. It is clear that we can take any straight line at the distance $O(\varepsilon)$ from the rigid parts as an interface. Hence it remains to prove that the law by Beavers and Joseph doesn't depend on the position of the interface. We have the following auxiliary result

Lemma 8. *(see [47]) Let $a < 0$ and let $\beta^{a,bl}$ be the solution for (5.57)–(5.61) with S replaced by $S_a = (0,1) \times \{a\}$, Z^+ by $Z_a^+ = (0,1) \times (a, +\infty)$ and $Z_a^- = Z_{BL} \setminus (S_a \cup Z_a^+)$. Then we have*

$$C_1^{a,bl} = C_1^{bl} - a.$$

This simple result implies that the constant in the law by Beavers and Joseph depends on the position of the boundary. That dependence is linear. Nevertheless, the following result proves that the change of the constant is compensated by the change of the position of the interface and the effective solution changes at order ε^2 :

Lemma 9. *(see [47]) Let $\Omega_{a\varepsilon} = (0,b) \times (a\varepsilon, h)$ for $a < 0$ and let $\{u^{a,eff}, p^{a,eff}\}$ be a solution for (5.79) – (5.84) in $\Omega_{a\varepsilon}$, with (5.84) replaced by*

$$u_2^{a,eff} = 0 \text{ and } u_1^{a,eff} + \varepsilon C_1^{a,bl} \frac{\partial u_1^{a,eff}}{\partial x_2} = 0 \text{ on } \Sigma_a = (0,b) \times a\varepsilon. \quad (5.88)$$

The unique solution $\{u^{a,eff}, p^{a,eff}\}$ for (5.79) – (5.83), (5.88) is given by

$$u^{a,eff} = \left(\frac{p_b - p_0}{2b\nu}\left((x_2 - a\varepsilon)^2 - (x_2 - a\varepsilon - \varepsilon C_1^{a,bl})\frac{(h - a\varepsilon)^2}{h - a\varepsilon - \varepsilon C_1^{a,bl}}\right), 0\right)$$

for $a\varepsilon \leq x_2 \leq h$ and

$$p^{a,eff} = p^0 = \frac{p_b - p_0}{b} x_1 + p_0 \quad \text{for } 0 \leq x_1 \leq b.$$

By the preceding lemma, $C_1^{a,bl} = C_1^{bl} - a$ and

$$u^{a,eff}(x) = u^{eff}(x) + O(\varepsilon^2).$$

Therefore, a perturbation of the interface position for an $O(\varepsilon)$ implies a perturbation in the solution of $O(\varepsilon^2)$. Consequently, there is a freedom in fixing position of Σ. It influences the result only at the next order of the asymptotic expansion.

Remark 6. Therefore we have found that the effective velocity field satisfies the interface condition of Beavers and Joseph in the sense of the estimates (5.85) – (5.87). It is important to point out that the parameter α from the law (5.47) is determined through the corresponding Navier's boundary layer (5.57) – (5.61) and (5.67) by $\alpha = -\frac{1}{\varepsilon C_1^{bl}} > 0$.

In what concerns the pressure approximation, after extending it, we get the uniform bound on it. Hence the effective pressure in Ω_2 is $\nu C_\omega^{bl} \frac{\partial v_1^0}{\partial x_2}(0)$. This indicates the discontinuity of the pressure field at Σ and doesn't confirm the law proposed in [27].

5.2 Interactions Flow-Structures

5.2.1 Introduction

Structure of the porous media met in the nature is frequently very complicated. In order to model their behavior it is necessary to make some hypothesis on it. In general it is supposed that there are two phases, a solid and a fluid one. Solid phase could be rigid or deformable. Furthermore, the porous medium is supposed to be heterogeneous at the microscopic (pore) level but statistically homogeneous at macroscopic level. Then we introduce the characteristic length of the non-homogeneities ℓ. Since the theory for the physical velocities, pressures and other quantities is too complicated, one prefers working with the averaged quantities over characteristic volumes being of order ℓ^3.

For a rigid porous medium, averaging leads to an upscaled filtration law, known as Darcy's law. In this chapter we study small deformations of a deformable porous matrix containing an incompressible fluid.

Modeling of elastic porous materials is a classical issue, undertaken in the extensive works on the general three-dimensional continuum theory by M. Biot. Biot's theory is used to predict the propagation of elasto-acoustic waves through porous media. The Biot's effective equations take into account the displacement of both the fluid in the pores and the solid skeleton and the coupling between these. For details of Biot's theory we refer to his collected works edited by I. Tolstoy [57].

We start with the case of a porous medium saturated by an inviscid fluid. Materials of this type are actually used in aerospace applications for reducing the noise transmission. Corresponding theory was originally developed by M. Biot in fifties. The characteristic length of the non-homogeneities was ℓ and in [15] Biot introduced the averaged displacements $\mathbf{u}$ and $\mathbf{U}$ of the solid and fluid phase, respectively. They correspond to averages over volumes proportional to ℓ^3 of the microscopic displacements. Then by generalizing the classical procedure from the linear theory of elasticity, he introduced the generalized stress-strain relations through the elastic potential energy

$$2E_p = A_1 D(\mathbf{u}) : D(\mathbf{u}) + A_2 (\operatorname{div} \mathbf{u})^2 + B(\operatorname{div} \mathbf{U})^2 + C(\operatorname{div} \mathbf{u})(\operatorname{div} \mathbf{U}),$$

where D is the symmetrized gradient. The stress components are then given by the partial derivatives of E_p.

Next, in [16] Biot obtained dynamical relations in the absence of dissipation. It was argued that compressible fluids behave as incompressible ones if the wavelength of the elastic waves was much bigger than the pore size and that the microscopic flow pattern of the fluid, relative to the solid, depended only on the direction of the flow and not on its magnitude. Then the kinetic energy $\mathcal{E}_c$ of the statistically isotropic system is

$$2\mathcal{E}_c = \rho_{11}|\frac{\partial \mathbf{u}}{\partial t}|^2 + 2\rho_{12}\frac{\partial \mathbf{u}}{\partial t}\frac{\partial \mathbf{U}}{\partial t} + \rho_{22}|\frac{\partial \mathbf{U}}{\partial t}|^2. \tag{5.89}$$

Through invariance considerations, it was found that

$$\rho_{11} = (1-\varphi)\rho_s + \rho_a, \quad \rho_{12} = -\rho_a \quad \text{and} \quad \rho_{22} = \varphi\rho_f + \rho_a,$$

where φ is the porosity, ρ_s and ρ_f are the mass densities for the solid and the fluid, respectively, and ρ_a is an additional mass due to the fluid.

After considering the Lagrangian $\mathcal{L} = \mathcal{E}_c - E_p$, the following system of equations is obtained

$$\begin{cases} \rho_{11}\dfrac{\partial^2 \mathbf{u}}{\partial t^2} + \rho_{12}\dfrac{\partial^2 \mathbf{U}}{\partial t^2} = A_1\,\mathrm{div}(D(\mathbf{u})) + A_2\nabla\,\mathrm{div}\,\mathbf{u} + \dfrac{C}{2}\nabla\,\mathrm{div}\,\mathbf{U} \\ \rho_{12}\dfrac{\partial^2 \mathbf{u}}{\partial t^2} + \rho_{22}\dfrac{\partial^2 \mathbf{U}}{\partial t^2} = B\nabla\,\mathrm{div}\,\mathbf{U} + \dfrac{C}{2}\nabla\,\mathrm{div}\,\mathbf{u} \end{cases} \quad (\mathcal{B})$$

The more realistic dissipative case is modeled by adding the friction effects due to the relative motion between the fluid and the solid. Fundamental references are Biot's papers [16] and [18]. The coefficients in $(\mathcal{B})$ are not given and their determination is in general not clear. In this direction we refer to the paper [17] by Biot and Willis. Furthermore, it is not clear that coefficients in $(\mathcal{B})$ are scalars. All those open questions give motivation to undertake a derivation of the equations $(\mathcal{B})$ starting from the first principles, i.e. from the coupled system containing the linear elasticity system for the solid skeleton and the linearized Euler equations for the inviscid fluid. The homogenization limit, when the pore size ε tends to zero should give an upscaled model. The fluid could be supposed incompressible, the compressible case being simply the penalization of the incompressible case. We are going to present an outline of the rigorous derivation from [29].

In the case of moderate Reynolds numbers, Biot modeled the dissipative effects by arguing that they depend only on the relative motion between the fluid and the solid. This leads to the modified system $(\mathcal{B})$

$$\begin{cases} \rho_{11}\dfrac{\partial^2 \mathbf{u}}{\partial t^2} + \rho_{12}\dfrac{\partial^2 \mathbf{U}}{\partial t^2} + b\dfrac{\partial}{\partial t}(\mathbf{u}-\mathbf{U}) = A_1\,\mathrm{div}(D(\mathbf{u})) + A_2\nabla\,\mathrm{div}\,\mathbf{u} + \dfrac{C}{2}\nabla\,\mathrm{div}\,\mathbf{U} \\ \rho_{12}\dfrac{\partial^2 \mathbf{u}}{\partial t^2} + \rho_{22}\dfrac{\partial^2 \mathbf{U}}{\partial t^2} - b\dfrac{\partial}{\partial t}(\mathbf{u}-\mathbf{U}) = B\nabla\,\mathrm{div}\,\mathbf{U} + \dfrac{C}{2}\nabla\,\mathrm{div}\,\mathbf{u} \end{cases}$$

Biot claimed that $b = \mu\varphi^2 K^{-1}$, where K was the permeability, φ the porosity and μ the fluid viscosity. Then in his subsequent publications it was transformed to

$$\rho\dfrac{\partial^2 \mathbf{u}}{\partial t^2} + \rho_f\varphi\dfrac{\partial^2}{\partial t^2}(\mathbf{U}-\mathbf{u}) = \mathrm{Div}\,\tau$$

$\tau = A_1 D(\mathbf{u}) + (C/2+B)\,\mathrm{div}\,\mathbf{U} + (A_2+C/2)\nabla\,\mathrm{div}\,\mathbf{u}$ (total effective stress)

$p_f = -\dfrac{B}{\varphi}\,\mathrm{div}\,\mathbf{U} - \dfrac{C}{2\varphi}\,\mathrm{div}\,\mathbf{u}$ (effective fluid pressure in the pores)

$$\rho_f\dfrac{\partial^2 \mathbf{u}}{\partial t^2} + \bar{Y}(\dfrac{\partial}{\partial t})\varphi\dfrac{\partial^2}{\partial t^2}(\mathbf{U}-\mathbf{u}) = -\nabla p_f$$

where $\bar{Y}$ is the *viscodynamic operator* and $\rho = (1-\varphi)\rho_s + \varphi\rho_f$ is the effective mass density.

In the most general anisotropic case the expressions for τ and p_f where further generalized to

$$\tau_{ij} = \sum_{l,\nu} A_{ij}^{l\nu} D_{l\nu}(\mathbf{u}) + \varphi M_{ij}\,\mathrm{div}\,(\mathbf{u}-\mathbf{U})$$

$$p_f = \sum_{i,j} M_{ij} D_{ij}(\mathbf{u}) + \varphi m\,\mathrm{div}\,(\mathbf{u}-\mathbf{U})$$

where $\{A_{ij}^{l\nu}\}$ is a symmetric positive definite 4th order tensor, m is a parameter and $\{M_{ij}\}$ is a symmetric positive definite matrix. The viscodynamic operator $\bar{Y}$ depends in general on the frequency. Its exact form was subject of detailed investigations in Biot's papers and in the subsequent works. In the small frequency range, Biot found the representation

$$\bar{Y}(\frac{\partial}{\partial t}) = \mu K^{-1} + \frac{\rho_{22}}{\varphi^2}\frac{\partial}{\partial t}.$$

The derivation of the dissipative Biot's law by homogenization was studied by a number of authors: R. Burridge, J. B. Keller, T. Levy, E. Sanchez-Palencia, J. Sanchez-Hubert, J. L. Auriault, G. Nguentseng, J. Saint Jean Paulin and others. The recent exhaustive references are papers by Gilbert and Mikelić [30] and by Clopeau et al. [23], where the Biot's law describing the acoustic properties of the seabed is derived. They contain a number of references to the papers on dissipative Biot's law. Also derivation of a variant of the dissipative Biot's law was one of the first applications of the 2-scale convergence (see Nguetseng [51]).

To the contrary, only the paper [29] considers the non-dissipative case. The main difficulty is the absence of the space derivatives in the linearized incompressible Euler system. Consequently, the result depends strongly on the connectedness of the fluid part.

We discuss the homogenization of the non-dissipative case in section 5.2.2. The upscaled equations are derived and compared with the Biot's model.

The dissipative case is discussed in section 5.2.3.

5.2.2 Biot's Model Without Dissipation

In this section we study a porous medium obtained by a periodic arrangement of the pores, with connected fluid part. The formal description goes along the following lines:

Firstly we define the geometrical structure inside the unit cell $\mathcal{Y} =]0,1[^3$. Let $\mathcal{Y}_s$ (the solid part) be a closed subset of $\bar{\mathcal{Y}}$ and $\mathcal{Y}_f = \mathcal{Y}\backslash\mathcal{Y}_s$ (the fluid part). Now we make the periodic repetition of $\mathcal{Y}_s$ all over $\mathbb{R}^3$ and set $\mathcal{Y}_s^k = \mathcal{Y}_s + k$, $k \in \mathbb{Z}^3$. Obviously the set $E_S = \bigcup_{k \in \mathbb{Z}^3} \mathcal{Y}_s^k$ is a closed subset of $\mathbb{R}^3$ and $E_F = \mathbb{R}^3\backslash E_S$ is an open set in $\mathbb{R}^3$. Following Allaire [5] we make the following assumptions on $\mathcal{Y}_f$ and E_F:

(i) $\mathcal{Y}_f$ is an open connected set of strictly positive measure, with a Lipschitz boundary and $\mathcal{Y}_s$ has strictly positive measure in $\bar{\mathcal{Y}}$ as well.

(ii) E_F and the interior of E_S are open sets with the boundary of class $C^{1,1}$, which are locally located on one side of their boundary. Moreover E_F is connected and the solid part, E_S, is supposed connected in $\mathbb{R}^3$

Now we see that $\Omega =]0, L[^3$ is covered with a regular mesh of size ε, each cell being a cube $\mathcal{Y}_i^\varepsilon = \varepsilon(\mathcal{Y} + i)$, with $1 \leq i \leq N(\varepsilon) = |\Omega|\varepsilon^{-3}[1 + 0(1)]$. We define $\mathcal{Y}^\varepsilon_{s_i} = (\Pi^\varepsilon_i)^{-1}(\mathcal{Y}_s)$ and $\mathcal{Y}^\varepsilon_{f_i} = (\Pi^\varepsilon_i)^{-1}(\mathcal{Y}_f)$. For sufficiently small $\varepsilon > 0$ we suppose $L/\varepsilon \in \mathbb{N}$ and consider $T_\varepsilon = \{k \in \mathbb{Z}^3 | \mathcal{Y}^\varepsilon_{s_k} \subset \Omega\}$ and define

$$\Omega^\varepsilon_s = \bigcup_{k \in T_\varepsilon} \mathcal{Y}^\varepsilon_{s_k}, \quad \Gamma^\varepsilon = \partial \Omega^\varepsilon_s, \quad \Omega^\varepsilon_f = \Omega \setminus \Omega^\varepsilon_s.$$

Obviously, $\partial \Omega^\varepsilon_f = \partial \Omega \cup \Gamma^\varepsilon$. The domains Ω^ε_s and Ω^ε_f represent, respectively, the solid and fluid parts of a porous medium Ω. We suppose small deformations and the displacements are described by linearized equations in both media. More precisely, in Ω^ε_f we have the linearized momentum equation for the time derivative of the fluid deformation u^ε, in Eulerian variables. Ω^ε_s is the reference configuration of a deformable elastic body and the equation of linear elasticity for the deformation w^ε are in Lagrangean variables. In both domains all quadratic terms are neglected. The interface between two media changes with the deformation of the structure. If the pores contain an inviscid fluid then the kinematic interface condition is continuity of the normal velocities. It reads

$$\forall t, \forall x \in \Gamma^\varepsilon, \frac{\partial u^\varepsilon}{\partial t}(x + w^\varepsilon(x,t), t) \cdot \nu = \frac{\partial w^\varepsilon}{\partial t}(x,t) \cdot \nu.$$

We note that in the case of a viscous fluid, considered in section 5.2.3, one has continuity of the velocities on Γ^ε. after linearization and neglecting the term $\nabla_x \frac{\partial u^\varepsilon}{\partial t}(x,t) \cdot w^\varepsilon(x,t)$, we get continuity of the normal deformations at Γ^ε. In the viscous case this means continuity of the deformations.

Let $F \in C^\infty([0,T]; L^2(\Omega)^3)$ and curl $F \in C^\infty([0,T]; L^2(\Omega)^3)$. Then we consider the system

$$\rho_s \frac{\partial^2 w^\varepsilon}{\partial t^2} - \text{div}(\sigma(w^\varepsilon)) = F\rho_s \quad \text{and} \quad \sigma(w^\varepsilon) = AD(w^\varepsilon) \text{ in } \Omega^\varepsilon_s \times]0,T[, \quad (5.90)$$

$$\rho_f \frac{\partial^2 u^\varepsilon}{\partial t^2} + \nabla p^\varepsilon = F\rho_f \quad \text{and} \quad \text{div}\frac{\partial u^\varepsilon}{\partial t} = 0 \text{ in } \Omega^\varepsilon_f \times]0,T[, \quad (5.91)$$

$$(u^\varepsilon - w^\varepsilon) \cdot \nu = 0 \quad \text{and} \quad (p^\varepsilon I + \sigma(w^\varepsilon)) \cdot \nu = 0 \text{ on } \Gamma^\varepsilon \times]0,T[\quad (5.92)$$

$$\{u^\varepsilon, w^\varepsilon, p^\varepsilon\} \text{ are } L - \text{periodic.} \quad (5.93)$$

$$u^\varepsilon(x,0) = \frac{\partial u^\varepsilon}{\partial t}(x,0) = 0 \text{ in } \Omega^\varepsilon_f \quad (5.94)$$

$$w^\varepsilon(x,0) = \frac{\partial w^\varepsilon}{\partial t}(x,0) = 0 \text{ in } \Omega^\varepsilon_s \quad (5.95)$$

The above system has a unique variational solution in

$$V = \{\varphi \in L^2(\Omega)^3 ; \varphi \in H^1(\Omega^\varepsilon_s)^3, \text{div } \varphi = 0 \text{ in } \Omega^\varepsilon_f, \text{div } \varphi \in L^2(\Omega)$$
$$\text{and } \varphi \text{ is L-periodic}\}.$$

It is arbitrary smooth with respect to t.

After taking $u^\varepsilon \in H^3(0,T;L^2(\Omega_f^\varepsilon)^3) \cap H^2(0,T;H^1_{\text{per}}(\Omega_f^\varepsilon)^3)$, $w^\varepsilon \in H^3(0,T;L^2(\Omega_s^\varepsilon)^3)$ and $p^\varepsilon \in H^1(0,T;L^2(\Omega_f^\varepsilon))$ as the test function in (5.90) – (5.95) we obtain

$$\left\|\frac{\partial u^\varepsilon}{\partial t}\right\|_{L^\infty(0,T;L^2(\Omega_f^\varepsilon)^3)} + \left\|\frac{\partial w^\varepsilon}{\partial t}\right\|_{L^\infty(0,T;L^2(\Omega_s^\varepsilon)^3)} + \left\|\nabla w^\varepsilon\right\|_{L^\infty(0,T;L^2(\Omega_s^\varepsilon)^9)} \leq C, \tag{5.96}$$

The linearized incompressible Euler system doesn't involve derivatives of the velocity field with respect to x and an H^1-estimate for the velocity doesn't follow directly. One way to proceed is to use the $H(\text{div};\Omega_f^\varepsilon)$ estimate in the fluid part. Nevertheless, after taking the curl of the linearized Euler system, we get

$$\text{div } u^\varepsilon = 0 \quad \text{and} \quad \left\|\text{curl } u^\varepsilon\right\|_{L^\infty(0,T;L^2(\Omega_f^\varepsilon)^3} \leq C \tag{5.97}$$

and it is natural to estimate the L^2-norm of the ∇u^ε by the L^2-norms of div u^ε and curl u^ε. Such estimate requires a boundary condition at the interface. In our model it is a given normal component $u^\varepsilon \cdot \nu$ and then the H^1-estimate uniform with respect to ε holds if and only if the first Betti number of Ω_f^ε is zero. By assumptions, $\partial\Omega_f^\varepsilon$ is a connected 2D manifold. Each such manifold is homeomorphic to a sphere with "handles" and the first Betti number or the genus of $\partial\Omega_f^\varepsilon$ is the number of handles. For a simply connected domain, the first Betti number is zero. Our domain is multiply connected and the estimate requires some effort.

Proposition 6. *(see [29]) By supposing that $\partial \mathcal{Y}_f \in C^{1,1}$ we get*

$$\left\|\nabla u^\varepsilon\right\|_{L^\infty(0,T;L^2(\Omega_f^\varepsilon)^9)} \leq \frac{C}{\varepsilon}. \tag{5.98}$$

We note that for isolated fluid parts the velocity gradient is uniformly bounded, leading to different results. For more details we refer to [29]. The extension of the pressure field p^ε to $\Omega_s^\varepsilon \times]0,T[$ is given by

$$\tilde{p}^\varepsilon(x,t) = \begin{cases} p^\varepsilon(x,t) - \frac{1}{|\Omega|}\int_{\Omega_f^\varepsilon} p^\varepsilon(x,t)\,dx, & x \in \Omega_f^\varepsilon, \\ -\frac{1}{|\Omega|}\int_{\Omega_f^\varepsilon} p^\varepsilon(x,t)\,dx, & x \in \Omega_s^\varepsilon. \end{cases} \tag{5.99}$$

Then $\int_\Omega \tilde{p}^\varepsilon \text{div } \varphi = \int_{\Omega_f^\varepsilon} p^\varepsilon \text{div } \varphi$, $\forall \varphi \in H^1_{\text{per}}(\Omega)^3$, and

$$\left\|\tilde{p}^\varepsilon\right\|_{H^1(0,T;L^2_0(\Omega))} + \left\|\nabla \tilde{p}^\varepsilon\right\|_{H^1(0,T;H^{-1}_{\text{per}}(\Omega)^3)} \leq C. \tag{5.100}$$

In order to prove the main convergence results of this paper we use the notion of *two-scale convergence* which was introduced in [50] and developed further in [6].

Definition 1. *The sequence $\{z^\varepsilon\} \subset L^2(\Omega)$ is said to <u>two-scale converge</u> to a limit $z \in L^2(\Omega \times \mathcal{Y})$ iff for any $\sigma \in C^\infty(\Omega; C^\infty_{\text{per}}(\mathcal{Y}))$ ("per" denotes 1-periodicity) one has*

$$\lim_{\varepsilon \to 0} \int_\Omega z^\varepsilon(x)\sigma(x, \frac{x}{\varepsilon})\,dx = \int_\Omega \int_\mathcal{Y} z(x,y)\sigma(x,y)\,dy\,dx.$$

The main interest of this convergence lies in the compactness properties, being more precise that the standard weak compactness.

Lemma 10. *(see [50]) From each bounded sequence in $L^2(\Omega)$ one can extract a subsequence which two-scale converges to a limit $z \in L^2(\Omega \times \mathcal{Y})$.*

Lemma 11. *(i) Let z^ε and $\varepsilon\nabla_x z^\varepsilon$ be bounded sequences in $L^2(\Omega)$. Then there exists a function $z \in L^2(\Omega; H^1_{\text{per}}(\mathcal{Y}))$ and a subsequence such that both z^ε and $\varepsilon\nabla_x z^\varepsilon$ two-scale converge to z and $\nabla_y z$, respectively.*

(ii) Let z^ε and $\nabla_x z^\varepsilon$ be bounded sequences in $L^2(\Omega)$. Then there exists functions $z \in L^2(\Omega)$, $v \in L^2(\Omega; H^1_{\text{per}}(\mathcal{Y}))$ and a subsequence such that both z^ε and $\nabla_x z^\varepsilon$ two-scale converge to z and $\nabla_x z(x) + \nabla_y v(x,y)$, respectively.

Proof: See [6] and [50]. ■

If we have two different estimates for gradients in the solid and in the fluid part, then the classical way to proceed is by extending the deformation from Ω_f^ε to Ω and then passing to the limit $\varepsilon \to 0$. For precise recent results on homogenization of Neumann problems in perforated domains with Lipschitz perforations we refer to [1] and to the book Jikov, Kozlov, Oleinik [40]).

The a priori estimates (5.96) – (5.100) allow us to pass to the limit as $\varepsilon \to 0$. That is possible because of the following result

Theorem 6. *(see [29]) There exist subsequences such that, $\forall t \in [0,T]$,*

$$\chi_{\Omega_f^\varepsilon} u^\varepsilon + \chi_{\Omega_s^\varepsilon} w^\varepsilon \to u^0(x,t) + \chi_{\mathcal{Y}_f}(y)v(x,y,t) \text{ in the 2-scale sense,} \quad (5.101)$$

$$\chi_{\Omega_s^\varepsilon}\nabla w^\varepsilon \to \chi_{\mathcal{Y}_s}(y)[\nabla_x u^0(x,t) + \nabla_y u^1(x,y,t)] \text{ in the 2-scale sense,} \quad (5.102)$$

$$\chi_{\Omega_f^\varepsilon}\varepsilon\nabla u^\varepsilon \to \chi_{\mathcal{Y}_f}(y)\nabla_y v(x,y,t) \text{ in the 2-scale sense,} \quad (5.103)$$

$$\tilde{p}^\varepsilon \to \tilde{p}^0(x,y,t) \text{ in the 2-scale sense,} \quad (5.104)$$

with

$$v \in H^3(0,T; L^2(\Omega; H(\text{div}; \mathcal{Y}))^3) \cap H^1(0,T; L^2(\Omega; H^1_{\text{per}}(\mathcal{Y}_f))^3),$$

$$v = 0 \text{ on } \mathcal{Y}_s, \quad (5.105)$$

$$\text{div } \tfrac{\partial v}{\partial t} = 0 \quad \text{and in } \Omega \times \mathcal{Y}_f \times]0,T[\text{ and} \quad (5.106)$$

$$\text{div}_x \int_{\mathcal{Y}_f} \tfrac{\partial v}{\partial t}\,dy \in H^1(]0,T[, L^2(\Omega)) \quad (5.107)$$

$$u^0 \in H^3(0,T; H^1_{\text{per}}(\Omega)^3) \quad \text{and} \quad u^1 \in H^3(0,T; L^2(\Omega; H^1_{\text{per}}(\mathcal{Y}_s)/\mathbb{R})^3)$$

and

$$\begin{cases} \tilde{p}^0(x,y,t) = \chi_{\mathcal{Y}_f}(y)p(x,t) + B(t), \ p \in H^1(]0,T[;L^2(\Omega)) \\ B(t) = -\frac{|\mathcal{Y}_f|}{|\Omega|}\int_\Omega p(x,t)\,dx \end{cases} \quad (5.108)$$

It should be noted that the above convergence result relies on the connectivity of the solid part. The connectivity of the fluid part is not required.

Using standard procedures from the theory of 2-scale convergence it is straightforward to pass to the limit in the system (5.90) – (5.95). For details we refer to [29]. We obtain the following two-scale system for the effective displacement u^0, the effective relative pore displacement v between phases, the correction $D_y(u^1)$, the effective pressure p and the second pressure π

$$\rho_f \frac{\partial^2}{\partial t^2}(u^0 + v) + \nabla_x p + \nabla_y \pi = F\rho_f \quad \text{in } \Omega \times \mathcal{Y}_f \times]0,T[, \quad (5.109)$$

$$-\text{div}_y\{A(D_x(u^0) + D_y(u^1))\} = 0 \quad \text{in } \Omega \times \mathcal{Y}_s \times]0,T[, \quad (5.110)$$

$$\begin{cases} \rho_f \frac{\partial^2}{\partial t^2}\left\{|\mathcal{Y}_f|u^0 + \int_{\mathcal{Y}_f} v\,dy\right\} + \rho_s|\mathcal{Y}_s|\frac{\partial^2 u^0}{\partial t^2} + |\mathcal{Y}_f|\nabla_x p \\ -\text{div}_x\left\{\int_{\mathcal{Y}_s} A(D_x(u^0) + D_y(u^1))\,dy\right\} = F\rho \quad \text{in } \Omega \times]0,T[, \end{cases} \quad (5.111)$$

$$\text{div}_x\left\{|\mathcal{Y}_f|\frac{\partial u^0}{\partial t} + \int_{\mathcal{Y}_f}\frac{\partial v}{\partial t}\,dy\right\} = \int_{\mathcal{Y}_s} \text{div}_y \frac{\partial u^1}{\partial t}\,dy \quad \text{in } \Omega \times]0,T[, \quad (5.112)$$

$$A(D_x(u^0) + D_y(u^1))\nu = -p\nu \quad \text{on } \Omega \times (\partial\mathcal{Y}_s\setminus\partial\mathcal{Y})\times]0,T[, \quad (5.113)$$

$$v \cdot \nu = 0 \quad \text{on } \Omega \times (\partial\mathcal{Y}_f\setminus\partial\mathcal{Y})\times]0,T[, \quad (5.114)$$

$$\{u^0, u^1, v, p\} \text{ is L-periodic in } x, \quad \{u^1, v\} \text{ is 1-periodic in } y, \quad (5.115)$$

$$\begin{cases} u^0(x,0) = \frac{\partial u^0}{\partial t}(x,0) = 0 \quad \text{in } \Omega, \\ v(x,y,0) = \frac{\partial v}{\partial t}(x,y,0) = 0 \quad \text{in } \Omega \times \mathcal{Y}_f, \end{cases} \quad (5.116)$$

where $\rho = |\mathcal{Y}_f|\rho_f + |\mathcal{Y}_s|\rho_s$.

Theorem 7. *(see [29]) Let $\{u^0, p, u^1, v\}$ be given by (5.101) – (5.108). Then they define the unique variational solution for (5.109) – (5.116).*

Hence the two-scale system (5.109) – (5.116) is the complete upscaled problem. Nevertheless, it is too complicated for numerical calculations and it is necessary to eliminate the fast scale y. It is eliminated using the ansatz

$$u^1(x,y,t) = p(x,t)w^0(y) + \sum_{i,j} \left(D_x(u^0(x,t))\right)_{ij} w^{ij}(y) \tag{5.117}$$

where functions $\{w^{ij}\}$ are given by

$$\begin{cases} \operatorname{div}_y \left\{ A\left(\frac{e_i \otimes e_j + e_j \otimes e_i}{2} + D_y(w^{ij})\right) \right\} = 0 \text{ in } \mathcal{Y}_s, \\ A(\frac{e_i \otimes e_j + e_j \otimes e_i}{2} + D_y(w^{ij}))\nu = 0 \text{ on } \partial\mathcal{Y}_s \setminus \partial\mathcal{Y}, \\ \int_{\mathcal{Y}_s} w^{ij}(y)\,dy = 0, \ w^{ij} \text{ is } 1-\text{periodic}. \end{cases} \tag{5.118}$$

and w^0 by

$$\begin{cases} -\operatorname{div}_y \{AD_y(w^0)\} = 0 \text{ in } \mathcal{Y}_s, \\ AD_y(w^0)\nu = -\nu \text{ on } \partial\mathcal{Y}_s \setminus \partial\mathcal{Y}, \\ \int_{\mathcal{Y}_s} w^0(y)\,dy = 0, \ w^0 \text{ is } 1-\text{periodic}. \end{cases} \tag{5.119}$$

The auxiliary functions define the following effective coefficients:

$$A^H_{klij} = \left(\int_{\mathcal{Y}_s} A\left(\frac{e_i \otimes e_j + e_j \otimes e_i}{2} + D_y(w^{ij})\right)\right)_{kl}, \tag{5.120}$$

$$\mathcal{B}^H = \int_{\mathcal{Y}_s} AD_y(w^0)\,dy, \tag{5.121}$$

$$\int_{\mathcal{Y}_s} \operatorname{div}_y w^{ij}(y)\,dy = -\int_{\mathcal{Y}_s} AD_y(w^0) : D_y(w^{ij})\,dy = \mathcal{B}^H_{ij} \tag{5.122}$$

$$\int_{\mathcal{Y}_s} \operatorname{div}_y w^0\,dy = -\int_{\mathcal{Y}_s} AD_y(w^0) : D_y(w^0) < 0. \tag{5.123}$$

The tensors A^H and $\mathcal{B}^H$ defined by (5.120) and (5.121), respectively, are positive definite and symmetric.

Ansatz for v and π is more complicated. Here we write v as

$$\frac{\partial^2 v}{\partial t^2} = \frac{1}{\rho_f} \sum_j (e_j - \nabla_y \xi^j(y)) \left\{ \rho_f F_j(x,t) - \frac{\partial p}{\partial x_j}(x,t) - \rho_f \frac{\partial^2 u^0_j}{\partial t^2}(x,t) \right\}, \tag{5.124}$$

$$\pi = \sum_j \xi^j(y) \left\{ \rho_f F_j(x,t) - \frac{\partial p}{\partial x_j}(x,t) - \rho_f \frac{\partial^2 u^0_j}{\partial t^2}(x,t) \right\} \tag{5.125}$$

where ξ^i is the unique solution of

$$\begin{cases} -\Delta_y \xi^i = 0 & \text{in } \mathcal{Y}_f, \\ \dfrac{\partial \xi^i}{\partial \nu} = e_i \cdot \nu & \text{on } \partial \mathcal{Y}_f \setminus \partial \mathcal{Y}, \\ \xi^i \text{ is 1-periodic in } \mathcal{Y}, & \text{and } \int_{\mathcal{Y}_f} \xi^i \, dy = 0. \end{cases} \quad (5.126)$$

The effective coefficient is now

$$\mathcal{A}_{ij} = \int_{\mathcal{Y}_f} \left(\delta_{ij} - \frac{\partial \xi^j}{\partial y_i} \right) dy. \quad (5.127)$$

The matrix $\mathcal{A}_{ij}$, given by (5.127), is positive definite and symmetric. Therefore, we have the following system for $\{u^0, p\}$:

$$\left(\rho I - \rho_f \mathcal{A}\right) \frac{\partial^2 u^0}{\partial t^2} - \operatorname{div}_x \{A^H D_x(u^0)\} + \left(|\mathcal{Y}_f| I - \mathcal{B}^H - \mathcal{A}\right) \nabla_x p(x,t)$$

$$= (\rho I - \rho_f \mathcal{A}) F(x,t) \quad \text{in} \quad \Omega \times]0, T[\quad (5.128)$$

$$-\frac{\partial^2 p}{\partial t^2} \int_{\mathcal{Y}_s} \operatorname{div}_y w^0 \, dy - \frac{1}{\rho_f} \operatorname{div} \left(\mathcal{A} \nabla_x p(x,t)\right) + \operatorname{div} \left\{ (|\mathcal{Y}_f| I - \mathcal{A} - \mathcal{B}^H) \frac{\partial^2 u^0}{\partial t^2} \right\}$$

$$= -\operatorname{div} \left(\mathcal{A} F\right) \quad \text{in} \quad \Omega \times]0, T[\quad (5.129)$$

$$\{u^0, p\} \quad \text{is} \quad L-\text{periodic in } x \quad (5.130)$$

$$u^0(x,0) = \frac{\partial u^0}{\partial t}(x,0) = 0 \quad \text{in} \quad \Omega \quad (5.131)$$

We refer to [29] for the proof that $\{u^0, p\}$ is a unique solution for the system (5.128) – (5.131). The new system doesn't contain the fast scale and it is much easier to study.

It remains to compare it with the Biot's system ($\mathcal{B}$). First, we introduce the effective displacement of the solid part as $\mathbf{u} = u^0$ and the effective displacement of the fluid part as $\mathbf{U} = u^0 + \frac{1}{|\mathcal{Y}_f|} \int_{\mathcal{Y}_f} v \, dy$. Then, after averaging the equation (5.124), we get

$$p(x,t) = \frac{1}{\int_{\mathcal{Y}_s} \operatorname{div}_y w^0} \left\{ -\operatorname{div}_x \left(\mathcal{B}^H \mathbf{u}\right) + |\mathcal{Y}_f| \operatorname{div}_x \mathbf{U} \right\} \quad (5.132)$$

i.e., even an incompressible fluid gave rise to a compressible effective 2-phase medium, as noticed by M. A. Biot. This formula coincides with the formula proposed by Biot for a general anisotropic material.

Let us now obtain the system ($\mathcal{B}$). After some transformations we have

$$\{\rho_s|\mathcal{Y}_s|I - \rho_f|\mathcal{Y}_f|(I - |\mathcal{Y}_f|\mathcal{A}^{-1})\}\frac{\partial^2 \mathbf{u}}{\partial t^2} + \rho_f|\mathcal{Y}_f|(I - |\mathcal{Y}_f|\mathcal{A}^{-1})\frac{\partial^2 \mathbf{U}}{\partial t^2}$$

$$= \rho_s|\mathcal{Y}_s|F + \mathrm{div}(\mathcal{A}^H D(\mathbf{u})) + \frac{1}{-\int_{\mathcal{Y}_s}\mathrm{div}_y w^0}\mathcal{B}^H \nabla \, \mathrm{div}(\mathcal{B}^H \mathbf{u})$$

$$- \frac{|\mathcal{Y}_f|}{-\int_{\mathcal{Y}_s}\mathrm{div}_y w^0}\mathcal{B}^H \nabla \, \mathrm{div} \, \mathbf{U}.$$

$$\rho_f|\mathcal{Y}_f|(I - |\mathcal{Y}_f|\mathcal{A}^{-1})\frac{\partial^2 \mathbf{u}}{\partial t^2} + |\mathcal{Y}_f|^2 \rho_f \mathcal{A}^{-1}\frac{\partial^2 \mathbf{U}}{\partial t^2}$$

$$= \rho_f|\mathcal{Y}_f|F + \frac{|\mathcal{Y}_f|^2}{-\int_{\mathcal{Y}_s}\mathrm{div}_y w^0}\nabla \, \mathrm{div} \, \mathbf{U}$$

$$- \frac{|\mathcal{Y}_f|}{-\int_{\mathcal{Y}_s}\mathrm{div}_y w^0}\nabla \, \mathrm{div}(\mathcal{B}^H \mathbf{u}).$$

If the above system is compared with the original Biot's system ($\mathcal{B}$), then we see that they have the same structure.

The first of the above equations corresponds to the first equation in ($\mathcal{B}$) and the second one to the second in ($\mathcal{B}$). The difference is that we got matrices for ρ_{ij}. $\rho_{22} = |\mathcal{Y}_f|^2 \rho_f \mathcal{A}^{-1}$, $\rho_{12} = \rho_f|\mathcal{Y}_f|(I - |\mathcal{Y}_f|\mathcal{A}^{-1})$ and $\rho_{11} = \rho_s|\mathcal{Y}_s|I + |\mathcal{Y}_f|\rho_f(|\mathcal{Y}_f|\mathcal{A}^{-1} - I)$. The added mass is $\rho_a = \rho_f|\mathcal{Y}_f|(|\mathcal{Y}_f|\mathcal{A}^{-1} - I)$. The matricial coefficient $|\mathcal{Y}_f|\mathcal{A}^{-1}$ correspond to the tortuosity factor. Also in the compressibility terms ∇ div, we have $\mathcal{B}^H \mathbf{u}$ and not $\mathbf{u}$. Nevertheless, the four matricial coefficients corresponding to the effective stress-strain relations in the Biot's theory are reduced to only two: $\mathcal{A}^H$ and $\mathcal{B}^H$. To the contrary, the study of the purely elastic and dilatation waves for our system is much more complicated.

The advantage of our approach is that we are not only able to justify the Biot's model without ad hoc assumptions, but also we are able to calculate the coefficients from the first principles.

5.2.3 Biot's Model with Dissipation

In this subsection we undertake a rigorous derivation of the diphasic Biot's law. The fluid is supposed viscous and incompressible. After an appropriate scaling of the displacements, the lengths and the time, our starting equations are again (5.90) – (5.95), but with (5.91) replaced by

$$\rho_f \frac{\partial^2 u^\varepsilon}{\partial t^2} + \nabla p^\varepsilon - \mu\varepsilon^{-r}\Delta\frac{\partial u^\varepsilon}{\partial t} = F\rho_f \quad \text{and} \quad \mathrm{div}\,\frac{\partial u^\varepsilon}{\partial t} = 0 \text{ in } \Omega_f^\varepsilon \times]0, T[,$$
(5.133)

and (5.92) by

$$u^\varepsilon = w^\varepsilon \quad \text{and} \quad (p^\varepsilon I - 2\mu\varepsilon^{-r} D(\frac{\partial u^\varepsilon}{\partial t}) + \sigma(w^\varepsilon)) \cdot \nu = 0 \text{ on } \Gamma_\varepsilon \times]0, T[\quad (5.134)$$

The term $\mu\varepsilon^{-r}$ correspond to the *contrast of property number* $C = \dfrac{\mu}{a\lambda} = O(\varepsilon^{-r})$, where μ is the fluid viscosity, a is the characteristic size of the elasticity coefficients and λ is the characteristic length of the acoustic waves. It is customary to set "the dimensionless viscosity" to be $\mu\varepsilon^{-r}$ and then to study the corresponding 2-scale asymptotic expansion. Furthermore, the contrast of property number is related to the Strouhal number Sh, by $Sh = O(\varepsilon^{-2} C)$. For the modeling using the Strouhal number we refer to [11].

This leads to four cases of physical interest

- Model A: $C = O(\varepsilon^3), r = -3$, the acoustics of a fluid in a rigid porous matrix regime. This case was considered previously by Auriault in [11] and by Levy in [44].
- Model B: $C = O(\varepsilon^2), r = -2$, diphasic macroscopic behavior of the fluid and solid matrix. It was studied in [23] in great detail. The case of the slightly compressible flows through a visco-elastic matrix was considered by Nguetseng in [51].
- Model C: $C = O(\varepsilon), r = -1$, monophasic elastic macroscopic behavior.
- Model D: $C = O(1), r = 0$, monophasic viscoelastic macroscopic behavior. The case is discussed by Gilbert and Mikelić in [30] in great detail, along with the special case of a slightly compressible fluid component. In this case the fluid viscosity terms dominate the solid stress terms. Such a system can be compared to an unconsolidated, saturated marine sediment. Such a sediment possesses little skeletal rigidity.

Model B, the diphasic case corresponds to the Biot model [16, 18, 19] and has been considered by numerous authors. The meaning of $\mu\varepsilon^2$ is that the normal stress of the elastic matrix is of the same order as the fluid pressure. We concentrate at this case.

The a priori estimates are obtained after taking $\varphi = \dfrac{\partial u^\varepsilon}{\partial t}$ as the test function. We get

$$\|\frac{\partial u^\varepsilon}{\partial t}\|_{H^3(0,T;L^2(\Omega)^3)} + \|w^\varepsilon\|_{H^3(0,T;H^1(\Omega_s^\varepsilon)^3)} \leq C, \quad (5.135)$$

$$\|D\left(\frac{\partial u^\varepsilon}{\partial t}\right)\|_{H^3(0,T;L^2(\Omega_f^\varepsilon)^9)} \leq \frac{C}{\varepsilon}, \quad (5.136)$$

We extend the pressure field p^ε to $\Omega^\varepsilon \times]0, T[$ by

$$\tilde{p}^\varepsilon(x,t) = \begin{cases} p^\varepsilon(x,t) - \frac{1}{|\Omega|} \int_{\Omega_f^\varepsilon} p^\varepsilon(x,t)\, dx, & x \in \Omega_f^\varepsilon, \\ -\frac{1}{|\Omega|} \int_{\Omega_f^\varepsilon} p^\varepsilon(x,t)\, dx, & x \in \Omega_s^\varepsilon \end{cases} \quad (5.137)$$

Then $\int_\Omega \tilde{p}^\varepsilon \mathrm{div}\varphi = \int_{\Omega_f^\varepsilon} p^\varepsilon \mathrm{div}\varphi$, $\forall \varphi \in H^1_{\mathrm{per}}(\Omega)^n$, and

$$\|\tilde{p}^\varepsilon\|_{H^1(0,T;L^2_0(\Omega))} + \|\nabla \tilde{p}^\varepsilon\|_{H^1(0,T;H^{-1}_{\mathrm{per}}(\Omega)^n)} \leq C. \tag{5.138}$$

The two-scale convergence result is based now on the results from Allaire [6] and Fasano, Mikelić, Primicerio [28]. We have

Theorem 8. *(see [23])* There exist subsequences such that, $\forall t \in [0,T]$,

$$\chi_{\Omega_f^\varepsilon} u^\varepsilon + \chi_{\Omega_s^\varepsilon} w^\varepsilon \to u^0(x,t) + \chi_{\mathcal{Y}_f}(y) v(x,y,t) \text{ in the 2-scale sense,} \tag{5.139}$$

$$\chi_{\Omega_s^\varepsilon} \nabla w^\varepsilon \to \chi_{\mathcal{Y}_s}(y)[\nabla_x u^0(x,t) + \nabla_y u^1(x,y,t)] \text{ in the 2-scale sense,} \tag{5.140}$$

$$\chi_{\Omega_f^\varepsilon} \varepsilon \nabla u^\varepsilon \to \chi_{\mathcal{Y}_f}(y) \nabla_y v(x,y,t) \text{ in the 2-scale sense,} \tag{5.141}$$

$$\tilde{p}^\varepsilon \to \tilde{p}^0(x,y,t) \text{ in the 2-scale sense,} \tag{5.142}$$

with

$$\begin{cases} v \in H^1(0,T;L^2(\Omega;H^1_{\mathrm{per}}(\mathcal{Y}))^3), \ v = 0 \text{ on } \mathcal{Y}_s, \\ \mathrm{div}\,\frac{\partial v}{\partial t} = 0 \text{ in } \Omega \times \mathcal{Y}_f \times]0,T[\text{ and } \mathrm{div}_x \int_{\mathcal{Y}_f} \frac{\partial v}{\partial t}\, dy \in H^1(]0,T[,L^2(\Omega)) \end{cases} \tag{5.143}$$

$$u^0 \in H^3(0,T;H^1_{\mathrm{per}}(\Omega)^3) \quad \text{and} \quad u^1 \in H^3(0,T;L^2(\Omega;H^1_{\mathrm{per}}(\mathcal{Y}_s)/\mathbb{R})^3)$$

and

$$\begin{cases} \tilde{p}^0(x,y,t) = \chi_{\mathcal{Y}_f}(y) p(x,t) + B(t), \ p \in H^1(]0,T[;L^2(\Omega)) \\ B(t) = -\frac{|\mathcal{Y}_f|}{|\Omega|} \int_\Omega p(x,t)\,dx \end{cases} \tag{5.144}$$

Using standard procedures from the theory of 2-scale convergence it is straightforward to pass to the limit in the system (5.90), (5.133), (5.134), (5.93)–(5.95). For details we refer to [23]. We obtain once more the system (5.109) – (5.116), but with (5.109) replaced by

$$\rho_f \frac{\partial^2}{\partial t^2}(u^0 + v) + \nabla_x p + \nabla_y \pi - \mu \Delta_y \frac{\partial v}{\partial t} = F\rho_f \text{ in } \Omega \times \mathcal{Y}_f \times]0,T[, \tag{5.145}$$

and (5.114) by

$$v = 0 \text{ on } \Omega \times (\partial \mathcal{Y}_f \setminus \partial \mathcal{Y}) \times]0,T[. \tag{5.146}$$

In complete analogy with the subsection which precedes we have once more uniqueness.

Elimination of the fast scale y goes along the same lines. We have

$$u^1(x,y,t) = p(x,t) w^0(y) + \sum_{i,j} \left(D_x(u^0(x,t))\right)_{ij} w^{ij}(y) \tag{5.147}$$

where functions $\{w^{ij}\}$ are given by (5.118) and w^0 by (5.119).

Ansatz for v and π is more complicated.

We construct the solution v by first generating solutions to

$$\begin{cases} \dfrac{\partial w^i}{\partial t} - \Delta_y w^i + \nabla_y \pi^i = 0, \\ \text{div}_y w^i = 0, \quad w^i(y,0) = e_i, \\ w^i \mid_{\partial \mathcal{Y}_f} = 0, \quad \{w^i, \pi^i\} \text{ is 1-periodic.} \end{cases} \tag{5.148}$$

In terms of the $\{w^i, \pi^i\}$, v and π have the following representation

$$\begin{cases} \dfrac{\partial v_i}{\partial t} = \sum_j \int_0^t w_i^j(y, \dfrac{\rho_f}{\mu}(t-\tau))\phi_j(x,\tau)d\tau, \\ \pi = \mu \sum_j \int_0^t \pi^j(y, \dfrac{\rho_f}{\mu}(t-\tau))\phi_j(x,\tau)d\tau. \\ \phi(x,\tau) = F(x,\tau) - \dfrac{1}{\rho_f}\nabla_x p(x,\tau) - \dfrac{\partial^2 u^0}{\partial \tau^2} \end{cases} \tag{5.149}$$

The coefficients $\mathcal{A}$ are now replaced by the time dependent kernels

$$\begin{cases} \mathcal{A}_{ij}(t) = \int_{\mathcal{Y}_f} w_i^j(y, \dfrac{\rho_f}{\mu} t) dy \\ \mathcal{A}_{ij}(0) = \int_{\mathcal{Y}_f} w_i^j(y,0)\, dy = |\mathcal{Y}_f|\delta_{ij}. \end{cases} \tag{5.150}$$

$\mathcal{A}(t)$ is a symmetric positive definite matrix for all $t \geq 0$ and $|\mathcal{A}_{ij}(t)| \leq C \exp\{-c_0 t\}$. We refer to [45] for details. Furthermore, if $\hat{\mathcal{A}}(\gamma)$ denotes the Laplace transform of $\mathcal{A}$, then the matrix $\rho I - \gamma \rho_f \hat{\mathcal{A}}(\gamma)$ is positive definite (see [23]). Therefore, we have the following system for $\{u^0, p\}$:

$$\begin{cases} \rho I \dfrac{\partial^2 u^0}{\partial t^2} - \dfrac{d}{dt}\int_0^t \mathcal{A}(t-\tau)\rho_f \dfrac{\partial^2 u^0}{\partial \tau^2}(x,\tau)\, d\tau - \text{div}_x\left\{A^H D_x(u^0)\right\} \\ + \left(|\mathcal{Y}_f|I - \mathcal{B}^H\right)\nabla_x p(x,t) - \dfrac{d}{dt}\int_0^t \mathcal{A}(t-\tau)\nabla_x p(x,\tau) = \\ \rho F(x,t) - \dfrac{d}{dt}\int_0^t \mathcal{A}(t-\tau)\rho_f F(x,\tau)\, d\tau \quad \text{in} \quad \Omega \times]0,T[\end{cases} \tag{5.151}$$

$$\begin{cases} -\dfrac{\partial^2 p}{\partial t^2}\int_{\mathcal{Y}_s}\text{div}_y w^0\, dy - \dfrac{1}{\rho_f}\dfrac{d}{dt}\text{div}\int_0^t \mathcal{A}(t-\tau)\nabla_x p(x,\tau)\, d\tau + \\ \text{div}\left\{(|\mathcal{Y}_f|I - \mathcal{B}^H)\dfrac{\partial^2 u^0}{\partial t^2}\right\} - \dfrac{d}{dt}\text{div}\int_0^t \mathcal{A}(t-\tau)\dfrac{\partial^2 u^0}{\partial \tau^2}(x,\tau)\, d\tau \\ = -\dfrac{d}{dt}\text{div}\int_0^t \mathcal{A}(t-\tau)F(x,\tau)\, d\tau \quad \text{in} \quad \Omega \times]0,T[\end{cases} \tag{5.152}$$

$$\{u^0, p\} \quad \text{is} \quad L-\text{periodic in } x \tag{5.153}$$

$$u^0(x,0) = \dfrac{\partial u^0}{\partial t}(x,0) = 0 \quad \text{in} \quad \Omega \tag{5.154}$$

We refer to [23] for the proof that $\{u^0, p\}$ is a unique solution for the system (5.151) – (5.154). The new system doesn't contain the fast scale and it is much easier to study.

It remains to compare it with the modified Biot's system ($\mathcal{B}$).

For the effective pressure we get once more the formula (5.132) corresponding to Biot's modeling of the effective pressure in the general anisotropic case.

Concerning the decomposition formula for u^1, it was derived in Burridge and Keller [20]. They wrote it as

$$D_y(u^1) = Q(x,y)p + L(x,y)AD_x(u^0), \qquad (5.155)$$

where Q was a second order tensor and L a fourth order tensor, respectively. However, contrary to the paper [23] and to Nguetseng [51], determination of the tensor Q and L was not discussed there. For the effective stress tensor they have set

$$\overline{\tau}_0 = \frac{1}{|\mathcal{Y}_s|} \int_{\mathcal{Y}_s} A(D_x(u^0) + D_y(u^1)) = (A + \overline{ALA})D_x(u^0) + \overline{AQ}p, \qquad (5.156)$$

which differs slightly from our approach, since we got for the effective stress-strain relations

$$\begin{cases} \tau(x,t) = -\dfrac{1}{\int_{\mathcal{Y}_s} \mathrm{div}_y w^0}(|\mathcal{Y}_f|I - \mathcal{B}^H)\,\mathrm{div}\,_x\{(|\mathcal{Y}_f|I - \mathcal{B}^H)\mathbf{u}\} \\ +A^H D_x(u^0) + \dfrac{1}{\int_{\mathcal{Y}_s} \mathrm{div}_y w^0}(|\mathcal{Y}_f|I - \mathcal{B}^H)\,\mathrm{div}\,_x(\mathbf{u}-\mathbf{U}) \\ p(x,t) = \dfrac{1}{\int_{\mathcal{Y}_s} \mathrm{div}_y w^0}\{\mathrm{div}_x\left((|\mathcal{Y}_f|I - \mathcal{B}^H)\mathbf{u}\right) + |\mathcal{Y}_f|\mathrm{div}_x(\mathbf{U}-\mathbf{u})\} \end{cases}$$

$$(5.157)$$

i.e. we got Biot's relations with $M_{ij} = (|\mathcal{Y}_f|\delta_{ij} - \mathcal{B}^H_{ij})\dfrac{1}{\int_{\mathcal{Y}_s} \mathrm{div}_y w^0}$ (see Biot [18]).

Finally, the Biot's law describing relative motion of the fluid in the pores reads

$$-\nabla p_f - \rho_f \frac{\partial^2 \mathbf{u}}{\partial t^2} = \overline{Y}\left(\frac{d}{dt}\right)\frac{\partial w}{\partial t}, \qquad (5.158)$$

In Burridge and Keller [20] the Laplace's transform of v was written in the form

$$\hat{v} = W(x,y)(\nabla_x \hat{p} - \rho_f \omega^2 \hat{u}^0) \qquad (5.159)$$

but without giving hints about W, which also should depend on ω ! If one compares (5.158) with (5.149), we see that the Laplace transform of $\overline{Y}\left(\frac{d}{dt}\right)$ is $\rho_f|\mathcal{Y}_f|\hat{\mathcal{A}}^{-1}$. Hence we have not only justified (5.158), but also we are giving a method for determining the viscodynamic operator. It is interesting to point out the connection between (5.158) and the filtration of an inviscid incompressible fluid through a rigid porous medium in [46], where also we have (5.158) but with $\mathbf{u} = 0$ and a different $\overline{Y}$.

Acknowledgement: The author would like to thank Jose-Luis Ferrin for reading the second part of the review.

References

1. E. Acerbi, V. Chiado Piat, G. Dal Maso and D. Percivale. *An extension theorem from connected sets and homogenization in general periodic domains*, Nonlinear Anal., Theory Methods Appl. **18** (1992), 481–496.
2. Y. Achdou, O. Pironneau, *Domain decomposition and wall laws*, C. R. Acad. Sci. Paris, Série I, 320 (1995), p. 541–547.
3. Y. Achdou, O. Pironneau, F. Valentin, *Shape control versus boundary control*, eds F. Murat et al., Equations aux dérivées partielles et applications. Articles dédiés à J.L.Lions, Elsevier, Paris, 1998, p. 1–18.
4. Y. Achdou, O. Pironneau, F. Valentin, *Effective Boundary Conditions for Laminar Flows over Periodic Rough Boundaries*, J. Comp. Phys., 147 (1998), p. 187–218.
5. G. Allaire. *Homogenization of the Stokes flow in a connected porous medium*, Asympt. Anal. **2** (1989), 203–222.
6. G. Allaire. *Homogenization and two-scale convergence*, SIAM J. Math. Anal. **23.6** (1992), 1482–1518.
7. G. Allaire, M. Amar, *Boundary layer tails in periodic homogenization*, ESAIM: Control, Optimisation and Calculus of Variations 4(1999), p. 209–243.
8. Y. Amirat, J. Simon, *Influence de la rugosité en hydrodynamique laminaire*, C. R. Acad. Sci. Paris, Série I, 323 (1996), p. 313–318.
9. Y. Amirat, J. Simon, *Riblet and Drag Minimization*, in Cox, S (ed) et al., Optimization methods in PDEs, Contemp. Math, 209, p. 9–17, American Math. Soc., Providence, 1997.
10. Y. Amirat, D. Bresch, J. Lemoine, J. Simon, *Effect of rugosity on a flow governed by Navier-Stokes equations*, to appear in Quaterly of Appl. Maths 2001.
11. J.-L. Auriault. *Poroelastic media*, in Homogenization and Porous Media, U. Hornung, ed., Interdisciplinary Applied Mathematics, Springer, Berlin, (1997), 163–182.
12. G. S. Beavers, D. D. Joseph, *Boundary conditions at a naturally permeable wall*, J. Fluid Mech. 30 (1967), p. 197–207.
13. D. W. Bechert, M. Bartenwerfer, *The viscous flow on surfaces with longitudinal ribs*, J. Fluid Mech. 206(1989), p. 105–129.
14. D. W. Bechert, M. Bruse, W. Hage, J. G. T. van der Hoeven, G. Hoppe, *Experiments on drag reducing surfaces and their optimization with an adjustable geometry*, preprint, spring 1997.
15. M. A. Biot, *Theory of Elasticity and Consolidation for a Porous Anisotropic Solid*, J. Appl. Phys., **26**, 182–185 (1955).
16. M. A. Biot. *Theory of propagation of elastic waves in a fluid-saturated porous solid. I. Lower frequency range*, and *II. Higher frequency range*, J. Acoust Soc. Am. **28**(2) (1956), 168–178 and 179–191.
17. M. A. Biot and D. G. Willis, *The Elastic Coefficients of the Theory of Consolidation*, J. Appl. Mech., **24**, 594–601 (1957).

18. M. A. Biot. *Generalized theory of acoustic propagation in porous dissipative media*, Jour. Acoustic Soc. Amer. **34** (1962), 1254–1264.
19. M. A. Biot. *Mechanics of deformation and acoustic propagation in porous media*, Jour. Applied Physics **33** (1962), 1482–1498.
20. R. Burridge and J. B. Keller. *Poroelasticity equations derived from microstructure*, Jour. Acoustic Soc. Amer. **70** (1981), 1140–1146.
21. D. M. Bushnell, K. J. Moore, *Drag reduction in nature*, Ann. Rev. Fluid Mech. 23(1991), p. 65–79.
22. G. Buttazzo, R. V. Kohn, *Reinforcement by a Thin Layer with Oscillating Thickness*, Appl. Math. Optim. 16(1987), p. 247–261.
23. T. Clopeau, J. L. Ferrin, R.P. Gilbert and A. Mikelić, *Homogenizing the acoustic properties of the seabed: Part II*, Mathematical and Computer Modelling, 33(2001), p. 821–841.
24. I. Cotoi, *Etude asymptotique de l'écoulement d'un fluide visqueux incompressible entre une plaque lisse et une paroi rugueuse*, doctoral dissertation, Université Blaise Pascal, Clermont-Ferrand, January 2000.
25. G. Dagan, *The Generalization of Darcy's Law for Nonuniform Flows*, Water Resources Research, Vol. 15 (1981), p. 1–7.
26. R. Dautray and J.-L. Lions. *Mathematical Analysis and Numerical Methods for Science and Technology* **5** EVOLUTION PROBLEMS 1, Springer, Berlin, (1992).
27. H. I. Ene, E. Sanchez-Palencia, *Equations et phénomènes de surface pour l'écoulement dans un modèle de milieu poreux*, J. Mécan., 14 (1975), p. 73–108.
28. A. Fasano, A. Mikelić and M. Primicerio. *Homogenization of flows through porous media with grains*, Advances in Mathematical sciences and Applications, **8** (1998), 1–31.
29. J. L. Ferrin, A. Mikelić, *Homogenizing the Acoustic Properties of a Porous Matrix Containing an Incompressible Inviscid Fluid*, preprint, Université Claude Bernard Lyon 1, February 2001.
30. R. P. Gilbert and A. Mikelić. *Homogenizing the acoustic properties of the seabed: Part I*, Nonlinear Analysis, **40** (2000), 185–212.
31. V. Girault and P.-A. Raviart, *Finite Element Methods for Navier-Stokes Equations*, Springer Verlag, Berlin, 1986.
32. W. Jäger, A. Mikelić, *Homogenization of the Laplace equation in a partially perforated domain*, prépublication no. 157, Equipe d'Analyse Numérique Lyon-St-Etienne, September 1993, published in " *Homogenization, In Memory of Serguei Kozlov* ", eds. V. Berdichevsky, V. Jikov and G. Papanicolaou, p. 259–284, Word Scientific, Singapore, 1999.
33. W. Jäger, A. Mikelić, *On the Boundary Conditions at the Contact Interface between a Porous Medium and a Free Fluid*, Annali della Scuola Normale Superiore di Pisa, Classe Fisiche e Matematiche - Serie IV 23 (1996), Fasc. 3, p. 403–465.
34. W. Jäger, A. Mikelić, *On the effective equations for a viscous incompressible fluid flow through a filter of finite thickness*, Communications on Pure and Applied Mathematics 51 (1998), p. 1073–1121.
35. W. Jäger, A. Mikelić, *On the boundary conditions at the contact interface between two porous media*, in Partial differential equations, Theory and numerical solution, eds. W. Jäger, J. Nečas, O. John, K. Najzar, et J. Stará, π Chapman and Hall/CRC Research Notes in Mathematics no 406, 1999. pp. 175–186.

36. W. Jäger, A. Mikelić, *On the interface boundary conditions by Beavers, Joseph and Saffman*, SIAM J. Appl. Math., 60(2000), p. 1111–1127.
37. W. Jäger, A. Mikelić, *On the roughness-induced effective boundary conditions for a viscous flow*, J. of Differential Equations, 170(2001), p. 96–122.
38. W. Jäger, A. Mikelić, N. Neuß, *Asymptotic analysis of the laminar viscous flow over a porous bed*, SIAM J. on Scientific and Statistical Computing, 22(2001), p. 2006–2028.
39. W. Jäger, A. Mikelić, *Turbulent Couette Flows over a Rough Boundary and Drag Reduction*, preprint, Université Claude Bernard Lyon 1, september 2001.
40. V. V. Jikov, S. M. Kozlov and O. A. Oleinik. *Homogenization of Differential Operators and Integral Functionals.* Springer Verlag, New York, (1994).
41. J. L. Lions, *Some Methods in the Mathematical Analysis of Systems and Their Control*, Gordon and Breach, New York, 1981.
42. P. Luchini, F. Manzo, A. Pozzi, *Resistance of a grooved surface to parallel flow and cross-flow*, J. Fluid Mech. 228(1991), p. 87–109.
43. Th. Levy, E. Sanchez-Palencia, *On boundary conditions for fluid flow in porous media*, Int. J. Engng. Sci., Vol. 13 (1975), p. 923–940.
44. Th. Levy. *Acoustic phenomena in elastic porous media*, Mech. Res. Comm. **4** (4) (1977), 253–257.
45. A. Mikelić. *Mathematical derivation of the Darcy-type law with memory effects, governing transient flow through porous medium*, Glasnik Matematički **29 (49)** (1994), 57–77.
46. A. Mikelić, L. Paoli. *Homogenization of the inviscid incompressible fluid flow trough a 2D porous medium*, Proceedings of the AMS, vol. **17** (1999), 2019–2028.
47. A. Mikelić, *Homogenization theory and applications to filtration through porous media*, chapter in *Filtration in Porous Media and Industrial Applications*, by M. Espedal, A. Fasano and A. Mikelić, Lecture Notes in Mathematics Vol. 1734, Springer-Verlag, 2000, p. 127–214.
48. B. Mohammadi, O. Pironneau, F. Valentin, *Rough Boundaries and Wall Laws*, Int. J. Numer. Meth. Fluids, 27 (1998), p. 169–177.
49. C. L. M. H. Navier, *Sur les lois de l'équilibre et du mouvement des corps élastiques*, Mem. Acad. R. Sci. Inst. France, 369 (1827).
50. G. Nguetseng. *A general convergence result for a functional related to the theory of homogenization*, SIAM J. Math. Anal. **20** (1989), 608–623.
51. G. Nguetseng. *Asymptotic analysis for a stiff variational problem arising in mechanics*, SIAM J. Math. Anal. **20.3** (1990), 608–623.
52. O. A. Oleinik, G. A. Iosif'jan, *On the behavior at infinity of solutions of second order elliptic equations in domains with noncompact boundary*, Math. USSR Sbornik 40(1981), p. 527–548.
53. R. L. Panton, *Incompressible Flow*, John Wiley and Sons, New York, 1984.
54. P. G. Saffman, *On the boundary condition at the interface of a porous medium*, Studies in Applied Mathematics, 1(1971), p. 77–84.
55. E. Sanchez-Palencia, *Non-Homogeneous Media and Vibration Theory*, Springer Lecture Notes in Physics 127, Springer-Verlag, Berlin, 1980.

56. H. Schlichting, K. Gersten, *Boundary-Layer Theory*, 8th Revised and Enlarged Edition, Springer-Verlag, Berlin, 2000.
57. I. Tolstoy, ed., *Acoustics, elasticity, and thermodynamics of porous media.* Twenty-one papers by M.A. Biot, Acoustical Society of America, New York, 1992.
58. S. Vogel, *Life in Moving Fluids*, 2nd ed., Princeton university Press, Princeton, 1994.

6 From Molecular Dynamics to Conformation Dynamics in Drug Design

Peter Deuflhard

Konrad-Zuse-Zentrum Berlin, and Free University of Berlin, Dept. Mathematics and Computer Science, Takustraße 7, 14195 Berlin, Germany – deuflhard@zib.de

Dedicated to Good Bill Hunting,
Chief of Mount Highdle tribe,
on the occasion of his 60th birthday

6.1 Introduction

The design of pharmaceuticals, briefly called *drug design*, is a pyramidal multistage process, from a broad basis to an extremely narrow tip:

- molecular recognition studies
- intracellular impact studies
- physiological investigations
- animal experiments
- clinical tests
- market introduction

The basis level "molecular recognition studies", in turn, consists of two parts: studies in the chemical lab and studies in the *virtual lab* by means of the computer, often named as *computational drug design*. The impact of this rather new scientific field cannot be overestimated: The cost of identifying a marketable drug out of a huge set of promising chemical substances is commonly estimated as 500 million Euro. If, at the basis level, the number of promising drug candidates could be halved, then the cost per successful marketable pharmaceutical would also roughly be halved, not to mention the reduction of "time to market".

In computational biotechnology, algorithms from discrete mathematics or computer science already play a publicly visible role – for example, multiple alignment in the decoding of the human genome. These approaches primarily aim at a clarification of the *geometric form* of molecular systems. In view of the *biological function*, however, the *dynamics* of molecular systems need to be studied in detail. Here the situation is characterized by the fact that real times of pharmaceutical interest are in the region of *msec* up to *min*, whereas simulation times are presently in the region of *psec* up to *nsec* with *fsec* timesteps. Therefore some computational scientists advocate that the available computer power is the essential limiting factor for gaining insight into the dynamics of molecular systems.

Even though the dynamics of molecules is well recognized in its importance, its mathematical treatment seems to be still at an early phase of involvement. Up to now, classical numerical analysis essentially only enters via fast multipole methods (see Greengard and Rokhlin [17]) or via symplectic discretizations (cf. Sanz-Serna et al. [22]). However, the computation of molecular dynamics has a mathematical limitation, even stricter than the limitation by computer power: the arising trajectories are Hamiltonian and as such chaotic. Consequently, the traditional trajectory simulations give, at best, only information about time averages. Under some ergodic hypothesis, often carelessly a priori assumed, these averages are equivalent to statistical ensemble averages. Therefore an investigation of the dynamics of molecular systems over the time scales of interest will require a different mathematical approach.

In recent years the present author and Ch. Schütte have created some new mathematical model based on concepts of nonlinear dynamics (for early papers see, e.g., [7, 26, 25, 11]). This approach, now called *conformation dynamics*, will be worked out here together with its algorithmic implications and its scientific perspectives.

6.2 Classical Molecular Dynamics

In classical molecular dynamics the simplifying assumption is made that the motion of atoms and molecules can be described by Newtonian differential equations just as in classical mechanics, replacing mechanical potentials by special molecular potentials. Such an assumption obviously ignores the role of quantum mechanics, which actually provides the correct physical framework for these microscopic processes. Some part of the quantum-mechanical effects, at least, are introduced into the classical formalism via a parametrization of the potentials.

6.2.1 Hamiltonian Differential Equations

Let N atoms of a molecular system be specified in terms of their spatial coordinates (position variables) $q_j \in \mathbb{R}^3$, $j = 1, \ldots, N$, and their corresponding N generalized moments (momenta variables) $p_j \in \mathbb{R}^3$. Then the Hamilton function H has the form

$$H(q,p) = \frac{1}{2} p^T M^{-1} p + V(q).$$

The first, quadratic term, involving the symmetric, positive definite mass matrix M, is the kinetic energy, the second term is the potential energy or

6 From Molecular Dynamics to Conformation Dynamics in Drug Design

just potential, which is often highly nonlinear in the molecular context. From given H, the Hamiltonian differential equations are defined as

$$q_i' = \frac{\partial H}{\partial p_i}, \quad p_i' = -\frac{\partial H}{\partial q_i}, \quad i = 1, \ldots, N.$$

Of course, the quality of any molecular dynamics calculation is strongly dependent on the quality of the available potential data (we mostly use MMFF due to [18]). These potentials have the general form

$$V(q) = \sum_{k,l} V_{\text{bond}}(q_k, q_l) + \sum_{k,l,j} V_{\text{angle}}(q_k, q_l, q_j)$$
$$+ \sum_{k,l,j,m} V_{\text{out-of-plane}}(q_k, q_l, q_j, q_m) + \sum_{k,l,j,m} V_{\text{dihedral}}(q_k, q_l, q_j, q_m)$$
$$+ \sum_{k,l} V_{\text{Lennard-Jones}}(q_k, q_l) + \sum_{k,l} V_{\text{Coulomb}}(q_k, q_l)$$

or, in abbreviation,

$$V = V_B + V_A + V_T + V_{LJ} + V_Q,$$

where V_B describes the bond deformation, V_A the angle deformation, V_T the torsion angle deformation (two parts), V_{LJ} the van-der-Waals interaction in terms of the Lennard-Jones potential, and V_Q the electrostatic interaction in terms of Coulomb forces between charges Q.

The numerical solution of the initial value problem for these differential equations first requires the selection of an efficient nonstiff discretization scheme – consult, e.g., the specialized textbook of Sanz-Serna [22] or Section 4.3.4 in the more recent textbook [5, 6]. In the context of numerical integration an efficient evaluation of the right sides is needed. The above potential terms V_B, V_A, V_T, and V_{LJ} contribute a cost of order $O(N)$ operations. The direct evaluation of the long-range Coulomb potential V_Q appears to require $O(N^2)$ operations and hence constitutes a problem of its own, at least for realistic molecules. An efficient algorithm requiring only $O(N)$ operations is the *fast multipole method* of L. Greengard and V. Rokhlin [17].

In order to speed up the numerical computations, T. Schlick and followers suggested to skip the adaptive control of the numerical integrators and just run them with step sizes at the border of stability of the numerical schemes. Such an approach has an interpretation only in terms of some sampling based on the ergodic theorem – see, e.g., [23].

6.2.2 Condition of Molecular Initial Value Problems

Formally speaking, the above solution of the initial value problem is *unique*, which can be written in terms of the flow Φ as

$$x(t) = (q(t), p(t)) = \Phi^t x_0 .$$

For the purpose of numerical analysis, we additionally have to study the corresponding *condition number* κ, which characterizes the sensitivity of the unique solution under perturbation of the initial values. By virtue of first order perturbation theory such a quantity can be defined as (cf. Section 3.1.2 in [5, 6])

$$\|\delta x(t)\| \dot{\leq} \kappa(t) \|\delta x_0\| , \quad \kappa(t) = \|\partial \Phi^t / \partial x_0\| .$$

As already discovered by H. Poincaré, Hamiltonian systems are *chaotic*. In general mathematical terms, this means a characterization of the asymptotic behavior – in the present notation $\kappa(\infty) = \infty$. In the context of numerical analysis, this means that an ever so slight perturbation of the initial values will induce a resulting perturbed trajectory deviating markedly from the unperturbed trajectory after some characteristic critical time. The question is: How long is that "critical time"? Detailed examination shows that for the subclass of *integrable* Hamiltonian systems (such as the popular Kepler problem) the condition number grows linearly – see, e.g. V. I. Arnold [2]. In real life molecular dynamics problems, however, the growth is exponential, i.e.

$$\kappa(t) \sim \exp(t/t_{\text{crit}}) , \tag{6.1}$$

where the critical times t_{crit} are typically no longer than a few ps.

Example: Trinucleotide ACC. We illustrate the effect for the small biomolecule ACC – compare Section 1.2 in [5, 6]. This molecule is a short RNA segment consisting of 94 atoms; the genetic letters in its acronym stand for adenine (A) and cytosine (C). Figure 6.1 shows simulation snapshots at the times $t = 0.0$ ps, $t = 0.5$ ps, and $t = 20$ ps (picoseconds: 1 ps $= 10^{-12}$ sec).

As can be seen, the two molecular configurations are almost identical at the start, but differ completely after only 20 ps. The resulting configurations (left a spherical shape, right a stretched shape) remain essentially the same over quite long time spans. They are therefore called *metastable conformations*. These mathematical objects typically occur in nearly all molecular systems and should be directly computed as such.

6.3 Metastable Conformations as Almost Invariant Sets

The observations of the preceding section force severe changes in the mathematical modelling of molecular dynamics. Instead of the *point concept* of classical mechanics based on deterministic trajectories we need to derive a *set concept* based on the above mentioned metastable conformations. This is the key idea of conformation analysis to be presented here.

6 From Molecular Dynamics to Conformation Dynamics in Drug Design 273

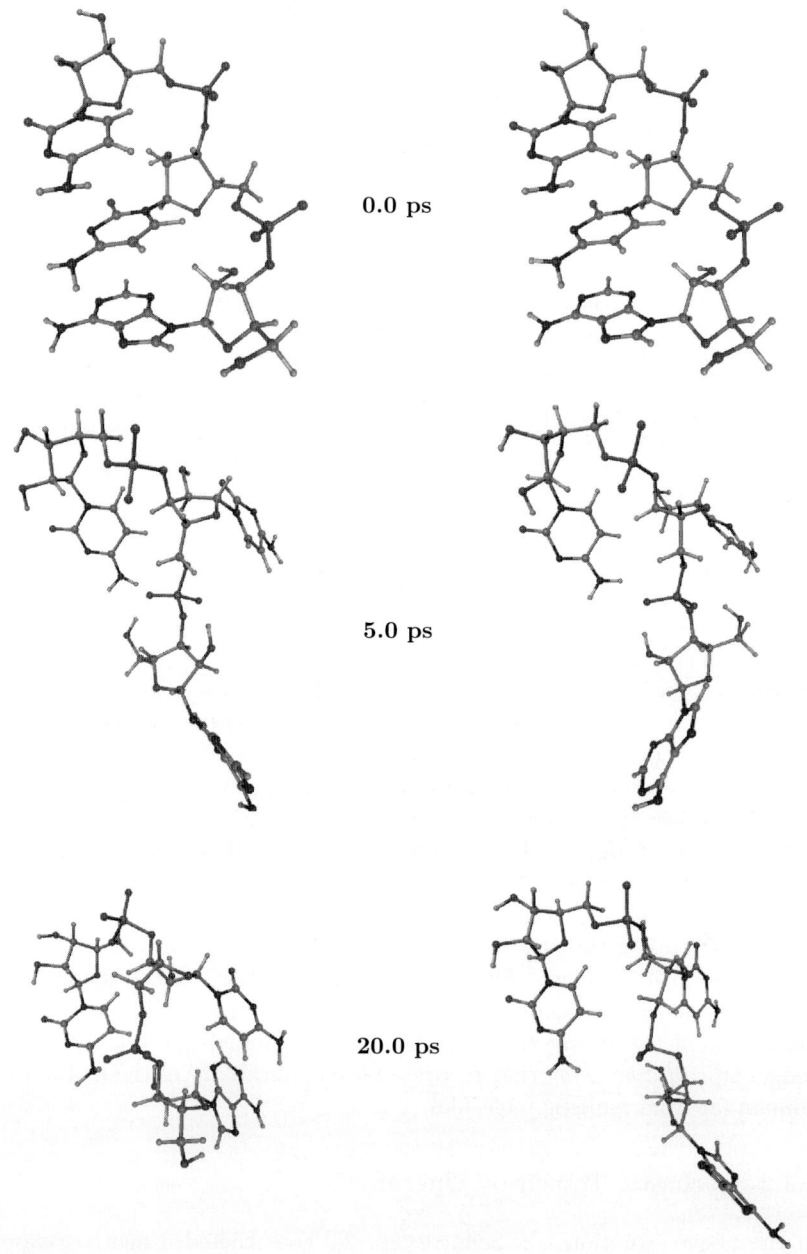

Fig. 6.1. ACC molecule: Development of distinct conformations from nearly identical initial configurations

6.3.1 Perron–Frobenius Operator

Starting point for the new approach was the pioneering work of Dellnitz and co-workers [4, 3] based on the Perron–Frobenius operator U. This operator (dating back to Ulam) is defined via measures in phase space $x = (p,q) \in \Gamma \subset \mathbb{R}^{6N}$ as

$$U\mu(B) = \mu(\Phi^{-\tau}(B)), B \subset \Gamma$$

An invariant measure $\bar\mu$ and the corresponding invariant set $\bar B$ are characterized by

$$\bar\mu(B) = \bar\mu(\Phi^{-\tau}(B)), \quad \bar B = \Phi^{-\tau}(\bar B),$$

which lead to the eigenvalue problem

$$U\bar\mu(\bar B) = \bar\mu(\bar B) \qquad (6.2)$$

for the Perron eigenvalue $\lambda = 1$. On this basis, these authors computed (relatively) global attractors by some adaptive multilevel box discretization. Moreover they found that (a) eigenvalues $\lambda \neq 1$ on the unit circle permit an interpretation in terms of cyclic dynamics, and (b) eigenvalues close to the Perron eigenvalue inside the unit cicle (due to discretization effects) seem to have an interpretation in terms of *almost invariant sets*.

The success of that approach was intimately linked to *hyperbolic* dynamics which is known to collapse asymptotically to some dynamics on a low-dimensional manifold. Being well aware of this restriction, the present author nevertheless risked to extend that basic scheme to *Hamiltonian* dynamics known not to collapse, but to remain on some high-dimensional energy surface. A first attempt in this direction, as published in [7], suffered from two important disadvantages. First, for a *deterministic* Hamiltonian system, the operator U is *unitary* in $L^2(\Gamma)$ so that real eigenvalues inside the unit circle cannot exist. But such eigenvalues had been computed and could be interpreted in detail within the model! The reason for that has been that the discretization had allowed for *stochastic* perturbations of the deterministic system so that such eigenvalues could, in fact, occur and did contain information about almost invariant sets. Second, the subdivision technique caused some *curse of dimension* that restricted the applicability of the method to a domain far from realistic molecules.

6.3.2 Stochastic Transition Operator

In the above situation Ch. Schütte [26, 25] constructed a new *self-adjoint* stochastic operator T. Starting point of his construction is the fact that in a chemical lab with constant temperature and constant volume the deterministic model should be embedded into a canonical or Boltzmann distribution f_0. With β the inverse temperature and for separable Hamiltonian $H = \frac{1}{2}p^T M^{-1} p + V(q)$ we may factorize this distribution according to

$$f_0 = \frac{1}{Z} exp(-\beta H) , \quad Z = \int exp(-\beta H) dq \, dp$$
$$= \frac{1}{Z_p} exp(-\frac{\beta}{2} p^T M^{-1} p) \frac{1}{Z_q} exp(-V(q)) \quad (6.3)$$

$$f_0 = \mathcal{P}\mathcal{Q}, \quad Z = Z_p Z_q, \quad \int \mathcal{P}(p) dp = \int \mathcal{Q}(q) dq = 1$$

The key idea is now that the mathematical objects of interest, the metastable conformations, are objects in *position space* $q \in \Omega \subset \mathbb{R}^{3N}$ rather than in the whole phase space $\Gamma = \Omega \times \mathbb{R}^{3N}$. Let $A, B \subset \Omega$ be subsets in the position space and define cylinders $\Gamma(A) := A \times \mathbb{R}^{3N}$ – see Fig. 6.2. Let $\chi(A)$ denote the characteristic function of a set A (a function which is 1 inside A and 0 outside). In this setting the probability for the dynamical system to *be* within A can be written as

$$\pi(A) = \int_{\Gamma(A)} f_0(p,q) dq \, dp = \int_A \mathcal{Q}(q) dq = \int_\Omega \chi_A^2 \mathcal{Q}(q) dq =: \langle \chi_A, \chi_A \rangle_\mathcal{Q}, \quad (6.4)$$

where we introduced some inner product with weighting $\mathcal{Q}$.

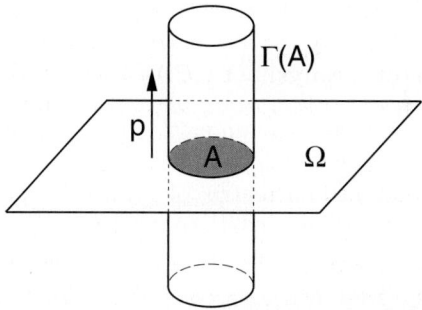

Fig. 6.2. Position space fibre (here: cylinder) $\Gamma(A)$ in phase space

The operator T is then constructed as the restriction of the Perron-Frobenius operator U to position space via averaging over the momentum part of the canonical distribution, which means integrating U over the cylinders $\Gamma(\cdot)$. The conditional probability for the system to *move* during time τ from A to B during time τ can then be defined by virtue of the new operator T as

$$w(A, B, \tau) = \frac{\langle \chi_A, T\chi_B \rangle_\mathcal{Q}}{\langle \chi_A, \chi_A \rangle_\mathcal{Q}}. \quad (6.5)$$

In the same manner, the probability for the system to *stay* in A during time τ comes out as

$$w(A, A, \tau) = \frac{\langle \chi_A, T\chi_A \rangle_Q}{\langle \chi_A, \chi_A \rangle_Q} \,. \tag{6.6}$$

The operator T is defined over the weighted spaces

$$L_Q^r(\Omega) = \{u : \Omega \to \mathcal{C}, \int_\Omega |u(q)|^r Q \, dq < \infty\}, \qquad r = 1, 2 \,.$$

Obviously, the Hilbert space $L_Q^2(\Omega)$ is associated with the above introduced weighted inner product $\langle \cdot, \cdot \rangle_Q$. With this notation, the properties of T can be listed as follows (due to Schütte [25]):

(i) T is bounded in $L_Q^r(\Omega)$: $\|Tu\|_Q \leq \|u\|_Q$, for $r = 1, 2$.
(ii) T is a Markov operator on $L_Q^1(\Omega)$.
(iii) T is *self-adjoint* in $L_Q^2(\Omega)$. Hence, the spectrum $\sigma(T)$ is real-valued and bounded: $\sigma(T) \subset [-1, 1]$.
(iv) There exists a cluster of eigenvalues close to the Perron eigenvalue well-separated from the remaining (continuous) part of the spectrum.
We call it the *Perron cluster*.

In summary, the operator T arises as the transition operator of a *reversible* Markov chain. We will use this basic structure for the discretization of the operator – see the subsequent Section 6.4. As a result of this kind of discretization we will obtain a stochastic and sparse matrix T which, due to the reversibility of the Markov chain, is also symmetric in a generalized sense.

6.3.3 Perron Cluster Analysis (PCCA)

The newly introduced name "Perron cluster analysis" characterizes a cluster analysis technique based on some analysis of the arising Perron cluster of eigenvalues of the transition matrix of a Markov chain. For this reason it should more correctly be named **Perron Cluster Cluster Analysis**, possibly abbreviated PCCA to distinguish it clear enough from the principal component analysis (PCA).

The PCCA method requires an input in terms of a stochastic (general symmetric) matrix $T = T_N$ of dimension N. The method analyzes the spectrum of such a matrix with respect to the possible existence of a Perron cluster of eigenvalues, say $\lambda_1 = 1, \lambda_2 \approx 1, \ldots, \lambda_k \approx 1$. The task is to identify k almost invariant sets corresponding to k metastable chemical conformations. Note that the number k is unknown in advance and must be identified as well. Here we will only sketch the main ideas behind the algorithm. For a broader introduction into the topic we refer to Section 5.5 in the recent editions of the textbook [9, 10], for more details to the original paper [11].

Just as in (6.2), we here obtain the (discrete) eigenvalue problem

$$\pi^T T = \pi^T, \quad Te = e, \quad \pi^T e = 1, \tag{6.7}$$

where the left eigenvector $\pi^T = (\pi_1, \ldots, \pi_N)$ represents the discrete invariant measure and the right eigenvector $e^T = (1, \ldots, 1)$ is the discrete invariant set – each corresponding to the Perron eigenvalue $\lambda_1 = 1$. Assume now that the total index set $\mathcal{S} = \{1, 2, \ldots, N\}$ can be decomposed into k disjoint index subsets

$$\mathcal{S} = \mathcal{S}_1 \oplus \cdots \oplus \mathcal{S}_k$$

such that there exist k *uncoupled* Markov chains, each of which is running "infinitely long" within one of the index subsets. Then, for a reversible Markov chain, the total transition matrix T is strictly block diagonal with block submatrices $\{T_1, \ldots, T_k\}$ – see, e.g., [21]. Each of these submatrices is stochastic and gives rise to a single Perron eigenvalue $\lambda(T_i) = 1$, $i = 1, \ldots, k$. Let the submatrices be primitive. Then, due to the Perron-Frobenius theorem, each block T_i possesses a unique right eigenvector $e_i = (1, \ldots, 1)^T$ of length $\dim(T_i)$ corresponding to its Perron root. Therefore, in terms of the total transition matrix T, the eigenvalue $\lambda = 1$ is k–fold and the corresponding eigenspace is spanned by the vectors

$$\chi_{\mathcal{S}_i} = (0, \ldots, 0, e_i^T, 0, \ldots, 0)^T, \qquad i = 1, \ldots, k .$$

In view of the identification problem to be treated, our notation deliberately emphasizes that these eigenvectors can be interpreted as *characteristic functions* of the invariant index subsets (see Fig. 6.3, left). In general, any basis $\{X_i\}_{i=1,\ldots,k}$ of the eigenspace corresponding to $\lambda = 1$ can be written as a linear combination of the characteristic functions $\chi_{\mathcal{S}_i}$ with coefficients $\alpha_{ij} \in \mathbb{R}$ such that

$$X_i = \sum_{j=1}^{k} \alpha_{ij} \, \chi_{\mathcal{S}_j}, \qquad i = 1, \ldots k .$$

As a consequence, eigenvectors corresponding to $\lambda = 1$ are *constant on each index subset* (see Fig. 6.3, bottom).

In reality, the block diagonal form will not be apparent due to unknown index permutations. We therefore need some elementwise criterion that is independent of any index permutation.

Lemma 6.3.1 *[11] Given a block-diagonal transition matrix T consisting of reversible, primitive blocks, a left eigenvector $\pi > 0$ and a basis $\{X_i\}_{i=1,\ldots,k}$ of its eigenspace corresponding to $\lambda = 1$. Associate with every state s_i its sign structure*

$$s_i \longmapsto (\text{sign}((X_1)_i), \ldots, \text{sign}((X_k)_i)).$$

Then

(i) invariant index subsets are collections of states with common sign structure,
(ii) different index subsets exhibit different sign structures.

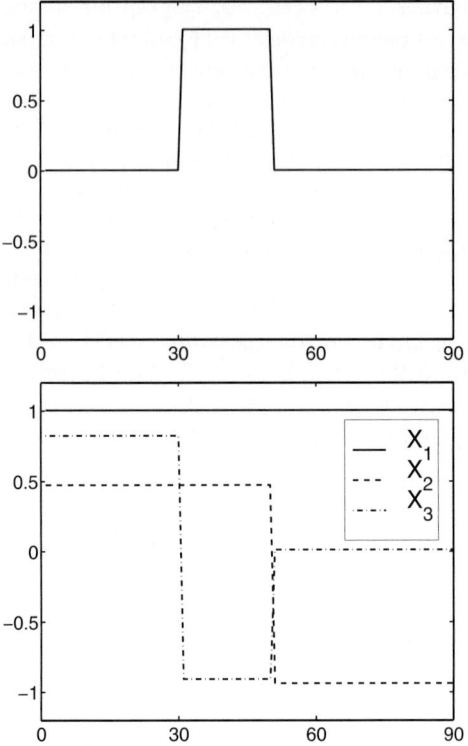

Fig. 6.3. Uncoupled Markov chain over $k = 3$ disjoint index subsets. The state space $\mathcal{S} = \{s_1, \ldots, s_{90}\}$ divides into the subsets $\mathcal{S}_1 = \{s_1, \ldots, s_{29}\}$, $\mathcal{S}_2 = \{s_{30}, \ldots, s_{49}\}$ and $\mathcal{S}_3 = \{s_{50}, \ldots, s_{90}\}$. *Top*: Characteristic function $\chi_{\mathcal{S}_2}$. *Bottom*: Eigenbasis corresponding to the 3-fold eigenvalue $\lambda = 1$. Observe that each eigenvector is constant on each subset. The sign structure for state s_{69}, for example, is $(+, -, 0)$ in the sense of Lemma 6.3.1.

Next suppose that we have k *nearly uncoupled* Markov chains, each of which is staying "for a long time" in one of the index subsets $\mathcal{S}_i$. For the transition probabilities (6.5) and (6.6) this means that

$$w(\mathcal{S}_i, \mathcal{S}_i, \tau) = 1 - O(\epsilon), \qquad w(\mathcal{S}_i, \mathcal{S}_j, \tau) = O(\epsilon), \quad i \neq j \qquad (6.8)$$

in terms of some not further specified perturbation parameter that indicates the *metastability* of the index subsets. In this case the transition matrix T is (after some unknown permutation) block diagonally *dominant*. Moreover, a *Perron cluster*

$$\lambda_1 = 1, \quad \lambda_2 = 1 - O(\epsilon), \ \ldots, \lambda_k = 1 - O(\epsilon)$$

arises as a perturbation of the k-fold Perron root in the uncoupled case $\epsilon = 0$. Upon applying Kato's perturbation theory [19] we obtain the following results for the corresponding eigenvectors:

Theorem 6.3.2 *[11] Let $T(\epsilon)$ be a family of matrices satisfying certain regularity conditions not specified here (for details see [11]). Let Π_j denote the projection on the eigenspace spanned by the eigenvector X_j of the unperturbed transition matrix $T(0)$. Then, for real ϵ, there exist π-orthonormal eigenvectors $X_1(\epsilon), \ldots, X_k(\epsilon)$ of the following form:*

(i) An eigenvector corresponding to the Perron root $\lambda_1(\epsilon) \equiv 1$ given by

$$X_1(\epsilon) \equiv e,$$

(ii) A set of $k-1$ eigenvectors corresponding to the eigenvalue cluster $\lambda_2(\epsilon), \ldots, \lambda_k(\epsilon)$ close to $\lambda = 1$ of the form

$$X_i(\epsilon) = \sum_{j=1}^{k}(\alpha_{ij} + \epsilon\beta_{ij})\chi_{\mathcal{S}_j} + \epsilon\sum_{j=k+1}^{n}\frac{1}{1-\lambda_j}\Pi_j T^{(1)}X_i + \mathcal{O}(\epsilon^2)$$

for appropriate coefficients $\alpha_{ij}, \beta_{ij} \in \mathbb{R}$ and index subsets $\mathcal{S}_1, \ldots, \mathcal{S}_k$ corresponding to the block-diagonal form of $T(0)$.

The theorem nicely indicates that we can essentially use the tools from the unperturbed case also for the perturbed case. As an illustration, see Fig. 6.4 where the locally constant pattern over each of the index subsets is still visible even under perturbation. Upon applying Lemma 6.3.1 and carefully observing perturbations of the strict zero, we again have an elementwise criterion independent of any permutation.

Summarizing, we finally have the desired k *metastable chemical conformations* (in spatial box discretization) as the k almost invariant subsets $\mathcal{S}_1, \ldots, \mathcal{S}_k$. In the true spirit of scientific computing these objects must be appropriately visualized – a scientific topic of its own right, which, however, cannot be touched upon here. For these conformations the algorithm supplies the following information:

- the probabilities $\pi(\mathcal{S}_i)$ for the system to *be* within the subset $\mathcal{S}_i$, as defined in (6.4),
- the probabilities $w_{ii} = w(\mathcal{S}_i, \mathcal{S}_i, \tau)$ for the system to *stay* during time τ in the subset $\mathcal{S}_i$, as defined in (6.6), and
- the probabilities $w_{ij} = w(\mathcal{S}_i, \mathcal{S}_j, \tau)$, $i \neq j$, for the system to *move* from the subset $\mathcal{S}_i$ to the subset $\mathcal{S}_j$, as defined in (6.5).

In other words: The Perron cluster analysis supplies the number, the life times, and the decay pattern of the metastable chemical conformations. As for the parameter ϵ used above without specification, we naturally arrive at the definition

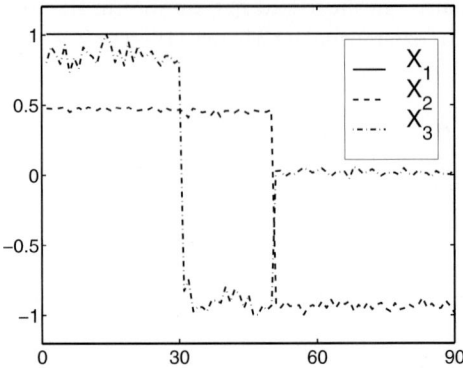

Fig. 6.4. Eigenbasis X_1, X_2, X_3 corresponding to Perron cluster $\lambda = 1, 0.75, 0.52$ of the transition matrix associated with $k = 3$ nearly uncoupled Markov chains. Observe the nearly constant level pattern on each of the index subsets $\mathcal{S}_1, \mathcal{S}_2$ and $\mathcal{S}_3$ – to be compared with Fig. 6.3 for the uncoupled case.

$$\epsilon = \max_{i=1,\ldots,k} (1 - w_{ii}) = 1 - \min_{i=1,\ldots,k} w_{ii} \qquad (6.9)$$

For each of the $\mathcal{S}_i$ the characteristic life times are roughly found to be

$$\tau_{\mathcal{S}_i} \approx \frac{\tau}{1 - w_{ii}}.$$

The blow-up from $\tau \ll t_{\text{crit}}$, the deterministic time scale as defined in (6.1), to the time scales $\tau_{\mathcal{S}_i}$ of the metastable conformations is significant. This relation documents in a nutshell the telescoping of the deterministic model, based on short term trajectories, and the statistical model, based on the eigenvalue problem for the (discretized) transition operator, to obtain a long term model.

Above all it is clear that the whole Perron cluster analysis will only work, if the stochastic transition operator T can be discretized avoiding the curse of dimension – which is the topic of the next section.

6.4 Approximation of the Transition Operator

The spatial stochastic transition operator T as discussed in Section 6.3 is associated with an underlying Markov chain. Upon introducing the projection π on the position variables via $\pi(q, p) = q$, we may write this Markov chain as

$$q_{k+1} = \pi \varPhi^\tau(q_k, p_k), \qquad p_k : \quad \mathcal{P}\text{ – distributed}. \qquad (6.10)$$

As shown schematically in Fig. 6.5, it combines a short term deterministic model, characterized by the flow $\varPhi^\tau$, with a statistical model, characterized by the $\mathcal{P}$-distribution, the momentum part of the Boltzmann distribution – see (6.3).

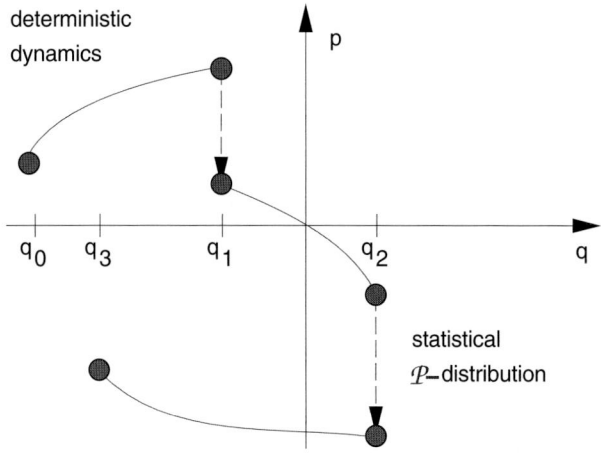

Fig. 6.5. Hybrid Monte Carlo process and Markov chain (6.10)

Hybrid Monte Carlo method. Given a discretization of the position space Ω in terms of boxes $\{B_1, \ldots, B_N\}$, the elements of the transition matrix $T = (T_{ij})$ can be computed by virtue of

$$T_{ij} = \frac{\#\{q_{k+1} \in B_j \wedge q_k \in B_i\}}{\#\{q_k \in B_i\}} \qquad i,j = 1, \ldots N \ .$$

By construction, the evaluation of the matrix elements thus leads to some *hybrid Monte Carlo* process – see again Fig. 6.5. If we run M samples within such a process, then we obtain an approximation $T^{(M)}$ with an approximation error

$$|T - T^{(M)}| \leq \gamma/\sqrt{M} \ .$$

As in all Monte Carlo type processes, *trapping* within local minima will occur, unless we take special precautions. In particular, if the spectral gap at the Perron root approaches 0, then the above constant γ blows up to ∞. However, this is just the case treated here, since we want to analyze Perron clusters! In this situation a technique of temperature embedding has been developed, which circumvents critical slowing down of the MC process in the case under consideration. Unlike simulated annealing this method can "heat" the momenta of the system separately in a nonphysical fashion – compare the factorization in (6.3). First results have been published in the early paper [12] by A. Fischer et al., an improvement in the direction of a hierarchical coupling-uncoupling method can be found in [13].

Spatial box discretization. The number N of spatial boxes is also the dimension of the arising transition matrix T. In order to avoid the *curse of dimension* we must assure that N remains of moderate size even for larger molecular systems.

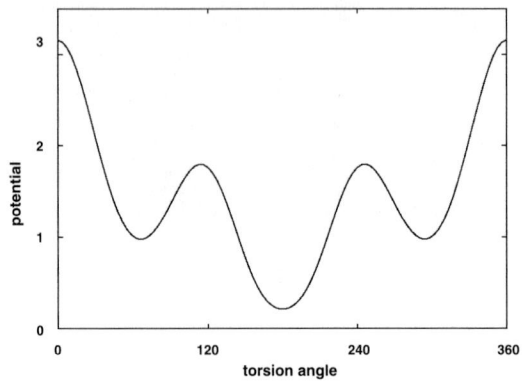

Fig. 6.6. Molecular torsion potential with triple well ($s = 3$)

From chemical insight into the problem, different conformations are caused by the double or triple well structure in the torsion potentials – see Fig. 6.6. Let s be the number of minima in the torsion potential ($s = 2$ or $s = 3$) and m the number of torsion angles ($m \approx 7$ per nucleotide), then we obtain a number
$$N \approx s^m$$
of boxes. For the above example molecule ACC we have $m = 37$ and would therefore arrive at some $N > 10^{11}$ – which is certainly intolerable for such a small system!

As a first remedy we adopted the technique of identification of *essential degrees of freedom* originally suggested by Berendsen et al. [1]. Generally speaking, this method is based on a principal component analysis (PCA) of fluctuations of the time series obtained by molecular dynamics calculations. We modified the method such that it only works on the torsion angles, i.e. on a preassigned subset of the variables – see [26]. Note that in this case the cylinder $\Gamma(A)$ reduces to a fibre associated with this subset – see Fig. 6.2. For the ACC molecule, which was just mentioned above, the method suggests only $m_{\text{ess}} = 4$ generalized torsion coordinates and a number of
$$N_{\text{ess}} = 36$$
of boxes. For a while we were quite content with this approach, until we found out experimentally that it is also not efficient enough for larger molecules.

6 From Molecular Dynamics to Conformation Dynamics in Drug Design   283

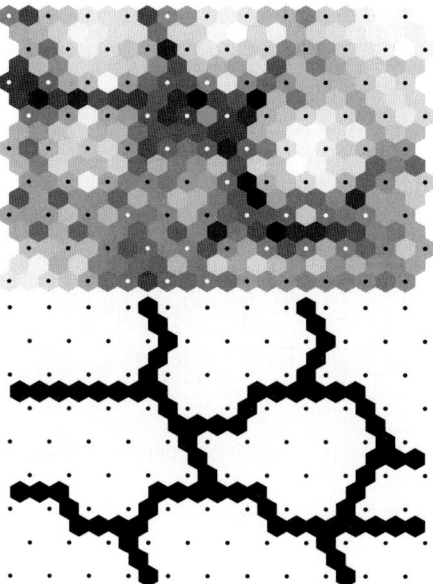

Fig. 6.7. Results of cluster analysis. *Top*: SOM. *Bottom*: SOM combined with new Perron cluster analysis

In a further step of the development the MD group at ZIB recurred to *neural networks*, especially to self-organizing maps (SOM) as suggested by Kohonen [20]. Upon combining SOM with the Perron cluster analysis as discussed in Section 6.3, T. Galliat et al. managed to develop some much more efficient tool for box discretization – see [16]. In Fig. 6.7 we illustrate the improvement achieved by the addition of the Perron cluster analysis to SOM using a typical SOM representation in terms of hexagonal topology. The result on the right in the figure was obtained a lot faster than the result on the left (a quarter of an hour on a work station as compared to about a week). In [15, 14] the idea has been further developed toward an adaptive multilevel box discretization called *self-organizing box maps* (SOBM) extending techniques from numerical partial differential equations to neural networks.

Example: Tri–nucleotide ACC. This example has been used several times before for illustration purposes. From the neural network approach to box discretization we obtain $N = 54$ boxes. In Fig. 6.8 the sparse pattern of the associated $(54, 54)$-matrix is given, representing the discretization of the stochastic transition operator over the given 54 boxes.

In Table 6.1 we list the first eigenvalues of the transition matrix (ordered according to modulus). As can be observed, there are gaps at $k = 2$ and at $k = 8$, which can both be analyzed. Note that by construction via Lemma 6.3.1 a larger value of k just leads to some substructuring of the conformations: the

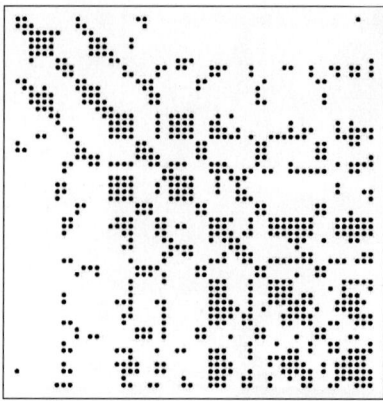

Fig. 6.8. ACC: sparse transition matrix of dimension $N = 54$

Table 6.1. ACC: eigenvalues of transition matrix

k	1	2	3	4	5	6	7	8	9	...
λ_k	1.000	0.999	0.995	0.993	0.980	0.972	0.961	0.930	0.874	...

Table 6.2. ACC: probabilities for metastable conformations

conformations	S_1	S_2	S_3	S_4	S_5	S_6	S_7	S_8
$\pi(S_i)$	**0.325**	0.097	0.009	0.037	0.107	0.105	**0.273**	0.046
w_{ii}	0.995	0.992	0.919	0.966	0.964	0.991	0.987	0.969

extended sign structure of the eigenvectors just adds more sign information to the already existing one. In Table 6.2 we list the computed probabilities $\pi(S_i)$ to *be* within and $w_{ii} = w(S_i, S_i, \tau)$ to *stay* for $\tau = 50$ fsec within one of the conformations S_i for $i = 1, \ldots, 8$. All elements w_{ii} in the second row are close below 1, which indicates that the computed conformations are in fact *metastable*. Moreover, the numbers in the first row clearly indicate that the conformations S_1 and S_7 dominate the dynamics, which explains the first eigenvalue gap at $k = 2$. From the numbers w_{ii} and definition (6.9) we here obtain the perturbation parameter $\epsilon = 0.081$.

Example: HIV protease inhibitor VX-478. This molecule is the basis for the anti-AIDS drug Agenerase distributed by Glaxo Wellcome. Generally speaking, the HIV is hard to attack directly by drugs, since it is a so-called retrovirus that mutates faster than any molecular recognition can take place. As a consequence, any HIV pharmaceutical will attack the supporting en-

6 From Molecular Dynamics to Conformation Dynamics in Drug Design   285

zymes. One of them is the HIV protease, which regulates the passage of HIV through the cell membrane. The here selected molecule has been exactly designed (by Vertex) to inhibit this passage. The molecular data were taken from the public domain Protein Data Bank (PDB).

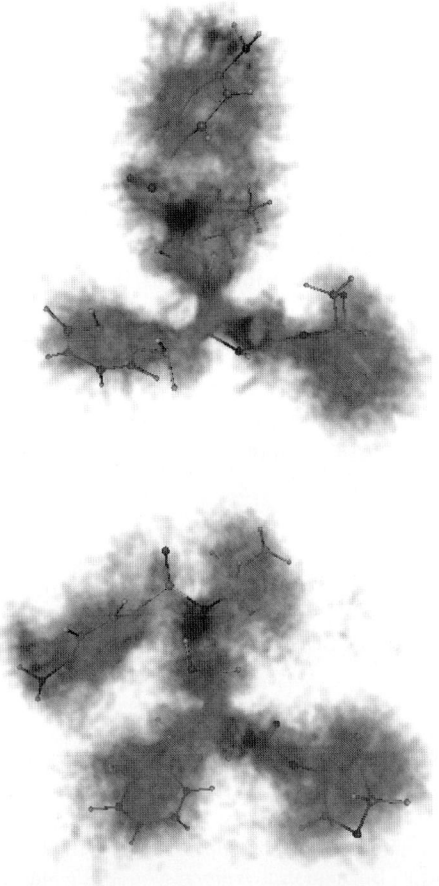

Fig. 6.9. HIV protease inhibitor: T-bone and double T conformations

We started the conformation analysis at a virtual temperature of $1400K$ (to avoid trapping in the HMC process, see above). At this level there arose $k = 3$ metastable conformations. At the next lower level ($1000K$), these conformations could be analyzed in terms of substructures. In Fig. 6.9 two out of these substructures are shown. In view of drug design it is important to understand which of the conformations (of the same molecule!) actually exhibits the desired pharmaceutical effect. Questions of this kind can be studied in terms of the probabilities as exemplified above in the tables for the ACC molecule – assuming, of course, that the input potentials give a reliable description of the physics of the molecule.

6.5 Perspectives

Conformation analysis opens the door to an understanding of molecular dynamics on time scales of pharmaceutical interest. Even though the essential structure of the mathematical model and its algorithmic realization seem to be quite clear at this time, further progress is needed to allow for the successful analysis of larger biomolecules. In the opinion of the author, the new mathematical concepts of conformation dynamics have a real chance to play an important role in drug design in the near future.

Acknowledgements: The author is greatly indebted to all members of the ZIB and FU molecular dynamics groups who have joined their efforts to improve the mathematical models and to develop efficient algorithms for computational drug design.

References

1. A. Amadei, A.B.M. Linssen, and H.J.C. Berendsen. *Essential dynamics on proteins.* Proteins **17**, pp. 412-425 (1993).
2. V. I. Arnold. *Mathematical Methods of Classical Mechanics.* Second edition. Springer, Heidelberg, New York (1989).
3. M. Dellnitz and A. Hohmann. *A subdivision algorithm for the computation of unstable manifolds and global attractors*, Numer. Math. **75**, pp. 293–317 (1997).
4. M. Dellnitz and O. Junge. *On the approximation of complicated dynamical behavior*, SIAM J. Num. Anal. **36**, pp. 491–515 (1999).
5. P. Deuflhard and F. Bornemann. *Numerische Mathematik II. Gewöhnliche Differentialgleichungen.* 2. Auflage. Walter de Gruyter, Berlin, New York (2002).
6. P. Deuflhard and F. Bornemann. *Scientific Computing with Ordinary Differential Equations.* Springer, Berlin, Heidelberg, New York (2002).
7. P. Deuflhard, M. Dellnitz, O. Junge, and Ch. Schütte. *Computation of essential molecular dynamics by subdivision techniques.* In [8], pp. 98–115 (1999).
8. P. Deuflhard, J. Hermans, B. Leimkuhler, A. E. Mark, S. Reich, and R. D. Skeel, editors. *Computational Molecular Dynamics: Challenges, Methods, Ideas*, volume 4 of *Lecture Notes in Computational Science and Engineering*. Springer, Berlin, Heidelberg, New York (1999).
9. P. Deuflhard and A. Hohmann. *Numerische Mathematik I. Eine algorithmisch orientierte Einführung.* 3. Auflage. Walter de Gruyter, Berlin, New York (2002).

10. P. Deuflhard and A. Hohmann. *Introduction to Scientific Computing.* 2nd edition. Springer, Berlin, Heidelberg, New York (2002).
11. P. Deuflhard, W. Huisinga, A. Fischer, and Ch. Schütte. *Identification of almost invariant aggregates in reversible nearly uncoupled Markov chains.* Lin. Alg. Appl. **315**, pp. 39–59 (2000).
12. A. Fischer, F. Cordes, and C. Schütte. *Hybrid Monte Carlo with adaptive temperature in mixed-canonical ensemble: Efficient conformation analysis of RNA.* J. Comput. Chem. **19**, pp. 1689–1697 (1998).
13. A. Fischer, Ch. Schütte, P. Deuflhard, and F. Cordes. *Hierarchical uncoupling-coupling of metastable conformations.* In [24], (2002).
14. T. Galliat. *Adaptive Multilevel Cluster Analysis by Self-Organizing Box Maps.* Submitted as PhD thesis, Department of Mathematics and Computer Ccience, Free University of Berlin, (March 2002).
15. T. Galliat, P. Deuflhard, R. Roitzsch, and F. Cordes. *Automatic identification of metastable conformations via self-organized neural networks.* In [24], (2002).
16. T. Galliat, W. Huisinga, and P. Deuflhard. *Self-organizing maps combined with eigenmode analysis for automated cluster identification.* In H. Bothe and R. Rojas, editors, *Proceedings of the 2nd International ICSC Symposium on Neural Computation*, Academic Press, pp. 227–232 (2000).
17. L. Greengard, and V. Rokhlin. *On the evaluation of electrostatic interactions in molecular modeling.* Chem. Ser. **29A**, pp. 139–144 (1989).
18. T.A. Halgren. *Merck molecular force field. I-V.* J. Comp. Chem., **17**, pp. 490–641 (1996).
19. T. Kato. *Perturbation Theory for Linear Operators.* Springer, Berlin, Heidelberg, New York (1995).
20. T. Kohonen. *Self-Organizing Maps.* Springer, Berlin, Heidelberg, New York, 3rd edition (2001).
21. C. D. Meyer. *Stochastic complementation, uncoupling Markov chains, and the theory of nearly reducible systems.* SIAM Rev., **31**, pp. 240–272 (1989).
22. J. Sanz-Serna, and M. Calvo. *Numerical Hamiltonian Problems.* Chapman and Hall, London, UK (1994).
23. T. Schlick. *Some Failures and Successes of Long-Time Approaches to Biomolecular Simulations.* In [8], pp. 227–262 (1999).
24. T. Schlick and H. H. Gan, editors. *Computational Methods for Macromolecules: Challenges and Applications — Proc. of the 3rd Intern. Workshop on Algorithms for Macromolecular Modelling, New York, 2000.* Springer, Berlin, Heidelberg, New York, 2002, in press.
25. Ch. Schütte. *Conformation Dynamics: Modelling, Theory, Algorithm, and Application to Biomolecules.* Habilitation thesis, Department of Mathematics and Computer Science, Free University of Berlin, 1998. Available as ZIB-Report SC-99-18 via http://www.zib.de/bib/pub/pw/.
26. Ch. Schütte, A. Fischer, W. Huisinga, and P. Deuflhard. *A direct approach to conformation dynamics based on hybrid Monte Carlo.* J. Comput. Phys., Special Issue on Computational Biophysics, **151**, pp. 146–168 (1999).

7 A Posteriori Error Estimates and Adaptive Methods for Hyperbolic and Convection Dominated Parabolic Conservation Laws

Dietmar Kröner, Marc Küther, Mario Ohlberger, Christian Rohde

Institut für Angewandte Mathematik, Universität Freiburg,
Hermann-Herder-Straße 10, 79104 Freiburg, Germany –
dietmar@mathematik.uni-freiburg.de, mario@mathematik.uni-freiburg.de

Abstract: In this contribution we will give a survey on rigorous a posteriori error estimates and adaptive methods for finite volume approximations of hyperbolic and convection dominated parabolic conservation laws. Scalar problems are considered as well as weakly coupled systems where the coupling is only due to lower order terms. In the context of scalar hyperbolic conservation laws error estimates are obtained for cell centered finite volume schemes and for the staggered Lax–Friedrichs scheme in multi dimensions. In the case of weakly coupled convection dominated parabolic equations we get a posteriori error estimates for a vertex centered finite volume scheme which are uniform in the lower bound of the diffusion. Numerical experiments underline the applicability of the theoretical results in adaptive computations.

7.1 Introduction

We are concerned with the following Cauchy problem for the unknown function $u : \mathbb{R}^2 \times [0, \infty) \to \mathbb{R}$:

$$\partial_t u + \operatorname{div} \mathbf{f}(u) + \lambda(u) = \varepsilon \Delta u \quad \text{in } \mathbb{R}^d \times (0, \infty), \tag{7.1}$$
$$u(., 0) = u_0 \quad \text{in } \mathbb{R}^2. \tag{7.2}$$

Here $\mathbf{f} \in C^1(\mathbb{R}, \mathbb{R}^2)$ is some flux function and $\lambda \in C^0(\mathbb{R})$ a source term. The problem (7.1) can be seen as a model for the evolution of small amplitude waves in compressible fluid flow [30]. We shall consider the cases

$$\varepsilon = 0 \quad \text{and} \quad 0 < \varepsilon << 1,$$

corresponding to ideal, respectively low diffusion convection dominated flow models.

In order to approximate the exact solution u of the problem (7.1), (7.2) by a numerical method finite volume type methods on unstructured grids are considered. Our goal is to develop *self adaptive* numerical algorithms which give approximative solutions u_h within a prescribed error tolerance by local refinement/coarsening of the underlying grid.

To be more specific, fixing some error tolerance $tol > 0$, a compact set $K \subset \mathbb{R}^2$, and $T > 0$ we want to have

$$||u - u_h|| \leq tol.$$

Here $||.||$ denotes some appropriate norm for functions defined on $K \times [0, T]$. The first step is to derive an a-posteriori error estimate of the form

$$||u - u_h|| \leq C\,\eta(u_h, h) + R_h,$$

where η depends only on the known approximate solution u_h and the discretization. $C > 0$ is some computable quantity and R_h stands for the approximation error of the data. To adjust the grid accordingly we need the error estimator to be local in the sense that we have

$$\eta(u_h, h) = \sum_j \eta_j(u_h, h_j), \qquad (7.3)$$

where η_j corresponds in some appropriate manner to the error on *one* element of the discretization with the local discretization parameter h_j.

In this paper we will concentrate on the derivation of rigorous a posteriori error estimators for exact solutions of (7.1), (7.2) which are localizable in the sense described above and uniform in ε.

In the sequel this contribution will be organized as follows. In Section 7.2 we will present a posteriori results concerning finite volume approximations of scalar hyperbolic conservation laws in several space dimensions, i.e. the case $\varepsilon \equiv 0$ in (7.1). In particular, cell centered finite volume schemes (Section 7.2.1) and a very general class of Lax-Friedrichs schemes on staggered grids (Section 7.2.2) are considered.

Extending these results we will present error estimators for the convection dominated case $\varepsilon > 0$ in Section 7.3. In fact the somewhat more general problem of weakly coupled systems is considered. As we use the techniques developed in the purely hyperbolic case we obtain error estimators that are robust in ε. Finally in Section 7.4 recent results concerning numerical experiments of adaptive finite volume schemes deduced from the a posteriori error estimates are presented.

To conclude the introduction we want to relate our results to the known literature in the field of self adaptive algorithms for PDE's, in particular for conservation laws.

For elliptic and parabolic problems there exists a well developed theory which can be used to accelerate numerical codes considerably (cf. [1, 4, 5, 9, 10, 11, 12, 15, 18]).

Concerning linear symmetric hyperbolic systems a posteriori error estimates in the $H^{-1,2}$-norm for the error and in the L^2-norm for the local error have been proven in [14]. In [27] nonlinear scalar conservation laws in 1D have been considered. In that paper the author could estimate the error in the Lip'-norm by the residual.

In order to demonstrate the main difficulties when analyzing convection dominated problems let us quote the following result from [13]. For $T > 0$, $\Omega \subset \mathbb{R}^2$ bounded, and $u_0 : \Omega \to \mathbb{R}$, consider the boundary value problem

$$\begin{aligned} \partial_t u + \operatorname{div} \mathbf{f}(u) - \epsilon \Delta u &= g & &\text{in } \Omega \times (0,T), \\ u &= 0 & &\text{in } \partial\Omega \times (0,T), \\ u(\cdot, 0) &= u_0 & &\text{in } \Omega, \end{aligned} \qquad (7.4)$$

with the small diffusion parameter $\varepsilon > 0$ and $g : \Omega \times (0,T) \to \mathbb{R}$. To obtain a numerical approximation u_h of the exact solution u of (7.4) the authors present a mixed finite volume finite element method which shares a lot of properties with the methods considered here. Using typical energy based techniques they prove the following a priori error bound.

Theorem 1. *For $u_0 \in L^\infty(\mathbb{R}^2) \cap W^{1,2}(\mathbb{R}^2)$, there exist constants $c_1, c_2 > 0$ independent of ε such that*

$$\|u(\cdot,t) - u_h(\cdot,t)\|_{L^2(\Omega)} \leq c_1 \, h \, e^{c_2 T/\epsilon} \quad (t \leq T). \qquad (7.5)$$

This estimate strongly depends on the small parameter ε and will blow up if ε tends to zero.

In [28] for the stationary linear problem a posteriori error estimates have been proven for stabilized finite element approximations. The estimate is uniform only if the grid size h is of the order of the parameter ε. There are a number of results mainly in 1D where similar estimates uniformly in ε have been proven, [10], [17]. While in [28], [13] and many other papers the estimates are based on L^2- or energy methods, in this contribution we will give an overview on results concerning a posteriori error estimates which are based on L^1-techniques due to Kuznetsov [21].

7.2 A Posteriori Error Estimates for Scalar Hyperbolic Conservation Laws

We consider the following scalar hyperbolic conservation law in multi dimensions.

$$\partial_t u + \text{div}\, \mathbf{f}(u) = 0 \quad \text{in} \quad \mathbb{R}^2 \times \mathbb{R}^+, \tag{7.6}$$
$$u(x,0) = u_0(x) \quad \text{in} \quad \mathbb{R}^2. \tag{7.7}$$

For the data we have to assume the following conditions. $u_0 \in L^\infty(\mathbb{R}^d) \cap BV(\mathbb{R}^2)$ with constants A and B such that $A \leq u_0 \leq B$ a.e. and $\mathbf{f} \in C^1(\mathbb{R}; \mathbb{R}^2)$.

We are now going to discuss a posteriori error estimates which are obtained for cell centered finite volume schemes [19] and for the staggered Lax–Friedrichs scheme [20].

7.2.1 Cell Centered Finite Volume Approximations

Let $\mathcal{T} = \{T_j | j \in I\}$ be a mesh of $\mathbb{R}^d$ such that the interface of two neighboring cells T_j, T_l of $\mathcal{T}$ is included in a hyperplane (see also [19]). The joint edge of T_j and T_l will be denoted by S_{jl}. We assume that there exists an $\alpha > 0$ such that we have for all $j \in I$

$$\alpha h_j^2 \leq |T_j|, \quad \alpha |\partial T_j| \leq h_j, \quad h_j := \text{diam}(T_j). \tag{7.8}$$

Moreover we define $h := \sup_{j \in I} h_j$ and $h_{\min} := \infty_{j \in I} h_j$. For any $j, l \in I$ there is a numerical upwind flux $g_{jl} \in C^1(\mathbb{R}^2, \mathbb{R})$ which satisfies the following conditions for all $u, v, u', v' \in [A, B]$.

$$\partial_u g_{jl}(u,v) \geq 0, \quad \partial_v g_{jl}(u,v) \leq 0, \tag{7.9}$$

and

$$g_{jl}(u,v) = -g_{lj}(v,u), \quad g_{jl}(u,u) = n_{jl}|S_{jl}|\mathbf{f}(u), \tag{7.10}$$
$$|g_{jl}(u,v) - g_{jl}(u',v')| \leq L|S_{jl}|(|u-u'| + |v-v'|). \tag{7.11}$$

Here $\Delta t > 0$ is the timestep, $t^n := n\Delta t$, $(n \in \mathbb{N})$ and n_{jl} is the outer unit normal to S_{jl}. L is some positive constant. Now the cell centered upwind finite volume scheme for computing approximate solutions to (7.6), (7.7) is given by

Definition 1. *(Finite volume scheme)* Let

$$u_j^0 := \frac{1}{|T_j|}\int_{T_j} u_0, \quad u_j^{n+1} := u_j^n - \frac{\Delta t}{|T_j|}\sum_{l \in N(j)} g_{jl}(u_j^n, u_l^n) \tag{7.12}$$

for all $n \in \mathbb{N}$ and $j \in I$. Here $N(j)$ denotes the indices of the neighboring triangles of T_j.

For the time step we assume the following CFL-condition $\Delta t \leq (1-\xi)\alpha^2 h_{\min}/L$ for a given $\xi \in]0,1[$ and α as defined in (7.8), where L is the Lipschitz constant from (7.11). Let us denote

$$u_h(x,t) := u_j^n \quad \text{if} \quad x \in T_j, \quad t^n < t \leq t^{n+1}. \tag{7.13}$$

Let u be the exact solution of (7.6), (7.7) and u_h be the discrete solution as defined in (7.13). In [6], [29], [8] it was shown that under the assumption, mentioned above, we have for any compact set $K \subset \mathbb{R}^2 \times \mathbb{R}^+$

$$\int_K |u(x,t) - u_h(x,t)|\, dx\, dt \leq ch^{1/4} \tag{7.14}$$

where the constant c depends only on K and the given data.

Now let us present the corresponding a posteriori error estimate. Let R, ω, T be given and

$$I_0 := \{n \mid 0 \leq t^n \leq \min\{\frac{R+1}{\omega}, T\}\},$$
$$D_R := \{(x,t) \mid |x - x_0| + \omega t < R + 1\}, \tag{7.15}$$
$$M(t) := \{j \mid \text{there exists } x \in T_j \text{ such that } (x,t) \in D_{R+1}\}. \tag{7.16}$$

Theorem 2. *[19] Assume the conditions as mentioned above. Let $K \subset\subset \mathbb{R}^2 \times \mathbb{R}^+$, $\omega = \sup_{A \leq s \leq B} |\mathbf{f}'(s)|$ and choose T, R and x_0 such that $T \in\,]0, \frac{R}{\omega}[$ and (see Figure 7.1)*

$$K \subset \cup_{0 \leq t \leq T} B_{R-\omega t}(x_0) \times \{t\}. \tag{7.17}$$

Then we have

$$\int_K |u - u_h| \leq T a_0 \left[\int_{|x-x_0|<R+1} |u_0 - u_h(\cdot,0)| + aQ + 2\sqrt{bcQ} \right], \tag{7.18}$$

where

$$Q := \sum_{n \in I} \sum_{j \in M(t^n)} \Delta t^n h_j^2 |u_j^{n+1} - u_j^n|$$
$$+ 2L\Delta t^n \sum_n \sum_{E(t_n)} (\Delta t + h_{jl}) h_{jl} |u_j^n - u_l^n|)$$

and $E(t_n)$ is the set of all edges, which lie in $M(t^n)$. In the sum over $E(t_n)$ the indices j, l refer to the triangles T_j, T_l such that $T_j \cap T_l$ is the corresponding edge and $h_{jl} := diam(T_j \cup T_l)$.

Remark 7. The constants a_0, a, b, c are known explicitly. A corresponding result also holds for higher space dimensions. Under suitable conditions this theorem can be generalized if we replace $\mathbf{f}(u)$ by $\mathbf{f}(x, t, u)$ (see [19]).

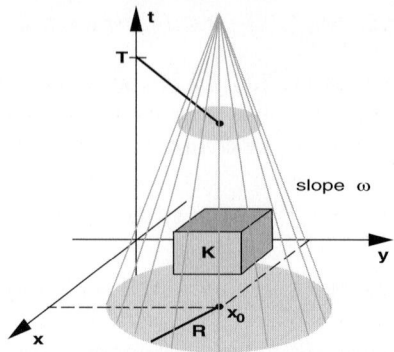

Fig. 7.1. Cone of dependence.

Example 1. Now let us use the a posteriori error estimate of Theorem 2 for the following numerical experiment. We want to solve (7.6), (7.7) with $\mathbf{f}(u) := (u^2, u^2)^t$ and

$$u_0(x) = 2, \quad \text{if} \quad \frac{x_1 + x_2}{2} - 0.5 \leq 0$$

$$u_0(x) = 1, \quad \text{if} \quad \frac{x_1 + x_2}{2} - 0.5 > 0.$$

We want to get an approximate solution such that the error measured in

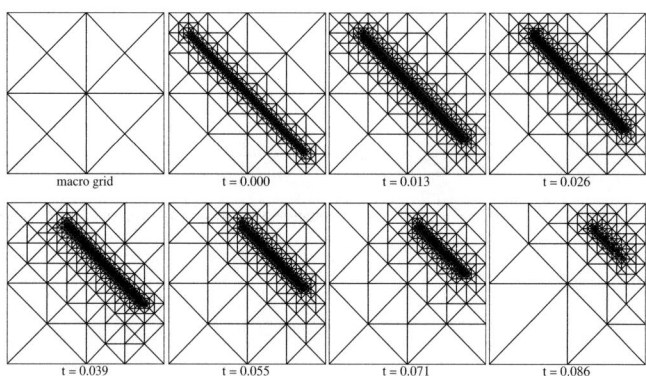

Fig. 7.2. Adaptive grid of a Burgers shock-type problem.

the L^1-norm at time $t = 0.086$ in the circle $B_{0.1}(0.75, 0.75)$ is within a given tolerance. The corresponding grids which are produced by the error estimator of Theorem 2 for different times can be seen in Figure 7.2.

7.2.2 Staggered Lax-Friedrichs Approximations

In this subsection we shall introduce a staggered Lax–Friedrichs scheme on a class of very general unstructured grids in arbitrary space dimensions. The idea to use such staggered grids in combination with some higher order scheme is due to Nessyahu and Tadmor [23]. Special unstructured staggered grids in two and three spatial dimensions were introduced by Arminjon et al. [3, 2].

We consider a sequence of unstructured grids $(\mathcal{T}_h^n)_{n\in\mathbb{N}}$ of $\mathbb{R}^d$ which do not need to be nested but satisfy the regularity assumption (7.8) with a parameter $\alpha > 0$ independent from n. Here n corresponds to the time level. In order to link two consecutive grids $\mathcal{T}_h^n$ and $\mathcal{T}_h^{n+1}$ we have to introduce some notations.

- $\mathcal{T}_h^n = (T_i^n)_{i \in I^n}$, $I^n \subset \mathbb{N}$, consists of open and non-overlapping polyhedrons of finite diameter. Set $h_i^n := \mathrm{diam}\,(T_i^n)$ and $h_n := \max_{i \in I^n} h_i^n$.
- Let $i \in I^{n+1}$ be given.

$$K^{n,n+1}(i) := \{j \in I^n \mid T_j^n \cap T_i^{n+1} \neq \emptyset\},$$
$$K_\partial^{n,n+1}(i) := \{j \in I^n \mid S_{ij}^{n,n+1} := T_j^n \cap \partial T_i^{n+1} \neq \emptyset\}.$$

The scaled outer normal $\nu_{ij}^{n,n+1}$ to $S_{ij}^{n,n+1}$ is piecewise defined (as the sum of the scaled outer normals on each part of $S_{ij}^{n,n+1}$ contained in a $(d-1)$ dimensional hyperplane) and has the length of $S_{ij}^{n,n+1}$.

These notations are illustrated in Figure 7.3.

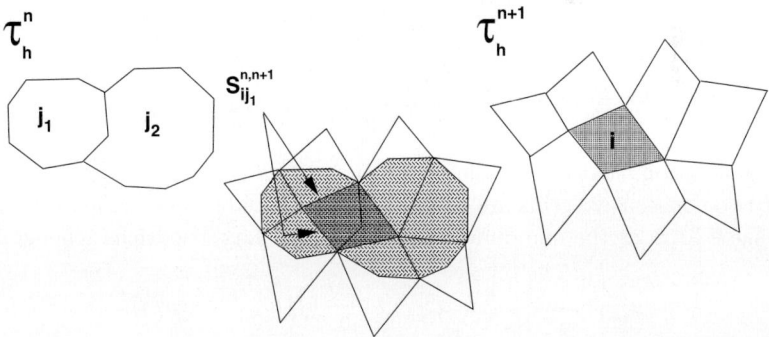

Fig. 7.3. Two consecutive staggered grids (*left* and *right*), with $K^{n,n+1}(i) = K_\partial^{n,n+1}(i) = \{j_1, j_2\}$ (*middle*).

Furthermore, two consecutive grids have to satisfy the following overlap assumption.
There exists a constant $C_{\text{ov}} > 0$ such that for all $n \in \mathbb{N}$, $i \in I^{n+1}$, $j \in K^{n,n+1}(i)$ the estimate holds

$$C_{\text{ov}} \leq \frac{|T_j^n \cap T_i^{n+1}|}{|T_i^{n+1}|} \leq 1. \tag{7.19}$$

Furthermore, suppose that for all $n \in \mathbb{N}$, for all $i \in I^{n+1}$ and all $j \in K^{n,n+1}(i)$

$$\partial T_i^{n+1} \cap \partial T_j^n \quad \text{has dimension at most } d-2. \tag{7.20}$$

Now, we can define a Lax-Friedrichs scheme on such unstructured staggered grids.

Definition 2. *(Staggered Lax-Friedrichs scheme)* Define an approximation u_h to the solution of (7.6), (7.7) by the following scheme.
For $i \in I^0$ set

$$u_i^0 := \frac{1}{|T_i^0|} \int_{T_i^0} u_0(x) \, \mathrm{d}x. \tag{7.21}$$

For given values u_i^n, $i \in I^n$ define values u_i^{n+1}, $i \in I^{n+1}$ by

$$u_i^{n+1} = \sum_{j \in K^{n,n+1}(i)} r_{ij}^{n,n+1} u_j^n - \frac{k^n}{|T_i^{n+1}|} \sum_{j \in K_\partial^{n,n+1}(i)} \mathbf{f}(u_j^n) \nu_{ij}^{n,n+1}, \tag{7.22}$$

with

$$r_{ij}^{n,n+1} := \frac{|T_j^n \cap T_i^{n+1}|}{|T_i^{n+1}|}.$$

Set

$$u_h(x,t) = u_i^n \quad \text{for } t \in [t^n, t^{n+1}[\text{ and } x \in T_i^n. \tag{7.23}$$

Let us mention, that due to the overlap condition (7.20) we exclude the case that two consecutive grids are equal. The structure of the scheme (7.22) is of the same form as the one dimensional staggered Lax-Friedrichs scheme [22]

$$u_{i+1/2}^{n+1} = \frac{1}{2}(u_i^n + u_{i+1}^n) - \frac{\Delta t}{\Delta x}(f(u_{i+1}^n) - f(u_i^n)).$$

Note, that in the definition of the scheme (7.22) no numerical flux is involved. That means that one does not need any information about Riemann problems which is useful, if one considers hyperbolic systems where the solution of the Riemann problem is not known explicitly or hard to compute.
However, this staggered Lax-Friedrichs scheme is linked to a finite volume scheme on a fixed mesh (see Definition 1) which is quite useful for the error analysis of it.

Remark 8. The scheme (7.22) can be decomposed into three steps (prolongation, evolution and averaging). From two consecutive grids $\mathcal{T}_h^n$ and $\mathcal{T}_h^{n+1}$ construct the intersection grid $\mathcal{T}_h^{n,n+1}$ by intersecting each element of $\mathcal{T}_h^n$ by each element of $\mathcal{T}_h^{n+1}$. At time t^n piecewise constant values are given on $\mathcal{T}_h^n$. In a first step, these are trivially prolongated to $\mathcal{T}_h^{n,n+1}$. Secondly, on $\mathcal{T}_h^{n,n+1}$ a finite volume step of the form given in Definition 1 is performed. Finally, these values are projected onto $\mathcal{T}_h^{n+1}$ which is nothing else than an averaging step. Considering the complete algorithm, consisting of the aforementioned three steps, it turns out that numerical fluxes appearing in the second step either cancel due to conservation or reduce to continuous flux evaluations due to consistency and the overlap assumption of two consecutive grids. It can be shown that the schemes introduced in [2, 3] are special cases of the scheme in Definition 2. Details can be found in [20].

We are now ready to state the second main theorem of this contribution.

Theorem 3 (A posteriori error estimate).
Let the assumptions of Theorem 2 be fulfilled and assume that the overlap assumption (7.19), (7.20) and the following CFL-condition hold

$$\Delta t^n L \leq \frac{1}{2}(1-\xi)\alpha^2 h_{min}^{n,n+1} \quad \xi \in]0,1[, \tag{7.24}$$

where $h_{min}^{n,n+1}$ is the minimal diameter of the intersection grid $\mathcal{T}_h^{n,n+1}$. Let K, ω, T, R and x_0 be defined as in Theorem 2 and u_h from (7.23), then we have

$$\int_K |u - u_h| \leq T a_0 \left[\int_{|x-x_0|<R+1} |u_0 - u_h(.,0)| + aQ + \sqrt{bcQ} \right], \tag{7.25}$$

where the constants a_0, a, b and c are the same as in Theorem 2, and

$$Q := \frac{1}{2} \sum_{n=0}^{N_0} \sum_{i \in I_D^{n+1}} h_i^{n+1} |T_i^{n+1}| \sum_{j,l \in K^{n,n+1}(i)} r_{ij}^{n,n+1} r_{il}^{n,n+1} |u_j^n - u_l^n|$$

$$+ \sum_{n=0}^{N_0-1} \Delta t^n \sum_{i \in I_D^{n+1}} |T_i^{n+1}| \left| u_i^{n+1} - \sum_{j \in K^{n,n+1}(i)} r_{ij}^{n,n+1} u_j^n \right|$$

$$+ 6L \sum_{n=0}^{N_0} \Delta t^n \sum_{i \in I_D^{n+1}} (h_i^{n+1} + \Delta t^n) \sum_{(j,l) \in \mathcal{E}^{n,n+1}(i)} |S_{jl}^{n,n+1}(i)| |u_j^n - u_l^n|$$

and $\mathcal{E}^{n,n+1}(i)$ is the set of all edges in $\mathcal{T}_h^n$ which have a nonempty intersection with T_i that is denoted by $S_{jl}^{n,n+1}(i)$ and I_D^n denotes the set of all indices associated to elements lying in the domain of dependence D_{R+1}, see (7.15).

Remark 9. Under suitable conditions this theorem can be generalized if one replaces $f(u)$ by $f(x,t,u)$. In addition it can be shown that the estimator is of the order $h^{1/4}$, with $h = \max_n h^n$, if some further conditions on the mesh and the time step size are imposed (see [20]).

7.3 A Posteriori Error Estimates for Convection Dominated Weakly Coupled Systems

The results of Section 7.2 can be extended to the case where the diffusion parameter ε in (7.1) does not vanish.

Here we consider in fact the following *weakly coupled system* of $M \in \mathbb{N}$ equations for the unknown function $\mathbf{u} = (u^1, \ldots, u^M)^T : \mathbb{R}^2 \times [0, T] \to \mathbb{R}^M$, $T > 0$:

$$u_t^i + \nabla \cdot \mathbf{f}^i(u^i) + \lambda^i(\mathbf{u}) = \varepsilon \Delta u^i \quad (i = 1, \ldots, M). \tag{7.26}$$

$\mathbf{f}^i : \mathbb{R} \to \mathbb{R}^2$ is a nonlinear flux function, $\varepsilon > 0$ a diffusion coefficient, and $\lambda^i : \mathbb{R}^M \to \mathbb{R}$ the source term. Note that this system is coupled weakly in the sense that coupling appears only in the zero order terms.

As we are interested in the Cauchy problem for (7.26) we impose for $\mathbf{u}_0 \in L^\infty(\mathbb{R}^2, \mathbb{R}^M) \cap W^{1,1}(\mathbb{R}^2, \mathbb{R}^M)$ the condition

$$\mathbf{u}(\cdot, 0) = \mathbf{u}_0 \text{ in } \mathbb{R}^2. \tag{7.27}$$

An application of (7.26) in the context of subsurface flow models will be considered in detail in the next section.

To perform the analysis we impose some properties on the data of the problem. Assume for all $i \in \{1, \ldots, M\}$ and $\Lambda^i > 0$

$$\mathbf{f}^i \in C^2(\mathbb{R}), \; \mathbf{f}^{i\prime} \text{ bounded}, \lambda^i \in C^1(\mathbb{R}^M), \; \|\nabla \lambda^i(\mathbf{u})\|_{L^\infty(\mathbb{R}^M)} \leq \Lambda^i. \tag{7.28}$$

As we intend to apply the techniques from the hyperbolic case considered in the preceding section we adjust the concept of entropy solutions. An *entropy solution* of (7.26), (7.27) is a function $\mathbf{u} \in L^1(\mathbb{R}^2 \times [0,T]) \cap L^2(0,T; H^1(\mathbb{R}^2))$ such that

(i) $\mathbf{u}$ is a distributional solution of (7.26),
(ii) $\partial_t u^i \in L^2(0,T; H^{-1}(\mathbb{R}^2))$, $f(u^i) \in L^2(\mathbb{R}^2 \times [0,T])$ for $i = 1, \ldots, M$,
(iii) for all $\varphi \in \mathcal{D}(\mathbb{R}^2 \times [0,T))$ and $\kappa \in \mathbb{R}$ we have for $i = 1, \ldots, M$

$$\int_{\mathbb{R}^2 \times [0,T]} U(u^i - \kappa) \partial_t \varphi + \int_{\mathbb{R}^2} U(u_0^i - \kappa) \varphi(., 0)$$
$$+ \int_{\mathbb{R}^2 \times [0,T]} \mathbf{F}^i(u^i, \kappa) \cdot \nabla \varphi - \varepsilon U'(u^i - \kappa) \nabla u^i \cdot \nabla \varphi$$
$$- \int_{\mathbb{R}^2 \times [0,T]} \varepsilon U''(u^i - \kappa) |\nabla u^i|^2 \varphi + \lambda^i(\mathbf{u}) U'(u^i - \kappa) \varphi = 0.$$

Here U is some convex entropy function (cf. [25] for details) and $\mathbf{F}^i(v, \kappa) = \int_\kappa^v \mathbf{f}^i(w) U(w - \kappa) dw$.

Existence and uniqueness of entropy solutions $\mathbf{u}$ is proven in [26], for instance.

The finite volume scheme. In this subsection we define the implicit vertex centered finite volume approximation of problem (7.26), (7.27). In order to do so we consider a primal triangulation $\mathcal{T} := \{T_j | j \in I\}$ as in Section 7.2 with an associated set of vertices $\{p_k | k \in K\}$ for some index set K. We now construct the set of dual cells $\{\Omega_k | k \in K\}$.

The (directed) connecting line of two neighboring vertices p_k and $p_{k'}$ will be denoted by $\Gamma_{kk'}$, the corresponding oriented normal vector by $m_{kk'}$. (cf. Figure 7.4). Define for each vertex p_k the corresponding dual cell Ω_k by connecting the centers of gravity of the surrounding triangles with the midpoints of the edges $\Gamma_{kk'}$ (cf. Figure 7.4).

For some vertex p_k we denote by $E(k)$ the set of local indices l of all edges S_{kl} of Ω_k. The corresponding outer normals are denoted by n_{kl}. Finally we let $\beta = \beta(k, l)$ be some mapping that maps (k, l) to the unique global number $k' \in K$ such that $S_{kl} \cap \Omega_{k'} \neq 0$.

We define for $k, k' \in K$

$$h_k := \mathrm{diam}\,(\Omega_k), \quad h_{kk'} := \mathrm{diam}\,(\Omega_k \cup \Omega_{k'}).$$

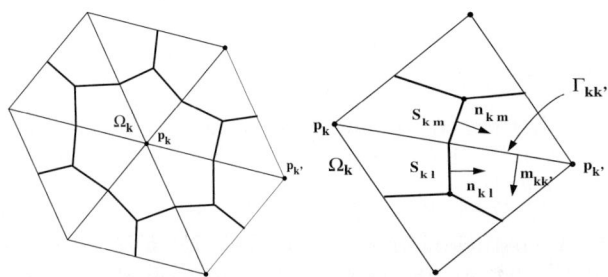

Fig. 7.4. Dual cells and grid notations.

With these notations we define the numerical fluxes of the problem.

Convective fluxes: For any $i \in \{1, \ldots, M\}, k \in K, l \in E(k), n \in \{1, \ldots, N\}$ let $g_{kl}^{n,i} \in C^1(\mathbb{R}^2, \mathbb{R})$ be a numerical flux satisfying the conditions (7.9), (7.10), (7.11) mentioned in Section 7.2 accordingly.

In contrast to the definition of the numerical fluxes in Section 7.2 the number l denotes the local number of an edge of some dual cell and *not* a global number of a neighboring element. This notation is more convenient for the finite volume scheme considered here and should not cause any confusion.

Diffusive fluxes: Let $V_h \subset H^1(\mathbb{R}^2)$ be the space of continuous functions that are linear on each triangle $T_j \in \mathcal{T}$.
For any $i \in \{1, \ldots, M\}, k \in K, l \in E(k)$, and $n \in \{1, \ldots, N\}$ the function $d_{kl}^{n+1,i} : V_h \to \mathbb{R}$, given by

$$d_{kl}^{n+1,i}(w_h) = \varepsilon |S_{kl}| \nabla w_h|_{T_{kl}} \cdot n_{kl} \quad (w_h \in V_h), \tag{7.29}$$

is called *diffusive numerical flux*. Here T_{kl} is the element of $\mathcal{T}$ that contains S_{kl}.

Now the upwind finite volume scheme for computing the approximate solutions to (7.26), (7.27) is given by

Definition 3 (Implicit finite volume scheme). *For $n \in \{0, \ldots, N-1\}$ and $i \in \{1, \ldots, M\}$ the approximate solution $\mathbf{u}_h^n := (u_h^{n,1}, \ldots, u_h^{n,M})$, $u_h^{n,i} \in V_h$ is given by the nodal basis coefficients $u_k^{n,i}$, $k \in K$, defined as:*

$$u_k^{0,i} := \frac{1}{|\Omega_k|} \int_{\Omega_k} u_0^i,$$

$$u_k^{n+1,i} + \frac{\Delta t^n}{|\Omega_k|} \sum_{l \in E(k)} \left\{ g_{kl}^{n+1,i}(u_k^{n+1,i}, u_{\beta(k,l)}^{n+1,i}) - d_{kl}^{n+1,i}(u_h^{n+1,i}) \right\} \tag{7.30}$$
$$+ \Delta t^n \lambda(\mathbf{u}_j^{n+1}) = u_k^{n,i}.$$

Furthermore we define the space-time function $\mathbf{u}_h$ by

$$\mathbf{u}_h(\cdot, 0) := \mathbf{u}_h^0, \quad \mathbf{u}_h(\cdot, t) := \mathbf{u}_h^{n+1} \text{ for } t \in (t^n, t^{n+1}] \text{ and } n \in \{0, \ldots, N-1\}.$$

Note that for the vertex centered schemes the dual cells are the finite volumes. We now state the third main theorem of this contribution (For more details and the proof see [25]).

Theorem 4 (A posteriori error estimate). *Let $\mathbf{u}$ be the entropy solution of (7.26), (7.27) and $\mathbf{u}_h$ the approximate solution, defined in (7.30). Let the underlying triangulation be weakly acute. Then the following a posteriori error estimate holds for any $\Lambda \geq \sum_{i=1,\ldots,M} \Lambda^i$ and $\Delta t^n \Lambda \leq 1$:*

$$\sum_{i=1,\ldots,M} \|u_h^i - u^i\|_{L^1(\mathbb{R}^2 \times [0,T])} \leq \eta,$$

where η is given by

$$\eta := \frac{1}{\Lambda} \sum_{i=1,\ldots,M} \left(a(\eta_0^i + \eta_\lambda^i) + b(\eta_t^i + \eta_c^i + \eta_d^i) + c\sqrt{\eta_{\bar{U}}^i} + d\sqrt{\eta_t^i + \eta_c^i + \eta_d^i} \right).$$

The positive constants a, b, c and d are computable and the error estimator terms are given as:

$$\eta_0^i = \int_{\mathbb{R}^2} |u_h^i(x,0) - u_0^i(x)|\,dx,$$

$$\eta_t^i = \sum_{n=0}^{N-1} \Delta t^n \sum_{k \in K} \left(|\Omega_k| |u_k^{n+1,i} - u_k^{n,i}| + h_k \int_{\Omega_k} |\nabla u_h^i(x,t^{n+1})|\,dx \right),$$

$$\eta_c^i = \sum_{n=0}^{N-1} \Delta t^n \sum_{\text{all edges } S_{kl}} (h_{k(\beta(k,l))} + \Delta t^n) Q_{kl}^{n+1,i} |u_k^{n+1,i} - u_{\beta(k,l)}^{n+1,i}|$$

$$+ \sum_{n=0}^{N-1} \Delta t^n \sum_{k \in K} h_k \int_{\Omega_k} |\nabla u_h^i(x,t^{n+1})|\,dx,$$

$$\eta_d^i = \varepsilon \sum_{n=0}^{N-1} \Delta t^n \sum_{\text{all edges } \Gamma_{kk'}} |\Gamma_{kk'}| [\nabla u_h^{n+1,i} \cdot m_{kk'}]_{\Gamma_{kk'}} (h_{kk'} + \Delta t^n),$$

$$\eta_\lambda^i = \sum_{n=0}^{N-1} \Delta t^n \sum_{k \in K} \int_{\Omega_k} |\lambda^i(\mathbf{u}_h^{n+1}(x)) - \lambda^i(\mathbf{u}_k^{n+1})|\,dx,$$

$$\eta_{\bar{U}}^i = \|\bar{U}''\|_{L^\infty} \sum_{n=0}^{N-1} \Delta t^n \sum_{\text{all edges } \Gamma_{kk'}} \varepsilon |\Gamma_{kk'}| [\nabla u_h^{n+1,i} \cdot m_{kk'}]_{\Gamma_{kk'}} |u_k^{n+1,i} - u_{\beta(k,l)}^{n+1,i}|$$

$$+ \sum_{n=0}^{N-1} \Delta t^n \sum_{k \in K} h_k |\lambda^i(\mathbf{u}_h^{n+1})| \int_{\Omega_k} |\nabla u_h^{n+1,i}(x)|\,dx.$$

Here $[\cdot]_{\Gamma_{kk'}}$ denotes the absolute value of the jump of the given value across the edge $\Gamma_{kk'}$ and $Q_{kl}^{n+1,i}$ is defined as

$$Q_{kl}^{n+1,i} := \frac{2 g_{kl}^{n+1,i}(v,w) - g_{kl}^{n+1,i}(v,v) - g_{kl}^{n+1,i}(w,w)}{(v-w)},$$

where $v := u_k^{n+1,i}$ and $w := u_{\beta(k,l)}^{n+1,i}$.

Note that the estimates in Theorem 4 are robust for $\varepsilon \to 0$. In the limit we recover the terms of the a posteriori error estimate for the purely hyperbolic case (cf. Theorem 2). In this form the result does not cover situations where the diffusion can degenerate locally depending on the solution. For the numerical treatment of this case we refer to [16]. It would be interesting to extend the results in this direction.

In these notes we have restricted ourselves to the most simple setting of the weakly coupled system (7.26) for the sake of simplicity. The more general case for which $\mathbf{f}^i, \lambda^i$ and ε are allowed to depend on space and time is considered in [25].

7.4 Numerical Experiments

In the case of a scalar convection-diffusion-reaction equation we derived in [24] an adaptive solution strategy from the corresponding a posteriori error estimate and gave some numerical experiments to underline the efficiency and applicability of the resulting numerical method. The adaptive strategy was derived by localizing the global error estimator η of Theorem 4 into local error indicators η_j^n, where n denotes the time step and j the triangle number of the underlying mesh. Finally an equal distribution strategy of the local indicators led us to a space–time adaptive algorithm.

Particularly in [24] it was shown that, at least for a certain class of problems, the error estimator η converges with the same rate as the error itself and that the derived adaptive numerical scheme is optimal.

Thus, we want to demonstrate here rather the applicability to weakly coupled systems for some realistic problems, than studying convergence rates.

7.4.1 Transport of Contaminants with Degradation

In the following numerical experiment we want to focus on a simplified model of contaminant transport with degradation in a porous medium. A contaminant (substrate) with concentration c^s is injected into an aquifer and advected within an underlying flow field with given velocity $\mathbf{u}$. In the presence of oxygen with concentration c^o, the contaminant and oxygen react to some harmless third component. Thereby the contaminant, as well as the oxygen is degraded. The reaction taking place is modeled by Monod kinetics. We point out that the interaction of the diffusion and reaction process is very critical in such applications and may lead to an overestimation of the degradation, when standard numerical schemes with large numerical viscosity are used for the simulation. The adaptivity of our finite volume method highly resolves the solution within the critical regions of the computational domain and thereby reduces the artificial numerical viscosity, where it is necessary.

Let $\Omega \subset \mathbb{R}^2$ be our bounded computational domain, which may represent a horizontal cut through the aquifer.

The mathematical model for the degradation of a contaminant consists of a weakly coupled system for the concentrations c^o and c^s of oxygen and substrate respectively (cf. [7] for a detailed description of the model).

$$\phi \partial_t c^o + \mathrm{div}\,(\mathbf{u} c^o - \phi D \nabla c^o) = -\nu^o k_{gr}(c^o, c^s) \text{ in } \Omega \times (0, T),$$
$$\phi \partial_t c^s + \mathrm{div}\,(\mathbf{u} c^s - \phi D \nabla c^s) = -\nu^s k_{gr}(c^o, c^s) \text{ in } \Omega \times (0, T). \quad (7.31)$$

Here, the transport of oxygen and substrate are linear and are given by the known velocity $\mathbf{u}$. The porosity $\phi \in [0,1]$ is a given constant. The diffusion D is assumed to be a small constant and the reaction rates are assumed to be of the Monod form

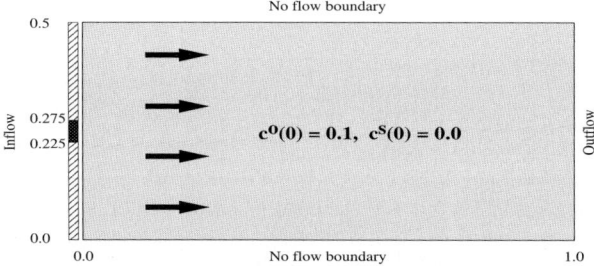

Fig. 7.5. Geometry, initial, and boundary conditions of the model problem.

$$k_{gr}(c^o, c^s) := \frac{c^o}{c^o + K^o} \frac{c^s}{c^s + K^s} B. \tag{7.32}$$

The constants ν^o and ν^s denote the corresponding stoichiometric coefficients, K^o, K^s the Monod coefficients and B a prescribed concentration of biomass, which acts as a catalyst for the reaction.

For our model problem we choose $\Omega = (0, 1) \times (0, 0.5)$, $\phi = 0.2$, $\mathbf{u} = (1, 0)^T$, $D = 0.0001$, $\nu^o = 0.5$, $\nu^s = 5.0$, $K^o = 0.1$, $K^s = 0.1$, $B = 1.0$. As initial conditions we choose $c^o(\cdot, 0) = 0.1$ and $c^s(\cdot, 0) = 0.0$ in Ω. The substrate is injected at the middle part of the inflow boundary. We therefore prescribe $c^s = 1.0, c^o = 0.0$, on $\{0\} \times (0.225, 0, 275) \times (0, T)$ and $c^s = 0.0, c^o = 1.0$, on $\{0\} \times (0, 0.5) \setminus (0.225, 0, 275) \times (0, T)$. On the upper and lower boundary we prescribe no flow boundary conditions and on the left boundary outflow conditions (cf. Figure 7.5) .

In Fig. 7.6 the numerical solution obtained with the adaptive finite volume method is compared with the numerical solution on a uniform grid. The left and middle column show the adaptive solution at $t = 0.15$ and $t = 5$, where the solution has reached the stationary state. The right column shows the solution on a uniform mesh at $t = 5$, which also corresponds to the stationary state. In all these columns the concentration of the contaminant is shown at top, the concentration of the oxygen in the middle and the underlying computational grid at the bottom. The adaptive algorithm detects the critical regions where diffusion and reaction interacts very well and thus the degradation of the contaminant is not overestimated. This leads to a fan of the contaminant through the whole domain in the stationary state. The calculation on the uniform grid however introduces to much numerical viscosity which leads to an overestimation of the degradation. Thus the resulting fan in the stationary state stays within the computational domain and does not reach the right hand boundary.

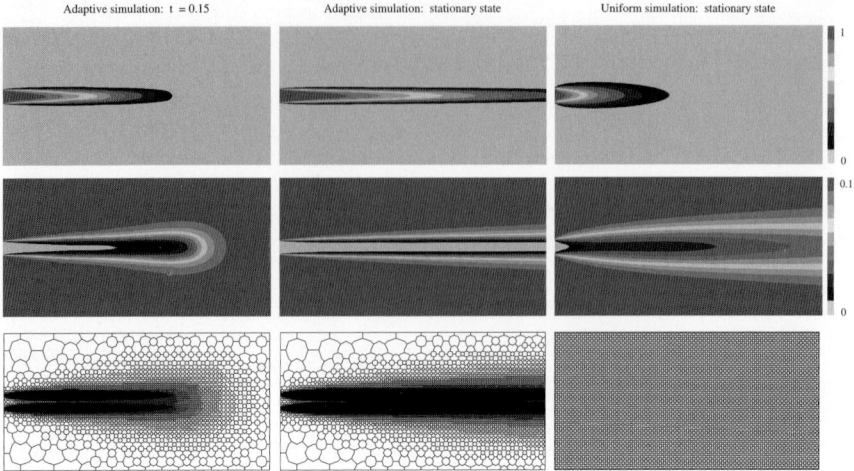

Fig. 7.6. Comparison of the concentrations of substrate (*top*), oxygen (*middle*) and the underlying computational grid (*bottom*) for an adaptive simulation at t = 0.15 (*left*) and t = 5.0 (*middle*) with an uniform calculation at t = 5.0 (*right*). See Plate 12 in the Appendix for a version of this figure in colour.

7.5 Conclusion

In this paper we have collected some recent results which show that a posteriori error estimates can be rigorously derived for finite volume methods and for staggered methods on unstructured grids for nonlinear conservation laws, weakly coupled systems of conservation laws and convection dominated diffusion equations. The theory mainly relies on the doubling of variables technique of Kruzkov and the stability estimates of Kuznetsov and does not use energy estimates. In the case of diffusion, the estimates are uniform in the lower bound of the diffusion.

In this context there are still many open questions. For the scalar case in Section 7.2 we can obtain the a priori error estimate (7.14) from (7.18) on a fixed grid and similarly on staggered grids from (7.25). At the moment this is not clear how to do it analogously in Section 7.3. Some terms in the error estimator in Theorem 4 seem to be technical and we expect that these terms can be avoided. In many applications one has to replace $\varepsilon\Delta u^i$ in (7.26) by $\text{div}(D^i(x,t)\nabla u^i)$ and $D^i(x,t)$ degenerates on an open set and may also depend on u. This case should be taken into account too. The problem will become much harder if $D^i(x,t)$ will depend on ∇u. It is also not clear how to prove the efficiency of the a posteriori error estimators in Section 7.2 and 7.3. By this we mean that the true error can be estimated by η from below with a different constant. This estimate can be used to avoid an unnecessary amount of numerical costs during the adaption process. All the results in this paper refer to the initial value problem. It remains to generalize these results to the initial boundary value problem.

References

1. L. Angermann, P. Knabner, and K. Thiele. An error estimate for a finite volume discretization of density driven flow in porous media. *Appl. Numer. Math.*, 26:179–191, 1998.
2. P. Arminjon, A. Madrane, and A. St-Cyr. Numerical simulation of 3-d flows with a non-oscillatory central scheme on unstructured tetrahedral grids. Talk presented at the eight international conference on hyperbolic problems, Magdeburg, 2000.
3. P. Arminjon, D. Stanescu, and M.C. Viallon. A two-dimensional finite volume extension of the Lax-Friedrichs and Nessyahu-Tadmor schemes for compressible flows. In *6th Int. Symposium on Comp. Fluid Dynamics*, volume 4, pages 7–14, Lake Tahoe, Septempber 4-8, 1995.
4. Becker, R. and Rannacher, R. A feed-back approach to error control in finite element methods: Basic analysis and examples. *East-West J. Numer. Math.*, 4:237–264, 1996.
5. D. Braess and R. Verfürth. A posteriori error estimators for the Raviart-Thomas element. *SIAM J. Numer. Anal.*, 33:2431–2444, 1996.
6. C. Chainais-Hillairet. Finite volume schemes for a nonlinear hyperbolic equation. Convergence towards the entropy solution and error estimates. *M2AN Math. Model. Numer. Anal.*, 33:129–156, 1999.
7. O. Cirpka, E. Frind, and R. Helmig. Numerical simulation of biodegradation controlled by transverse mixing. *Journal of Contaminant Hydrology*, 40:159–182, 1999.
8. B. Cockburn, F. Coquel, and P.G. Lefloch. An error estimate for finite volume methods for multidimensional conservation laws. *Math. Comput.*, 63:77–103, 1994.
9. W. Dörfler. A convergent adaptive algorithm for poisson's equation. *SIAM J. Num. Anal.*, 33:1106–1124, 1996.
10. W. Dörfler. Uniformly Convergent Finite–Element Methods for Singularly Perturbed Convection–Diffusion Equations. Habilitationsschrift, Mathematische Fakultät, Freiburg, 1998.
11. K. Eriksson and C. Johnson. Adaptive streamline diffusion finite element methods for stationary convection-diffusion problems. *Math. Comput.*, 60:167–188, 1993.
12. K. Eriksson and C. Johnson. Adaptive finite element methods for parabolic problems. V: Long-time integration. *SIAM J. Numer. Anal.*, 32:1750–1763, 1995.
13. M. Feistauer, J. Felcman, M. Lukáčová-Medvid'ová, and G. Warnecke. Error estimates for a combined finite volume-finite element method for nonlinear convection-diffusion problems. *SIAM J. Numer. Anal.*, 36(5):1528–1548, 1999.
14. P. Houston, J.A. Mackenzie, E. Süli, and G. Warnecke. A posteriori error analysis for numerical approximation of Friedrichs system. *Numer. Math.*, 82(3):433–470, 1999.

15. P. Houston and E. Süli. Adaptive Lagrange-Galerkin methods for unsteady convection-dominated diffusion problems. Report 95/24, Numerical Analysis Group, Oxford University Computing Laboratory, 1995.
16. W. Jäger and J. Kačur. Solution of porous medium type systems by linear approximation schemes. *Numer. Math.*, 60:407–427, 1991.
17. C. Johnson and A. Szepessy. Adaptive finite element methods for conservation laws based on a posteriori error estimates. *Commun. Pure Appl. Math.*, 48:199–234, 1995.
18. Claes Johnson, Rolf Rannacher, and Mats Boman. Numerics and hydrodynamic stability: Toward error control in computational fluid dynamics. *SIAM J. Numer. Anal.*, 32(4):1058–1079, 1995.
19. D. Kröner and M. Ohlberger. A-posteriori error estimates for upwind finite volume schemes for nonlinear conservation laws in multi dimensions. *Math. Comput.*, 69:25–39, 2000.
20. M. Küther. Error estimates for the staggered Lax-Friedrichs scheme on unstructured grids. *To appear in SIAM J. Numer. Anal.*, 2001.
21. N.N. Kuznetsov. Accuracy of some approximate methods for computing the weak solutions of a first-order quasi-linear equation. *USSR, Comput. Math. and Math. Phys.*, 16(6):159–193, 1976.
22. P. D. Lax. Weak solutions of nonlinear hyperbolic equations and their numerical computation. *Commun. Pure Appl. Math.*, 7:159–193, 1954.
23. H. Nessyahu and E. Tadmor. Non-oscillatory central differencing for hyperbolic conservations laws. *J. Comput. Phys.*, 87:408–463, 1990.
24. M. Ohlberger. A posteriori error estimates for vertex centered finite volume approximations of convection-diffusion-reaction equations. *M2AN Math. Model. Numer. Anal.*, 35(2):355–387, 2001.
25. M. Ohlberger and C. Rohde. Adaptive finite volume approximations for weakly coupled convection dominated parabolic systems. Preprint, Mathematische Fakultät, Freiburg, April 2001.
26. C. Rohde. Entropy solutions for weakly coupled hyperbolic systems in several space dimensions. *Z. Angew. Math. Phys.*, 49(3):470–499, 1998.
27. Tadmor, E. Local error estimates for discontinuous solutions of nonlinear hyperbolic equations. *SIAM J. Numer. Anal.*, 28:891–906, 1991.
28. R. Verfürth. Robust a posteriori error estimators for a singularly perturbed reaction-diffusion equation. *Numer. Math.*, 78:479–493, 1998.
29. J.P. Vila. Convergence and error estimates in finite volume schemes for general multi-dimensional scalar conservation laws. I Explicit monotone schemes. *RAIRO, Modelisation Math. Anal. Numer.*, 28:267–295, 1994.
30. G.B. Whitham. *Linear and nonlinear waves*. John Wiley & Sons Inc., New York, 1999. Reprint of the 1974 original, A Wiley-Interscience Publication.

8 On Anisotropic Geometric Diffusion in 3D Image Processing and Image Sequence Analysis

Karol Mikula[1], Tobias Preußer[2], Martin Rumpf[2], and Fiorella Sgallari[3]

[1] Department of Mathematics, Slovak University of Technology, Bratislava, Slovakia – *mikula@vox.svf.stuba.sk*
[2] Faculty for Mathematics, University of Duisburg, Duisburg, Germany – *preusser@math.uni-duisburg.de* or *rumpf@math.uni-duisburg.de*
[3] Department of Mathematics and CIRAM, University of Bologna, Bologna, Italy – *sgallari@dm.unibo.it*

Abstract: A morphological multiscale method in 3D image and 3D image sequence processing is discussed which identifies edges on level sets and the motion of features in time. Based on these indicator evaluation the image data is processed applying nonlinear diffusion and the theory of geometric evolution problems. The aim is to smooth level sets of a 3D image while preserving geometric features such as edges and corners on the level sets and to simultaneously respect the motion and acceleration of object in time. An anisotropic curvature evolution is considered in space. Whereas, in case of an image sequence a weak coupling of these separate curvature evolutions problems is incorporated in the time direction of the image sequence. The time of the actual evolution problem serves as the multiscale parameter. The spatial diffusion tensor depends on a regularized shape operator of the evolving level sets and the evolution speed is weighted according to an approximation of the apparent acceleration of objects. As one suitable regularization tool local L^2–projection onto polynomials is considered. A spatial finite element discretization on hexahedral meshes, a semi-implicit, regularized backward Euler discretization in time, and an explicit coupling of subsequent images in case of image sequences are the building blocks of the algorithm. Different applications underline the efficiency of the presented image processing tool.

8.1 Introduction

Processing three dimensional images and image sequences is a task of growing interest in various applications. Especially in medical imaging different image generation hardware such as CT or MRI devices, and more recently also 3D ultrasound devices deliver large image data at high resolution for further post processing. Based on that data anomalies can be analyzed and the progress of deseases can be studied. Furthermore, physical experiments can be recorded via MRI or other 3D measurement devices. Thus comparisons with 3D simulations became possible. Frequently the resulting images and image sequences are characterized by a rather unsatisfying signal to noise ratio, which leads to serious difficulties in the further post processing. Especially in 3D many features are hidden and the essential structure or the involved motions and developments are hard to catch visually. Frequently, one is interested in the extraction of certain level surfaces from the data, which bound volumes or separate regions of interest. Often the actual intensity value is of minor importance and dependent on the modality in the image generation process. Methods which behave invariant under transformations of the intensity or gray scale are called morphological. They only effect the morphology of the image, which coincides with the geometry of the level sets. The aim of this paper is to combine recent results on anisotropic geometric diffusion for the denoising of 3D images and a smoothing method for 3D image sequences which takes into account feature motion and acceleration. The peculiarity of the method is, that it is able to preserve edges and corners on level sets while still allowing tangential smoothing along the edges. Furthermore, a suitable acceleration quantity — the apparent acceleration — is used to modulate the speed of propagation.

The core of the method is an evolution driven by *anisotropic geometric diffusion* of level surfaces. In case of image sequences the diffusion processes are coupled on different frames of the sequence in time and then simultaneously applied to every frame. Thus, an anisotropic diffusion tensor depending on a presmoothed shape operator and thus on presmoothed principal curvatures and principal directions of curvature, is sensitive to the identification of the important surface features. Furthermore, the speed of diffusion is modulated based on the measured motion of level sets in image sequences. In the identification of curvature and motion quantities, a build-in regularization and projection on prototype shapes turns out to be essential to make the proposed method robust and mathematically well-posed.

The paper is organized as follows. First, in Section 8.2 we discuss some background work on image and image sequence processing. Section 8.3 briefly introduces the anisotropic geometric diffusion method on still images and in Section 8.4 we discuss the generalization to image sequences via a suitable coupling of the diffusion problems on different frames of the sequence. Afterwards, in Section 8.5 we sketch how to extract the required curvature and motion quantities and in Section 8.6 we present the actual discretization with finite elements.

8.2 Review of Related Work

Let us consider a noisy image given by an intensity map $\phi_0 : \Omega \to \mathbb{R}; x \mapsto \phi_0(x)$ on some image domain $\Omega \subset \mathbb{R}^3$ or a continuous image sequence

$\phi_0 : [0,T] \times \Omega \to \mathbb{R}$; $(s,x) \mapsto \phi_0(s,x)$. Scale space methods define an evolution operator $E(t)$ which acts on the initial data ϕ_0 and delivers a family of representations $\{E(t)\phi_0\}_{t\geq 0}$ on successively coarser scales. Here, the time parameter t acts as a scale parameter, leading form a fine, but noisy representation for time $t = 0$ to successively smoother and coarser representation for increasing time parameter t. To avoid any confusion we will always use t for the time scale of the smoothing evolution and s for the sequence parameter, which represents time in the image sequence data base. One of the first successful methods along this concept was presented by Perona and Malik [24]. For a given initial image ϕ_0 they considered the evolution problem

$$\partial_t \phi - \operatorname{div}(G(\|\nabla \phi\|)\nabla \phi) = 0$$

For increasing time t - the scale parameter - the original image at the initial time is successfully smoothed and image patterns are coarsened. Simultaneously edges - indicated by steep image gradients - are enhanced if one chooses a diffusion coefficient $G(.)$ which suppresses diffusion in areas of high gradients. A suitable choice is $G(\alpha) = \left(1 + \frac{\alpha^2}{\lambda^2}\right)^{-1}$ for a positive constant λ. Catté et al. [6] proposed a regularization method where the diffusion coefficient is no longer evaluated on the exact intensity gradient. Instead they suggested to consider the gradient evaluation on a prefiltered image, i.e., they consider the equation

$$\partial_t \phi - \operatorname{div}(G(\|\nabla \phi_\sigma\|)\nabla \phi) = 0$$

where $\phi_\sigma = K_\sigma * \phi$ with a suitable local convolution kernel K_σ of width σ. Compared to the original Perona–Malik method this model turns out to be well-posed and edges are still retained. Indead, the prefiltering avoids the detection and pronouncing of artificial edges, which are due to the initial noise.

Weickert [34] improved this method taking into account anisotropic diffusion, where the Perona–Malik type diffusion is concentrated in one direction, for instance the direction perpendicular to the level set or feature direction. This leads to an additional tangential smoothing on level sets and enables to amplify intensity correlations along lines or on level sets. The geometry of this evolution problem especially influences our investigations on anisotropic diffusion. In the axiomatic work by Alvarez et al. [1] general nonlinear evolution problems based on the scale space idea where derived from a set of axioms. Especially including the axiom of gray value invariance they end up with a curvature evolution model, i. e.

$$\partial_t \phi - \|\mathcal{D}\phi\| \left(t \operatorname{div}\left(\frac{\mathcal{D}\phi}{\|\mathcal{D}\phi\|}\right)\right)^{\frac{1}{3}} = 0.$$

Curvature motions has been studied for a long time in geometry and in physics, where interfaces are driven by surface tension. In dimensions higher than two, singularities may occur in the evolution. Existence of generalized viscosity solutions has been proved by Evans and Spruck [13]. Anisotropic curvature flow has been studied for instance by Bellettini and Paolini [5]. In case of planar curves Kačur and Mikula [19] considered an evolution equation for the curvature, from which one can recover the shape of the curves. Concerning the application this is closely related to the preferability of certain interface orientations in the crystalline structure of material (cf. [3, 31]). Starting with the above mentioned axiomatic results curvature motion proved to be a successful ingredient in segmentation and image enhancement methods, e. g. compare Pauwels et al. [23]. Sapiro [28] proposed a modification of Mean-Curvature-Motion (MCM) considering a diffusion coefficient which depends on the image gradient.

In [7] a parametric anisotropic curvature motion was applied to the smoothing of noisy triangulated surfaces. It preserves edges on the surfaces and incorporates diffusion solely along the edge and not perpendicular to it. In [26] a corresponding level set formulation has been discussed and compared to the parametric model. This method will be presented in detail below and enter our image sequence smoothing scheme via the involved spatial operator.

Motion detection in image sequences is already a classical research area in computer vision. Various approaches have been presented to extract object velocities from movie data and the motion or deformation of objects therein. Either one asks for a deformation controlled by elastic stresses where the elastic properties may depend on knowledge about the material or one considers flow fields which give rise for the deformation. For details we refer to [32, 18, 8]. Alternatively, optical flow techniques can be implied, which are based on suitable regularization of the inverse problem to identify the deformation [22, 11, 2, 27, 34]. An axiomatic scale space theory for continuous image sequences has been developed by Guichard [14].

Mikula et al. [21] presented an extension of the original Perona–Malik approach to image sequences via a modulation of the propagation speed depending on a measured acceleration quantity. The speed modulation presented here will be based on these results. In their model they consider a quantity introduced by Guichard [15] which assumes that points preserve their intensity along the smooth (*lambertian*) motion trajectories, i.e. they proposed a finite volume scheme for the scale space model for image sequences $\phi : \mathbb{R}^+ \times [0, T] \times \Omega \to [0, 1]$

$$\partial_t \phi - \mathrm{clt}(\phi_\sigma) \mathrm{div}(G(\|\nabla \phi_\sigma\|) \nabla \phi) = 0$$

where the index σ indicates a usual regularization. The so called *curvature of lambertian trajectories* $\mathrm{clt}(\phi)$ at time s for scale t is defined by

$$\mathrm{clt}(\phi)(t,s,x) := \lim_{\Delta s \to 0} \min_{w_1,w_2 \in M} \frac{1}{(\Delta s)^2} \Big(|<\nabla \phi(t,s,x), w_1 - w_2>|$$
$$+ |\phi(t, s - \Delta s, x - w_1) - \phi(t,s,x)|$$
$$+ |\phi(t, s - \Delta s, x + w_2) - \phi(t,s,x)|\Big),$$

where M is a small ball around x. It measures the coherence of the moving structures in time. The first part of the clt(ϕ) definition actually measures the so called *apparent acceleration* and the second and third term evaluate gray value coherences. In the implementation one confines with Δs being the time offset between two frames of the given discrete image sequence and considers M as a discrete ball of pixels around x.

Concerning the general numerical implementation of PDE methods in image processing among others Weickert proposed finite difference schemes [34] and Kačur and Mikula [16] suggested a semi-implicit finite element implementation for the isotropic model by Catté et al.[6]. Adaptive finite element methods in image processing are discussed by Bänsch and Mikula [4], Schnörr [30] and in [25]. Kimmel [17] generalizes scale space methodology to textures on surfaces, considering the appropriate intrinsic differential operators. A finite volume implementation has been studied by Mikula and Ramarosy [20].

The numerical approximation of curvature motion in level set form has recently been investigated by Deckelnick and Dziuk [9]. They have analyzed a corresponding fully discrete finite element method and proved convergence toward viscosity solutions.

8.3 Anisotropic Geometric Diffusion on Still Images

At first, we confine to the processing of still images. Thus, we consider a noisy initial image $\phi_0 : \Omega \to \mathbb{R}$; $x \mapsto \phi_0(x)$ with $\Omega \subset \mathbb{R}^3$ and ask for a scale of images $\{\phi(t, \cdot) \,|\, t \geq 0\}$ with $\phi(0, \cdot) = \phi_0$. Throughout this paper Ω will always be the unit cube $[0,1]^3$. We will define a level set formulation for a generalized anisotropic curvature motion of the isosurfaces.

Here, as long as we derive the model we assume $\phi(\cdot, \cdot)$ to be sufficiently smooth and $\nabla \phi(t,x) \neq 0$ for all $(t,x) \in \mathbb{R}_0^+ \times \Omega$. Indeed, due to the implicit function theorem the corresponding level sets are actually smooth surfaces. To keep our method invariant under gray scale transformations we confine to curvature quantities as the driving forces for the corresponding evolution of the level sets. The simplest morphological smoothing model would be to consider mean curvature motion of the level sets (cf. Section 8.2). But in addition to the smoothing of the level sets our aim is to maintain or even enhance edges on these surfaces. Edge type features on a smooth level set are characterized by a small curvature in the direction along the feature and a sufficiently large curvature in the perpendicular direction in the tangent space. For implicit surfaces these curvature quantities are represented by the

Fig. 8.1. As a test case for the anisotrpic geometric diffusion for still images we consider the function $\phi(x) = |x_1| + |x_2| + |x_3|$ whose level sets are octahedrons. This function was perturbed and then taken as initial data for the anisotropic geometric diffusion method for single images. From left to right an original perturbed level set and the corresponding first, second, and fifth time step of its evolution on a 64^3 grid are depicted. In the bottom row we visualize the dominant curvature on the level sets from the left column. A color ramp from blue to red indicates the dominant curvature value. See Plate 13 in the Appendix for a version of this figure in colour.

shape operator $S = S(\phi) := DN$, where $N = \frac{\nabla \phi}{\|\nabla \phi\|}$. In the vicinity of an edge there will be a small and a large eigenvalue κ^1 and κ^2 respectively. The corresponding eigenvectors v^1 and v^2 point in tangential direction. The evaluation of the shape operator on a level set of a noisy image might be misleading with respect to the true but unknown level sets and edges. E. g. noise might be identified as features. Thus we have to consider a regularization in advance and prefilter the current image $\phi(t, \cdot)$ before evaluating the shape operator. We take into account a local L^2 projection of the image intensity on the space of quadratic polynomials. The width σ of the projection stencil is considered to be the parameter steering this regularization. The evaluation of the shape operator on this polynomials is straightforward. We end up with the following type of nonlinear parabolic problem. Given the initial 3D image ϕ_0 on a domain Ω, we ask for a scale of images $\{\phi(t, \cdot)\}_{t \geq 0}$ which obey the anisotropic geometric evolution equation:

$$\partial_t \phi(t, x) - \|\nabla \phi(t, x)\| \operatorname{div}\left(a^\sigma(t, x) \frac{\mathcal{D}\phi}{\|\mathcal{D}\phi\|}(t, x)\right) = 0 \qquad (8.1)$$

on $\mathbb{R}^+ \times \Omega$ and satisfy the initial condition

$$\phi(0,\cdot) = \phi_0(\cdot).$$

Furthermore, we suppose natural boundary conditions on $\partial\Omega$, i. e.

$$a^\sigma(t,x)\frac{\partial \phi}{\partial \nu}(t,x) = 0$$

where ν denotes the outer normal on $\partial\Omega$. The diffusion tensor a^σ is supposed to depend on the regularized shape operator S^σ

$$a^\sigma(t,x) := \mathcal{A}(S^\sigma(t,x))$$

where $\mathcal{A} : \text{Sym}(\mathbb{R}^3) \to \text{Sym}(\mathbb{R}^3)$. Here $\text{Sym}(\cdot)$ denotes the space of symmetric maps. Finally $S^\sigma(x)$ is the shape operator at the position x evaluated on the local L^2 projection on quadratic polynomials with respect to a stencil ball $B_\sigma(x)$ of radius σ. As a suitable choice for this mapping $\mathcal{A}$ we consider the scalar function G from the basic image processing model, with $G(\alpha) = (1 + \lambda^{-2}\alpha^2)^{-1}$, now acting on the shape operator of the projection. Mapping the normal space to the 0 we trivially expand it to $\text{Sym}(\mathbb{R}^3)$. Here λ serves as a steering parameter for the identification of edges. The shape operator S^σ diagonalizes with respect to the basis $\{v^{1,\sigma}, v^{2,\sigma}, N^\sigma\}$, where $v^{1,\sigma}, v^{2,\sigma}$ are eigenvectors corresponding to principal curvatures $\kappa^{1,\sigma}, \kappa^{2,\sigma}$ on the level set and N^σ is the corresponding normal. Hence we obtain the matrix representation

$$\mathcal{A}(S^\sigma) = B_\sigma^T \begin{pmatrix} G(\kappa^{1,\sigma}) & & \\ & G(\kappa^{2,\sigma}) & \\ & & 0 \end{pmatrix} B_\sigma.$$

Here $B_\sigma \in SO(3)$ is the basis transformation from the regularized frame of principal directions of curvature and the normal $\{v^{1,\sigma}, v^{2,\sigma}, N^\sigma\}$ onto the canonical basis $\{e_1, e_2, e_3\}$.

In Figure 8.2 a 3D echocardiographical image of a human heart is taken as initial data. Here, different time steps of the evolution under anisotropic geometric diffusion are shown. A second example is concerned with true measurement data. The salt concentration in a density driven flow through a porous media filled with fresh and salt water is measured in a laboratory experiment by an MRI device. In Figure 8.3 level sets of the salt concentration are drawn, whereas Figure 8.4 shows slices through the 3D data set at different stages of the experiment. In both cases we compare the original measurement data with smoothing results obtained by our method.

Concerning the further details and an analysis of this model we refer to [26]. We especially observe that the underlying evolution is equivalent to the propagation of the level sets with speed f in normal direction N, i. e. $\partial_t x = f N$ holds for

$$f := \text{tr}(a^\sigma (S_\sigma - S)) + (\text{div } a^\sigma)(N^\sigma - N).$$

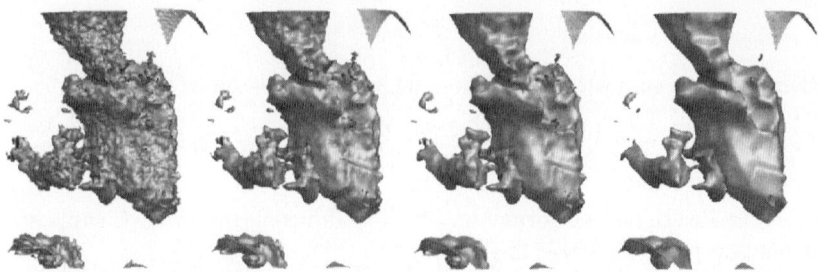

Fig. 8.2. From left to right a certain level set - visualizing the shape of one ventricel of the human heart - is extracted from the anisotropic geometric evolution process for still images. Here successive steps of the smoothing process are shown. The computation was performed on a 128^3 grid.

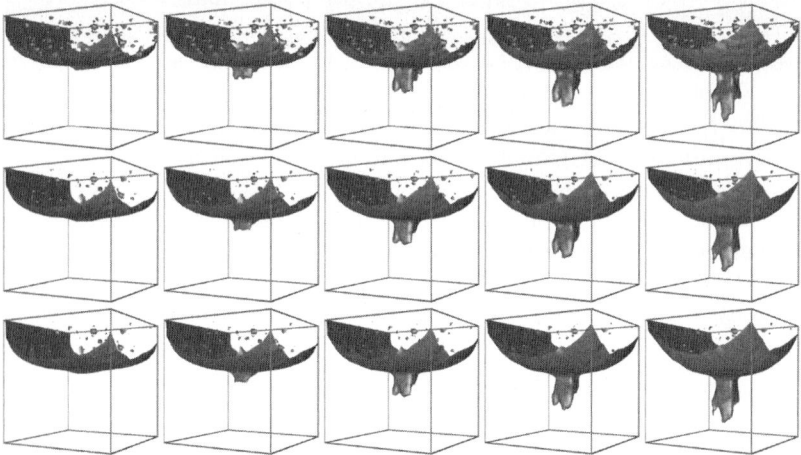

Fig. 8.3. The anisotropic geometric level set method is applied to noisy data from a fingering experiment in a two phase porous medium flow of fresh and salt water. During the experiment the salt concentration was measured using an MR imaging device. From left to right different stages of the experiment (corresponding to different frames in the image sequence) are depcited. In the top row the original noisy data is shown, whereas in the second and third row the scale steps 2 respectively 3 of the coupled anisotropic evolution on a 64^3 grid are depicted (cf. also Fig. 8.4).

Here we define $S_\sigma = DN^\sigma$, where N^σ is the normal of the locally projected image. This implies that our method is invariant on images ϕ_0 which are quadratic polyonomials.

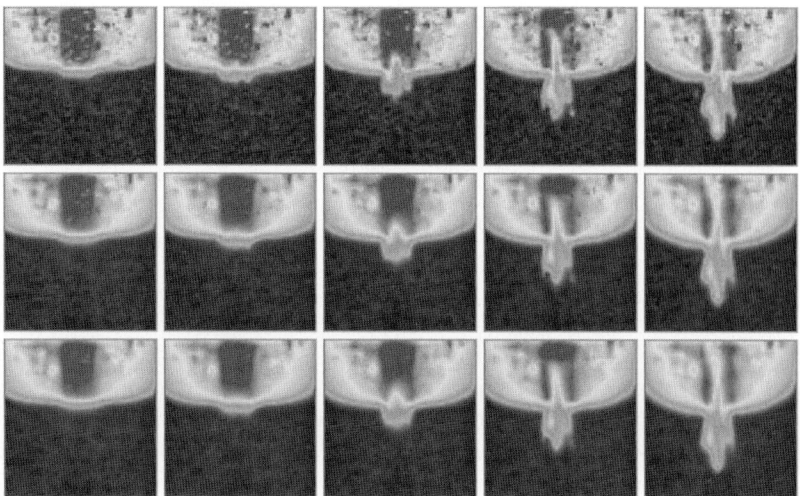

Fig. 8.4. From the experimental data shown in Figure 8.3 vertical slices are extracted. Again from left to right different frames of the sequence and from top to bottom different scales from the evolution are shown. See Plate 14 in the Appendix for a version of this figure in colour.

8.4 Processing Image Sequences via Coupled Anisotropic Geometric Diffusion

Let us now focus on the multiscale evolution of image sequences via an anisotropic geometric diffusion method. We therefore consider a grey valued image sequence $\phi_0 : [0,T] \times \Omega \to [0,1]$ where $[0,T]$ is the time interval of the given sequence. Our approach here is based on a combination of the ansitropic geometric diffusion method for still images (cf. Section 8.3) applied to time slices of the sequence and a coupling of these initially seperate processes in the sequence parameter based on the approach presented in [14, 15, 29]. The coupling consists of the speed modulation via the clt-term which measures the curvature of a motion trajectory in the plane spanned by the normal on the level set of the image intensity and the apparent velocity [14, 15]. The curvature is proved to coincide with the acceleration of the normal component of the velocity in the direction of the apparent velocity. Thus, noisy motion trajectories will lead to a faster diffusion, which induces a smoothing. In Section 8.6 we will outline that the proposed model is still handsome with respect the computation complexity also for large 3D image sequence. This holds true although we end up with a PDE in 5 dimensions: three spatial coordinates x, the sequence parameter s, and the scale parameter t.
I. e. we obtain the following problem:
For each frame $\phi_0(s,\cdot)$ of the image sequence find a scale of images $\{\phi(t,s,\cdot)\}_{t\geq 0}$ such that for $(t,s,x), \in \mathbb{R}^+ \times \{s\} \times \Omega$

$$\partial_t \phi(t,s,x) - \text{clt}^\sigma(\phi)(t,s,x) \|\nabla \phi(t,s,x)\| \, \text{div}\left(a^\sigma(t,s,x) \frac{D\phi}{\|D\phi\|}(t,s,x)\right) = 0$$

and in Ω

$$\phi(0, s, \cdot) = \phi_0(s, \cdot).$$

Furthermore, we suppose natural boundary conditions on $\partial \Omega$, i. e.

$$a^\sigma(t,s,x) \frac{\partial \phi}{\partial \nu}(t,s,x) = 0$$

where ν denotes the outer normal on $\partial \Omega$. Here, we take into account the same anisotropic diffusion tensor $a^\sigma(t, s, x)$ as in the case of still images, except the involved regularized shape operator $S^\sigma(t, s, x)$ is now evaluated for a time slice s in the image sequence $\phi(t, \cdot, \cdot)$. Furthermore, the curvature of lambertian trajectory $\text{clt}(\phi)$ is evaluated on the spatially regularized image sequence. Indead we apply the same local projection of the image sequence time slices on quadratic polynomials and evaluate the operator $\text{clt}(\cdot)$ for fixed position x and scale t on the family of resulting polynomials in s. Hence, the involved spatial regularization is indicated by an upper index σ.

In general we can not guarantee that $\nabla \phi \neq 0$ and $\text{clt}(\phi) \neq 0$ even if the initial data fulfills this regularity assumptions. Thus, we have to regularize the problem replacing the norm $\|\cdot\|$ in equation (8.1) and the $\text{clt}(\cdot)$-term by a regular approximation. I. e. we choose

$$\|v\|_\epsilon := \sqrt{\epsilon^2 + \|v\|^2},$$
$$\text{clt}^\sigma(\psi)_\epsilon := \max\{\text{clt}^\sigma(\psi), \epsilon\}.$$

The corresponding regularized variational formulation is then given by ($\zeta = (t, s, x)$)

$$\left(\frac{\partial_t \phi(\zeta)}{\text{clt}^\sigma(\phi)_\epsilon(\zeta) \|\nabla \phi(\zeta)\|_\epsilon}, \vartheta\right) + \left(a^\sigma(\zeta) \frac{\nabla \phi(\zeta)}{\|\nabla \phi(\zeta)\|_\epsilon}, \nabla \vartheta\right) = 0,$$

for all test functions $\vartheta \in C^\infty(\Omega)$. Here the brackets $(\cdot, \cdot)$ indicate the L^2 product on Ω. Considering a finite element implementation we will pick up this formulation in Section 8.6.

8.5 Local Curvature and Motion Evaluation

In our model above we have made extensive use of a regularized shape operator S^σ and the curvature of lambertian trajectory term $\text{clt}(\cdot)$, on which we base the computation of the anisotropic diffusion tensor and the speed modulation coupling the evolution problems on different time steps.

To this end let us fix a scale $t \in \mathbb{R}^+$. For each frame $\psi(s,\cdot) := \phi(t,s,\cdot)$ of the image sequence scale we consider a local L^2 projection in space of $\psi(s,\cdot)$ onto a subspace of the space of quadratic polynomials $\mathcal{P}_2$. Suppose we have fixed a point $x \in \Omega$ and we denote by

$$\mathcal{Q} := \operatorname{span}\{x_1^2, x_2^2, x_3^2, \, x_1x_2, x_1x_3, x_2x_3, \, x_1, x_2, x_3, 1\}$$

this subspace of $\mathcal{P}_2$. The local L^2 projection $\Pi_{x,\sigma}\psi(s,\cdot) \in \mathcal{Q}$ of the intensity $\psi(s,\cdot)$ onto $\mathcal{Q}$ is then defined via the orthogonality relation

$$\int_{\mathcal{B}_\sigma(x)} (\psi(s, y-x) - (\Pi_{x,\sigma}\psi)(s,y)) \, q \, dy = 0 \qquad \forall q \in \mathcal{Q},$$

where $\mathcal{B}_\sigma(x)$ is a ball of radius σ around x. For the ease of presentation we write $\psi_{s,x}^\sigma(y)$ instead of $(\Pi_{s,x,\sigma}\psi)(y)$ for fixed $(s,x) \in [0,T] \times \Omega$ in case of the local L^2 projection applied to an image frame. Now we define in analogy to the non-regularized case the shape operator $S^\sigma(s,x) = S(\psi_{s,x}^\sigma(\cdot))(x) := (D_y N^\sigma)(x)$ (cf. Section 8.3), with $N^\sigma(y) = \frac{\nabla_y \Psi_{s,x}^\sigma(y)}{\|\nabla_y \Psi_{s,x}^\sigma(y)\|}$. Thus, S^σ is charaterized by its eigenvalues 0, $\kappa^{j,\sigma}$, $j=1,2$ and the eigenvectors $\{v^{1,\sigma}, v^{2,\sigma}, N^\sigma\}$. Therefore, with an appropriate basis transformation $B_\sigma \in SO(3)$, we obtain

$$S^\sigma = B_\sigma^T \begin{pmatrix} \kappa^{1,\sigma} & & \\ & \kappa^{2,\sigma} & \\ & & 0 \end{pmatrix} B_\sigma.$$

Furthermore we will base the evaluation of the curvature of the lambertian trajectory on the locally projected images. Hence, in anticipation of a later discrete sequence of frames we consider the local projections of a previous $(s - \Delta s)$, the actual (s) and a next frame $(s + \Delta s)$, respectively. Then we compute the clt-term by the following formula

$$\operatorname{clt}^\sigma(\psi)(t,s,x) := \min_{w_1, w_2 \in \mathcal{B}_\sigma(x)} \frac{1}{(\Delta s)^2} \Big(\langle \nabla \psi_{s,x}^\sigma(0), w_1 - w_2 \rangle$$
$$+ |\psi_{s-\Delta s,x}^\sigma(w_1) - \psi_{s,x}^\sigma(0)|$$
$$+ |\psi_{s+\Delta s,x}^\sigma(w_2) - \psi_{s,x}^\sigma(0)| \Big),$$

since we have shifted the projections locally such that $\psi_{s,x}^\sigma(0) = \phi(t,s,x)$. The evaluation of the clt-term can be done via testing a lattice of different $w_i \in \mathcal{B}_\sigma(x)$ or by a continous minimization of the resulting polynomial. In our implementation we currently apply the lattice approach.

8.6 Finite Element Discretization

Up to now we have considered an image intensity $\phi(t,s,x)$ which has been a sufficiently smooth function on $\mathbb{R}^+ \times [0,T] \times \Omega \subset \mathbb{R}^3$. Concerning the implementation of the proposed multiscale method and its actual application to digital image sequences we now have to discretize our model in space and in the scale parameter t. First, as already mentioned we consider our image sequence to consist of $m+1$ frames at time steps $\Delta s := \frac{T}{m}$. In applications these image frames typically arise as arrays of pixels. We interpret pixel values as nodal values on a uniform hexahedral mesh $\mathcal{C}$ covering the whole image domain Ω and consider the corresponding trilinear interpolation on cells $C \in \mathcal{C}$ to obtain discrete intensity functions in the accompanying finite element space. To clarify the notation we will always denote spatially discrete quantities with upper case letters to distinguish them from the corresponding continuous quantities in lower case letters. A sub- or superscript h indicates the grid size, an upper index the time step, and a lower index the processed frame, i.e. $\Phi_i(t,x) = \Phi(t, i\Delta s, x)$. Let us define the space of piecewise trilinear, continuous functions

$$V^h = \{\Phi \in C^0(\Omega) \,|\, \Phi|_C \in \mathcal{P}_1 \otimes \mathcal{P}_1 \otimes \mathcal{P}_1 \,\forall C \in \mathcal{C}\},$$

where $\mathcal{P}_1$ denotes the space of linear polynomials and $\otimes$ the tensor product. Discretizing first only in space we obtain a variational finite element formulation of our level set evolution problem:

For $i = 0, \ldots, m$ find $\Phi_i : \mathbb{R}_0^+ \to V^h$ with initial data $\Phi_i(0) = \mathcal{I}_h \phi_0(i\Delta s)$, such that

$$\left(\frac{\partial_t \Phi_i}{\mathrm{clt}^\sigma(\Phi)_\epsilon \|\nabla \Phi_i\|_\epsilon}, \theta\right)^h + \tau \left(A_i^\sigma \frac{\nabla \Phi_i}{\|\nabla \Phi_i\|_\epsilon}, \nabla \theta\right) = 0$$

for all $\theta \in V^h$.

Here, $\mathcal{I}_h : C^0(\Omega) \to V^h$ is the Lagrange interpolation on the grid $\mathcal{C}$ and the diffusion tensor A_i^σ is supposed to be a suitable approximation of $a^\sigma(t, i\Delta s, \cdot)$. Finally, we have used the lumped mass scalar product $(\cdot, \cdot)^h$, which is defined by

$$(U, V)^h := \sum_{C \in \mathcal{C}} \int_C \mathcal{I}_h(U\,V)\,\mathrm{d}x$$

for discrete functions $U, W \in V^h$ (cf. [33]). As an immediate consequence the corresponding nonlinear mass matrix $M_h(\Phi)$ is diagonal. This simplifies the resulting scheme significantly. We end up with a system of ordinary differential equations for the nodal values of the intensity function Φ. Following Dziuk and Deckelnick [10] we select $\epsilon \approx h$. Furthermore we consider a stencil width $\sigma = C\,h$ for integers C equal to $2, 3$ or 4.

Next, we have to discretize in time, which includes the choice of some time stepping scheme and the decision which term to be handled implicitly and which explicitly. Here we choose a semi-implicit backward Euler discretization. Expressed in geometric terms we consider the metric and the regularized shape operator explicitly for each frame(cf. [12]). Let τ be a selected scale step size and let Φ_i^n to be an approximation of $\Phi_i(n\tau)$. Then we obtain the time and space discrete problem:

For $i = 0, \ldots, m$ find a sequence of discrete intensity functions $\{\Phi_i^n\}_{n=0,\cdots}$ with $\Phi_i^n \in V^h$ and $\Phi_i^0 = \mathcal{I}_h \phi_0(i\Delta s)$, such that

$$\left(\frac{\Phi_i^{n+1} - \Phi_i^n}{\text{clt}^\sigma(\Phi^n)_\epsilon \|\nabla \Phi_i^n\|_\epsilon}, \theta\right)^h + \tau \left(A_i^{\sigma,n} \frac{\nabla \Phi_i^{n+1}}{\|\nabla \Phi_i^n\|_\epsilon}, \nabla \theta\right) = 0$$

for all $\theta \in V^h$.

Finally, in each step of the discrete evolution, for each frame in the image sequence we have to solve a single system of linear equations. In terms of nodal vectors indicated by a bar on top of the corresponding discrete function we can rewrite the scheme and get

$$\left(M_h(\Phi_i^n) + \tau L_h(\Phi_i^n)\right) \bar{\Phi}_i^{n+1} = M_h(\Phi_i^n) \bar{\Phi}_i^n$$

for the new vector of nodal values $\bar{\Phi}_i^{n+1}$ at time $t_{n+1} = (n+1)\tau$. Here, we have applied the nonlinear lumped mass and stiffness matrices

$$M_h(\Phi_i^n) = \left(\left(\frac{\Psi_\mu}{\text{clt}^\sigma(\Phi^n)_\epsilon \|\nabla \Phi_i^n\|}, \Psi_\nu\right)^h\right)_{\mu\nu},$$

$$L_h(\Phi_i^n) = \left(\left(A_i^{n,\sigma} \frac{\nabla \Psi_\mu}{\|\nabla \Phi_i^n\|_\epsilon}, \nabla \Psi_\nu\right)\right)_{\mu\nu},$$

where $\{\Psi_\mu\}_\mu$ is the nodal basis of V^h.

In each time step the discrete diffusion tensor $A_i^{n,\sigma}$ and $\text{clt}(\Phi^n)_\epsilon$ are evaluated for every grid node and every image frame separately. Then, on cells $C \in \mathcal{C}$ we use trilinear interpolation to define $A_i^{n,\sigma}$. Furthermore, in the numerical application we have replaced the integration of $(A_i^{n,\sigma} \frac{\nabla \Psi_\mu}{\|\nabla \Phi_i^n\|_\epsilon}, \Psi_\nu)$ by the one point numerical quadrature which refers only to the value at the element's center of mass.

Concerning the proper choice of the stencil parameter σ with respect to the grid size we refer to [26].

Acknowledgements: The echocardiographical data show in Figure 8.2 was provided by TomTec Imaging Systems and C. Lamberti from DEIS, Bologna University. Furthermore we acknowledge Sascha Oswald from Sheffield University, who performed the fresh and salt water experiment shown in Figures 8.3 and 8.4.

References

1. L. Alvarez, F. Guichard, P. L. Lions, and J. M. Morel. Axioms and fundamental equations of image processing. *Arch. Ration. Mech. Anal.*, 123(3):199–257, 1993.
2. L. Alvarez, J. Weickert, and J. Sánchez. A scale–space approach to nonlocal optival flow calculations. In M. Nielsen, P. Johansen, O. F. Olsen, and J. Weickert, editors, *Scale-Space Theories in Computer Vision. Second International Conference, Scale-Space 1999, Corfu, Greece, September 1999*, Lecture Notes in Computer Science; 1682, pages 235–246. Springer, 1999.
3. S. B. Angenent and M. E. Gurtin. Multiphase thermomechanics with interfacial structure 2, evolution of an is othermal interface. *Arch. Rational Mech. Anal.*, 108:323–391, 1989.
4. E. Bänsch and K. Mikula. A coarsening finite element strategy in image selective smoothing. *Computing and Visualization in Science*, 1:53–63, 1997.
5. G. Bellettini and M. Paolini. Anisotropic motion by mean curvature in the context of finsler geometry. *Hokkaido Math. J.*, 25:537–566, 1996.
6. F. Catté, P.-L. Lions, J.-M. Morel, and T. Coll. Image selective smoothing and edge detection by nonlinear diffusion. *SIAM J. Numer. Anal.*, 29(1):182–193, 1992.
7. U. Clarenz, U. Diewald, and M. Rumpf. Nonlinear anisotropic diffusion in surface processing. In *Proc. Visualization 2000*, pages 397–405, 2000.
8. C. A. Davatzikos, R. N. Bryan, and J. L. Prince. Image registration based on boundary mapping. *IEEE Trans. Medical Imaging*, 15(1):112–115, 1996.
9. K. Deckelnick and G. Dziuk. Discrete anisotropic curvature flow of graphs. *Mathematical Modelling and Numerical Analysis, to appear*, 2000.
10. K. Deckelnick and G. Dziuk. A fully discrete numerical scheme for weighted mean curvature flow. Technical Report 30, Mathematische Fakultät Freiburg, 2000.
11. R. Deriche, P. Kornprobst, and G. Aubert. Optical–flow estimation while preserving its discontinuities: A variational approach. In *Proc. Second Asian Conf. Computer Vision (ACCV '95, Singapore, December 5-8, 1995)*, volume 2, pages 290–295, 1995.
12. G. Dziuk. An algorithm for evolutionary surfaces. *Numer. Math.*, 58:603–611, 1991.
13. L. Evans and J. Spruck. Motion of level sets by mean curvature I. *J. Diff. Geom.*, 33(3):635–681, 1991.
14. F. Guichard. *Axiomatisation des analyses multi-échelles d'images et de films*. PhD thesis, University Paris IX Dauphine, 1994.
15. F. Guichard. A morphological, affine, and galilean invariant scale–space for movies. *IEEE Transactions on Image Processing*, 7(3):444–456, 1998.
16. J. Kačur and K. Mikula. Solution of nonlinear diffusion appearing in image smoothing and edge detection. *Appl. Numer. Math.*, 17 (1):47–59, 1995.
17. R. Kimmel. Intrinsic scale space for images on surfaces: The geodesic curvature flow. *Graphical Models and Image Processing*, 59(5):365–372, 1997.
18. F. Maes, A. Collignon, D. Vandermeulen, G. Marchal, and P. Suetens. Multi-modal volume registration by maximization of mutual information. *IEEE Trans. Medical Imaging*, 16(7):187–198, 1997.

19. K. Mikula and J. Kačur. Evolution of convex plane curves describing anisotropic motions of phase interfaces. *SIAM Journal on Scientific Computing*, 17(6):1302–1327, 1996.
20. K. Mikula and N. Ramarosy. Semi–implicit finite volume scheme for solving nonlinear diffusion equations in image processing. *Numerische Mathematik*, 2001.
21. K. Mikula, A. Sarti, F. Sgallari, and C. Lamberti. *Nonlinear multiscale analysis models for filtering of 3D + time biomedical images*. Lectures Notes in Computational Science and Eng. Springer Verlag, 2001.
22. H. H. Nagel and W. Enkelmann. An investigation of smoothness constraints for the estimation of displacement vector fields from images sequences. *IEEE Trans. Pattern Anal. Mach. Intell.*, 8:565–593, 1986.
23. E. Pauwels, P. Fiddelaers, and L. Van Gool. Enhancement of planar shape through optimization of functionals for curves. *IEEE Trans. Pattern Anal. Mach. Intell.*, 17:1101–1105, 1995.
24. P. Perona and J. Malik. Scale space and edge detection using anisotropic diffusion. In *IEEE Computer Society Workshop on Computer Vision*, 1987.
25. T. Preußer and M. Rumpf. An adaptive finite element method for large scale image processing. *Journal of Visual Comm. and Image Repres.*, 11:183–195, 2000.
26. T. Preußer and M. Rumpf. A level set method for anisotropic diffusion in 3D image processing. *SIAM J. Appl. Math.*, 2001, to appear.
27. E. Radmoser, O. Scherzer, and J. Weickert. Scale–space properties of regularization methods. In M. Nielsen, P. Johansen, O. F. Olsen, and J. Weickert, editors, *Scale-Space Theories in Computer Vision. Second International Conference, Scale-Space '99, Corfu, Greece, September 1999*, Lecture Notes in Computer Science; 1682, pages 211–220. Springer, 1999.
28. G. Sapiro. Vector (self) snakes: A geometric framework for color, texture, and multiscale image segmentation. In *Proc. IEEE International Conference on Image Processing, Lausanne*, September 1996.
29. A. Sarti, K. Mikula, and F. Sgallari. Nonlinear multiscale analysis of 3D echocardiography sequences. *IEEE Transactions of Medical Imaging*, 18(6):453–466, 1999.
30. C. Schnoerr. A study of a convex variational diffusion approach for image segmentation and feature extraction. *J. Math. Imaging Vis.*, 8(3):271–292, 1998.
31. J. E. Taylor, J. W. Cahn, and C. A. Handwerker. Geometric models of crystal growth. *Acta metall. mater.*, 40:1443–1474, 1992.
32. J. P. Thirion. Image matching as a diffusion process: An analogy with maxwell's demon. *Medical Imag. Analysis 2*, pages 243–260, 1998.
33. V. Thomee. *Galerkin–Finite Element Methods for Parabolic Problems*. Springer, 1984.
34. J. Weickert. *Anisotropic diffusion in image processing*. Teubner, 1998.

9 Population Dynamics: A Mathematical Bird's Eye View

Odo Diekmann[1] and Markus Kirkilionis[2]

[1] Universiteit Utrecht, Mathematisch Instituut, P.O. Box 80010, 3508 TA Utrecht, The Netherlands – *O.Diekmann@math.uu.nl*
[2] University of Warwick, Mathematics Department & Centre for Scientific Computing, Coventry CV4 7AL, United Kingdom – *mak@maths.warwick.ac.uk*

Abstract: The aim of this chapter is to provide interested outsiders with a brief (and therefore incomplete) overview of the kind of questions and insights concerning the dynamics of biological populations that can be formulated in mathematical language and derived by mathematical methods. Ideally the chapter should serve as an invitation to further reading and hence we give many pointers to the extensive literature. In order to highlight the ideas, we sacrifice the precise statement of assumptions (implying that some of our statements are sloppy from the point of view of the pedantic mathematician). Likewise we shall focus on the simplest examples that illustrate a key issue and not strive for generality. Our aim is to enlighten, not to impress. Moreover, we have not tried to hide our bias deriving from taste and experience, so the views we present are somewhat idiosyncratic.

9.1 The Chemostat

In this section we introduce the chemostat, a simple laboratory device to culture microorganisms continuously. We use it to illustrate basic ecological or biotechnological principles, exploiting the fact that some parameters are "tunable" by the experimentalist to motivate bifurcation analysis. For our audience the basic reference is Smith & Waltman [58], the classical first reference to a mathematical description of the chemostat is Monod [44]. (Here and in the following our references are most of all a convenient starting point for tracing lots of relevant references. We do not aim to give deserved credit to all various authors that have contributed to the subject.)

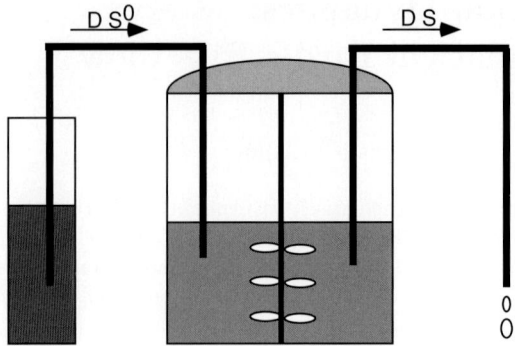

Fig. 9.1. In a vessel of volume $V = 1$ bacteria or algae are grown. All they need to grow is provided in excess except for one substance (e.g. phosphate, nitrate, vitamin B12,...) which accordingly is called the limiting substrate or resource. Fresh medium containing substrate with concentration S^0 is pumped into the vessel at rate D. The vessel is well-stirred. The concentration of the substrate within the vessel is denoted by S. The volume is kept constant by pumping liquid out at the same rate D.

9.2 Consumer-Resource Interaction

In the absence of consumers the substrate concentration obeys

$$\frac{dS}{dt} = [S^0 - S]D, \tag{9.1}$$

and so converges to the concentration S^0 defined by the inflowing medium. Now let's introduce a consumer species. From the point of view of an individual consumer the substrate concentration S is the key aspect of the environmental condition under which it has to live. When S is constant in time, it lives in a constant world. In a constant world one can, as an equivalent alternative to considering changes in numbers (or concentrations/densities) in the course of real time, study population growth or decline on a generation basis.

The *basic reproduction ratio (number)* R_0 is, by definition, the expected number of offspring produced by a newborn individual. Here we assume for the time being, that individuals are identical at birth (but see sections 9.5 and 9.6). In the case individuals are not only identical at birth, but in fact throughout their life (i.e., their relevant properties do not change at all in the course of their life) R_0 is the product of the expected life time and the rate at which an individual produces offspring while being alive.

Let us assume that consumer individuals cannot die in the chemostat and let us declare them dead when they are washed out. Then the length of their life is an exponentially distributed variable with mean D^{-1} (note that we have scaled volume such that the vessel has volume 1; more generally one should interpret D as the fraction of the vessel volume that is replaced per unit of time).

How much offspring an individual produces is related to how much substrate it consumes. Let us assume the two are simply proportional with constant of proportionality η, which is called the conversion efficiency (this assumption is of course debatable, especially if we measure consumers in terms of number of individuals per unit of volume and less so if we measure consumers in terms of biomass per unit of volume).

How much substrate an individual consumes per unit of time depends on the substrate concentration. Let us denote this quantity by $g(S)$. Various submodels for the short-time-scale consumption process (such as the Holling time budget model, explained in, e.g., Metz & Diekmann[40] or the Michaelis-Menten enzyme reaction model as presented in, e.g., Segel[55], yield expressions of the form

$$g(S) = \frac{mS}{S+k}, \qquad (9.2)$$

where m and k are positive constants. The most relevant features are that g increases linearly with S at low concentrations (the output of a factory is proportional to the rate at which raw material is provided when this rate is low) but approaches a limit k as concentrations get higher and higher (the output of the factory is determined by its capacity, not by the input). Below we shall explain how one can actually use the chemostat to measure g.

We conclude that the basic reproduction ratio depends on the environmental condition S and is given explicitly by

$$R_0(S) = \frac{1}{D}\eta g(S). \qquad (9.3)$$

If $R_0(S^0) < 1$, the organism is not able to sustain itself in the "virgin" (and best possible) environment. If, on the other hand, $R_0(S^0) > 1$, it will start growing when introduced in the virgin environment. While growing, it exerts influence on the environmental condition by feedback (which in this case is consumption). If, as an end result, the environmental condition becomes constant again then necessarily we should have that the steady substrate concentration $\bar{S}$ is such that

$$R_0(\bar{S}) = 1, \qquad (9.4)$$

i.e., the organism "sets" the environmental condition such that it neither grows nor declines (in general this *must* be the case, though not necessarily pointwise in time, only on average over time).

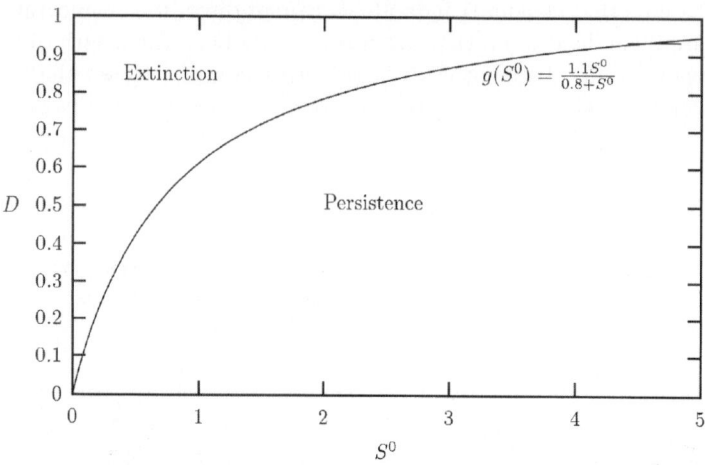

Fig. 9.2. The curve $D = \eta g(S^0)$ (corresponding to $R^0(S^0) = 1$) separates in the plane of the two tunable parameters S^0 and D the regions of consumer extinction and consumer persistence. The constant η is chosen to be equal to one.

That S does indeed have to stabilise at a constant level is most easily concluded from the differential equations

$$\begin{aligned}\frac{dS}{dt} &= [S^0 - S]D - g(S)X, \\ \frac{dX}{dt} &= [\eta g(S) - D]X,\end{aligned} \quad (9.5)$$

where both substrate S and the consumer concentration denoted by X have nonnegative initial conditions, i.e., $S_0 \geq 0$ and $X_0 \geq 0$ as required by the biological interpretation. Indeed, note that the "total amount of substrate" (including that incorporated into consumers) in the chemostat defined by

$$M = S + \frac{1}{\eta}X \quad (9.6)$$

satisfies the linear scalar equation

$$\frac{dM}{dt} = [S^0 - M]D, \quad (9.7)$$

since the chance of washout is the same for the dissolved (in the culture medium) and incorporated substrate and nothing is lost or gained at the transformation. It follows from Eq. (9.7) that $M(t) \to S^0$ as $t \to \infty$. Thus, in view of Poincaré-Bendixson theory, there is no other possibility than that both S and X have a limit. The conclusion of the analysis is summarised in Figure 9.2 (but note that even in the persistence part of the parameter space

$X(t) = 0$ for all $t > 0$ when $X_0 = 0$). Experimentally one can easily vary D and measure both $\bar{S}$ and $\bar{X}$. A first check on the model is to verify whether the quotient $\frac{\bar{X}}{S^0 - \bar{S}}$ is, within experimental accuracy, constant. Next one plots the points $(\bar{S}, \frac{D(S^0 - \bar{S})}{\bar{X}})$ in the plane to find the graph of g.

To end the section, we make the following remark: As long as we stay in the persistence part of the parameter space, $\bar{S}$ is independent of S^0. So increasing S^0 benefits the consumer, not the substrate itself. This is a first aspect of the so-called "Paradox of Enrichment".

9.3 Competition for Substrate in the Chemostat

We introduce a second type of micro-organism that is competing for the same substrate. To distinguish the two organisms, we use an index i taking the values 1 and 2. The dynamical changes are now described by the equations

$$\frac{dS}{dt} = [S^0 - S]D - g_1(S)X_1 - g_2(S)X_2,$$
$$\frac{dX_1}{dt} = [\eta g_1(S) - D]X_1, \qquad (9.8)$$
$$\frac{dX_2}{dt} = [\eta g_2(S) - D]X_2.$$

By setting

$$1 = R_0^i(S) \stackrel{\text{def}}{=} \frac{1}{D}\eta_i g_i(S) \quad \text{for} \quad i \in \{1, 2\}, \qquad (9.9)$$

as before, we define the two levels $\bar{S}_i$ at which the respective consumers would set the substrate level when in isolation from one another: we assume that $R_0^i(S^0) > 1$ for $i = 1, 2$, (to avoid that either species is doomed to go extinct even in the absence of competition). Suppose the indices are chosen such that $\bar{S}_2 < \bar{S}_1$. Then, in view of the monotone dependence of $R_0^i(S)$ on S, species 2 can grow in the environment as set by species 1 but *not* vice versa (we say that species 2 can invade successfully when species 1 is the resident while species 1 will fail to invade when species 2 is the resident). In the framework of Adaptive Dynamics (see Metz et. al. [41], Geritz et al. [23], Geritz et al. [24], Rand et al. [51], SpIss [18]) it is then concluded (by assumption; but see Geritz et al. [25] for a justification when the differences between the two species are sufficiently small) that species 2 will outcompete (i.e., drive to extinction) species 1 and become the new resident, whenever it is introduced (e.g. by mutation) while species 1 is the resident. If the species differ in a particular trait, we say that a trait substitution took place. Thus one can think about long term phenotypic evolution in terms of trait substitution sequences (the ecological interactions of the residents determine an attractor and hence the environmental conditions in which a mutant has to try its luck;

if a mutant invades successfully, a change in the composition of the residents takes place and a new attractor is approached; then the game is repeated; note that the underlying assumption is that the time scale of convergence to the ecological attractor is much shorter than the time scale at which mutants arise; this is expressed by saying that evolution is mutation limited). Adaptive Dynamics is concerned with the systematic study of such trait substitution sequences (but it acknowledges the fact that populations need not necessarily stay monomorphic in the process: (repeated) branching may occur).

Clearly then competition for limiting substrate in the chemostat will lead, on the long evolutionary time scale, to a consumer that *minimises*, as far as allowed by physiological constraints, the steady state substrate level $\bar{S}$. This is "Verelendung": evolution favours the "type" that sets the (one-dimensional!) environmental quality index at the worst possible level. It is the other side of the coin of optimal adaptation to the environment: if there is feedback to the environment and the environment is accurately described by a scalar then the environment deteriorates in the course of evolution (see Mylius & Diekmann [45], Diekmann [19], Metz et al. [43].)

Fortunately, the one-dimensionality of the environment resulted from the experimental setup and many simplifying assumptions (see Huisman & Weissing [31]). Yet one can rightfully wonder why there exist so very many species, despite ubiquitous competitive exclusion in models? Could it be that evolution has as an emerging property that the complexity of the environmental conditions increases?

For the present model one can rigorously establish competitive exclusion by reducing the three-dimensional system to a two-dimensional one. This is done by constructing from Eq. (9.8) a so-called asymptotically autonomous system, where the fact that

$$M \stackrel{\text{def}}{=} S + \frac{1}{\eta_1} X_1 + \frac{1}{\eta_2} X_2 \to S^0 \qquad (9.10)$$

as $t \to \infty$ is used. See Thieme [59] for warnings and precise results. In the next step the competitive exclusion is proven by phase plane arguments (see Smith & Waltman [58], section 1.5).

9.4 A Chemostat Containing a Food-Chain

Consider now an additional trophic level, a predator feeding on the microorganism, a classical prey-predator relationship. This can be described by the system of equations

$$\frac{dS}{dt} = [S^0 - S]D - g(S)X,$$
$$\frac{dX}{dt} = [\eta g(S) - D]X - h(X)Y, \qquad (9.11)$$
$$\frac{dY}{dt} = [\xi h(X) - D]Y.$$

It can be reduced to a two-dimensional system since

$$M \stackrel{\text{def}}{=} S + \frac{1}{\eta}X + \frac{1}{\xi\eta}Y \to S^0, \qquad (9.12)$$

as $t \to \infty$. From the point of view of the predator the environmental condition is captured by the consumer concentration X. Its basic reproduction ratio equals

$$\tilde{R}_0(X) \stackrel{\text{def}}{=} \frac{1}{D}\xi h(X), \qquad (9.13)$$

when X is constant. The curve $\tilde{R}_0(X) = 1$ subdivides the region of consumer persistence in the parameter plane into two subregions, one in which the predator is doomed to go extinct and one in which the predator persists as well. The latter subregion is further subdivided by a Hopf bifurcation curve into a subregion in which consumer exploitation by the predator leads to a steady state and one in which it leads to oscillatory coexistence. More formally we have

(i) If $R_0(S^0) < 1$ then prey and predator die out, i.e., $S \to S^0$, $X \to 0$ and $Y \to 0$ for $t \to \infty$.
(ii) If $R_0(S^0) > 1$ and $\tilde{R}_0(\frac{D(S^0-\bar{S})}{g(\bar{S})}) = \tilde{R}_0(\eta(S^0 - \bar{S})) < 1$, then only the prey population survives, i.e., $S \to \bar{S}$, $X \to \eta(S^0 - \bar{S})$ and $Y \to 0$ for $t \to \infty$.
(iii) If $R_0(S^0) > 1$ and $R_0(\eta(S^0 - \bar{S})) > 1$, then both species persist, i.e., either $X \to \bar{X}$ and $Y \to \bar{Y}$ for $t \to \infty$, or the concentrations oscillate.

Again we find a "Paradox of Enrichment" phenomenon: in region III the consumer level is fixed at $\bar{X}$ such that

$$\tilde{R}_0(\bar{X}) = 1, \qquad (9.14)$$

i.e., by the predator, and so it doesn't depend at all on S^0 (in region III, however, the steady state S level increases with S^0 since the consumers are held in check by the predators). In region IV it is even worse: an increase of S^0 leads to enhanced "violence" of the predator-consumer fluctuations. This means that consumer concentrations go through extreme troughs, in which the danger of extinction by demographic stochasticity (i.e., the fact that populations consist of discrete entities, individuals, which either do reproduce or do not reproduce, but never produce one tenth of an offspring) is real. In

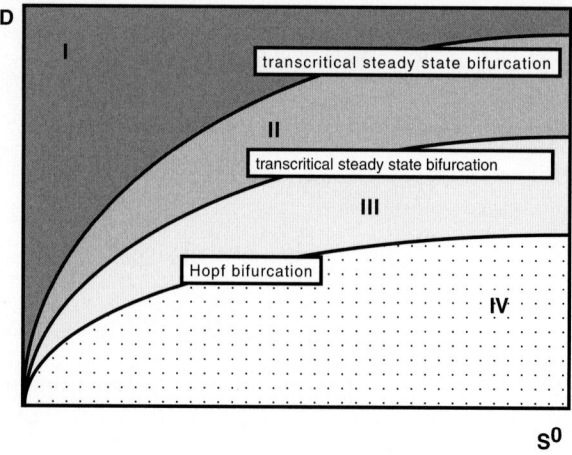

Fig. 9.3. Schematic operating diagram of the simplest food chain chemostat. In region I both consumer and predator go extinct; in region II the consumer reaches a positive steady state but the predator goes extinct; in region III both species attain a positive steady state while in region IV they also both persist, but in an oscillatory manner. To show uniqueness of the limit cycle in region IV is a tough problem, but it can be solved, see Yang [62] . We encourage the reader to verify this scheme numerically with bifurcation packages like CONTENT (see [13]) or AUTO (see [2]) for appropriately chosen strictly monotone functions f and g, see Fig.9.2.

other words, increasing the supply of nutrient, which in principle is beneficial for the consumer, may have the counterintuitive effect that it goes extinct, see Boer et al. [4].

Some famous lab populations (e.g. Huffaker's oranges, mites and Nicholson's blowflies [see Nisbet & Gurney [48]]) do indeed show violent oscillations, but, as a rule, preys and predators in the field seem to fluctuate, but not so very extreme. One theory (see Jansen & de Roos [33]) ascribes this fact to the influence of spatial structure: since local oscillations are out of phase they average out (to some extent) over larger domains. We shall return to this point in subsection 9.6.2.

One can extend the chain by adding a super-predator, i.e., a species that feeds on the predator. We then speak about a system with three trophic levels (the substrate, which doesn't reproduce, is not counted as a trophic level). In recent years the operating diagram of such systems has been studied in detail (see again Boer et al. [4] for related phenomena) and many of its complications and complexities have been revealed. When one keeps on increasing the number of species under consideration, it becomes less likely that they form a chain. Many ecosystems constitute complicated food webs. To unravel the relationships between food web structure and qualitative as well as quantitative aspects of population dynamics constitutes a major challenge (Yodzis [63], Neutel et al. [47]).

For the consumer, the environmental condition is now characterised by the couple (S, Y), i.e., by food availability and predation pressure. This creates the possibility of predator mediated coexistence: if one species is better in reducing S, but the other is less vulnerable to predation, they may be able to coexist in a stable steady state, thus forming a rather simple food web (see Butler et al. [7], Butler & Wolkowicz [9]). An ecological analysis ends with such a conclusion but in the framework of Adaptive Dynamics there is a natural next question: is such a dimorphism contracting (meaning the traits of the two types get closer and closer together as a result of substitutions) or expanding? As far as we know this has not been analysed for the present system (but see Pugliese [50] for a somewhat similar situation that could be analysed in detail).

9.5 Infectious Agents and the Art of Averaging

Next we switch from the chemostat model to a simple description of an infectious disease, an SIR-model. A key feature of infectious agents is that reproduction occurs at two levels, within hosts (see Nowak & May [49] for an inspiring account) and via transmission between hosts, Anderson & May [1], Diekmann & Heesterbeek [21]. So the population is more of a metapopulation (see Hanski [27]) with each infected host carrying a local (but moving!) subpopulation of agents. (The analogy in an ecological situation would be a landscape consisting of patches providing suitable habitat for some species, but where outside the patches life is impossible. Patches may appear or dissappear (for example patches may just consist of a tree)).

When the "size" of the local subpopulation is not of much influence on the transmission probability, one forgets about it and just works with the number or density of infected hosts. A representative model takes the form

$$\begin{aligned}\frac{dS}{dt} &= B - \mu S - \alpha \frac{I}{N} S, \\ \frac{dI}{dt} &= -\mu I + \alpha \frac{I}{N} S - \gamma I, \\ \frac{dR}{dt} &= -\mu R + \gamma I,\end{aligned} \qquad (9.15)$$

where now S denotes "susceptibles", I denotes infectives, and R denotes the recovered individuals, while by definition $N = S + I + R$, the total population of hosts. The infectious period is exponentially distributed with parameter γ. The parameter α is a contact intensity times a transmission probability, μ is a given death rate. All these parameters are positive. As $N(t) \to \frac{B}{\mu}$ for $t \to \infty$ (and so we might as well assume that $N = \frac{B}{\mu}$ to begin with), the fact that we incorporated a factor $\frac{1}{N}$ in the transmission term seems to affect only the interpretation of α. But when comparing populations of different sizes or when considering growing or declining populations, the issue deserves serious attention. In chemical kinetics the reaction rates are, according to the principle of mass action, directly proportional to the concentrations of the two substances that react, simply because this is true for the collision rate. A similar reasoning would suggest to eliminate the $\frac{1}{N}$ factor in (9.15) when the variables are densities. But what if we consider a sexually transmitted disease (STD)? Does the number of sexual contacts an individual has per unit of time, increase with population density? Next, consider Phocine Distemper Virus circulating in a seal colony, see de Koeijer et al. [36]. Transmission takes place when the seals gather at rocks and beaches to rest and sun-bath. Whenever space is not in short supply, they adhere to typical distances to each other. So when population numbers go down, as a result of mortality due to the virus, the contact intensity does not change (this example also illustrates that "density" is not always "number divided by area"; also think of animals living in herds or schools or flocks).

The basic reproduction ratio is given explicitly by

$$R_0(S) = \frac{\alpha}{\gamma + \mu} \frac{S}{N}, \qquad (9.16)$$

and we find that S stabilises at $N = \frac{B}{\mu}$ if $R_0(N) < 1$, and at $\bar{S}$ if $R_0(N) > 1$, where $\bar{S}$ is defined implicitly by $R_0(\bar{S}) = 1$. Evolution will tend to minimise $\bar{S}$, so to maximise $\frac{\alpha}{\gamma+\mu}$. There are trade-offs that complicate this simple picture: can α (say, the production of virus) be increased without increasing the additional death rate γ as well? Moreover, cross immunity may not be complete, so it may be necessary to consider superinfection (i.e., simultaneous infection with two different strains). Unfortunately, not that much is known about the within host competition of bacteria and viruses (but see Nowak & May [49]). For example, the modelling of the emergence of anti-biotic resistance in hospital infections is severely hampered by lack of such detailed knowledge, leading to ill-defined concepts such as "costs of resistance" (see the book edited by Dieckmann et al.[17].)

So far we have assumed that the relevant (for transmission) host behaviour is uniform, i.e., we ignore heterogeneity. As a rule, averaging over differences in infectivity is safe, but with differences in susceptibility one should be more careful. For example, let us calculate R_0 for a sexually transmitted disease in a heterogeneous population. More specifically, assume there are two types of individuals, labeled "1" and "2", which occur in the proportion $N_1 : N_2 = 1 : 100$ and for which the number of effective contacts (i.e., contacts multiplied by the transmission probability for a contact between a susceptible and an infective) are in the proportion $c_1 : c_2 = 10 : 1$ (so there is a minority of relatively active individuals). Now assume that the types of two individuals have an *independent* influence on the probability that they meet (in more technical jargon, assume *separable* (also called *proportional*) mixing in the sense that for any type the proportion of its partners that are of type i is given by $\frac{c_i N_i}{c_1 N_1 + c_2 N_2}$). Then

$$R_0 = c_1 \frac{c_1 N_1}{c_1 N_1 + c_2 N_2} + c_2 \frac{c_2 N_2}{c_1 N_1 + c_2 N_2} \approx 1.8 \ c_2,$$

since the probability distribution of the type of newly infected individuals is independent of the type of the individual that is responsible for the infection and given explicitly by $\frac{c_i N_i}{c_1 N_1 + c_2 N_2}$. This should be contrasted with the naieve (but wrong!) guess that $R_0 = \bar{c}$, where $\bar{c}$ is the average number of effective contacts, i.e., $\bar{c} = \frac{c_1 N_1 + c_2 N_2}{N_1 + N_2} \approx 1,1 \ c_2$. The point is that type 1 individuals are overrepresented among the newly infected individuals because of their greater susceptibility. When there are more than two types the result is more conveniently expressed by

$$R_0 = \text{mean} + \frac{\text{variance}}{\text{mean}}$$

Such an explicit formula is possible whenever susceptibility of the susceptible and the infectivity of the infective are *independent* factors for the probability of their contact (whence the name "separable mixing"). When there is dependence, R_0 can be characterised as the dominant eigenvalue of a positive (so-called next-generation) matrix or, even more general, as the spectral radius of a bounded and positive linear operator. The point is that the distribution of types among newly infected individuals can no longer be found a priori (to determine this distribution becomes part of the problem; it is now given as the normalised eigenvector corresponding to R_0) We refer to Diekmann & Heesterbeek [21] for a systematic exposition and a wealth of examples.

9.6 Heterogeneity

When biological processes are considered, there is always a certain amount of heterogeneity in the objects being considered. No individual is the same as another individual, no environment is really the same at two different spatial positions. The important question for modelling is when heterogeneities can be neglected? And if so, how can the variability of properties in the considered objects be averaged?

9.6.1 Heterogeneity Deriving from Physiological Differences

Individuals may differ from one another in many ways. The information which captures the essential (for population dynamics) properties we call the *i*-state (*i* for individual). Apart from changes in numbers due to death and reproduction we then also have to model

- the distribution of the i-state-at-birth of offspring.
- changes in i-state corresponding to maturation/growth/development or, in the case of spatial position, movement.

In particular we should specify how such quantities depend on the i-state itself and on the prevailing environmental condition. (Here "dynamic energy budget" theory by Kooijman [37] provides guidelines.) In addition one has to model the feedback loop: how is, in turn, the environmental condition influenced by the population(s)?

The aim of such elaborate modeling exercises is to unravel how mechanisms at the i-level lead to phenomena at the p-level. We give two examples:

- Cannibalism can be part of the adult foraging strategy, as small individuals may have access to food that cannot be consumed directly by big individuals (see Claessen et al.[12]). This is relevant for lakes with *only* piscivorous fish, like pike or perch.
- Variable maturation delay (i.e., the fact that the length of the period between birth and the onset of reproduction depends on food availability which, in turn, depends on overall consumption) can create demographic cycles which differ markedly from the usual predator-consumer cycles (as described in Section 9.4), as periods in which the population consists of many small individuals alternate with periods in which there are only a few individuals which, however, are very large (deRoos & Persson[53], deRoos & Persson[54], this is relevant for Daphnia feeding on algae as well as for various species of fish).

We refer to Tuljapurkar & Caswell [61], DeAngelis [15], Metz & Diekmann [40], for biological motivation and surveys, and to Cushing [14], Calsina & Saldaña [11], Tucker & Zimmermann [60], Diekmann et al. [20], Diekmann et al. [22] for the still somewhat underdeveloped mathematical theory, and to deRoos [52], Kirkilionis et al. [35] for numerical methods.

9.6.2 Heterogeneity Deriving from Spatial Position

The preceding section focussed on "internal" differences among individuals where as in the present section we shall focus on "external" differences, i.e., differences in the environmental condition as perceived by individuals inhabiting different positions in an extended domain.

A first phenomenon is that now population growth may actually correspond to expansion of the inhabited domain like, e.g., in the case of a species newly introduced into another continent. The art is to derive the asymptotic speed of propagation c_0 from the ingredients of a model and to show rigorously that a local disturbance does indeed trigger a change that propagates with speed c_0 (see Metz et al. [42]).

A second phenomenon is that spatial pattern may alter the dynamics of interacting populations. We refer to Dieckmann et al.[16] for a comprehensive survey of the state-of-the-art. A striking example is how the hypercycle is saved from parasites by spatial pattern formation, in particular spirals, as shown by Boerlijst [5]. As another example, recall the damping of prey-predator cycles due to phase differences between local oscillations (but as a warning see Huang and Diekmann[30]: incorporation of a natural Holling type II correction factor in the diffusion term has a strong diminishing influence on this mechanism).

The gradostat and the creation of niches Finally it should be said that the ecological situation of having unequally distributed resources can also be studied with the extension of the chemostat as described in Sections 9.1 - 9.4. If we build a chain with two or more chemostats, with fresh medium flowing into the first vessel, which itself has its outflow directed to the second vessel etc. (an outflow away from the system is thus only from the last vessel) and if all this "communication" between the vessels is with the same rate D as before, this chain of vessels called the *gradostat"* (see Lovitt & Wimpenny[38]) will produce a gradient of resource levels. More precisely the first vessel will have the relatively highest and the last the lowest resource concentrations. The result is that species can "adapt" to different resource concentrations and coexist in competition, which they could not have done in a homogeneous environment (Jäger et al. [32], Smith & Tang[56]). In general a gradostat can have maximally as many coexisting species as there are number of vessels. Instead of having different resource concentrations in space, it is also possible to create variations of resource levels in time. Again several competitors may survive, see Butler et al.[10], Smith [57], Hale & Somolinos [26].

9.7 The Pecularities of Semelparity

A semelparous organism reproduces only once in its life. Examples for this type of behaviour are:

- Annual or biennial plants
- Salmon populations. In this specific example there is nursery competition.
- Cicadas. There are examples of species waiting 13 respectively 17 years before a year class reproduces again. See Behnke [3] and the references given there.

Going to reproduce is like signing ones own death sentence. If reproduction is restricted to a small time window in the year and life span has a fixed length of, say, k years, a population splits into reproductively isolated year classes according to the year of birth counted modulo k. The population dynamics can be described by a Ricker equation:

$$\begin{aligned} x_{n+1} &= R_0 x_n \exp^{-I_n} \quad \text{in case of annual reproduction.} \\ x_{n+k} &= R_0 x_n \exp^{-I_n} \quad \text{in case of multiennial reproduction.} \end{aligned} \quad (9.17)$$

In the first case we can take $I_n = x_n$, but in the second case there is a multitude of possible (and, depending on species, plausible) choices. The year classes may coexist or they may outcompete each other. Bulmer[6] calls an insect "periodical" if it consists of a single year class, i.e., if all but one year classes are missing. The Cicadas are a famous example.

Competitive success or failure now depends on how the period k of the life cycle compares to the period of the environmental cycle that results by feedback from the population dynamics. In particular *resonance* can have a large impact. Each year class is characterised by its phase in the life cycle relative to the phase of the environmental oscillation. Thus one year class may have a systematic advantage while another has a systematic disadvantage. First of all this creates the possibility of resonance mediated coexistence. But it also forms the basis for a "resident strikes back" phenomenon in which an initially successfully invading year class is in the long run outcompeted since it triggers a phase shift in the resident attractor which turns its advantage into a disadvantage. Moreover, alternating temporary "explosions" of several year classes may form a heteroclinic cycle. See Mylius & Diekmann [46] and the references given there.

9.8 Concluding Sermon

The guiding principle for meaningful mathematical population dynamics should come from modeling considerations: what phenomena do we try to understand and what do we know about potentially underlying mechanisms? On the basis of such considerations one formulates a model and then derives a mathematical representation of it, usually in the form of differential - or integral equations. The next challenge is to analyse the equations and to arrive at conclusions concerning the qualitative and/or quantitative behaviour of their solutions which, hopefully, allow a biological interpretation that yields insight (not infrequently, in fact, by falsifying a hypothesis one had; we quote from J. Maynard Smith, Mathematical Ideas in Biology, [39], p.92: "It is characteristic of mathematical ideas in biology that they are most suggestive when they are contradicted by experiment", and from D.G. Kendall, [34]: "To know the consequences may be to reject the theory, but this is one way in which truth prevails"; it is a pity that the publication culture stimulates people to value positive results stronger than negative results).

Naturally the applied analyst often has a quite different perspective: aware of his/her toolbox, (s)he "hunts" for equations that can be handled with some, but not too much, difficulty. As long as the results add to our understanding of the complicated relationship between mechanisms and phenomena, this is allright. But it becomes a meaningless activity when the equations themselves are "massaged" without biological motivation (e.g., introducing delay in diffusion equations easily leads to nonsense when the delay concerns moving individuals, as those who are now at the same position have followed quite different paths in the past).

Cooperation between (theoretical) population biologists and ecologists on the one hand and applied mathematicians on the other, is the safest way to avoid such fallacies and, instead, to address the major challenge of unravelling cause and effect in the extremely complex network of interacting populations.

References

1. Anderson, R.M. and R.M. May, *Infectious diseases of humans*. Oxford University Press, **2000**.
2. AUTO continuation package, see URL: http://indy.cs.concordia.ca/auto/main.html
3. Behnke, H., *Periodical cicadas*. J. Math. Biol., Vol. 40, p. 413–431 (**2000**).
4. Boer, M.P., B.W. Kooi and S.A.L.M. Kooijman, *Multiple attractors and boundary crisis in a tri-trophic food chain*. Math. Biosciences, Vol. 169, p. 109–128 (**2001**).
5. Boerlijst, M.C., *Spirals and spots: novel evolutionary phenomena through spatial self-structuring*. In: (eds), The geometry of ecological interaction. Simplifying spatial complexity. Cambridge University Press, **2000**.
6. Bulmer, *Periodical insects*. Am. Nat., Vol. 3?, p. 1099–1117 (**1977**).
7. Butler, G.J., S.B. Hsu and P. Waltman, *Coexistence of competing predators in a chemostat*. J. Math. Biol., Vol. 17, p. 133–151 (**1983**).
8. Butler, G.J. and G.S.K. Wolkowicz, *A mathematical model of the chemostat with a general class of functions describing nutrient uptake*. SIAM J. Appl.Math., Vol. 45, 1, p. 138–151 (**1985**).
9. Butler, G.J. and G.S.K. Wolkowicz, *Predator-mediated competition in the chemostat*. J. Math. Biol, Vol. 24, p. 167–191 (**1986**).
10. Butler, G.J., S.B. Hsu and P. Waltman, *A mathematical model of the chemostat with periodic washout rate*. SIAM J.Appl.Math., Vol. 45, 3, p. 435–449 (**1985**).
11. Calsina, À. and J. Saldaña, *A model of physiologically structured population dynamics with nonlinear individual growth rate*. J. Math. Biol., Vol. 33, p. 335–364 (**1995**).
12. Claessen, D., A.M. deRoos and L. Persson, *Dwarfs and giants: cannibalism and competition in size-structured populations*. Am. Nat., Vol. 155, p. 219–237 (**2000**).
13. CONTENT continuation package for different computing platforms, see URL: http://www.math.uu.nl/people/kuznet/index.html
14. Cushing, J.M., *An introduction to structured population dynamics*. CBMS-NSF Regional Conference Series in Applied Mathematics 71. SIAM, **1998**.
15. DeAngelis, D. and L. Gross, *Individual-based models and approaches in ecology: Populations, communities and ecosystems*. Chapman & Hall, New York. **1992**.
16. Dieckmann, U., R. Law and J.A.J. Metz (eds.), *The geometry of ecological interactions*. Cambridge University Press, (**2000**).
17. Dieckmann, U., J.A.J. Metz, M.W. Sabelis and K. Sigmund, *Adaptive dynamics of infectious diseases: in pursuit of virulence management*. Cambridge Studies in Adaptive Dynamics Cambridge University Press, **2002**.

18. Diekmann, O. F.B. Christiansen and R. Law (eds.) *Special Issue on Evolutionary Dynamics*. J. Math. Biol., Vol. 34, Issues 5/6, (**1996**).
19. Diekmann, O., *The many facets of evolutionary dynamics*. J. Biol. Systems, Vol. 5, p. 325–339 (**1997**).
20. Diekmann, O., M. Gyllenberg, J.A.J. Metz and H.R. Thieme, *On the formulation and analysis of general deterministic structured population models. I. Linear theory*. J. Math. Biol., Vol. 36, 4, p. 349–388 (**1998**).
21. Diekmann, O. and H. Heesterbeek, *Mathematical epidemiology of infectious diseases*. Wiley, **2000**.
22. Diekmann, O., M. Gyllenberg, H. Huang, M. Kirkilionis, J.A.J. Metz and H.R. Thieme, *On the formulation and analysis of general deterministic structured population models. II. Nonlinear theory*. J. Math. Biol., Vol. 43, 2, p. 157–189 (**2001**).
23. Geritz, S.A.H., . Kisdi, G. Meszna and J.A.J. Metz, *Evolutionarily singular strategies and the adaptive growth and branching of the evolutionary tree.*. Evol. Ecol., Vol. 12, 1, p. 35–57 (**1989**).
24. Geritz, S.A.H., J.A.J. Metz, . Kisdia and G. Meszna, *Dynamics of adaptation and evolutionary branching*. Phys. Rev. Letters, Vol. 78, p. 2024–2027 (**1997**).
25. Geritz, S.A.H., M. Gyllenberg, F. J. A. Jacobs and K. Parvinen, *Invasion dynamics and attractor inheritance*. J. Math. Biol., Vol. 44, p. 548–560 (**2002**).
26. Hale, J.K. and A.S. Somolinos, *Competition for fluctuating nutrient*. J. Math. Biol., Vol. 18, p. 255–280 (**1983**).
27. Hanski, I.A. and M.E. Gilpin, *Metapopulation biology. Ecology, genetics, and evolution*. Academic Press, San Diego. **1997**.
28. Hsu, S.B., *A competition model for a seasonally fluctuating nutrient*. J. Math. Biol., Vol. p. (**1980**).
29. Hsu, S.B. and P. Waltman, *On a system of reaction-diffusion equations arising from competition in an unstirred chemostat*. SIAM J. Appl. Math., Vol. 53, p. 1026–1044 (**1993**).
30. Huang, Y. and O. Diekmann, *Predator migration in resonse to prey density: what are the consequences?*. J. Math. Biol., Vol. 43, p. 561–581 (**2001**).
31. Huisman, J. and F.J. Weissing, *Oscillations and chaos generated by competition for interactively essential resources*. Ecological Research, Vol. 17, p. 175–181 (**2002**).
32. Jäger, W., B. Tang and P. Waltman, *Competition in the Gradostat*. J. Math. Biol., Vol. 25, p. 23–42 (**1987**).
33. Jansen, V.A.A. and A.M. de Roos, *The role of space in reducing prey-predator cycles*. In: U. Diekmann, R. Law and M. a. J. Metz (eds), The geometry of ecological interaction. Simplifying spatial complexity. Cambridge University Press, **2000**.
34. Kendall, D.G., *Mathematical models of the spread of infections*. In: Mathematics and computer sciences in biology and medicine. Medical Research Council, London, **1965**.
35. Kirkilionis, M., O. Diekmann, B. Lisser, M. Nool, A. deRoos and B. Sommeijer, *Numerical continuation of equilibria of physiologically structured population models. I. Theory*. Mathematical Models and Methods in Applied Sciences, Vol. 11, 6, p. 1–27 (**2001**).
36. de Koeijer, A. , O. Diekmann and P. Reijnders, *Modelling the spread of Phocine Distemper Virus among harbour seals.*. Bull. Math. Biol., Vol. 60, 3, p. 585–596 (**1998**).

37. Kooijman, S.A.L.M., *Dynamic energy budgets in biological systems. Theory and applications in ecotoxicology.*. Cambridge University Press, Cambridge, **1993**.
38. Lovitt, R.W. and J.W.T. Wimpenny, *The gradostat: a bidirectional compound chemostat and its application in microbiological research.* J. Gen. Microbiol., Vol. 127, p. 261–268 (**1981**).
39. Maynard Smith, J., *Mathematical Ideas in Biology.* Cambridge Univ. Press, **1968**.
40. Metz, J.A.J., O. Diekmann and (Editors), *The dynamics of physiologically structured populations.* Lecture Notes in Biomathematics 68, Springer-Verlag, Berlin and Heidelberg **1986**.
41. Metz, J.A.J., S.A.H. Geritz, G. Meszna, F.J.A. Jacobs and J.S. v. Heerwaarden, *Adaptive dynamics: A geometrical study of the consequences of nearly faithfull reproduction.* In: S. J. v. Strien and S. M. Verduyn-Lunel (eds), Stochastic and spatial structures of dynamical systems. North Holland, Elsevier, **1996**.
42. Metz, J.A.J., D. Mollison and F. van den Bosch, *The dynamics of invasion waves.* In: U. Dieckmann, R. Law and M. A. J. Metz (eds), The geometry of ecological interaction. Simplifying spatial complexity. Cambridge University Press, **2000**.
43. Metz, J.A.J., S. Mylius and O. Diekmann, *When does evolution optimise? On the relation between types of density dependence and evolutionary stable life history parameters.* Report from: IIASA, Nr.: WP-96-04, **1996**.
44. Monod, J., *La technique de culture continue, theorie et application.* Ann.Inst.Pasteur, Vol. 79, p. 390–410 (**1950**).
45. Mylius, S. and O. Diekmann, *On evolutionary stable life histories, optimization and the need to be specific about density dependence.* Oikos, Vol. 74, p. 218–284 (**1995**).
46. Mylius, S.D. and O. Diekmann, *The resident strikes back: invasion induced switching of resident attractor.* J. theor. Biol., Vol. 211, p. 297–311 (**2001**).
47. Neutel, A.-M., J.A.P. Heesterbeek and P.C. de Ruiter, *Stability in real food webs : weak links in long loops.* Science, Vol. 296, p. 1120–1123 (**2002**).
48. Nisbet, R.M. and W.S.C. Gurney, *Modelling fluctuating populations.* John Wiley & Sons, Singapore, **1982**.
49. Nowak, M.A. and R.M. May, *Virus dynamics.* Oxford University Press, **2000**.
50. Pugliese, A., *On the evolutionary coexistence of parasites.* Mathematical Biosciences, Vol. 177/178, p. 355–375 (**2002**)
51. Rand, D. A., H. B. Wilson and J. M. McClade, *Dynamics and evolution: evolutionarily stable attractors, invasion exponents and phenotype dynamics.* Phil. Trans. R. Soc. Lond. B, Vol. 343, p. 261–283 (**1994**).
52. de Roos, A.M. *Numerical methods for structured population models: the escalator boxcar train.* Num. Meth. Part. Diff. Equations, Vol. 4, p. 173-195 (**1988**).
53. de Roos, A.M. and L. Persson, *Physiologically structured models - from versatile technique to ecological theory.* Oikos, Vol. 94, p. 51–71 (**2001**).
54. de Roos, A.M. and L. Persson, *Competition in size-structured populations: mechanisms inducing cohort formation and population cycles.* Theor. Pop. Biol., (**2002, in press**).
55. Segel, L.A., *Modeling dynamic phenomena in molecular and cellular biology.* Cambridge University Press, Cambridge, **1984**.
56. Smith, H. and B. Tang, *Competition in the gradostat: the role of the communication rate.* J.Math.Biol., Vol. 27, p. 139–165 (**1989**).

57. Smith, H. L., *Competitive coexistence in an oscillating chemostat.* SIAM J.Appl.Math., Vol. 40, 3, p. 498–522 (**1981**).
58. Smith, H.L. and P. Waltman, *The theory of the chemostat.* Cambridge Studies in Mathematical Biology 13. Cambridge University Press, Cambridge, **1995**.
59. Thieme, H.R., *Convergence results and a Poincaré-Bendixson trichotomy for asymptotically autonomous differential equations.* J. Math. Biol., Vol. 30, p. 755–763 (**1992**).
60. Tucker, S.L. and S.O. Zimmermann, *A nonlinear model of population dynamics containing an arbitrary number of continuous structure variables.* SIAM J. Appl. Math., Vol. 48, 3, p. 549–591 (**1988**).
61. Tuljapurkar, S. and H. Caswell, *Structured-population models in marine, terrestrial, and freshwater systems.* Population and Community Biology Series 18. Chapman & Hall, New York, **1997**.
62. Yang, K. and H.I. Freedman, *Uniqueness of limit cycles in Gause type models of predator-prey systems.* Math. Biosciences, Vol. 88, p. 67–84 (**1988**).
63. Yodzis, P., *Introduction to theoretical ecology.* Harper & Row, **1989**.

10 Did Something Change? Thresholds in Population Models

Frank Hoppensteadt[1] and Paul Waltman[2]

[1] 606 GWC - SSERC, ASU, Tempe, AZ 85287-7606, USA – *fchoppen@asu.edu*
[2] Department of Mathematics and Computer Science, Emory University Suite 148, 1784 N. Decatur Rd. Atlanta, GA 30322, USA – *waltman@mathcs.emory.edu*

Dedication: *This paper is dedicated to our good friend Professor Willi Jäger on the occasion of his 60th birthday. The three of us have collaborated since a momentous year (for us) at the Courant Institute of Mathematical Sciences in 1969-70.*

Abstract: The goal of this article is to illustrate several interesting bifurcations that can arise in population biology. These are of interest since it is often through bifurcation phenomena that changes significant enough to be measured occur. For example, a minor change in some environmental parameter can cause a system to change from being at rest to oscillating. We illustrate here the role of several canonical types of bifurcations in population modeling.

10.1 Introduction

The subject of mathematical population biology is a vast one, much too large to be surveyed in a single article. Since the thrust of this volume is to present material in a way to make the subject attractive to students, we have chosen to focus on specific examples and to illustrate the phenomenon of change. This forms one of the connections between theories and experiments since experiments often focus on when a moderate change in a parameter produces a major change in the outcome. Such changes in the model's solutions can reflect the occurrence of a bifurcation of solutions and with it an exchange of stabilities. For example, a stable static state can bifurcate into a new stable state or a stable oscillation as model parameters change. We present here several examples of where bifurcation phenomena have been observed in population biology and how they have been used to explain, predict and control outcomes.
A major proponent of this approach is Rene Thom [30] who used canonical forms of singularities to identify bifurcation phenomena in developmental biology. In [14] a systematic development of certain bifurcations that arise in neuroscience is presented, and there are many excellent texts that present bifurcation phenomena for mathematicians and engineers, for example [2, 8,

11, 21, 24, 33] to name a few. We summarize here some of the elementary canonical models of bifurcations into new rest states and new oscillations.

Mathematical biology involves the formulation and analysis of mathematical models, often followed by computer simulation and visualization. Modeling is itself an art where one tries to capture the essence of a biological phenomenon in the simplest possible mathematical terms. The value of this approach is attested to by the surprising applicability of simple mathematical models to describe complex systems. There are two approaches to modeling: One tries to incorporate all possible variables and parameters and formulate a model in ordinary language - these are called ordinary language models. The other tries to reduce the problem to a minimal set of variables and to obtain detailed mathematical results. The former creates a full description of the phenomenon and is precise where data are known; the latter provides a rigorous solution and how it depends on critical (often dimensionless) parameters. Both have their place and both make contributions to understanding biological phenomena. For example, ordinary language modeling is the standard laboratory approach, while major successes, such as population genetics, have been achieved with mathematical modeling. Mathematical models can take various forms: discrete time and space models involving difference equations and continuous time and space models involving differential equations, integro-differential equations, and stochastic processes.

In this article, we select standard models, using the minimal model approach, and we illustrate the mathematics needed to study them. Since details of all of the work presented here appears elsewhere in the literature, we describe the model and some results for it, but do not give proofs in each case. We focus on bifurcations and thresholds that occur in mathematical population biology. The selection of the problems certainly reflects the interests of the authors, but they are representative of more general questions discussed in mathematical biology.

The organization of the article is as follows. We begin with a description of mathematical structures of bifurcations that we use later. Then we describe several well-known mathematical models in epidemics, ecology and population dynamics to illustrate typical threshold problems. For example, we describe a standard continuous time predator-prey model and show both the creation of a stable coexistence steady state (a saddle-node bifurcation) and a stable coexistence limit cycle (a super-critical Andronov-Hopf bifurcation). We also give conditions for a sub-critical Andronov-Hopf bifurcation and consider the problem of coexistence of competing predators (a bifurcation from a limit cycle). Interesting chaotic phenomena, that are closely related to bifurcations, are described in the context of a discrete time model from fisheries that shows cascading period doubling bifurcations leading to chaotic behavior as, in this case, the reproductive rate increases. Thus, apparently random behavior can occur after a bifurcation event. Other interesting problems occur when a system that is near a bifurcation is exposed to perturbations by random noise. These are difficult problems for mathematicians, and we illustrate two aspects of them.

We attempt to provide the reader with a listing of many excellent books on modeling, bifurcation, and mathematical biology where mathematical and biological details can be found. We have not attempted here to give an exhaustive list of all relevant literature, however.

10.2 Mathematical Background on Bifurcations

We summarize some basic canonical models of bifurcation theory in this section.

The Fold: One wishes to determine *real* solutions of the equation

$$x^2 - A = 0$$

for x in terms of the parameter A. Obviously if $A < 0$ there is no such solution, but if $A > 0$, then there are two solutions, $x = \pm\sqrt{A}$. Thus, the equation changed from having no real solutions for $A < 0$, to having one for $A = 0$ (not counting multiplicity), to having two for $A > 0$. In plotting solutions x vs. the parameter A, we obtain a parabola going to the right as shown in Figure 10.1. Because of this, the bifurcation is referred to as being a *fold bifurcation*. It is the simplest one.

There are other ways to view this. For example, this equation describes the equilibria for the dynamical system

$$\dot{x} = -(\partial/\partial x)\Phi(x)$$

where

$$\Phi(x) = (x^3/3 - Ax + B).$$

is like a potential function for the system. (Here and below $\dot{x} = dx/dt$.) There are two equilibria for A positive, and the dynamics are described in Figure 10.1.

This bifurcation has one free parameter (A) and so is referred to as having *codimension one*.

An important example of the fold appears in problems where a saddle and node come together and disappear. For example, consider

$$\dot{x} = A - 1 + \cos x.$$

If $|A - 1| < 1$, there are two equilibria for this equation (modulo 2π), namely the two branches of $x = \arccos(1 - A)$. One equilibrium is stable (a node) and the other (a saddle) is unstable. When A is near zero, we have

$$A - 1 + 1 - x^2/2 + h.o.t. = A - x^2/2.$$

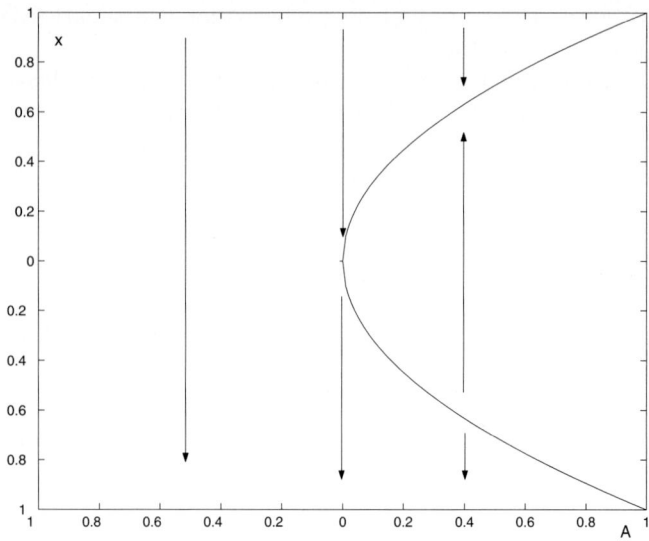

Fig. 10.1. The fold: Solutions of $\dot{x} = -x^2 + A$ for $-1 \leq A \leq 1$. The arrows indicate the behavior of solutions starting at various initial points.

so we apparently have a fold bifurcation at $A = 0$. The two equilibria are approximately $\pm\sqrt{2A}$ if $A > 0$, and they coalesce for $A = 0$. This local analysis indicates that solutions go to infinity for $A < 0$, but in this example, they are captured by the nonlinearity $(\cos x)$, and they can be viewed as describing an oscillation on a circle where x is the angle variable. (This example is referred to as being a saddle-node bifurcation on a limit cycle [14].)

The pitch fork: In this case, we consider

$$x(A - x^2) = 0$$

This is similar to the fold, except that $x = 0$ is always a solution for any choice of A. In this case, $\Phi(x) = x^4/4 - Ax^2/2$, and the dynamics are depicted in Figure 10.2. The name comes from the similarity between the bifurcation diagram in Figure 10.2 to a three-tined pitch fork. In population problems one might be interested only in solutions that are not negative, so such equations might have solutions that are spurious to a particular application.

Andronov-Hopf bifurcation: Consider the differential equation

$$\dot{z} = (A + i\omega - |z|^2)z$$

for a complex function $z(t)$ where ω is a fixed constant. Something interesting happens as A increases through $A = 0$: To see this we convert the problem

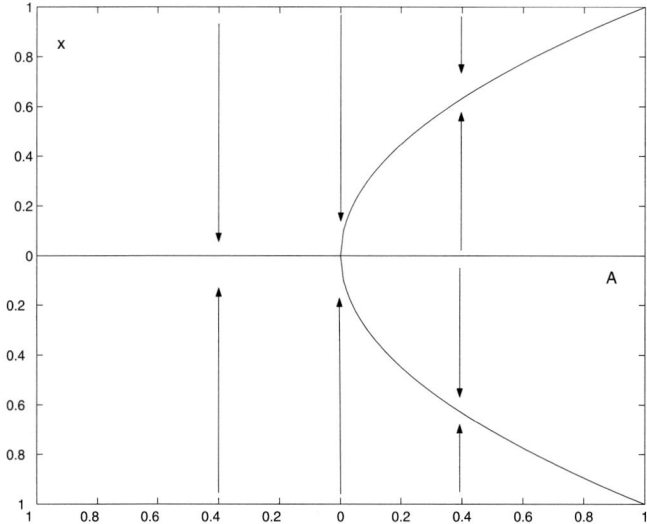

Fig. 10.2. Pitchfork bifurcation $x(A - x^2) = 0$. The arrows indicate the dynamics of solutions starting at various initial points for various values of A. Note that the dynamics are quite different than those in Figure 10.1.

to polar coordinates by setting $z = r \exp i\theta$ where $i = \sqrt{-1}$. A system of two real differential equations for the amplitude (r) and the phase (θ) results; namely,

$$\dot{r} = (A - r^2)r$$
$$\dot{\theta} = \omega.$$

For $A < 0$, all solutions for r approach 0. But, for $A > 0$, all solutions for r with $r(0) \neq 0$ approach $r = \sqrt{A}$ (or $-\sqrt{A}$ which indicates the anti-phase version of this solution.) Therefore, for $A > 0$ the solution for z approaches

$$z(t) \to \sqrt{A} \exp(i\omega t)$$

and so an oscillation has appeared that has frequency ω and amplitude $\sqrt{A}$. This case is referred to as being a *super-critical Andronov-Hopf bifurcation*. Figure 10.3 shows the solutions of this equation for 50 values of A.
The equation
$$\dot{z} = (A + i\omega + |z|^2)z$$
is similar. In polar coordinates it is

$$\dot{r} = (A + r^2)r$$
$$\dot{\theta} = \omega.$$

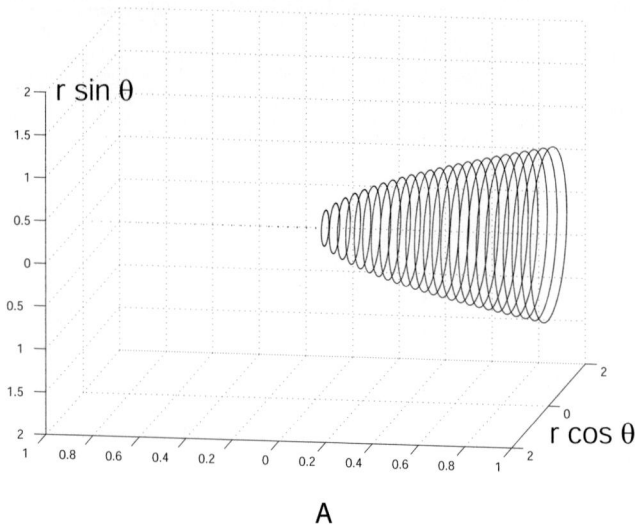

Fig. 10.3. Andronov-Hopf bifurcation for 50 values of λ

Now, for $A < 0$ there are solutions $r = \pm\sqrt{-A}$, but they are unstable while the solution $r = 0$ is stable. For $A > 0$, the only small solution is $r = 0$ which is unstable. This is shown in Figure 10.4.

Thus, we have seen two co-dimension one bifurcations that move a system from a stable rest state to a stable oscillation. In the first, a saddle-node bifurcation resulted in a periodic solution on a circle. In that case, the oscillation has constant amplitude and its frequency increases as the bifurcation parameter A increases. In the second, the super-critical Andronov-Hopf bifurcation moves the system from a stable rest point to a stable oscillation. In that case, the oscillation has amplitude increasing from 0 proportional to $\sqrt{A}$ and the frequency constant at ω. Each of these has co-dimension one (the free parameter is A).

The Cusp: The equation

$$x^3 - ax + b = 0$$

describes equilibria of the dynamical system

$$\dot{x} = -(\partial/\partial x)(x^4/4 - ax^2/2 + b).$$

It has two parameters a, b, and so it has co-dimension 2. The equilibria of this equation can be concisely described in a single three-dimensional plot of solutions x as functions of a and b, which we leave to the reader [11].

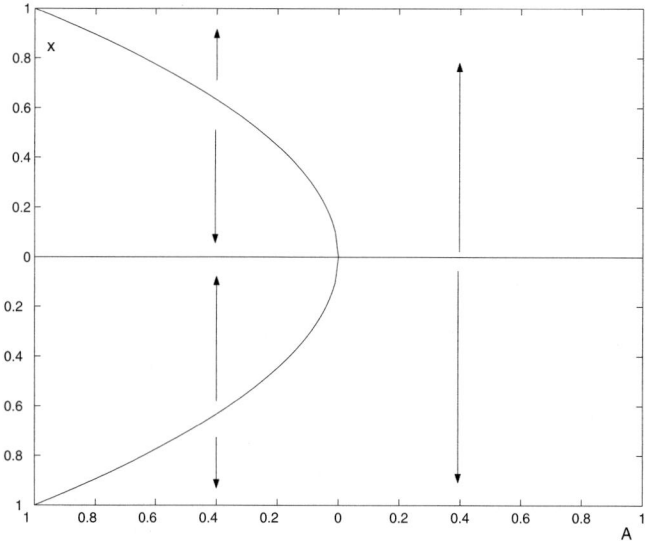

Fig. 10.4. Subcritical Andronov-Hopf bifurcation. For $A < 0$ there is an unstable oscillation but the equilibrium $z = 0$ is asymptotically stable. It loses its stability and the oscillation disappears as A increases through zero.

10.3 Disease Thresholds

10.3.1 Kermack-McKendrick

There are useful analogies between epidemics and chemical reactions. A theory of epidemics was derived by W.O. Kermack, a chemist, and A.G. McKendrick, a physician, who worked at the Royal College of Surgeons in Edinburgh between 1900 and 1930. They introduced and used many novel mathematical ideas in studies of populations. One important result of theirs is that an infection determines a threshold size for the susceptible population, above which an epidemic will propagate. Their theoretical epidemic threshold is observed in practice, and it measures to what extent a real population is vulnerable to the spread of an epidemic. At roughly the same time, V.I. Semenov, a Soviet chemist, derived a theory of combustion that identified explosion limits in terms of pressure and temperature beyond which chemicals begin explosive, chain-branched reactions. The two calculations are quite similar. Epidemic thresholds of population size beyond which there might be an epidemic are derived and discussed in this section. Propagation of infection is modeled to determine what aspects of a population might be controlled to reduce the risk of an epidemic. The calculation also enables one to estimate how severe an epidemic will be.

Suppose that there are functions $S(t)$, $I(t)$, and $R(t)$ and a small sampling interval h that describe the susceptible population (S), the population of infectives (I), and the population of those removed (R) from the process, respectively, at discrete times $t = nh$.

The notation introduced in [17] describes this model to be of the form

$$S \to I \to R$$

which indicates that susceptibles (S) can become infective (I), which in turn can be removed from the disease process (R).

We let the populations be described at discrete sampling times $t_0 = 0, t_1 = h, t_2 = 2h, t_3 = 3h, \ldots$ where the sampling interval h is small. The populations are described at these sampling times by $S(t_0), S(t_1), S(t_2), S(t_3), \ldots$, $I(t_0), I(t_1), I(t_2), I(t_3), \ldots$ and $R(t_0), R(t_1), R(t_2), R(t_3), \ldots$, respectively. The proportion of susceptibles that successfully avoid effective contact with infectives in the time interval $t_n \le t < t_{n+1}$ is taken to be

$$e^{-rhI(t_n)} S(t_n)$$

where r is called the contact rate per unit time. The Kermack-McKendrick model for S, I and R is:

$$\begin{aligned}
S(t_{n+1}) &= e^{-rhI(t_n)} S(t_n) \\
I(t_{n+1}) &= (1 - h\sigma) I(t_n) + (1 - e^{-rhI(t_n)}) S(t_n) \\
R(t_{n+1}) &= R(t_n) + (1 - (1 - h\sigma)) I(t_n),
\end{aligned} \qquad (10.1)$$

where r describes the rate of effective contact between an infective and a susceptible over the time step of size h and σ is the proportion of infectives that are removed from the process per unit time, for example through cure or quarantine [12].

Next, if h is small, we can rewrite this system using $t = hn$ as

$$\begin{aligned}
S(t+h) - S(t) &= (e^{-rhI(t)} - 1) S(t) \sim -rhI(t)S(t) + O(h^2) \\
I(t+h) - I(t) &= -h\sigma I(t) + (1 - e^{-rhI(t)}) S(t) \\
&\sim -h\sigma I(t) + rhI(t)S(t) + O(h^2) \\
R(t+h) - R(t) &= h\sigma I(t).
\end{aligned}$$

Dividing these equations by h, passing to the limit $h = 0$ and assuming that these limits exist gives three differential equations for approximations to $S(t)$, $I(t)$, and $R(t)$. We write these as

$$\begin{aligned}
\dot{S} &= -r I S \\
\dot{I} &= r I S - \sigma I \\
\dot{R} &= \sigma I.
\end{aligned}$$

This system of equations is referred to as being the continuous-time version of Kermack and McKendrick's model (10.2). Obviously, this model for epidemics depends on the assumption that the populations are thoroughly mixed as the process continues.

We can solve these differential equations and so determine the severity of an epidemic. Taking the ratio of the first two equations gives

$$\frac{dS}{dI} = -\frac{rIS}{rIS - \sigma I} = \frac{-rS}{rS - \sigma}.$$

Therefore,

$$dI = \left(\frac{\sigma}{rS} - 1\right) dS.$$

Integrating this equation yields

$$I = (\sigma/r) \ln S - S + C,$$

where C is a constant of integration that is determined by the initial conditions; namely,

$$C = I_0 - \frac{\sigma}{r} \ln S_0 + S_0.$$

Typical trajectories are shown in Figure 10.5, where $S^* = 1, I_o = 0.1$ and $S_o = S^* + k/2, k = 0, \ldots, 8$, respectively.

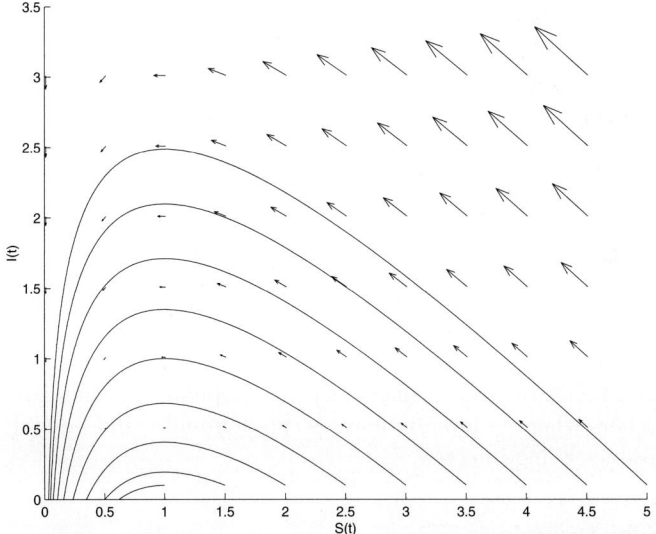

Fig. 10.5. $I(t)$ vs. $S(t)$ in nine cases. $I_o = 0.1$ in each case. The arrows show the lineal diagram for this system (i.e., the local vector field defined by the derivatives.)

The severity of the an epidemic can be predicted as follows: First,

$$I/S_0 = (\sigma/rS_0)\ln(S/S_0) + 1 - S/S_0 + I_0/S_0$$

Since $I_\infty = 0$, we have

$$0 = (\sigma/rS_0)\ln(S_\infty/S_0) + 1 - S_\infty/S_0 + I_0/S_0$$

Setting the final proportion $F = S_\infty/S_0$, the initial dose to be $\varepsilon = I_0/S_0$ and $\lambda = r/(\sigma S_0)$, we have

$$0 = \ln F + \lambda(1 - F + \varepsilon)$$

Expanding this equation about $F = 1$ and $\lambda = 1$, say $x = F - 1$ and $\mu = \lambda - 1$ gives

$$0 = \varepsilon + \varepsilon\mu - x^2 - \mu x + \ldots$$

where ... indicates terms that are of higher order in x, μ and ε. Therefore, we see that there is a fold bifurcation at $F = 1, \lambda = 1$.

The threshold level for the susceptible population is

$$S^* = S_0 \sigma/r.$$

S^* is the value of S for which $dI/dt = 0$ and below which the infectives will not replace themselves. We see in Figure 10.5 that trajectories starting near but above this value describe epidemics that end at a comparable distance below this value. Trajectories that start well above S^* end up near $S = 0$. However, in each case the final size of the susceptible population, $S(\infty)$, is where the trajectory approaches the S axis. Therefore, solving the equation

$$S - (\sigma/r)\ln S = C$$

for its smaller of two roots gives the final size. This is not possible to do in a convenient form; however, it is easy to do using a computer. In this way, we can estimate an epidemic's severity once we have estimated the infectiousness (r) and the removal rate (σ). Note that this theory suggests that some susceptibles will always survive an epidemic, a phenomenon called herd immunity, and this is the case if the susceptible population is not driven to very small numbers where random influences can dominate, and possibly result in all susceptibles being infected.

10.3.2 Schistosomiasis

Schistosomiasis is a disease caused by a wormlike parasite called a helminth. Male and female helminths must mate in a host (e.g., humans, ducks, or swine). Thereafter, some of the fertilized eggs leave the host in its feces. When an egg comes in contact with fresh water, it hatches and attempts to find a snail that it can penetrate. Once a snail is infected, a large number of larvae are produced. They swim freely in search of a host to complete the reproduction cycle. It might penetrate the skin of a host or be ingested with water or food grown in the water.

We sample the populations at regular intervals, and we denote by H_n the mean number of worms infecting each host and by I_n the number of infected snails. The minimal model for this is

$$H_{n+1} = (1-\mu)H_n + c\, I_n$$
$$I_{n+1} = (1-\delta)I_n + b\frac{(S-I_n)H_n^2}{1+H_n},$$

where μ and δ are death probabilities of hosts and snails, respectively. c is the number of helminths becoming established in a host due to each infected snail, S is the number of snails (assumed to be fixed) and the term $bH_n^2/(1+H_n)$ gives the mean number of paired worms per infected host, and so is proportional to the number of eggs produced. The last term describes the interaction between eggs and susceptible snails.

Consider the steady state distributions of helminths and infected snails: Say $H_n = H$ and $I_n = I$. The equations for these are

$$H = (1-\mu)H + c\, I$$
$$I = (1-\delta)I + b\frac{(S-I)H^2}{1+H}.$$

The solutions of this system can be determined easily by eliminating I and solving the resulting cubic for H. In this case, $H \equiv 0$ is one solution and the two remaining solutions can be determined using the quadratic formula. Figure 10.6 shows the results.

The schistosomiasis model illustrates a cusp bifurcation surface [12]. In particular, if $\delta < 0.5$ in that simulation, the disease dies out of the population. At $\delta = 0.5$, there is a fold bifurcation at which a new stable endemic state (top branch) and a new unstable state (middle branch) appear. This diagram suggests possible control measures. For example, if δ can be reduced through treatment of the water to kill snails, the system could be driven into the regime where the disease would die out by effectively reducing δ below 0.5. On the other hand, if the system is in the endemic state, a one time treatment of the host population could drive the helminths below the unstable state, when they would die out of the system. In the latter case, there remains the potential for new helminths introduced from outside to push the system back to the endemic state. In the former case, if $\delta > 0.5$, but the one time treatment brings the effective value of δ below 0.5, then after the treatment, the population would also reside in a vulnerable condition lying under the unstable branch.

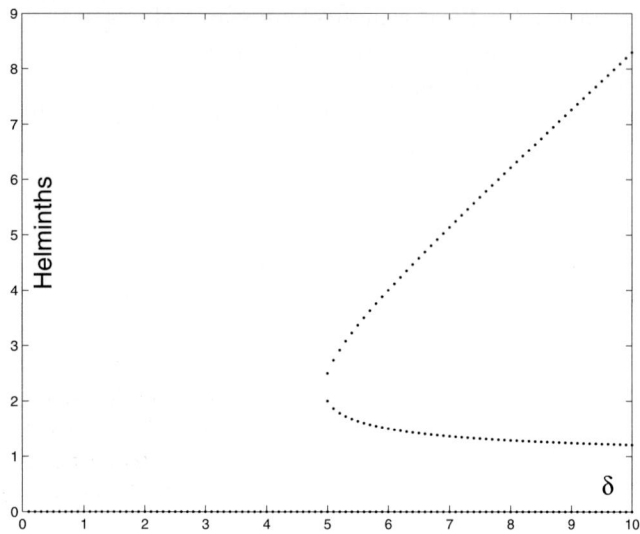

Fig. 10.6. Static states of the schistosomiasis model with $\mu b/c = 1, bS = 0.5$ as a function of the parameter 10δ. For $\delta < 0.5$ there is only the no-disease state ($H = 0$), and it is stable. For $\delta > 0.5$ there are three static states, both $H = 0$ and the top endemic state are stable; the intermediate state is unstable.

10.4 Predator-Prey Systems

Before attempting to describe the interacting populations that go by the general classification of being Predator-Prey systems we begin with a model of simple growth. Let $x(t)$ denote a population evolving with time, t. The basic quantity to be modeled is the intrinsic growth rate (or the per capita growth rate) given by $\dot{x}(t)/x(t)$ where dot here denotes the time derivative. Prescribing this quantity yields a growth model, i.e.,

$$\frac{\dot{x}(t)}{x(t)} = \text{MODEL}.$$

Two standard choices for the model are a constant, m, exponential growth, and $m(1 - \frac{x(t)}{K})$, logistic growth. In general one writes $\dot{x}(t) = x(t)f[x(t)]$.

To describe two interacting populations, x, y, one prescribes the intrinsic growth rates as a function of both variables so the model becomes

$$\dot{x}(t) = x(t)f[x(t), y(t)]$$
$$\dot{y}(t) = y(t)g[x(t), y(t)].$$

Note that $x(t) \equiv 0$ solves the first equation and $y(t) \equiv 0$ solves the second. Consequently, $\dot{y}(t) = y(t)f[0, y(t)]$ and $\dot{x}(t) = x(t)f[x(t), 0]$ describe one population when the other population is missing. Enough smoothness to guarantee unique solutions also allows one to conclude that populations with positive initial conditions, remain positive for all time. The functions f and g describe the nature of the interactions between populations. Models in this form have a long history of investigation, for example, having been studied by Kolmogoroff, [20], Rescigno and Richardson, [26],[27], May, [23], and Albrecht, Gatzke, and Haddad and Wax, [1]. See Chapter 5 of Freedman, [6], for a thorough discussion of such models. The discussion here follows that in [32].

10.4.1 The Basic Model

We first consider equations of the form (dropping the notation (t) for the independent variable)

$$\dot{x} = xg(x) - yp(x) \qquad (10.2)$$
$$\dot{y} = y(-s + cp(x))$$

which we refer to as a Gause system. We assume that the functions $g(x)$ and $p(x)$ are continuously differentiable. $g(x)$ describes the intrinsic growth rate (comprising birth and natural death) of the prey, and $p(x)$ reflects the capture of the prey by the predator. Predator growth depends on prey capture and the model accounts for this as being the only source of growth. s is the intrinsic "death" rate, and states that in the absence of prey, the predator will die out (that is, the predator population tends to zero as $t \to \infty$).

Reasonable assumptions for these general functions are

$g(0) > 0$, but there exists a $K > 0$ such that $g(x) < 0$ for $x > K$.
$p(0) = 0$; $p'(x) > 0$, $p(K) > s/c$.

The system then has three possible equilibria:

$$E_0 = (0, 0),$$
$$E_1 = (K, 0),$$
$$E_c = (x^*, y^*),$$

where (x^*, y^*) satisfy both isocline equations $\dot{x} = 0$ and $\dot{y} = 0$:

$$xg(x) - yp(x) = 0$$
$$-s + cp(x) = 0.$$

From the second equation the x isocline is a vertical line, $x = x^*$, and, if $x^* > 0$, then y^* is given by $y^* = x^* g(x^*)/p(x^*) > 0$. Since we focus on the bifurcation, we restrict the model further to known functions for $p(x)$ and

$g(x)$: We take $g(x) = m(1 - \frac{x}{K})$ (logistic growth) and $p(x) = \beta x/(a+x)$ (a typical Holling type II function, also referred to as Jacob-Monod or Michaelis-Menten functions). The resulting model is

$$\dot{x} = x\left(m(1 - \frac{x}{K}) - \frac{\beta y}{a+x}\right) \quad (10.3)$$
$$\dot{y} = y\left(-s + c\frac{\beta x}{a+x}\right).$$

The units of x, y, a and K are concentrations, m, β, s are rates (reciprocal time), and c is a dimensionless number. We re-scale the problem to non-dimensional variables by defining

$$\bar{x} = \frac{x}{K}, \bar{y} = \frac{y}{Kc}, \bar{a} = \frac{a}{K},$$
$$\bar{s} = s/m, \bar{\beta} = \frac{\beta c}{m}, \tau = tm$$

Making the changes and then dropping the bars results in a non-dimensional system where now $\dot{x}$ denotes the derivative with respect to τ:

$$\dot{x} = x[1 - x - \frac{\beta y}{a+x}] \quad (10.4)$$
$$\dot{y} = y[-s + \frac{\beta x}{a+x}].$$

The Jacobian matrix for (10.4) is

$$J = \begin{bmatrix} 1 - 2x + \frac{\beta a y}{(a+x)^2} & \frac{-\beta x}{a+x} \\ \frac{a \beta y}{(a+x)^2} & -s + \frac{\beta x}{a+x} \end{bmatrix}. \quad (10.5)$$

There are always two rest points, given by

$$E_0 = (0, 0)$$
$$E_1 = (1, 0).$$

At E_0 the Jacobian matrix is

$$J = \begin{bmatrix} 1 & 0 \\ 0 & -s \end{bmatrix}, \quad (10.6)$$

which has one positive and one negative eigenvalue. Thus, the origin is a saddle point with its stable manifold lying on the y-axis and its unstable manifold lying on the x-axis. The origin is not an omega limit point for any trajectory whose initial position is in the first quadrant.
Define, for $\beta > s$,

$$\lambda = \frac{as}{\beta - s}.$$

(If $\beta < s$, then $\dot{y} < 0$, and so $\lim_{t\to\infty} y(t) = 0$. In this case, the prey grows to its carrying capacity.)

At E_1, the Jacobian matrix is

$$J = \begin{bmatrix} -1 & \frac{-\beta}{1+a} \\ 0 & \frac{(\beta-s)(1-\lambda)}{1+a} \end{bmatrix}, \qquad (10.7)$$

whose eigenvalues lie on the main diagonal. If $\lambda > 1$, then E_1 is stable node. If we fix a and β and view λ as being a function of s, we write $\lambda = \lambda(s)$. Then $\lambda(s)$ is a monotone increasing function of s. As s decreases, λ decreases until $\lambda(s_0) = 1$ ($s_0 = \frac{\beta}{1+a}$) beyond which E_1 is a saddle point. For these values an asymptotically stable rest point $E_c = (x^*, y^*) = (\lambda, \frac{(a+\lambda)(1-\lambda)}{\beta})$ exists in the first quadrant.

The Jacobian matrix at E_c is

$$J = \begin{bmatrix} \lambda(-1 + \frac{1-\lambda}{a+\lambda}) & \frac{-\beta\lambda}{a+\lambda} \\ \frac{a(1-\lambda)}{(a+\lambda)} & 0 \end{bmatrix}. \qquad (10.8)$$

The eigenvalues of J can be determined by solving the characteristic equation

$$\mu^2 - tr(J)\mu + \det J = 0$$

where $tr(J)$ is the trace of J (the sum of the main diagonal elements) and $\det J$ is the determinant of J. The roots are

$$\mu = \frac{tr(J)}{2} \pm \sqrt{\left(\frac{tr(J)}{2}\right)^2 - \det J}.$$

The determinant is always positive so the sign of the trace determines the stability. When the trace is zero, the eigenvalues are pure imaginary. After some algebra, one has that $tr(J) = \frac{s}{\beta(\beta-s)}(\beta(1-a) - s(1+a))$. Let $\hat{s} = \frac{\beta(1-a)}{1+a}$, then for $s = \hat{s}$, the trace is zero and so the eigenvalues are purely imaginary. Moreover, the derivative of the trace with respect to s is negative at $\hat{s}$. Thus, an Andronov-Hopf bifurcation occurs and for $s < \hat{s}$, there is a periodic orbit [22]. In [4], it is shown that this periodic orbit is unique. More accurately, it was shown that any periodic orbit is locally asymptotically stable, and hence there can be only one. In [18], it is shown that for $s > \hat{s}$, there are no periodic orbits and hence the interior rest point is globally asymptotically stable. (This follows from the Poincarè-Bendixson Theorem.) Figure 10.7 shows the periodic orbit with two approaching trajectories: One initial condition is outside the periodic orbit and one inside; the dark region contains the periodic orbit.

Mathematica 4.1 was used to solve the equations and to plot the graphs in all of the illustrations for predator-prey equations.

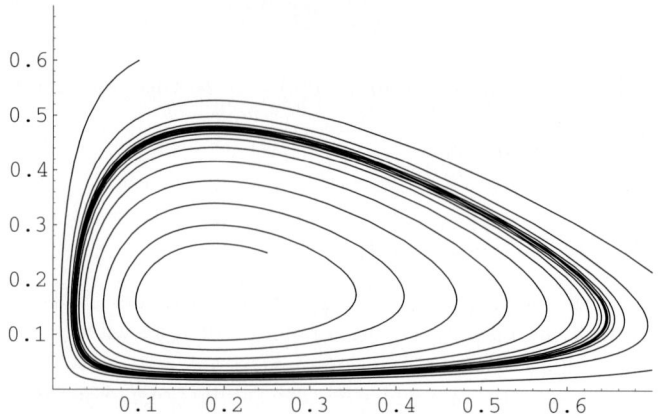

Fig. 10.7. Two orbits approaching a stable limit cycle for system (10.4). One trajectory starts at (0.1,0.6) and the other at (2.5,2.5).

10.4.2 Subcritical Bifurcation

Hofbauer and So [9] have shown that a predator prey system (10.2) can exhibit multiple limit cycles. We follow their presentation. Consider

$$\dot{x} = xg(x) - yp(x) \qquad (10.9)$$
$$\dot{y} = y(-s + cp(x)).$$

where

(i) $(x - K)g(x) < 0$ for $x \geq 0$, for some $K > 0$,

(ii) $p(0) = 0$, $p(K) > \frac{s}{c}$, $p'(x) > 0$ for $x \geq 0$,

(iii) If $h(x) \equiv xg(x)/p(x)$, then $h''(x) < 0$ for $x \geq 0$.

(Here $p' = dp/dx$, etc.)

The system (10.4) satisfies all three of these assumptions since $g(x) = (1 - x)$, $p(x) = \frac{\beta x}{a+x}$ and $c = 1$ in the equations for y. The restrictions guarantee the existence of an interior coexistence rest point $E^* = (x^*, y^*)$, and that this point is unique.

Multiply the vector field defined by the right hand side of system (10.9) by $y^{\alpha-1}/p(x)$ where α is to be determined. The result is the following system that will have the same trajectories:

$$\dot{x} = h(x)y^{\alpha-1} - y^\alpha$$
$$\dot{y} = y^\alpha(\frac{-s}{p(x)} + c).$$

Let $\hat{x}$ be the value where $h(x)$ reaches its maximum. In this case the Jacobian matrix is

$$J = \begin{bmatrix} y^{\alpha-1}h'(x) & (\alpha-1)y^{\alpha-2}[h(x)-y] - y^{\alpha-1} \\ y^\alpha \frac{sp'(x)}{p^2(x)} & \alpha y^{\alpha-1}(\frac{-s}{p(x)} + c) \end{bmatrix}. \tag{10.10}$$

E^* is stable if $h'(x^*) < 0$ (equivalently, if $x^* > \hat{x}$) and unstable if $h'(x^*) > 0$ (equivalently, if $x^* < \hat{x}$). The eigenvalues of J are pure imaginary if $x^* = \hat{x}$. Define $\hat{s} = p(\hat{x})$. At $\hat{s}$, an Andronov-Hopf bifurcation occurs (essentially as discussed above). The following discussion shows that for $s > \hat{s}$, but close to it, there are two limit cycles, one attracting and one repelling, and the bifurcation is sub-critical.

Now, we fix c in (10.10) and again regard s as being a bifurcation parameter. The coordinates of E^* are also functions of s, so we write

$$E^*(s) = (x^*(s), y^*(s)).$$

The divergence of the vector field (10.10) is given by

$$y^{\alpha-1}h'(x) + \alpha y^{\alpha-1}(\frac{-s}{p(x)} + c) = y^{\alpha-1}D(s,x),$$

and one has that

$$D(s, x^*) = h'(x^*).$$

At $s = \hat{s}$, $h'(x^*(\hat{s})) = h'(\hat{x}) = 0$ and the real part of the eigenvalues of (10.10) are zero. ($h'(x^*)$ is a multiple of the trace since $\frac{-s}{p(x^*)} + c = 0$.) Moreover, $\frac{\partial \dot{x}(s)}{\partial s} = h''(\hat{x}) \frac{\partial x^*(s)}{\partial s} < 0$ since $x^*(s)$ is monotone increasing function, so the eigenvalues cross the imaginary axis transversely.
Since

$$\frac{\partial D}{\partial x} = h''(x) + \alpha s \frac{p'(x)}{p^2(x)},$$

choosing α to be

$$\alpha = -\frac{(p^2(\hat{x})h''(\hat{x})}{\hat{s}p'(\hat{x})}$$

gives

$$\frac{\partial D}{\partial x} = 0.$$

Then

$$\frac{\partial^2 D}{\partial x^2}(\hat{s}, \hat{x}) = h'''(\hat{x}) - h''(\hat{x})\left(\frac{p''(\hat{x})}{p'(\hat{x})} - \frac{2p'(\hat{x})}{p(\hat{x})}\right). \tag{10.11}$$

If the right hand side of (10.11) is positive, then it will also be positive for x close to x^*. Then there are no limit cycles close to E^* by Dulac's criterion. One can also argue on grounds of increasing area under the flow that $E(\hat{s})$ is an unstable spiral. Then the limit cycle, which must exist for $s = \hat{s}$ by the Poincarè-Bendixson Theorem, exists for $s > \hat{s}$, i.e., the bifurcation is subcritical.

An example that illustrates assumptions was given in [9]; namely,

$$\dot{x} = x(1+x)(1 - \frac{x}{3})(3 - 4x - 2x^2) - yx(3 - 4x + 2x^2)$$
$$\dot{y} = y(-1.1 + x(3 - 4x + 2x^2))$$

which corresponds to the choices $s = 1.1$, $c = 1$, $g(x) = (1+x)(1 - \frac{x}{3})(3 - 4x + 2x^2)$, and $p(x) = x(3 - 4x + 2x^2)$.

Figure 10.8 shows two limit cycles associated with this subcritical Andronov-Hopf bifurcation. Figure 10.8 was computed by taking points far from E^*

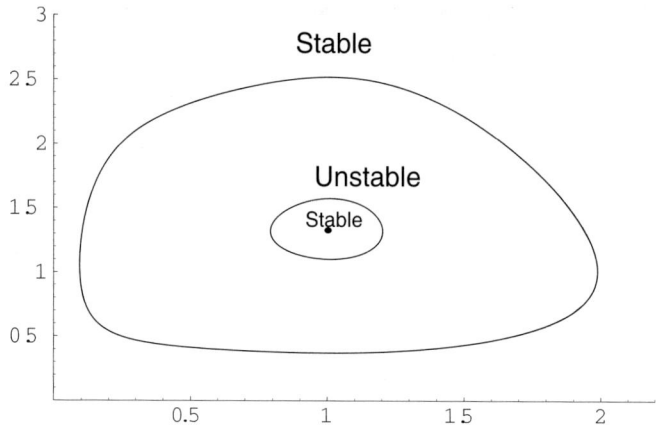

Fig. 10.8. Subcritical Bifurcation. In this case, the outside (larger) orbit is stable and the inside one is unstable. There is a stable equilibrium in the center of the smaller orbit.

and solving forward, taking points close to E^* and solving for negative time, and (having located the cycles) taking a point in the annulus and solving in both directions (which numerically verifies that there are no other periodic orbits between them). In each case, the last 10 time units were plotted for each of the four trajectories. The presence of only two closed curves indicates that the forward and backward orbits were identical.

10.4.3 Bifurcation from a Limit Cycle

Consider the equations investigated in section 10.4.1,

$$\dot{x} = x\left[1 - x - \frac{\beta y}{a+x}\right]$$
$$\dot{y} = y\left[-s + \frac{\beta x}{a+x}\right],$$
(10.12)

where a, β and s are fixed so that there exists a unique periodic orbit (with Floquet multiplier insider the unit circle). Now, suppose that there is a second predator in this system: We relabel the constants and write the new system as

$$\dot{x} = x\left[1 - x - \frac{\beta_1 y}{a_1+x} - \frac{\beta_2 z}{a_2+x}\right]$$
$$\dot{y} = y\left[-s_1 + \frac{\beta_1 x}{a_1+x}\right]$$
$$\dot{z} = z\left[-s_2 + \frac{\beta_2 x}{a_2+x}\right].$$
(10.13)

(10.12) is contained within (10.13) in the plane $z = 0$ – part of the boundary of E^3 where the trajectories of (10.13) lie. For s_2 sufficiently large, $\lim_{t\to\infty} z(t) = 0$, so (10.13) is an asymptotically autonomous system with limiting system (10.12) and the periodic orbit will be the omega limit set of at least some trajectories with initial conditions in the interior of the positive octant.

Is it possible that the limit cycle can bifurcate in such a way that a limit cycle lies interior to the positive octant? If so, there is a change of behavior from competitive exclusion, where no more than one predator can persist, to coexistence of both. We investigate this in this section. The procedure is essentially that considered above to obtain coexistence via a saddle-node bifurcation, but now the bifurcation can be viewed as one for a Poincarè mapping. We omit the details, but the principal consideration for bifurcation from an asymptotically stable limit cycle is that a Floquet exponent changes sign (i.e., a Floquet multiplier passes through the unit circle).

Label the periodic orbit in the $z = 0$ plane as Γ, its period as T, and linearize the system about this trajectory. The Jacobian matrix is

$$J = \begin{bmatrix} 1 - 2x + \frac{\beta_1 a_1 y}{(a_1+x)^2} & \frac{-\beta_1 x}{a_1+x} & -\frac{\beta_2 x}{a_2+x} \\ \frac{a_1 \beta_1 y}{(a_1+x)^2} & -s_1 + \frac{\beta_1 x}{a_1+x} & 0 \\ 0 & 0 & -s_2 + \frac{\beta_2 x}{a_2+x} \end{bmatrix}.$$
(10.14)

The Floquet exponents are $(0, -\mu, \nu)$ where μ is the exponent for (10.12) and

$$\nu = \int_0^T \frac{\beta_2 x(t)}{a_2 + x(t)} dt - s_2 T.$$

Consider β_2 and a_2 as being fixed and treat s_2 as a bifurcation parameter. (The other parameters were fixed above.) The critical value, s^*, is given by

$$s^* = \frac{1}{T} \int_0^T \frac{\beta_2 x(t)}{a_2 + x(t)} dt.$$

The principle result in [3] is: For $s_2 > s^*$ and $s_2 - s^*$ small, there is a periodic orbit of (10.13) interior to the positive octant. Figure 10.9 shows the periodic orbit. Trajectories are computed for several initial conditions in the positive octant and the last 25 times steps plotted. Coexistence of competing predators occurs in the form of a stable oscillation. To illustrate the location of the interior limit cycle relative to the limit cycles on the boundary, we plot all three simultaneously in Figure 10.9.

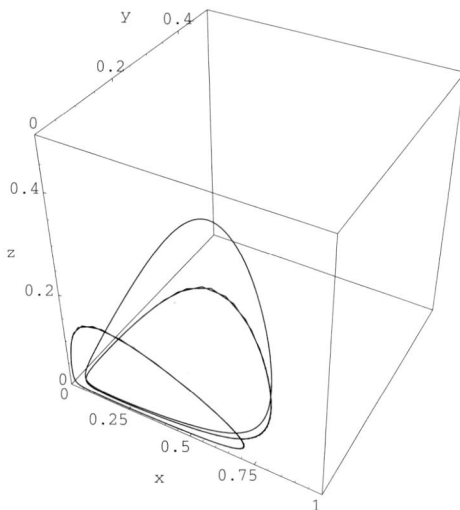

Fig. 10.9. Relative Location of the Limit Cycles. The outside orbits lie in the xz and xy planes, and the orbit extending into the interior of the first octant, which is stable, describes oscillatory coexistence of all three species.

A general reference for Hopf bifurcation is [22]. See also [18], [19], and [29], for work on competing predators.

10.5 Chaos

Some models have no randomness in them, yet their solutions are highly irregular. Such irregularities were known in the nineteenth century, but some new ideas emerged in the 1970's that re-ignited interest in them and where they occur in nature. The term 'chaos' was coined to describe this, and it has been described in the popular press [7]. One of the early descriptions of chaos was given in the context of an ecological system in terms of a discrete time model used in management of fisheries.

10.5.1 Iterating Reproduction Curves

An interesting way to study a reproduction function is to iterate it using a computer [16]. For example, suppose that the discrete time model is

$$B_{n+1} = F(B_n).$$

The population two generations ahead is given by

$$B_{n+2} = F(B_{n+1}) = F(F(B_n)) = F^{[2]}(B_n),$$

the population three generations ahead is given by

$$B_{n+3} = F(B_{n+2}) = F(F(B_{n+1})) = F(F(F(B_n))) = F^{[3]}(B_n),$$

and the population in generation 10 is described by the iterated reproduction function

$$F^{[10]}(B) = F(F(\cdots F(B))),$$

where there are $10 F$'s, etc.

Ricker's model, was used to study fish populations in the 1950's [25]. It happens that reproduction can actually decrease as population sizes become large and crowding occurs. For example, adult guppies will feed on their young, so the offspring of an extremely large adult population will face little chance of survival. Although populations of intermediate size may more than reproduce themselves, large ones may not. A reproduction function that describes this is

$$F(B) = r \exp(-B/K) B,$$

where r gives the population's geometric growth rate when the population is small (viz., $\exp(-B/K) \approx 1$ when $B/K \approx 0$). When $B = K$, the reproduction rate is $r \exp(-1) \approx r/2$. Some iterates of this function are shown in Figure 10.10.

Ricker's computer simulations of the model

$$B_{n+1} = r \exp(-B_n/K) B_n. \tag{10.15}$$

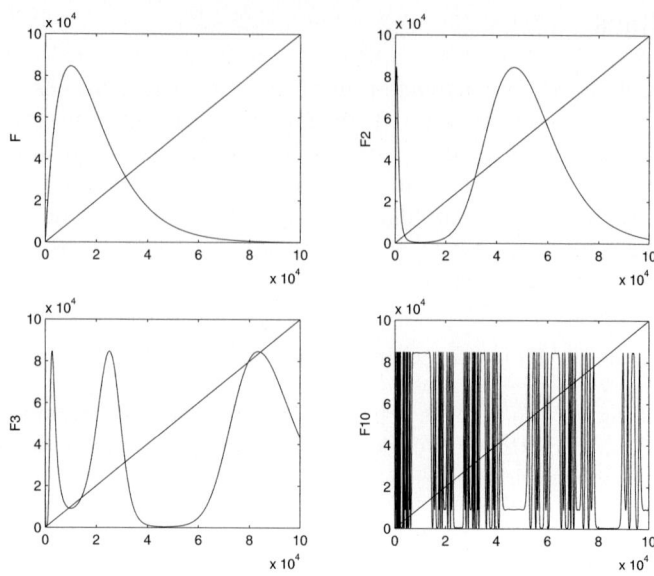

Fig. 10.10. Iterates of Ricker's reproduction function: *Upper left, F; Upper right* $F^{[2]}$; *Lower left* $F^{[3]}$; *Lower right* $F^{[10]}$. In each figure, the one-to-one line is drawn for reference.

showed that for some values of r and K the iterates are highly irregular. If $r < 1$, then the population will less than replace itself, and so will die out. However, if $r > 1$, a small population will more than replace itself, and the equilibrium $B^* = K \log r$ (where $F(B^*) = B^*$) is a likely eventual outcome for the population. This might or might not be the case depending on the value of r. If r is near 1, then the iterates will approach B^*. However, for larger values of r, the solutions can do other things. Table 10.1 illustrates four cases.

These sample paths are also drawn in Figure 10.11. When $r = 0.4$, the population dies out; when $r = 2.0$, it approaches 3466. When $r = 10.0$, the population numbers settle to a simple oscillation between two numbers 4673 and 18353. (This is referred to as being an orbit of period two.) However, when $r = 20.0$, the population numbers seem to be chaotic. A histogram method described in [13] provides a way for capturing this strange behavior. For example, for each value of r in Table 10.1, we can describe where the iterates lie by breaking the interval $0 \leq P \leq 40,000$ into $1,000$ equal cells and putting a marker in a cell each time an iterate hits that cell. The results in this case are shown in Figure 10.12.

Specifically, as in Table 10.1, an initial point is iterated many times (5000 there). Each time a cell is hit, a marker is placed in it. At the end of the iteration, a histogram depicting the distribution of markers is presented and it describes where and how often the iterates visited each cell. For example,

Table 10.1. Four sample paths generated by Ricker's model for four different values of r. Population numbers are $\times 10^4$. (Reprinted from [16] with permission.)

n	$B_n(r=0.4)$	$B_n(r=2.0)$	$B_n(r=10.0)$	$B_n(r=20.0)$
0	0.5000	0.5000	0.5000	0.5000
1	0.0736	0.3679	1.8394	3.6788
2	0.0254	0.3525	0.4645	0.0469
3	0.0097	0.3484	1.8345	0.8544
4	0.0038	0.3471	0.4678	3.0944
5	0.0015	0.3467	1.8354	0.1270
6	0.0006	0.3466	0.4672	1.9704
7	0.0002	0.3466	1.8353	0.7658
8	0.0001	0.3466	0.4673	3.3111
9	0.0000	0.3466	1.8353	0.0881
10	0.0000	0.3466	0.4673	1.4774
11	0.0000	0.3466	1.8353	1.5390
12	0.0000	0.3466	0.4673	1.4174
13	0.0000	0.3466	1.8353	1.6649
14	0.0000	0.3466	0.4673	1.1921
15	0.0000	0.3466	1.8353	2.1974
16	0.0000	0.3466	0.4673	0.5424
17	0.0000	0.3466	1.8353	3.6663
18	0.0000	0.3466	0.4673	0.0479
19	0.0000	0.3466	1.8353	0.8712

in Table 10.1 in the case $r = 10.0$, the cells containing 9346 and 36706 would eventually receive all of the hits, so after many iterations the total iterates would be equally be divided between these two, except for the first 10 transient iterates, as shown in the lower left figure. Therefore, we eliminate the first few iterates as being transients and plot the rest in a histogram in Figure 10.12.

A full-scale simulation of this model is described in Figure 10.13. There, for each of 400 values of r between 0 and 20, the same experiment as described above is performed. Fifty initial iterates at each r value are ignored as being transients, and thereafter whenever a cell is hit, it is darkened so dark cells indicate at least one hit. The results described in Figure 10.13 show how the collection of iterates of Ricker's model depend on r. This striking diagram shows that as r increases the behavior of iterates becomes successively more complicated. For example, as r increases through a value near eight, a stable equilibrium splits into an orbit of period two. Near twelve, that splits into an orbit of period four, etc. These splittings are *bifurcations*, and the diagram is the bifurcation diagram describing how things change. It shows to some extent how chaotic behavior emerges as r increases.

Similar behavior can be observed in continuous time models. For example, the Volterra integral equation

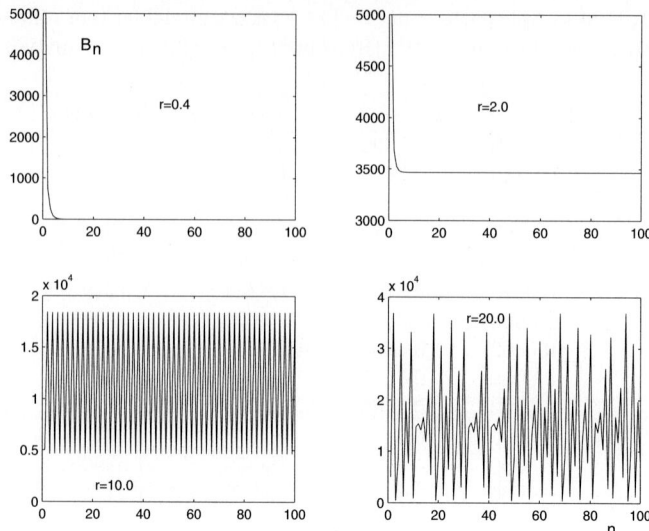

Fig. 10.11. Graphs of sample paths described in Table 10.1. *Upper left* $r = 0.4$; *Upper right* $r = 2.0$; *Lower left* $r = 10.0$; *Lower right* $r = 20.0$.

$$x(t) = \frac{r}{h} \int_{t-1-h}^{t-1} x(s) \exp\left(-x(s)\right) ds$$

can exhibit a cascade bifurcations as r increases that is similar to that of Ricker's model [10, 15]. In fact, as $h \to 0$, this model reduces to Ricker's.

The theory and applications of chaos are still being developed. There are interesting connections between chaotic population dynamics and random processes - in some senses, random processes, like Brownian motion, are more random than chaotic processes.

10.6 Random Perturbations of Ecological Systems

Chaotic systems remain difficult to analyze. However, scientists and engineers have studied highly irregular systems using ideas from probability theory that were introduced and developed by Norbert Wiener and others. The work in this section is described in greater detail in [28]. The general idea is to allow the parameters in one of our non-random models to depend on a random process, and then to determine the impact of this randomness on solutions of the system. To illustrate the method, we first study the model of Lotka and Volterra. First, we describe the non-random case in a way that sets up our random analysis.

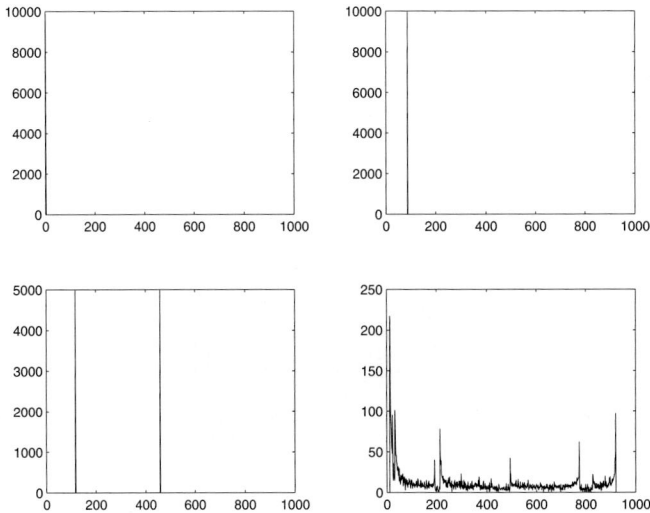

Fig. 10.12. Histograms of iterates described in Table 10.1. *Upper left* $r = 0.4$ (all hits are at $x = 0$); *Upper right* $r = 2.0$ (all hits are at $x = 3466$); *Lower left* $r = 10.0$; *Lower right* $r = 20.0$.

10.6.1 Lotka-Volterra Model with Random Perturbations

As in equation (10.2), we consider two interacting species, say prey and predators. Denote by $x(t)$ the number of prey and by $y(t)$ the number of predators at time t. The Lotka-Volterra model is described by the system of ordinary differential equations:

$$\begin{aligned} \dot{x} &= \alpha x - \beta xy \\ \dot{y} &= -\gamma y + \delta xy \end{aligned} \quad (10.16)$$

where $\alpha, \beta, \gamma, \delta$ are positive constants. Thus, in (10.2) we take $g(x) = \alpha$, $p(x) = x$, $s = \gamma$, and $\delta = c\beta$. Note that the growth rate for the prey, $\dot{x}/x$, is $\alpha - \beta y$, which decreases as the number of predators increases and becomes negative if $y > b = \alpha/\beta$. The growth rate of the predators, $\dot{y}/y$, is $\delta x - \gamma$, which increases as the number of prey increases, but is negative if $x < a = \gamma/\delta$ and is positive for $x > a$. The distribution $x = a, y = b$ is an equilibrium of coexistence for this system, and $x = 0, y = 0$ is the equilibrium corresponding to extinction. (It is unstable and cannot be reached if $x(0) > 0$.)

The first quadrant ($x \geq 0, y \geq 0$) is relevant to the biological system. In it, we see that orbits beginning with $x(0) > 0$ and $y(0) > 0$ revolve around the coexistence equilibrium point. Obviously, if $x(0) = 0$, then $y(t) = y(0)e^{-\gamma t}$, so $y(t) \to 0$ as $t \to \infty$, and in the absence of prey, the predators die out. And, if $y(0) = 0$, then $x(t) = x(0)e^{\alpha t}$ and $x(t) \to \infty$ as $t \to \infty$, and in the absence of predation the prey grow without bound.

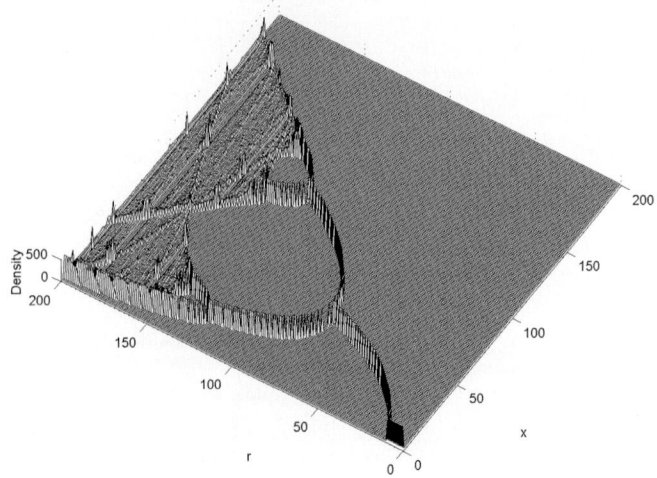

Fig. 10.13. Bifurcation diagram for Ricker's model. The scale on the r axis corresponds to $0 \leq r \leq 20$ (note that r moves from right to left here) and on the x-axis to $0 \leq x \leq 8$. This simulation performed 10000 iterations of Ricker's model for each of 200 r-values, and iterates were placed among 200 cells on the x-axis. For example, at $r = 5.0$ (50 on this plot), this plot shows that there is one cell occupied (at about $x = 2.0$ or 50 on this plot), and so there is a stable static state. Successive bifurcations can be traced as r increases. This plot was generated in approximately 30 secs. on a workstation. It took several hours when first done in 1975 on a CDC6600.

The behavior of system (10.16) is described by the following theorem which we state without proof.

Denote the coexistence equilibrium by $x = a, y = b$. Then
(i) System (10.16) has a first integral
$$\Phi(x,y) = x^\gamma y^\alpha \exp\{-\delta x - \beta y\}.$$

(ii) For $c \in (0, \Phi(a,b)]$ the orbits of the system with
$$\Phi(x,y) = c,$$
are closed curves.

(iii) If $\Phi(x_0, y_0) \in (0, \Phi(a,b))$, then the solution of system (10.16) is a periodic function determined in the following way:

Denote by $y_1(c) < y_2(c)$ the solutions of the equation
$$\Phi(a, y_k) = c, c \in (0, \Phi(a,b)) \quad \text{for} \quad k = 1, 2$$

and by $\phi_1(c,y) < \phi_2(c,y)$, the solutions for $k = 1, 2$, of the equations
$$\Phi(\phi_k(c,y), y) = c, y \in (y_1(c), y_2(c)).$$

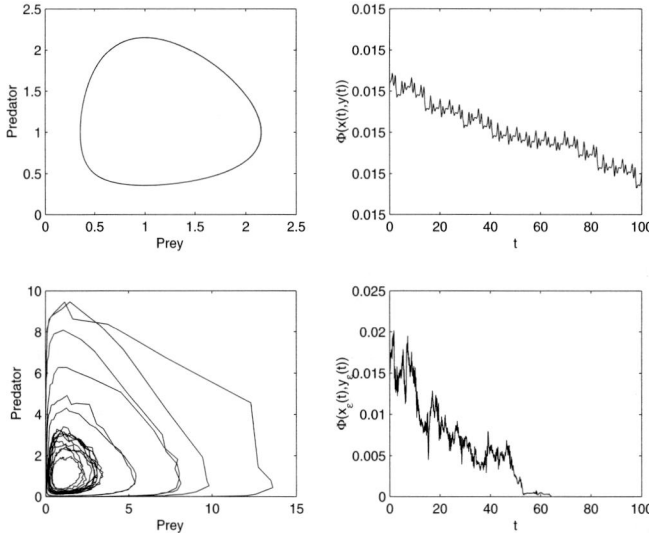

Fig. 10.14. Simulation of the Lotka-Volterra system with noise for $0 \le t \le 100$. Upper left: The solution of the averaged system. Upper right: $\Phi(x,y)$ evaluated on this trajectory. This function is essentially constant at 0.15. Lower right: Φ evaluated along a solution of the random system. Φ appears to be wandering toward $\Phi = 0$ like a diffusion process. Lower left: A sample path of the system with random perturbations. Noise drives this system into a situation where Φ is small. The predator populations for this simulation are shown in Figure 10.15.

$$\alpha(z_1) = \alpha_0 + z_1(t)(\alpha_1 - \alpha_0).$$

The average of $\alpha(z_1)$ with respect to ρ is

$$\alpha = \alpha_0 + (\rho_1 * 0 + \rho_2 * 1)(\alpha_1 - \alpha_0) = \alpha_0 + 0.5385(\alpha_1 - \alpha_0).$$

We define the other data $\beta(z_2), \gamma(z_3), \delta(z_4)$ similarly, where each z_j, for $j = 1, 2, 3, 4$, is an independent sample path of the process z described above. The averages of these functions are given by the formulas

$$\beta = \beta_0 + (\rho_1 * 0 + \rho_2 * 1)(\beta_1 - \beta_0) = \beta_0 + 0.5385(\beta_1 - \beta_0),$$

$$\gamma = \gamma_0 + (\rho_1 * 0 + \rho_2 * 1)(\gamma_1 - \gamma_0) = \gamma_0 + 0.5385(\gamma_1 - \gamma_0),$$

and

$$\delta = \delta_0 + (\rho_1 * 0 + \rho_2 * 1)(\delta_1 - \delta_0) = \delta_0 + 0.5385(\delta_1 - \delta_0).$$

Figure 10.14 shows the solution of the averaged system and one solution (a sample path) of the perturbed system. This indicates to what extent the system can be described by the first integral Φ. In this simulation, the stopping

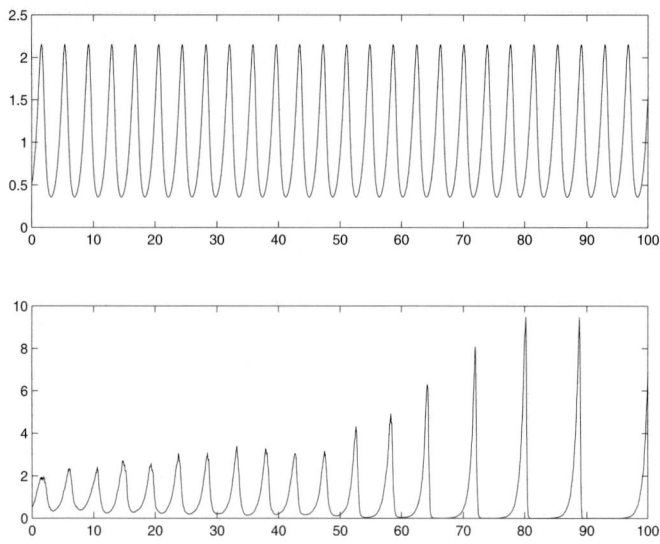

Fig. 10.15. Simulation of the Lotka-Volterra system as in Figure 10.14. Upper: The predator population of the averaged system. Lower: A sample path of the predator population with random perturbations. Noise drives this system into a situation where Φ is small and the result is a synchronized pattern of predation, resembling a sub-harmonic oscillation.

times are to have equal spacing. This shows that for the averaged system, Φ is essentially constant, but in this simulation, it grows indicating that the noise is moving the system toward orbits of higher and higher Φ values. In other simulations of the same system, Φ evaluated along the trajectory of the perturbed system moves toward zero, which indicates that the system is moving toward orbits that have successively lower Φ values, that is, the solutions spend most of their time near the coordinate axes. the result in that case is a synchronized appearance of the population where one species appears to be alone, except for fairly rapid transients in which the population distribution moves toward the other axis.

The connections between chaotic systems and random systems as just described is still not clear. Some insight can be gained by thinking of chaotic evolutions as being irregular, but not as irregular as random processes. The relevance of this example to the topics of this paper is that systems with noise can exhibit what appear to be oscillations that eventually deviate substantially from the solutions of the averaged equation. If the system has a first integral, then it provides a useful way to describe the solutions of the perturbed system.

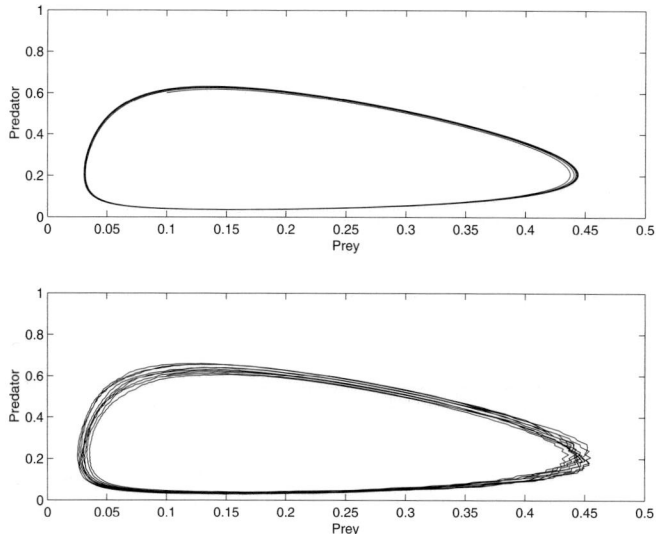

Fig. 10.16. This is a computer simulation of system (10.19). (*Top*) shows the orbit of the averaged system beginning at $x_0 = 0.2, y_0 = 0.1$, and (*Bottom*) a sample path of the orbit to the system with random coefficients as described in the text. Note that the randomly perturbed system seems also to execute a regular oscillation, but of larger amplitude and different phase.

10.6.2 The Basic Model with Random Perturbations

A second illustration of the influence of randomness is for the stable oscillation in equation (10.3). Let us consider the system

$$\dot{x} = x\left[1 - x - \frac{\beta y}{a + x}\right]$$
$$\dot{y} = y\left[\frac{\beta x}{a + x} - s\right] \qquad (10.19)$$

Figure 10.17 shows the differences between the predators in the averaged and randomly perturbed models 10.19 by comparing the amplitudes and phases as calculated from a point lying interior to both orbits in Figure 10.16. When $\beta = 2.8, a = 0.4$ and $s = 0.95$, this system has a unique stable oscillation as illustrated in Figure 10.7. Consider equation (10.19) with the noise function described as earlier for the Lotka-Volterra system where

$$a = 0.3 + 0.3 * z_1(t/\varepsilon), \beta = 2.7 + 0.1 * z_2(t/\varepsilon).$$

In this case, $\bar{a} = 0.452$ and $\bar{\beta} = 2.776$. It is shown in [28] that for system (10.19) with these data, the difference between the solution of the averaged

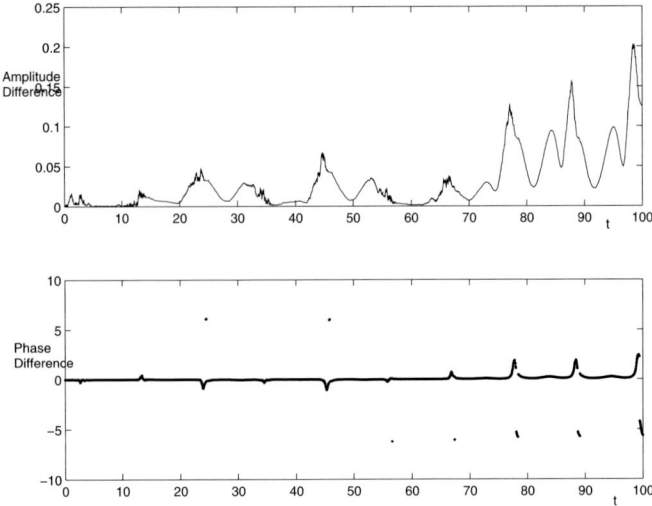

Fig. 10.17. This shows difference between the predator populations of the averaged and random systems (10.19) beginning at $x_0 = 0.2, y_0 = 0.1$. (*Top*) The difference of amplitudes between the two solutions. (*Bottom*) The difference of the phases between the two solutions. The growing deviation here indicates that the two solutions are still oscillating, but that they are diverging.

system and the solution of the randomly perturbed system, both starting at the same point, is approximately (to within order $O(\varepsilon)$) a diffusion process whose amplitude is order $\sqrt{\varepsilon}$. The results of a simulation are shown in Figure 10.16. The two orbits appear to be quite similar, but their amplitudes and phases are different as shown in Figure 10.17.

10.7 Summary

We have demonstrated here several mathematical mechanisms that can occur in population models resulting in interesting and non-trivial changes in behavior. The models are canonical models of the phenomena in the sense that they represent any model that can be mapped continuously into one of these. The models presented here have appeared in numerous separate investigations by us and others: Interesting and useful bifurcations for static states are described by the fold and cusp singularities, and bifurcations of a rest state to an oscillation by the saddle-node on limit cycle and Andronov-Hopf models. We have listed several references to point the reader to further reading on these topics, but our listing was not intended to be exhaustive.
Chaotic dynamics occur in quite simple appearing models, as shown for Ricker's model here. The relation between chaos and random noise is still

being investigated, but in rough terms they are similar with random noise being 'noisier'.

The impact of noise in systems that are near a bifurcation is difficult to analyze. We illustrated one method here using a first integral of the system, which provides a scalar function of the solution whose random aspects we can quantify. Although there were no bifurcation phenomena exhibited by the averaged system in that example, the system with noise exhibited quite different behavior. This suggests that the presence of noise can be responsible for a significant change in behavior. When no scalar function is available for us to monitor, we are left with describing the solutions of the full system using standard perturbation results, like the averaging method used in the second example here. That example illustrated how the averaged system can provide a useful approximation on a large time interval, but eventually the approximation is expected to fail.

Acknowledgements: This work was supported in part by the National Science Foundation under grants (FCH) DMS 98-05544 and (PW) DMS 98-01622.

References

1. F. Albrecht, H. Gatzke, A. Hadad, and N. Wax, The dynamics of two interacting populations, J.Math. Anal. Appl. 46 (1974), 658–670
2. V.I. Arnol'd, ed., Dynamical Systems V. Bifurcation Theory and Catastrophe Theory. Springer-Verlag, New York, 1994.
3. G. J. Butler and P. Waltman, Bifurcation from a limit cycle in a two predator-one prey ecosysstem modeled on a chemostat, J. Math. Bio. 12 (1981), 295–310
4. Cheng, K. -S., Uniqueness of limit cycles for a predator-prey system, SIAM J. Math. Analysis 12 [1981], 541–548
5. W. Feller, An Introduction to Probability Theory and its Applications, vols. I, II, Wiley-Interscience, New York.
6. H. I. Freedman, Deterministic Mathematical Models in Population Ecology, Marcel Dekker, New York, 1980
7. J. Glieck, Chaos: Making of a New Science, Viking, 1987.
8. J. Guckenheimer, P. Holmes, Nonlinear Oscillations, Dynamical Systems, and Bifurcation of Vector Fields, Springer-Verlag, New York, 1983.
9. J. Hofbauer and J.W.-H. So, Multiple Limit Cycles for Pedator-Prey Models, Math. Biosciences 99 (1990), 71–75
10. F. C. Hoppensteadt, A nonlinear renewal equation with periodic and chaotic solutions, Proc. AMS-SIAM Conf. Appl. Math., New York, April, 1976.
11. F.C. Hoppensteadt, Analysis and Simulation of Chaotic Systems, 2nd ed., Springer Verlag, New York, 2000.
12. F.C. Hoppensteadt, Mathematical Methods of Population Biology, Cambridge University Press, Cambridge, 1986.

13. F.C. Hoppensteadt and J.M. Hyman, Periodic solutions to a discrete logistic equation, SIAM Appl. Math. 32(1977)73–81.
14. F.C. Hoppensteadt, E.M. Izhikevich, Weakly Connected Neural Networks, Springer-Verlag, New York, 1997.
15. F.C. Hoppensteadt, Z. Jacekewicz, Numerical solution of nonlinear Volterra equations, in preparation.
16. F.C. Hoppensteadt, C.S. Peskin, Modeling and Simulation in Medicine and the Life Sciences, Springer-Verlag, in press.
17. F.C. Hoppensteadt, P. Waltman, A problem in the theory of epidemics (1970, 1971)Math. Biosci. 9,71-91, 12,133–145.
18. S. B. Hsu, S.P. Hubbell, and P. Waltman, Competing Predators, SIAM J. Appl. Math. 35 (1978), 617–625
19. S. B. Hsu, S.P. Hubbell, and P. Waltman, A contribution to the theory of competing predators, Ecol. Monographs 48 (1978), 337–349
20. A. Kolmorgoroff, Sulla teoria di Volterra della latta per l'esistenza, Gi. Inst. Ital. Attuari 7 (1936), 74–80
21. Yu. Kuznetzov, Elements of Applied Bifurcation Theory, Springer-Verlag, New York, 1995.
22. J. E. Marsden and M. McCracken, The Hopf bifurcation and its applications, Springer-Verlag, New York, 1976
23. R. M. May, Limit cycles in predator-prey communities, Science 177 (1972), 900–902.
24. J. D. Murray, Mathematical Biology, Springer-Verlag, New York, 1989.
25. W.E. Ricker, Stock and recruitment, J. Fish. Res. Bd. Canada, 11 (1954) 559–623.
26. A. Riscigno and I.W. Richardson, On the competitive exclusion principle, Bulletin Math. Biophysics, 27 (1965),85–89
27. A. Riscigno and I. W. Richardson,The struggle for life I:two species, Bulletin Math. Biphysics, 29 (1967), 377–388
28. A.V. Skorokhod, F.C. Hoppensteadt, H.S. Salehi, Random Perturbations of Dynamical Systems, Springer-Verlag, 2002.
29. H. L. Smith, The interaction of steady state and Hopf bifurcation in a two-predator, one prey competition model, SIAM J. Applied Math. (1982)27–43.
30. R. Thom, Structural stability and Morphogenesis: An outline of a general theory of models, W.A. Benjamin, Reading, MA, 1975.
31. P. Waltman, Deterministic Threshold Models in the Theory of Epidemics, Springer-Verlag, New York, 1974.
32. P. Waltman, Competition Models in Population Biology, Soc. Ind. Appl. Math., Philadelphia, 1983.
33. S. Wiggins, Introduction to Applied Nonlinear Dynamical Systems and Chaos, Springer-Verlag, New York, 1990.

11 Multiscale Modeling of Materials – the Role of Analysis

Sergio Conti[1], Antonio DeSimone[1], Georg Dolzmann[2], Stefan Müller[1], and Felix Otto[3]

[1] Max–Planck–Institute for Mathematics in the Sciences, Inselstraße 22–26, 04103 Leipzig, Germany – *conti@mis.mpg.de, desimone@mis.mpg.de, sm@mis.mpg.de*
[2] Mathematics Department, University of Maryland College Park, MD 20742-4015, USA – *dolzmann@math.umd.edu*
[3] Institut für Angewandte Mathematik, Wegelerstraße 10, 53115 Bonn, Germany – *otto@iam.uni-bonn.de*

Abstract: We present two case studies how analysis can be used to derive a hierarchy of models to capture multiscale behavior of materials. The determination, via Γ–convergence, of the thin film limit of micromagnetism delivers a reduced two–dimensional model for soft ferromagnetic films which justifies previously known theories for small fields and extends them to the regime of field penetration. The analytic evaluation of the quasiconvex envelope of the microscopic energy density of nematic elastomers allows efficient numerical computations with finite elements and shows the existence of a new "smectic" phase. In both cases, the numerical solution of the coarse–grained model is complemented by a reconstruction of the microscopic pattern associated with the reduced field.

11.1 Introduction

The behavior of natural and artificial materials is often determined by complex internal patterns spanning several different length scales, from the atomic to the macroscopic one. Proper understanding of the microstructure, and of ways to influence it, has often led to progress in the development of new materials. Classical approaches are based on intuition and experimental evidence, but recently increasing importance has been given to a complementary approach, based on modeling and simulation. Progress in atomistic molecular dynamics and *ab initio* quantum–mechanical simulations deliver precise information on the behavior of individual molecules, or of systems where a relatively small group of atoms is present. However, in many cases the relevant structures arise through collective effects, which involve huge numbers of atoms. A full quantum–mechanical, or even atomistic, computation of such systems is clearly unfeasible, and also not desirable from the point of view of qualitative understanding. It is therefore important to develop and apply methods which allow one to systematically transform the information available on the small scales into an effective model which describes the large scales alone.

In the last decades powerful mathematical tools have been developed to pass from models on microscopic or mesoscopic scales to macroscopic ones, within the general framework of weak–convergence methods. The ideas of Γ–convergence [19, 20, 18], relaxation [15], homogenization [49, 39, 10], quasiconvexification [46, 16, 47], Young–measures and compensated compactness [63, 55], H–measures [56, 33] and their variants provide general techniques which allow one, in principle, to derive rigorously large–scale effective models from the microscopic ones. The spectrum of methods is not yet complete, for example quantum–mechanical effects are still not properly included in the theory, and even the coupling of mechanical and electromagnetic fields has not yet been completely clarified. The development of such methods is a very active and interesting field of research (see, e.g. [57, 1, 44, 31]), which we do not discuss here.

Successful application of the general abstract methods to specific concrete model problems is a much more recent development. In the field of solid–solid phase transitions in crystals, the tools of quasiconvexity deliver sharp conditions on the compatibility of different phases, and on the possibility of finding microstructures which realize a given macroscopic average deformation [4, 13, 47]. Whereas explicit computations have been possible only in very few cases, the method has proven to have a wider field of applicability than phase transitions in crystals, see for example the work on nematic elastomers discussed below. Application of the same ideas to magnetostrictive materials has led to a simple interpretation of experimental data, which in turn furnishes the theoretical understanding needed to devise improved experimental devices [22, 58]. Even solids with simple (convex) elastic properties have a surprisingly complex behavior in nontrivial geometries. The blistering patterns of compressed thin films can be in a first approximation modelled by the scalar eikonal equation [34], and fascinating self–similar folding patterns emerge in a more refined vectorial plate theory [6, 41, 5]. The search for dimensionally reduced theories for the elasticity of thin films has attracted considerable attention for more than a century. It has been recently possible to rigorously derive, via Γ–convergence, the limiting two–dimensional theory describing resistance to bending of an unstretched film [51, 30]. A very interesting problem which is still open is how to rigorously derive a two–dimensional theory which accounts both for bending and for stretching and provides, in a well–defined sense, an optimal approximation of the three–dimensional theory.

This paper presents two case–studies in which a mathematically rigorous result on the transition from one scale to another has led to simplified modeling, easier numerical computation, and improved understanding of the large–scale effective material behavior. Before presenting the specific examples we summarize the general approach.

- The separation of scales allows one to derive analytically a reduced (coarse–grained) problem. This is a mathematical step, based on assuming a microscopic model, and rigorously deriving the limiting macroscopic model when one parameter, representing the ratio between the micro and the macro scales, tends to zero.
- The reduced macroscopic problem can be solved numerically in a much more efficient fashion. This second step, which typically is of computational nature, requires explicit knowledge not only of the general form of the model, but also of the parameters, and allows one to compare quantitatively with experiment. Detailed understanding of the coarse–grained model and of the kind of convergence used in deriving it from the microscopic one indicates also which quantities can be computed robustly.
- From the solution of the coarse–grained problem one can gain some insight into the fine–scale structure by retracing the steps that led to its derivation. While this step is often affected by non–uniqueness of the microscopic pattern corresponding to a given macroscopic state, it can nevertheless give some heuristic understanding of the small scales, as well as valuable insight on corrections to the coarse–grained model coming from the finiteness of the small length scales.

Two specific examples where this strategy has been successfully applied to the derivation of a macroscopic model from a microscopic one are soft magnetic films and nematic elastomers. In both examples, the mathematical approach has not only permitted to justify rigorously existing reduced models, but also led to a better understanding of the material behavior. For soft magnetic films the rigorous analysis led to a reduced model which extends into the regime of field penetration, and for nematic elastomers to the discovery of the "smectic" phase.

11.2 Soft Magnetic Films

Soft ferromagnetic films are of great interest both for applications and as a model physical system. Their sensitive response to applied magnetic fields make them useful for the design of many devices, including sensors and magnetoelectronic memory elements. The presence of significant hysteresis with relatively simple domain structures, together with the availability of a wealth of experimental and numerical data, makes such films an ideal candidate for understanding the microscopic origins of magnetic hysteresis. Such phenomena are correctly described by micromagnetism, a nonlocal, nonconvex variational model which dates back to Landau and Lifshitz [43] and Brown [11], described in Section 11.2.1. Direct numerical simulations of micron–size specimens based on micromagnetics are however beyond the current reach of scientific computing, due to the broad spectrum of length scales which need to be resolved simultaneously (see Section 11.2.1). We focus on the response to weak external fields of soft films, with parameters scaling as specified in Section 11.2.2. Important progress in the conceptual understanding of the equilibrium patterns in such films was achieved through simplified ad–hoc models [7, 12]. These models, however, lack a variational formulation (hence hindering efficient numerical simulations) and their connection with the general theory of micromagnetism has not yet been fully understood. It is therefore unclear whether these models deliver good approximations to minimizers of the full energy, and if so in which range of material and geometric parameters this approximation is valid. The natural mathematical tool to attack these issues is Γ–convergence [19, 20, 18, 10]. In Section 11.2.2 we present a rigorous derivation, via Γ–convergence, of a dimensionally reduced model, and discuss in which sense the minimizers of the micromagnetic energy converge, as the film thickness tends to zero, to the solutions of the limiting problem. The Euler–Lagrange equations coming from the reduced variational problem lead to the models proposed in [7] and [12] for small fields. Numerical results are compared with experimental measurements in Section 11.2.3.

11.2.1 Micromagnetics

Ferromagnetic materials display a complex microstructure of magnetic domains, walls, Bloch lines and singular points ranging from 100 μm down to a few nm. The rich source of experimental data and the simple mathematical formulation make the analysis of magnetic microstructures an excellent model problem to develop new mathematical tools for the understanding of multiscale problems.

Somewhat surprisingly the huge variety of magnetic structures can often be understood through minimization of a simple energy functional, which only involves two material parameters. Let the open bounded set $\omega \subset \mathbb{R}^3$ represent a magnetic body and let $m : \omega \to \mathbb{R}^3$ denote its magnetization, which satisfies the saturation condition $|m| = 1$. For convenience we shall extend the magnetization by zero outside ω, so that $m : \mathbb{R}^3 \to \mathbb{R}^3$ satisfies $|m| = \chi_\omega$, where the characteristic function of ω is defined by $\chi_\omega(x) = 1$ if $x \in \omega$, $\chi_\omega(x) = 0$ if $x \in \mathbb{R}^3 \setminus \omega$. We now define the energy associated to such a magnetization, in suitable non–dimensional units [43, 11] (see also [36, 8]). The micromagnetic energy depends on the material parameters d and Q, which are real constants, and on the anisotropy function $\phi : S^2 \to \mathbb{R}$, which we assume to be smooth.

Definition 1 (Micromagnetics). *Given an open bounded set $\omega \subset \mathbb{R}^3$ (the magnetic body), a function $h_{\text{ext}} \in L^1(\omega, \mathbb{R}^3)$ (the external field), and a vector field $m \in L^\infty(\mathbb{R}^3, \mathbb{R}^3)$ (the magnetization), the three–dimensional micromagnetic energy is*

$$E_{d,Q}^{(3D)}(m, h_{\text{ext}}, \omega) = d^2 \int_\omega |\nabla m|^2 dx + Q \int_\omega \varphi(m)\, dx$$
$$+ \int_{\mathbb{R}^3} |h_{\text{dem}}|^2 dx - 2 \int_\omega h_{\text{ext}} \cdot m\, dx, \qquad (11.1)$$

if $|m| = \chi_\omega$, and $+\infty$ otherwise. The demagnetizing field $h_{\text{dem}} = -\nabla u$ is obtained via Maxwell's equations,

$$\text{div}(-\nabla u + m) = 0 \qquad \text{in } \mathbb{R}^3, \qquad (11.2)$$

where the divergence is understood in the sense of distributions. The four terms in (11.1) are referred to as the exchange, anisotropy, magnetostatic and external field (or Zeeman) energy, respectively.

To the naive mathematical eye the exchange energy $|\nabla m|^2$ is the highest–order term which makes (11.1–11.2) a (nonlocal) lower–order perturbation of the harmonic map problem. While this point of view is useful to understand local properties, such as regularity of minimizers, it does not provide much insight into the complexity of the observed magnetic microstructures. Indeed much of the microstructure formation is driven by the magnetostatic energy, and the exchange energy acts primarily as a limiting factor against infinite refinement. To get a better understanding of the energy functional it is useful to look at the different energy terms separately.

- The anisotropy energy $\varphi(m)$ favors special directions of the magnetization. Most materials have either uniaxial or cubic symmetry. The relative importance of the anisotropy term is measured by the quality parameter Q, which varies over five orders of magnitude between different materials. Materials with a low Q, where the magnetization can rotate easily, are termed *soft* ferromagnets. In the polycrystalline Permalloy thin films discussed below, the fine–scale fluctuations in the crystal orientation result in a very small effective Q. In the following we shall consider only the homogenized material, which is magnetically soft, and not discuss the grain structure further.
- The magnetostatic energy $|h_{\text{dem}}|^2$ tries to eliminate the (distributional) divergence of m (see (11.2)). Written out separately for the interior and the boundary (with outer normal ν) of ω this becomes

$$\text{div}\, m = 0 \quad \text{in } \omega \qquad m \cdot \nu = 0 \quad \text{on } \partial\omega. \qquad (11.3)$$

This is known as the "principle of pole avoidance". It favors a magnetization which is parallel to the boundary and in particular strongly disfavors uniform magnetization of the sample.
- The exchange energy $|\nabla m|^2$ favors uniform or at least slowly–varying magnetizations. It sets the finest length scale and the properties of the walls, which in turn influence the larger length scales.

11.2.2 Thin Film Limit

We now consider a thin film, i.e. take $\omega = \Omega \times (0, t)$ with $t \to 0$ and $|\omega|$ of order 1. Throughout this section upper–case letters denote quantities entering the reduced two–dimensional problem. We first outline the general argument to motivate the definition of the two–dimensional functional, then state our convergence result.

We consider external fields h_{ext} with vanishing out–of–plane component, which scale linearly with the thickness t, and assume that the material parameters are such that $Q \ll t$ and $t^3 \ll d^2 \ll t/\ln(1/t)$. These conditions are well satisfied in typical Permalloy films, where $Q = 2.5 \times 10^{-4}$, $t = 0.01$ and $d = 0.005$ for a typical disk with diameter 1 μm (the length units adopted here are such that $|\Omega|$ is of order 1). We remark that this range includes film thicknesses over which radically different wall types, from symmetric Néel to asymmetric Bloch, are to be expected [36].

Our Γ–convergence result is based on the heuristic observation that, in the limit $t \to 0$, a hierarchical structure emerges in the micromagnetic energy $E^{(3D)}$. We now discuss the scaling of the various contributions, which is summarized in Table 11.2.2. This will motivate our choice of the scaling of material parameters and the identification of the relevant energy scaling, leading to the study of the limit of $t^{-2} E^{(3D)}$.

An out–of–plane component m_3 of the magnetization of order one throughout the sample determines a magnetostatic contribution of order t. Analogously, magnetization changes along the thickness direction x_3 give rise to an exchange energy of order $d^2/t \gg t^2$. Both are of order lower than t^2, hence turn into sharp constraints in the limit. Therefore we can write $m = (M, 0)$, where $M : \Omega \to \mathbb{R}^2$ is the two–dimensional magnetization.

The component of M orthogonal to the lateral boundary $\partial \Omega$ of the film's cross section leads to a magnetostatic contribution of order $t^2 \ln \frac{1}{t}$ associated with "poles" proportional to $M \cdot \nu$, where ν is the outer unit normal to $\partial \Omega$. Precisely the same mechanism penalizes jumps $[M \cdot \nu]$ of the normal component of the magnetization across a line of discontinuity of M with normal ν. These lines of discontinuity arise by approximating domain walls as sharp interfaces, and since their energy is of order lower than t^2, they are also forbidden in the limit. This explains why in the limiting problem only in–plane magnetizations occur with discontinuity lines along which $M \cdot \nu$ does not jump (see Definition 2).

At order t^2 we find the magnetostatic energy due to "charges" proportional to the in–plane divergence Div M, while anisotropy, walls, vortices, Bloch lines and cross–ties contribute only at higher order. Thus, with an external field scaling linearly with the thickness $h_{\text{ext}} = (t H_{\text{ext}}, 0)$, hence contributing the energy $-t^2 \int_\Omega H_{\text{ext}} \cdot M$, we have that, in the limit $t \to 0$, minimization of $E^{(3D)}$ within the restricted class outlined above results in an energetic competition at order t^2 between the aligning effect of H_{ext} and the demagnetizing effects due to Div M. This motivates the following definition of the reduced energy.

Table 11.1. Scaling of the relevant energy terms in (11.1) and their physical origin.

t		
$d^2/t \gg t^2$	m_3	magnetostatic
	$\frac{\partial m}{\partial x_3}$	exchange
$t^2 \ln(\frac{1}{t})$	$[M \cdot \nu]$	magnetostatic
t^2	Div M	magnetostatic
$Qt \ll t^2$	$\varphi(m)$	anisotropy
higher order	walls, Bloch lines	all energies

Definition 2 (Reduced energy). *Given an open bounded set $\Omega \subset \mathbb{R}^2$ (the magnetic film) and a function $H_{\text{ext}} \in L^1(\Omega, \mathbb{R}^3)$ (the reduced external field), the reduced micromagnetic energy is*

$$E^{(2D)}(M, H_{\text{ext}}, \Omega) = \int_{\mathbb{R}^3} |H_{\text{dem}}(x)|^2 dx - 2 \int_\Omega H_{\text{ext}} \cdot M(x) dx \qquad (11.4)$$

if the reduced magnetization $M \in L^\infty(\mathbb{R}^2, \mathbb{R}^2)$ obeys

$$|M| \le \chi_\Omega, \qquad \text{Div } M \in H^{-1/2}(\mathbb{R}^2) \qquad (11.5)$$

and $+\infty$ otherwise. The reduced demagnetizing field H_{dem} is obtained by solving $H_{\text{dem}} = -\nabla U$, $\nabla^2 U = 0$ outside $\Omega \times \{0\}$, and

$$\left[\frac{\partial U}{\partial x_3}\right] = \text{Div } M \qquad \text{on } \Omega \times \{0\} \qquad (11.6)$$

where Div denotes the two–dimensional divergence.

The first condition in (11.5) corresponds to the convexification of the unit–length constraint present in the three–dimensional problem, and the second one to the requirement that the normal component of M does not jump across discontinuity lines and the boundary of Ω.

Guided by the heuristic argument illustrated above, we introduce the following notion of convergence.

Definition 3. *Given a sequence $t^{(n)} \to 0$, we say that the sequence $\{(m^{(n)}, h_{\text{ext}}^{(n)}, \Omega \times (0, t^{(n)}))\}$ of admissible arguments for $E^{(3D)}$ converges to the two–dimensional limit $(M, H_{\text{ext}}, \Omega)$, which is an admissible argument for $E^{(2D)}$, if*

$$\frac{1}{t^{(n)}} \int_0^{t^{(n)}} m^{(n)}(\cdot, x_3) dx_3 \rightharpoonup \binom{M}{0} \qquad \text{in } L^2 \qquad (11.7)$$

and

$$\frac{1}{t^{(n)}} h_{\text{ext}}^{(n)} \rightharpoonup H_{\text{ext}} \qquad \text{in } L^2. \qquad (11.8)$$

We are now ready to state the main result of this Section.

Theorem 1 (Γ–convergence). *Let $\{t^{(n)}\}$, $\{d^{(n)}\}$ and $\{Q^{(n)}\}$ be sequences such that*

$$t^{(n)} \to 0, \qquad \left(d^{(n)}\right)^2 \frac{\ln 1/t^{(n)}}{t^{(n)}} \to 0, \qquad \text{and} \qquad \frac{Q^{(n)}}{t^{(n)}} \to 0. \qquad (11.9)$$

Then the reduced two–dimensional energy $E^{(2D)}$ (Definition 2) is the Γ–limit of the rescaled full micromagnetic energy $t^{-2}E^{(3D)}_{d^{(n)},Q^{(n)}}$ (Definition 1) with respect to the notion of convergence stated in Definition 3.

Proof. As customary for Γ–convergence results, the proof consists of two parts. The first one is a lower semicontinuity result, ensuring that $E^{(2D)}$ evaluated at the limit of a converging sequence is less than or equal to the lower limit of the energy $E^{(3D)}$ evaluated along the sequence. The second part consists of a construction guaranteeing that $E^{(2D)}$ provides a sharp lower bound to the limiting values of $E^{(3D)}$, i.e., that for every admissible argument M of $E^{(2D)}$, there exists a sequence converging to it such that the upper limit of $E^{(3D)}$ computed along the sequence is bounded above by $E^{(2D)}(M)$. The physical intuition which builds upon Table 11.2.2 plays a crucial role in guiding the constructive part of the proof, which is given in [25].

Remark 10. In the thin film limit, the unit length constraint on the magnetization vector is lost. Indeed, the non–convex constraint $|m| = \chi_\omega$ of the full magnetostatic problem is replaced by the weaker, convex constraint

$$|M| \le \chi_\Omega \qquad (11.10)$$

in the reduced problem. This is due to the fact that, at small thicknesses, the energy cost of in–plane, divergence–free fluctuations of the magnetization vector is small (in fact, zero at order t^2).

Remark 11. If the external field H_{ext} is a smooth gradient (this includes e.g. the typical case $H_{\text{ext}} = const$ on Ω), then the functional $E^{(2D)}$ depends on M only via the surface charge $\sigma = -\operatorname{Div} M$, and it is strictly convex in σ. Indeed, $\int_{\mathbb{R}^3} |H_{\text{dem}}|^2\, dx$ is a quadratic functional of σ and an integration by parts shows that $\int_\Omega H_{\text{ext}} \cdot M\, dx$ is a linear functional of σ.

The stationary points M of the functional (11.4) satisfy the Euler–Lagrange equations

$$H_{\text{dem}} + H_{\text{ext}} = \lambda M \quad \text{and} \quad \lambda(1 - |M|) = 0 \qquad (11.11)$$

in ω, where $\lambda(x) \ge 0$ is the Lagrange multiplier associated with the pointwise constraint $|M(x)| \le 1$. At zero external field, minimization of (11.4) gives $H_{\text{dem}} = 0$ (pole avoidance), and hence $\operatorname{Div} M = 0$ (flux closure). This corresponds to the model proposed by van den Berg [7]. In this case, (11.11) implies $\lambda = 0$. Increasing the external field strength, one first encounters a regime in which $H_{\text{dem}} + H_{\text{ext}}$ remains zero. This is the field–expulsion regime, which corresponds to the model proposed by Bryant and Suhl [12]. At higher fields, no such solution is possible, and one obtains $\lambda > 0$, $|M| = 1$ at least in part of the domain. In these regions the magnetization M is uniquely determined by the Euler–Lagrange equation (11.11). The regions of Ω where $\lambda(x) > 0$ are those where the induced field is unable to cancel the external field (i.e., the external field penetrates the sample).

Remarks 10 and 11 indicate that the reduced problem $E^{(2D)}$, or equivalently (11.11), is a convex, quadratic problem, which can be efficiently solved numerically, obtaining a unique solution for Div M for each external field H_{ext}. Since Div M determines H_{dem} uniquely, also the regions of field penetration and the magnetization inside these regions are uniquely determined through (11.11). This would conclude the first part of our program: a simple limiting functional has been derived, which can be easily minimized numerically, and which delivers the robust quantities of the problem, namely, Div M throughout the sample, the region of field penetration, and the magnetization M *inside* this region.

We shall now move to the final part of our program, i.e., to the reconstruction of the non–robust quantities entering the original problem from the solution of the reduced one. In particular, the key quantity of interest if the magnetization M *outside* the region of field penetration.

Remark 12. The set of regular in–plane vector fields of unit length and with given surface charge is large in the following sense: For any regular M_0 which obeys (11.10) there exist many regular M of unit length with the same surface charge: Div M = Div M_0. Indeed, we may write $M = \nabla^\perp \psi + M_0$ where $\nabla^\perp \psi = (-\partial \psi/\partial x_2, \partial \psi/\partial x_1)$ and the continuous function $\psi(x)$ on Ω solves the boundary value problem

$$|\nabla^\perp \psi + M_0| = 1 \text{ in } \Omega, \tag{11.12}$$
$$\psi = 0 \text{ on } \partial\Omega. \tag{11.13}$$

Condition (11.10) ensures the existence of a solution to (11.12–11.13). One can generate many solutions by imposing the additional condition $\psi = 0$ on an arbitrary curve contained in Ω.

A selection criterion among the minimizers of (11.4) based on further minimizing suitably defined wall and Bloch line energies should emerge from an asymptotic development of the micromagnetic energy functional to order higher than two in the film thickness, but this is not attempted here (see Section 11.2.4).

11.2.3 Numerical Results and Comparison with Experiment

The simplified structure of the two–dimensional limiting problem permits an efficient numerical solution. Minimization of $E^{(2D)}$ only delivers the divergence of M, not the full vector field. A direct experimental determination

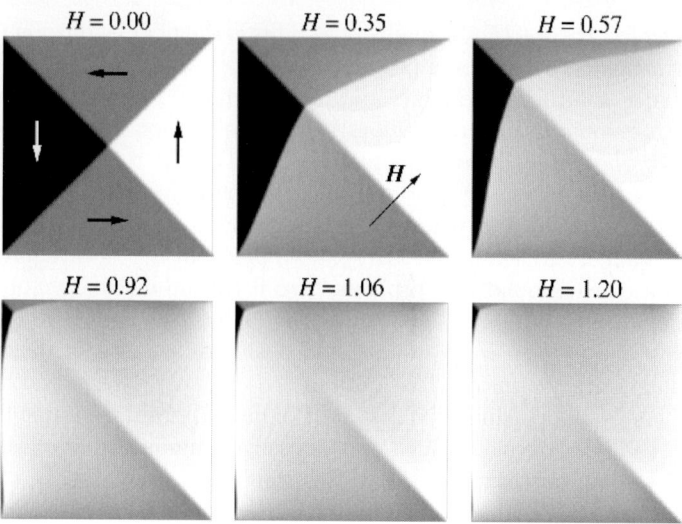

Fig. 11.1. Numerical results: gray–scale plots of the vertical component of magnetization for different values of the external field H_{ext}. The direction of the external field is indicated by the arrow in the second plot.

of the same quantity, for a comparison, is however difficult. Therefore, after having computed the divergence by computing one of the minimizers of $E^{(2D)}$ with length less than or equal to one, in a second step we determine a magnetization field which has unit length and the given divergence, using the heuristic selection criterion discussed below. The result is then compared with experiment.

The first computational step is a convex (though degenerate) variational problem. We solve it using an interior point method (see [27] for details). For the second step, we recall that the solution of (11.12–11.13) is not unique. However there is a unique viscosity solution (see e.g. [28]), which can be computed efficiently using the level–set method [53].

The numerical scheme above selects one of the many minimizers M. The selection principle implicit in this scheme is similar to the one proposed by Bryant and Suhl [12]. It appears to pick a minimizer with as few walls as possible. Thus it is not unlike the more physical selection mechanism of minimizing wall energy, which one can see as a higher–order correction to (11.4). Figure 11.1 shows the predictions of our numerical scheme for a square film of edge–length one, subject to a monotonically increasing uniform field applied along the diagonal. The comparison of our predictions with the response of two Permalloy ($\text{Ni}_{81}\text{Fe}_{19}$, $J_s = 1.0$ T) square samples of edge lengths $L = 30$ and 60 μm and thicknesses $D = 40$ and 230 nm, respectively, as observed in a digitally enhanced Kerr microscope (see Figures 11.2 and 11.3) shows a very good agreement between theory and experiment.

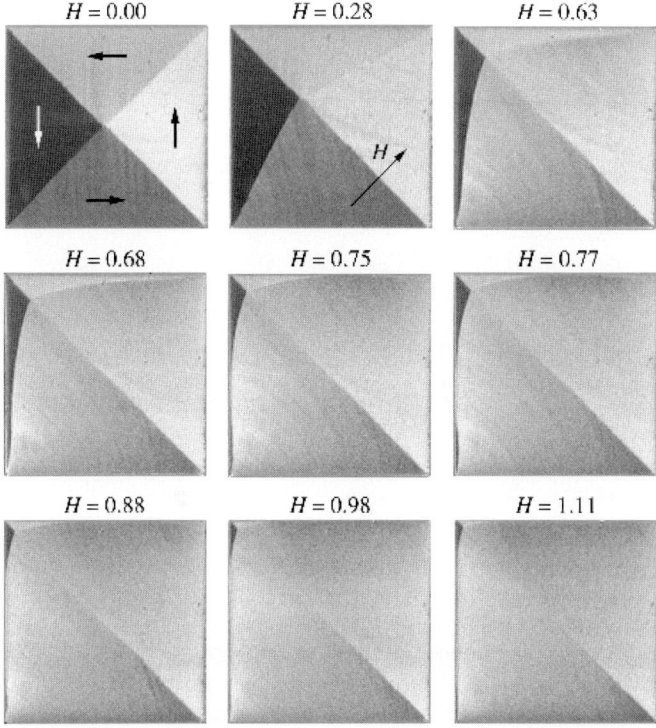

Fig. 11.2. Observed domain patterns on Permalloy square films with side 60 μm, and thickness 230 nm. The gray–scale represents the vertical component of the magnetization. Courtesy of R. Schäfer, IFW Dresden.

11.2.4 Discussion

The reduced model derived in the previous sections is able to capture the experimentally observed behavior in a broad range of material and geometric parameters. It determines the micromagnetic energy to principal order, and the associated robust physical quantities that are expected to have little or no hysteresis — the charge density, the region of field penetration, and the magnetization in the penetrated region.

A natural counterpart to the broad applicability of the model is its degeneracy, which limits the predictive power. The reconstruction procedure outlined above provides a specific magnetization pattern which is consistent with experimental observations, but a rigorous mathematical justification of the selection criteria has not yet been presented. The degeneracy is strictly connected to the disappearance of the exchange energy, and to the loss of the constraint of unit length. The derivation of a model in which this degeneracy is lifted, through inclusion of higher–order corrections to the energy, will most probably involve an analysis of the domain walls. As discussed in [36],

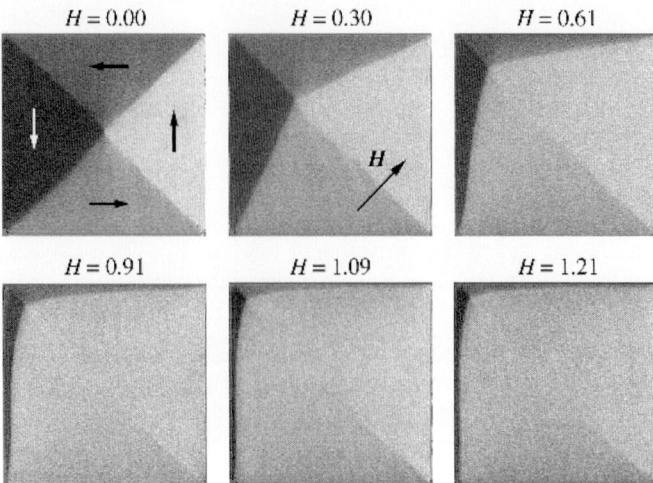

Fig. 11.3. Observed domain patterns on Permalloy square films with side 30 μm and thickness 40 nm. Courtesy of R. Schäfer, IFW Dresden.

different wall types are expected to be present in the range of thicknesses in which our theory applies. Therefore we expect that a reduction in the degeneracy of the theory will be accompanied by a partitioning of its range of applicability into a number of separate components, each requiring a different analytic treatment. Complete results of this nature are not yet available, but progress towards overcoming some of the key obstacles has been reported. On the one hand, the structure of the Néel walls has been elucidated, including the logarithmic tails, at least for the 180° case [24, 45]. Progress on the mathematical treatment of cross–tie walls is reported in [26]. Closely related variational problems concerning energies where the higher–gradient still survives as a singular perturbation have been studied in [3, 40, 52]. The related issue of the conservation of the constraint $|M| = 1$, which mathematically is a compactness issue, has also been successfully addressed in various singularly perturbed variational problems mimicking micromagnetism [2, 23, 38, 37]. A different class of refinements of the theory presented here would aim at including hysteresis and, in general, dynamic effects. The reduction to two dimensions of the Landau–Lifshitz–Gilbert equations of micromagnetic dynamics has been addressed, for example, in [32].

11.3 Nematic Elastomers

The elastic properties of weakly cross–linked nematic chains display in an experimentally accessible setting the fascinating consequences of the non–

quasiconvexity of an elastic energy density in nonlinear elasticity and its application to the mathematical modeling of rubber–like materials.

As first predicted by Golubović and Lubensky [35], isotropic gels prepared by cross–linking liquid polymers with a nearby nematic phase are, at least in some cases, close to a transition to an anisotropic phase, characterized by the coupling of elastic deformations to the alignment of the nematic director (see Figure 11.4). The elastic energy is in this case approximately minimized by all volume–preserving uniaxial deformations of given magnitude, independently of the orientation of the principal stretch directions [9, 59, 61]. These states form the zero set of a suitably defined microscopic energy W. Furthermore, all states with smaller stretches can be obtained by combining zero–energy states with different orientations in different parts of the domain, i.e., by using mixtures of pure states [59, 61].

The microscopic model involves therefore multiple energy–minimizing states, which can be combined on a small scale to obtain many more low–energy macroscopic deformations. In the effective macroscopic model the fine–scale oscillations are averaged out in the kinematics, but correctly accounted for in the energetics. Deducing such a macroscopic model amounts to determining the quasiconvexification of the microscopic energy. The macroscopic model, which is free from oscillations, is used to compute numerically a macroscopic deformation field. A possible microscopic representation of the macroscopic deformation is then recovered by "inverting" the quasiconvexification procedure. The organization of the following sections follows closely the mentioned line of thought. In Section 11.3.1 we present the derivation of the microscopic energy, whose quasiconvexification is then obtained in Section 11.3.2. Finite–element numerical computations are discussed in Section 11.3.3, together with the results on the microscopic deformation obtained with the mentioned inversion procedure. In Section 11.3.4 we discuss the issue of attainment, i.e. of whether structures on an infinitesimal scale in large parts of the domain are necessary in order to minimize the given energy, or if there is a configuration with the same energy where infinite refinement occurs only along the boundary, or not at all. Finally, in Section 11.3.5 we discuss critically our results and the validity of the adopted microscopic model.

11.3.1 Microscopic Model

Nematic elastomers are formed by cross–linking polymeric chains which are subject to nematic ordering in a certain temperature range. Nematic ordering, which is typical of elongated, rod–like molecules, means that the rotational invariance of the isotropic liquid is broken and molecules align their axis in some direction n. In an elastomer, where the mesogens cannot spatially rearrange, the orientation process is coupled to a uniaxial deformation of the polymeric chain along n (see Figure 11.4).

Elastomeric solids are characterized by a small shear stiffness, which derives mainly from entropic effects, whereas the resistance to compression is much

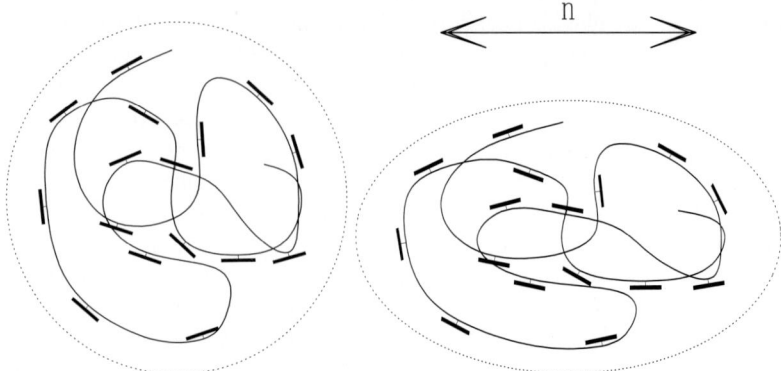

Fig. 11.4. Schematic representation of the deformation mechanism. In the isotropic phase (left panel) the polymer can be imagined as a Gaussian coil with attached side mesogens, whose orientation is random. In the nematic phase (right panel) the mesogens have a tendency to align, and this induces a uniaxial deformation of the coil. The director n, in non–chiral liquid crystals, does not have an orientation and should therefore be identified with $-n$.

larger. We follow common practice and model them as incompressible materials, i.e., from now on we assume that the elastic energy density is infinite for all deformations which are not volume–preserving. The free energy which governs the response to shears can be computed from statistical mechanics. The probability distribution giving the likelihood that an isotropic polymer chain has end–to–end span described by the vector R is, in the Gaussian approximation,

$$P_{\text{isotropic}}(R) = \left(\frac{3}{2\pi L}\right)^{3/2} \exp\left(-\frac{3}{2L}|R|^2\right), \quad (11.14)$$

where L is a measure of the the length of the polymer chains. Assume now that the polymer network is affinely deformed, and let F be the deformation gradient. The corresponding elastic free energy, of entropic nature, is obtained by averaging the logarithm of $P_{\text{isotropic}}(FR)$ with respect to $P_{\text{isotropic}}(R)$, i.e.

$$W_{\text{isotropic}}(F) = -k_B T \langle \ln P_{\text{isotropic}}(FR) \rangle_{P_{\text{isotropic}}(R)} = \frac{1}{2} k_B T |F|^2 + c \quad (11.15)$$

and corresponds to the standard neo–hookean elasticity of rubber. In (11.15) and below, c denotes a constant which does not depend on F, and hence has no effect on the elastic properties and can be safely ignored. Similarly, since we will be working at constant temperature, the Boltzmann factor $k_B T/2$ will be dropped in what follows.

In nematic elastomers, the chains have a preferred orientation, since extension parallel to n is more likely than in other directions. More precisely, one gets

$$P_n(R) = \left(\frac{3}{2\pi L}\right)^{3/2} \exp\left(-\frac{3}{2L} R_i (U_n^{-2})_{ij} R_j\right) \tag{11.16}$$

where

$$U_n = a^{1/6} \left[\mathrm{Id} + (a^{-1/2} - 1) n \otimes n\right] \tag{11.17}$$

corresponds to a uniaxial stretch along n. The coefficient $a \in (0,1)$ measures the effect of local ordering on the elastic properties, and clearly the isotropic case is recovered for $a = 1$ (this corresponds either to zero nematic ordering, or to zero coupling between the nematic ordering and the elastic deformation). Let n_0 denote the value of n at cross–linking, which we assume to be done in a perfectly ordered state, i.e. with n_0 and a constant across the sample (effects of weak disorder in the cross–linking configuration will be discussed in Section 11.3.5). We are interested in the elastic properties of the material after cross–linking. Since the chain distribution in the melt has been frozen, the probability distribution remains P_{n_0}. The free energy at given director n and deformation gradient $\tilde{F}$ is then obtained by the quenched average

$$\begin{aligned}
\tilde{W}_{\mathrm{BTW}}(\tilde{F}, n) &= -k_B T \langle \ln P_n(\tilde{F} R) \rangle_{P_{n_0}(R)} \\
&= \frac{1}{2} k_B T \operatorname{Tr} U_n^{-2} \tilde{F} U_{n_0}^2 \tilde{F}^T + c \\
&= \frac{1}{2} k_B T \operatorname{Tr} U_n^{-2} (\tilde{F} U_{n_0}) (\tilde{F} U_{n_0})^T + c.
\end{aligned} \tag{11.18}$$

The energy (11.18) was first obtained by Bladon, Terentjev and Warner (BTW) in [9]. Using this energy BTW predicted soft elastic response and microstructure (stripe domains) originating from the non–quasiconvexity of W_{BTW}; such predictions have been verified experimentally [42]. However, the experimental results do not show ideally soft response, but rather exhibit both a stress threshold and a small resistance at small stretches, on a scale much smaller than $k_B T$ but still measurable. This observation led to a theoretical search for corrections to (11.18), the main candidates being higher gradient energies (penalizing sharp changes in the director direction n) and disorder in the cross–linking configuration. Various corrections to W_{BTW} have been proposed in the literature [59, 62, 29, 60] and will be discussed in Section 11.3.5 below. The last expression in (11.18) emphasizes the high symmetry of the problem which can be obtained by choosing as reference configuration the stress-free state of the ideal isotropic "high-temperature" phase, which differs from the actual cross–linking configuration by a uniaxial stretch U_{n_0}. From now on we consider the strain with respect to this new reference configuration, $F = \tilde{F} U_{n_0}$. Choosing $k_B T / 2$ as energy units, and adding a suitable constant, one gets

$$W_{\mathrm{BTW}}(F, n) = \operatorname{Tr} U_n^{-2} F F^T - 3 = a^{-1/3} \left[|F|^2 - (1-a)|F^T n|^2\right] - 3 \tag{11.19}$$

for volume–preserving deformations F, and infinity if $\det F \neq 1$.

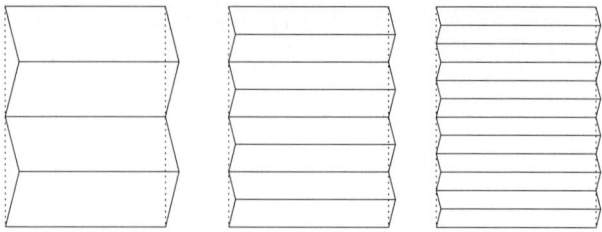

Fig. 11.5. Approximation of an affine deformation by a sequence of oscillating functions (see text).

Minimizing locally over unit vectors n corresponds to replacing $|F^T n|$ with the largest singular value of F. Indeed, due to the full rotational symmetry of this problem, the energy can only depend on the singular values of F, which we denote by $\lambda_1 \leq \lambda_2 \leq \lambda_3$ (their squares are the eigenvalues of $F^T F$ and $F F^T$). One then gets

$$W_{\text{ne}}(F) = \begin{cases} a^{-1/3}\lambda_1^2(F) + a^{-1/3}\lambda_2^2(F) + a^{2/3}\lambda_3^2(F) - 3 & \text{if } \det F = 1 \\ +\infty & \text{else} \end{cases} \quad (11.20)$$

which constitutes the starting point of our analysis.

We first present a simple argument which illustrates, in a special case, the origin of the experimentally–observed stripe domains from an energetic point of view. Consider a macroscopic deformation gradient $F = \text{diag}(a^{1/6}, \mu, a^{-1/6}/\mu)$, with $a^{1/6} < \mu < a^{-1/3}$. By definition, $W_{\text{ne}}(F) > 0$, but one can find a sequence of zero–energy deformations which converge (weakly in $W^{1,2}$) to $u(x) = Fx$. The construction uses a decomposition of the domain in equally spaced layers in which the deformation gradient takes the values

$$F_\pm = F \pm \delta e_2 \otimes e_3 = \begin{pmatrix} a^{1/6} & 0 & 0 \\ 0 & \mu & \delta \\ 0 & 0 & a^{-1/6}/\mu \end{pmatrix} \quad (11.21)$$

where δ is chosen so that $W_{\text{ne}}(F_\pm) = 0$. A short calculation shows that this is possible if and only if $a^{1/6} \leq \mu \leq a^{-1/3}$. We emphasize that it is possible to find a Lipschitz function whose gradient only takes the values F_+ and F_- precisely because $F_+ - F_-$ is a rank–one matrix. The boundary of the region where the gradient equals F_+ has normal e_3 (see Figure 11.5). It can be expected that for some deformation gradients more complicated constructions can lead to smaller energy than the one obtainable by a simple laminate. Indeed, we will prove below that a second iteration of the above construction is always optimal for this energy (Theorem 2), and then characterize the set of matrices F where a single iteration is sufficient (Theorem 3).

11.3.2 Quasiconvexification

In this section we give a mathematical formulation of the foregoing constructions. We start with some definitions. We say that a probability measure on 3×3 matrices $\nu = \mu_1 \delta_{F_1} + \mu_2 \delta_{F_2}$ is a first–order laminate with average F if $\mu_1 F_1 + \mu_2 F_2 = F$ and $\mathrm{rank}(F_1 - F_2) = 1$. Here δ_F denotes a Dirac delta concentrated on the matrix F. Laminates of order k with average F are then defined as the set of probability measures obtained from laminates of order $k-1$ replacing any δ_{F_j} with a first–order laminate with average F_j. We have seen above that laminates offer a natural way to reduce the energy by using complex deformation patters. The optimal energy which can be obtained with laminates is called the lamination convex envelope, and is defined as $\phi^{lc}(F) = \inf \langle \phi, \nu \rangle$, where the infimum is taken over all laminates with average F (we denote by ϕ a generic energy density, and use W for the specific expressions which concern nematic elastomers).

Laminates are very useful but also very special constructions, and it is therefore natural to ask whether other constructions can further reduce the energy. The optimal result is the quasiconvex envelope, defined by

$$\phi^{qc}(F) = \inf_{y \in W^{1,\infty}} \left\{ \frac{1}{|\Omega|} \int_\Omega \phi(\nabla y(x)) dx : y(x) = Fx \text{ on } \partial\Omega, \det \nabla y(x) = 1 \right\}. \quad (11.22)$$

It turns out that for the case of interest here $\phi^{qc} = \phi^{lc}$ and second–order laminates are sufficient. In order to better elucidate the mathematical structure of the problem, we consider a generalization of the above energy,

$$W(F) = \begin{cases} \left(\dfrac{\lambda_1(F)}{\gamma_1}\right)^p + \left(\dfrac{\lambda_2(F)}{\gamma_2}\right)^p + \left(\dfrac{\lambda_3(F)}{\gamma_3}\right)^p - 3 & \text{if } \det F = 1 \\ +\infty & \text{else} \end{cases} \quad (11.23)$$

where $\gamma_1 \leq \gamma_2 \leq \gamma_3$ are given positive constants, with $\gamma_1 \gamma_2 \gamma_3 = 1$ and $p \geq 2$. The energy (11.20) of nematic elastomers is recovered by choosing $p = 2$, $\gamma_1 = \gamma_2$. The energy W is nonnegative, and it vanishes only for matrices F which have singular values $(\gamma_1, \gamma_2, \gamma_3)$, i.e.

$$W(F) = 0 \text{ iff } F \in SO(3)\mathrm{diag}(\gamma_1, \gamma_2, \gamma_3)SO(3). \quad (11.24)$$

The quasiconvexification of W has been obtained by DeSimone and Dolzmann in [21], and later extended by Šilhavý in [54].

Theorem 2 ([21]). *Let W be given by (11.23). Then,*

$$W^{qc}(F) = \begin{cases} \tilde{g}(\lambda_3(F), \lambda_1^{-1}(F)) & \text{if } \det F = 1 \\ +\infty & \text{else} \end{cases} \quad (11.25)$$

where $\tilde{g} : (0, \infty)^2 \to \mathbb{R}$ is the convex nondecreasing function given in (11.32).

Proof. The proof is based on constructing a polyconvex function which lies below W, and therefore gives a lower bound to W^{qc}, and comparing with upper estimates obtained by constructing laminates. Here we outline the main ideas, using a slight variant of the original argument.

We first describe in general our strategy for evaluating the quasiconvex envelope of a function ϕ. The starting point is to obtain good candidates for the polyconvex envelope ϕ^{pc}, which is the largest polyconvex function less than or equal to ϕ. Recall that a function $\phi : \mathbb{R}^{3 \times 3} \to \mathbb{R}$ is polyconvex if there exists a convex function $h : \mathbb{R}^{19} \to \mathbb{R}$ such that $\phi(F) = h(F, \operatorname{cof} F, \det F)$, and that $\phi^{pc} \leq \phi^{lc}$. Given finitely many polyconvex functions $\{z_i\}$, all functions ϕ such that $\phi(F) = \eta(\{z_i(F)\})$, with η convex and nondecreasing in its arguments, are also polyconvex. An appropriate set of variables $\{z_i\}$ may be suggested by the structure of the problem. If one can write $\phi(F) = \eta(\{z_i(F)\})$, the largest convex and nondecreasing function $\tilde{\eta}$ less than or equal to η provides a polyconvex function $\tilde{\phi}(F) = \tilde{\eta}(\{z_i(F)\})$ not larger than $\phi(F)$, which is a lower bound for the quasiconvexification ϕ^{qc}. If one can further show that $\phi^{lc}(F) \leq \tilde{\phi}(F)$, then $\tilde{\phi}$ is the polyconvex and lamination convex envelope of ϕ.

We now apply this general strategy to our specific problem. Since the function

$$\psi(F) = \chi_1(\det F), \qquad \chi_1(x) = \begin{cases} 0 & \text{if } x = 1 \\ +\infty & \text{else} \end{cases} \qquad (11.26)$$

is polyconvex, $W^{qc}(F)$ is infinite unless $\det F = 1$. Therefore we only need to consider matrices with determinant equal to one. Due to its full rotational symmetry, the energy $W(F)$ only depends on the singular values of F, which are the natural variables in which the problem should be cast. These can in turn be expressed in terms of convex functions of F and $\operatorname{cof} F$, hence of polyconvex functions of F. Indeed, the largest singular value is given by

$$s(F) = \lambda_3(F) = \max_{|e|=1} |Fe| \qquad (11.27)$$

and is therefore a convex function of F. The inverse of the smallest can be written as

$$t(F) = \frac{1}{\lambda_1(F)} = \max_{|e|=1} |\operatorname{cof} Fe| \qquad (11.28)$$

and is a convex function of $\operatorname{cof} F$. The intermediate singular value is then recovered by the volume constraint, $\lambda_2 = t/s$. We thus write $W(F) = g(s(F), t(F))$, where $g : (0, \infty)^2 \to \mathbb{R}$ is given by

$$g(s,t) = \left(\frac{1}{t\gamma_1}\right)^p + \left(\frac{t}{s\gamma_2}\right)^p + \left(\frac{s}{\gamma_3}\right)^p - 3. \qquad (11.29)$$

By computing the Hessian matrix one can easily see that g is convex in s and t, however it is not increasing. More precisely,

$$\frac{\partial g}{\partial s} = ps^{p-1}\gamma_3^{-p} - pt^p s^{-p-1}\gamma_2^{-p} \tag{11.30}$$

is nonnegative if and only if $s^2 \geq t\gamma_3/\gamma_2$, and

$$\frac{\partial g}{\partial t} = pt^{p-1}s^{-p}\gamma_2^{-p} - pt^{-p-1}\gamma_1^{-p} \tag{11.31}$$

is nonnegative if and only if $t^2 \geq s\gamma_2/\gamma_1$. The significance of these conditions is best understood with reference to Figure 11.6. The range of s and t (as functions of F, with $\det F = 1$) is given by $0 < 1/t < t/s < s$, and it corresponds to the region between the two thick parabolas. The internal region between the two thin parabolas, denoted by S, is the one where both derivatives are positive. In the region I_1 we have $\partial_s g < 0$, and therefore the largest nondecreasing function below g is given by g evaluated on the common boundary between I_1 and S, i.e. $\tilde{g}(s,t) = g((t\gamma_3/\gamma_2)^{1/2}, t)$. Analogously in the region I_2 we get $\tilde{g}(s,t) = g(s, (s\gamma_2/\gamma_1)^{1/2})$. Finally, since $g(\gamma_3, 1/\gamma_1) = 0$ (and $g \geq 0$ everywhere), in the whole region L we get $\tilde{g} = 0$. We conclude that the largest nondecreasing function below g is given by

$$\tilde{g}(s,t) = \begin{cases} g(s,t) & \text{in } S, \text{ i.e. } s^2 \geq t\gamma_3/\gamma_2 \text{ and } t^2 \geq s\gamma_2/\gamma_1 \\ 2\left(\frac{t}{\gamma_2\gamma_3}\right)^{p/2} + \left(\frac{1}{\gamma_1 t}\right)^p - 3 & \text{in } I_1, \text{ i.e. } t \geq 1/\gamma_1 \text{ and } t \geq s^2\gamma_2/\gamma_3 \\ 2\left(\frac{1}{s\gamma_1\gamma_2}\right)^{p/2} + \left(\frac{s}{\gamma_3}\right)^p - 3 & \text{in } I_2, \text{ i.e. } s \geq \gamma_3 \text{ and } s \geq t^2\gamma_1/\gamma_2 \\ 0 & \text{in } L, \text{ i.e. } s \leq \gamma_3 \text{ and } t \leq 1/\gamma_1. \end{cases} \tag{11.32}$$

One can then easily check that $\tilde{g}$ is still convex (see [21] for details), and therefore $\tilde{\phi} = \tilde{g}(s(F), t(F))$ is polyconvex and it provides a lower bound for the quasiconvex envelope of W.

We now show that this bound is optimal, i.e. that the same energy can be reached by laminates. Again, the special choice of variables s and t makes the computation easy. Indeed, by rotational invariance it suffices to consider only diagonal matrices. Fix $F = \text{diag}(\mu_1, \mu_2, \mu_3)$, and consider $F_\delta = F + \delta e_1 \otimes e_2$. The matrices F_δ have $\det F_\delta = 1$, μ_3 as a singular value, and $|F_\delta|^2 = |F|^2 + \delta^2$. Let (ξ_1, ξ_2, μ_3) be the singular values of F_δ. Since $\xi_1^2 + \xi_2^2 = \mu_1^2 + \mu_2^2 + \delta^2$ and $\xi_1\xi_2\mu_3 = 1$, it is clear that by choosing a suitable δ one can reach a given pair (ξ_1, ξ_2) if and only if $|\xi_1 - \xi_2| \geq |\mu_1 - \mu_2|$, and the determinant constraint $\xi_1\xi_2\mu_3 = 1$ is satisfied. Then F is the average of a first-order laminate supported on $F_{\pm\delta}$, and $\phi^{lc}(F) \leq \phi(F_\delta) = \phi(F_{-\delta})$. Therefore in the (s,t) plane one can construct laminates along lines which keep one singular value constant, i.e. lines parallel to each coordinate axis, and lines which go through the origin as shown in Figure 11.6. The right panel shows the paths used to construct optimal laminates: constant s lines inside I_1, constant t lines inside I_3, and constant s/t lines inside L. Since the line segments used in L have one end point on the common boundary with I_1, I_3 and since

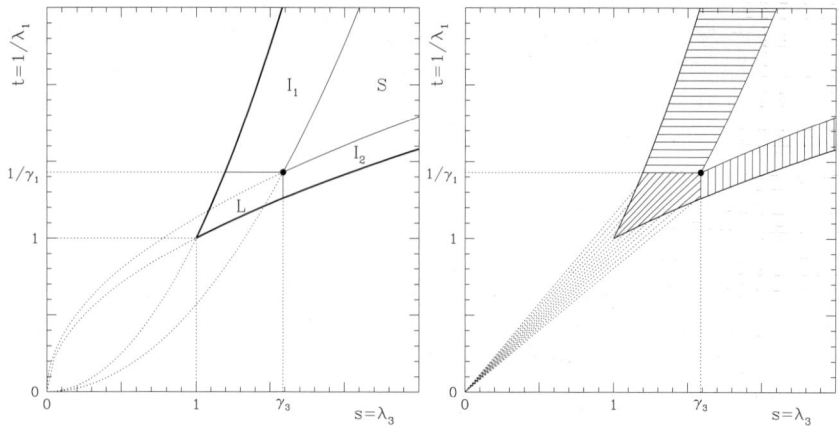

Fig. 11.6. The phase diagram for the relaxed energy. The left panel shows the different phases and phase boundaries (see Eq. (11.32)). The right panel shows the rank–one directions in the phase diagram which correspond to keeping one singular value fixed. First the energy in the regions I_1 and I_2 is constructed, by taking the value on the boundary with S (at fixed t in I_1, at fixed s in I_2). The construction in L is done by continuing from the boundaries of I_1 and I_2 along lines of constant $\lambda_2 = t/s$.

these points already represent laminates, the resulting construction in L is a second–order laminate. This concludes the proof that $W^{lc} = W^{pc}$. The fact that the quasiconvex envelope also coincides with them requires a delicate treatment of the volume constraint, for which we refer to [21].

Remark 13. For $p = 2$ and $\gamma_1 = \gamma_2 = a^{1/6}$, the relevant case for nematic elastomers, phase I_2 is absent, and (11.32) reduces to

$$\tilde{g}(s,t) = \begin{cases} \dfrac{a^{-1/3}}{t^2} + \dfrac{a^{-1/3}t^2}{s^2} + a^{2/3}s^2 - 3 & \text{in S, i.e. } s^2 \geq ta^{1/2} \\ 2a^{1/6}t + \dfrac{a^{-1/3}}{t^2} - 3 & \text{in Sm, i.e. } t \geq a^{1/6} \text{ and } t \geq a^{1/2}s^2 \\ 0 & \text{in L, i.e. } t \leq a^{1/6} \end{cases} \quad (11.33)$$

(see also Figure 11.10 below). We name Sm ("Smectic") the intermediate phase I_1 due to its peculiar mechanical behavior (see [14]).

In the construction above we used first laminates for I_1 and I_3, and second laminates for L. A natural question, which was posed and only partially answered in [21], is to ask whether the second laminates are really needed inside L. The following theorem gives a characterization of the set of gradients where first laminates are sufficient.

Theorem 3. Let F be a matrix in phase L, i.e. such that $\gamma_1 \le \lambda_1(F) \le \lambda_3(F) \le \gamma_3$. There is a first–order laminate with average F supported on the set $\{W(F) = 0\}$ if and only if
$$\lambda_1(F) \le \gamma_2 \le \lambda_3(F). \tag{11.34}$$

Proof. If F can be obtained as a simple laminate, we can write $F = M + a \otimes b$, with $W(M) = 0$. By rotational invariance we can assume M to be diagonal, and since it has zero energy $M = \operatorname{diag}(\gamma_1, \gamma_2, \gamma_3)$. We use an extension of the argument given in [21] for the $\gamma_1 = \gamma_2$ case, and consider a unit vector v orthogonal to e_3 and to b. Then $Fv = Mv$, but since v is a linear combination of e_1 and e_2, $|Mv| \le \gamma_2$, it follows that
$$\lambda_1(F) = \min_{|n|=1} |Fn| \le |Fv| = |Mv| \le \gamma_2. \tag{11.35}$$

Analogously, taking w as a unit vector orthogonal to e_1 and b, we get $\gamma_2 \le \lambda_3(F)$. This concludes the proof of the first implication.

It remains to show that every matrix in phase L obeying (11.34) can be obtained as a first–order laminate supported on zero–energy matrices. Consider $F = \operatorname{diag}(\mu_1, \mu_2, \mu_3)$ with $\gamma_1 \le \mu_1 \le \mu_2 \le \mu_3 \le \gamma_3$. The matrices
$$F_\delta = F + \delta(e_1 \cos\theta + e_3 \sin\theta) \otimes e_2 = \begin{pmatrix} \mu_1 & \delta\cos\theta & 0 \\ 0 & \mu_2 & 0 \\ 0 & \delta\sin\theta & \mu_3 \end{pmatrix} \tag{11.36}$$

all have unit determinant. The matrix F_δ has singular values $(\gamma_1, \gamma_2, \gamma_3)$ if and only if the two conditions $\operatorname{Tr} F_\delta^T F_\delta = \sum_i \gamma_i^2$, $\operatorname{Tr} \operatorname{cof} F_\delta^T \operatorname{cof} F_\delta = \sum_i \gamma_i^{-2}$ are satisfied. Equivalently,
$$\sum_i \mu_i^2 + \delta^2 = \sum_i \gamma_i^2 \tag{11.37}$$

and
$$\sum_i \mu_i^{-2} + \delta^2(\mu_3^2 \cos^2\theta + \mu_1^2 \sin^2\theta) = \sum_i \gamma_i^{-2}. \tag{11.38}$$

The first equation can always be solved for δ. The second one can then be solved for θ provided that
$$\mu_1^2 \le \frac{\sum \gamma_i^{-2} - \sum \mu_i^{-2}}{\sum \gamma_i^2 - \sum \mu_i^2} \le \mu_3^2. \tag{11.39}$$

The first inequality is equivalent to
$$0 \le -\mu_1^2 \sum_i \gamma_i^2 + \sum_i \gamma_i^{-2} + \mu_1^4 - \mu_1^{-2} = \mu_1^{-2} \prod_i (\mu_1^2 - \gamma_i^2), \tag{11.40}$$

which in turn gives $\mu_1 \le \gamma_2$. Analogously, the second inequality gives $\mu_3 \ge \gamma_2$.

Remark 14. For $\gamma_1 = \gamma_2$, the relevant case for nematic elastomers, the condition (11.34) is equivalent to $\lambda_1(F) = \gamma_1$. In turn, this indicates that first laminates are sufficient in phase Sm, up to the boundary with L, whereas second laminates are needed inside L.

11.3.3 Finite–Element Computations

In this section we present our numerical computations with the quasiconvexified energy W^{qc} on a model problem, whose geometry and material parameters correspond to experiments reported in the literature. The loading process consists in stretching in a direction orthogonal to the director at cross–linking $n_0 = e_3$ a thin, flat rectangular sheet of size $l_1 \times l_2 \times l_3$. In our numerical simulations, we choose a typical value $a = 0.5$, which gives a spontaneous stretch $a^{-1/3} = 1.26$. Moreover we considered a fixed thickness $l_1 = 0.1 l_3$ and an aspect–ratio $l_2/l_3 = 3$ (we also performed simulations with $l_2/l_3 = 1$, see below). Lengths are here measured in the reference configuration, which is the stress-free configuration of the high–temperature phase described by $a = 1$ [this corresponds to the change of variables mentioned before (11.19)]. The initial configuration is the stress–free one (in the low temperature phase, i.e. for $a = 0.5$) with director parallel to $n_0 = e_3$, hence it is deformed with respect to the reference configuration by the affine uniaxial strain U_{e_3}. The aspect ratio in the initial configuration is thus $3a^{1/2} \sim 2.12$. In the experiment, the two faces orthogonal to e_2 are glued to two pieces of rigid material (clamps), which are then pulled away from each other. In our simulation, on those faces we impose Dirichlet boundary conditions. The stretch s is the distance between the clamps (in direction 2) in units of its initial value. We always work with zero tractions on the unclamped part of the boundary. Computations have been performed on a SUN workstation by implementing a user–defined constitutive law in a commercial finite–element software package (ABAQUS). The high degeneracy of the energy (especially in phase L, see below) requires inclusion of a small regularizing perturbation. Details of the numerical procedure and of the regularization have been described in [14].

Figure 11.7 shows the reactive force exerted by the clamps plotted against the imposed displacements. These results show the existence of a "window" of liquid–like behavior, where the force is zero within the resolution of our simulation. We shall prove later (Theorem 6) that, at least in a smaller window, the energy (and hence the reaction force) is exactly zero.

Several experiments [42, 64] report formation of microstructures in the central part of the sample. This is explained by the fact that, in this region, the energy is reduced by forming small–scale oscillatory patterns with alternating shears, as discussed at the end of Section 11.3.1. To identify the part of the sample where this happens, we plot the so–called microstructure index

$$I_M = \frac{1}{\lambda_1} - a^{1/2} \lambda_3^2 \qquad (11.41)$$

which is positive in phases L and Sm, and negative in phase S. Hence, $I_M > 0$ characterizes regions where the macroscopic deformation is achieved by a mixture of different microscopic deformations. Figure 11.8 shows a density plot of I_M for different stretches. The striking result is that a region exhibiting

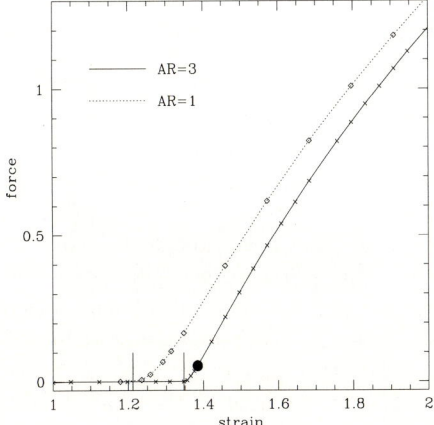

Fig. 11.7. Force versus strain curve (crosses joined by the full line). The diamonds joined by the dotted curve are results obtained for aspect–ratio 1 (i.e. $l_2 = l_3$). The black dot marks the configuration which is presented in more detail below. The two vertical lines mark the boundary of the liquid phase as obtained with the analytic construction in Theorem 6.

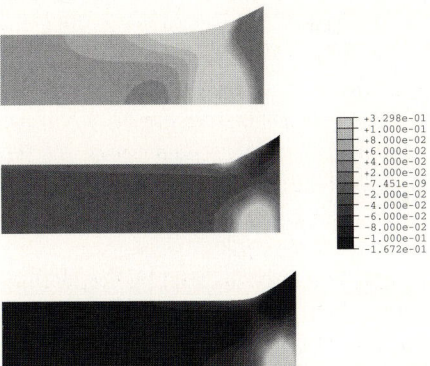

Fig. 11.8. Plot of the microstructure index I_M for different stretches. From top to bottom, $s = 1.31$, $s = 1.38$, and $s = 1.46$. Only the quarter of the sample in the first quadrant is plotted, the rest is symmetric. The clamped boundary is the one on the right side in this figure, the lower–left corner represents the center of the sample.

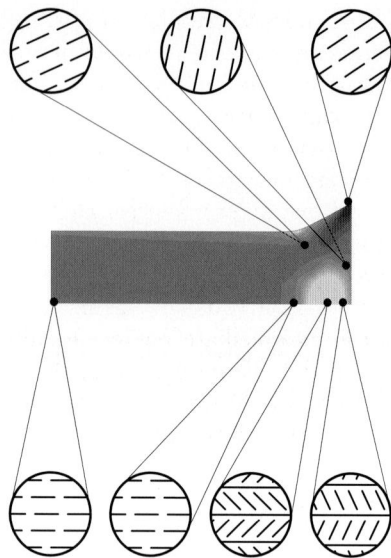

Fig. 11.9. Microstructure index I_M and resolution in microstructures at stretch 1.38.

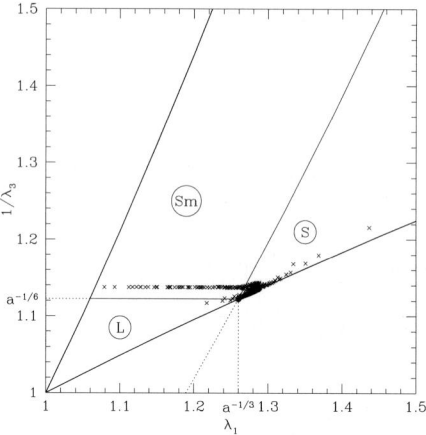

Fig. 11.10. Macroscopic phase diagram and phase distribution at $s = 1.38$. The three regions are defined in Remark 13. Only the region $\lambda_3^2 \geq 1/\lambda_1 \geq \lambda_3^{1/2}$ can be attained (due to the volume constraint).

microstructure survives even at large stretches, around the mid–point of the boundary. We now discuss in more detail the results for a representative value of the stretch ($s = 1.38$, see the central panel of Figure 11.8 for the resulting deformation).

In Figure 11.9 we show a possible resolution of the deformation gradient using first–order laminates (as done in the construction described in Section 11.3.2). We choose rank–one connected matrices $F_\pm$ such that $(F_+ + F_-)/2$ agrees with the local in–plane macroscopic deformation. The "sticks" in the figures indicate the local orientation of the nematic director n, and have been obtained by plotting the eigenvector of $F_+ F_+^T$ corresponding to the largest eigenvalue, i.e. the largest one among all vectors $F_+ e$ with $|e| = 1$. As was clear from the previous figures, microstructure is present only in the white region around the central part of the clamp. Indeed, the director n equals e_3 close to the clamp, whereas it equals e_2 in the center of the sample. The microstructure covers the intermediate regime. Finally, in Figure 11.10 we display the distribution in the phase diagram of the points explored by the numerically computed macroscopic deformation gradient. The row of points in phase Sm corresponds to the white region in the previous figures.

11.3.4 Attainment Results

The mathematical analysis of the microscopic energy of nematic elastomers allows one to understand and to reproduce in numerical simulations experimentally observed microstructures. The success of this approach motivates a further investigation of the mathematical origin of such microstructures. In particular, we consider two paradigmatic boundary–value problems, namely, affine Dirichlet conditions and the mixed boundary conditions which mimic the experimental setup. The implications of these results on the understanding of experiments on nematic elastomers are then discussed in Section 11.3.5.

Attainment and non–attainment for Dirichlet boundary conditions.
In this section we discuss the question of whether there exist Lipschitz functions u with affine boundary conditions $u(x) = Fx$ and with zero microscopic energy, i.e. whether for a given F such that $W^{qc}(F) = 0$ the infimum

$$\inf_{y \in W^{1,\infty}} \left\{ \frac{1}{|\Omega|} \int_\Omega W(\nabla y(x)) dx : y(x) = Fx \text{ on } \partial\Omega, \det \nabla y(x) = 1 \right\}, \tag{11.42}$$

which is zero by the definition of W^{qc}, is in fact attained [W was defined in (11.23)]. The answer only depends on the singular values of F, and not on the domain Ω. This issue was first investigated by Dacorogna and Tanteri in [17]. We give here an independent, complete characterization of the attainment set, based on a general construction by Müller and Šverák.

Theorem 4. *Let F be in the phase L, i.e.*

$$\gamma_1 \leq \lambda_1(F) \leq \lambda_3(F) \leq \gamma_3, \tag{11.43}$$

and such that $W(F) > 0$. Then the infimum in (11.42) is attained if and only if $\gamma_1 < \lambda_1(F)$ and $\lambda_3(F) < \gamma_3$.

Proof. By rotational invariance, we may assume that F is a diagonal matrix, $F = \text{diag}(\mu_1, \mu_2, \mu_3)$, with $\mu_1 \leq \mu_2 \leq \mu_3$. We first show that the infimum is not attained if $\mu_3 = \gamma_3$. Suppose otherwise. Then there exists a Lipschitz map u with $W(\nabla u) = 0$ a.e., and $u = Fx$ on $\partial(0,1)^3$. Then,

$$\gamma_3 = \int_{(0,1)^3} e_3 \cdot \nabla u e_3 \, dx \leq \int_{(0,1)^3} \lambda_3(\nabla u) \, dx = \gamma_3. \tag{11.44}$$

Therefore equality holds in this chain of inequalities, and since $|(\nabla u)e_3| \leq \lambda_3(\nabla u) = \gamma_3$ we get $(\nabla u)e_3 = \gamma_3 e_3$. It follows that

$$u(x_1, x_2, x_3) = F(x_1, x_2, 0)^T + \gamma_3 x_3 e_3 = Fx. \tag{11.45}$$

Hence $W(\nabla u) = W(F) > 0$, a contradiction. If instead $\mu_1 = \gamma_1$, we need to consider the cofactor matrix, which like the gradient itself is a null Lagrangian. Therefore its integral depends only on the boundary values, and

$$\frac{1}{\gamma_1} = \int_{(0,1)^3} e_1 \cdot (\text{cof } \nabla u) e_1 \, dx \leq \int_{(0,1)^3} \lambda_1^{-1}(\nabla u) \, dx \leq \frac{1}{\gamma_1}, \tag{11.46}$$

so that we can conclude as before.

To prove attainment in the case $\gamma_1 < \mu_1 \leq \mu_3 < \gamma_3$, we use a special case of a more general theorem by Müller and Šverák [48], which is stated below in Theorem 5. In our application, K is given by $K = SO(3)\text{diag}(\gamma_1, \gamma_2, \gamma_3)SO(3)$, and we only need to construct an in–approximation with $F \in U_1$. In order to do so, we choose

$$U_1 = \{F : \det F = 1, \gamma_1 < \lambda_1(F) \leq \lambda_3(F) < \gamma_3\} \tag{11.47}$$

and we define iteratively

$$U_{i+1} = U_i \cap \left\{ F : \lambda_1(F) < \gamma_1 + \frac{1}{i}, \lambda_3(F) > \gamma_3 - \frac{1}{i} \right\}. \tag{11.48}$$

The only thing that remains to be checked is that U_i is contained in the rank–one convex hull of U_{i+1}. This is however an immediate consequence of our construction in Section 11.3.2.

Before concluding this Section, we state the result regarding the constraint $\det F = 1$ which has been used in the construction above.

Theorem 5 ([48]). *Let $\Sigma = \{F \in M^{3\times 3} : \det F = 1\}$, and let K be a subset of Σ. Suppose that $\{U_i\}$ is an in–approximation of K, i.e. the U_i are open in Σ, uniformly bounded, U_i is contained in the rank–one convex hull of U_{i+1}, and U_i converges to K in the following sense: if $F_i \in U_i$ and $F_i \to F$, then $F \in K$. Then, for any $F \in U_1$ there exists a Lipschitz solution of the partial differential inclusion*

$$Du \in K \quad \text{a.e. in } \Omega$$
$$u(x) = Fx \quad \text{on } \partial\Omega.$$

Attainment for a Dirichlet–Neumann problem. We now explore the attainment question for the experimentally relevant case of Dirichlet boundary conditions prescribed only on two opposite faces of a thin rectangular sheet. We will show that, for a large range of imposed stretches, zero energy can be attained by deformations whose gradients take only finitely many values.

Theorem 6. *The problem*

$$W(\nabla u) = 0 \text{ a.e.}, \qquad u(x) = \mathrm{diag}(\gamma_1, t\gamma_2, \gamma_3)x \text{ for } x_2 = \pm l_2 \qquad (11.49)$$

has a piecewise affine solution in $\Omega = (-l_1, l_1) \times (-l_2, l_2) \times (-l_3, l_3)$ *such that* ∇u *takes only 4 different values for any t in the interval*

$$1 \leq t \leq 1 + \frac{\gamma_3 - \gamma_2}{\gamma_2} \frac{l_2 - AR_0 l_3}{l_2}, \qquad (11.50)$$

where

$$AR_0 = \frac{\gamma_2 + \gamma_3}{2\gamma_2}\left(\sqrt{1 + \frac{4\gamma_2\gamma_3}{(\gamma_2+\gamma_3)^2}} - 1\right). \qquad (11.51)$$

Proof. It is sufficient to construct $v(x_2, x_3)$ such that $v(\pm l_2, x_3) = \pm t\gamma_2 l_2 e_2 + \gamma_3 x_3 e_3$. Then one takes $u(x) = \gamma_1 x_1 e_1 + v(x_2, x_3)$. Figure 11.12 shows the regions where ∇v is constant, in the (x_2, x_3) plane. In the two regions denoted by A, which contain the portion of the boundary where the Dirichlet condition is imposed, we have $\nabla v = F_A = \mathrm{diag}(\gamma_2, \gamma_3)$. In the central region C, we have $\nabla v = F_C = \mathrm{diag}(\gamma_3, \gamma_2)$. In the regions denoted by B, $\nabla v = e^{i(\phi+\theta)}\mathrm{diag}(\gamma_2, \gamma_3)e^{-i\theta}$, where $e^{i\theta}$ denotes the matrix corresponding to a counterclockwise rotation by θ. The necessary conditions for the construction are that $F_A - F_B$ and $F_B - F_C$ are rank–one matrices, i.e.

$$\det(F_A - F_B) = \det(F_B - F_C) = 0, \qquad (11.52)$$

and that the corresponding rank–one directions have a suitable orientation (i.e. one should also check that they are compatible with the global geometry displayed in Figure 11.12). Eq. (11.52) is equivalent to

$$\cos(\phi + 2\theta) = 0, \qquad \cos\phi = \frac{4\gamma_2\gamma_3}{(\gamma_3 + \gamma_2)^2}, \qquad (11.53)$$

which has two pairs of opposite solutions, which we denote by $(\pm\phi, \pm\theta_1)$ and $(\pm\phi, \pm\theta_2)$. It remains to check that the corresponding rank–one direction have the relative orientation shown in the figure. Consider one of them, (ϕ, θ_1). Let ψ^{AB} and ψ^{BC} be the (uniquely defined) angles in $(0, \pi)$ such that $(F_A - F_B)(\cos\psi^{AB}, \sin\psi^{AB}) = (F_B - F_C)(\cos\psi^{BC}, \sin\psi^{BC}) = 0$ (ψ^{AB} and ψ^{BC} are the angles formed by the segments PR and PQ with the e_2 axis, respectively). If $\psi^{BC} \geq \psi^{AB}$ this solution is appropriate for B^+ (i.e. the part of

B which is contained in the first and third quadrant). If instead $\psi^{BC} \leq \psi^{AB}$, this solution is appropriate for B^-. In both cases, $(-\phi, -\theta_1)$ gives a solution for the other half of B. The same reasoning applies to $(\pm\phi, \pm\theta_2)$, hence we have two solutions.

We now focus on one of them, with sign chosen so that $\psi^{BC} \geq \psi^{AB}$. The boundary condition is satisfied if and only if

$$tl_2\gamma_2 = \gamma_3 P_2 + \gamma_2(l_2 - P_2) \tag{11.54}$$

where P_2 is the e_2 coordinate of the point P where the regions A and C touch. From the uniqueness of the elongation along e_2 (which is a consequence of the fact that ∇u is a gradient field),

$$\gamma_3 P_2 + \gamma_2(l_2 - P_2) = \gamma_3 Q_2 + (F_B)_{22}(R_2 - Q_2) + \gamma_2(l_2 - R_2), \tag{11.55}$$

we then get, since $|(F_B)_{22}| \leq \gamma_3$ and $Q_2 \leq R_2$, that $P_2 \leq R_2$, i.e. $\psi^{AB} \in (0, \pi/2)$ (a more refined analysis indicates that $0 \leq \psi^{AB} \leq \pi/2 \leq \psi^{BC} \leq \pi$ for both solutions).

It remains to check that the entire construction can be fitted inside the domain. The topology displayed in the first panel of Figure 11.11 can be realized if and only if B^+ lies entirely in the part of Ω in the first quadrant, i.e. if and only if

$$0 \leq \min(Q_2, P_2, R_2) \leq \max(Q_2, P_2, R_2) \leq l_2. \tag{11.56}$$

By a slight change in the connectivity of the domains, the first condition can actually be removed. Indeed, at least one of the two topologies of Figure 11.11 can be realized under the weaker condition

$$0 \leq P_2 \leq R_2 \leq l_2. \tag{11.57}$$

Since $R_2 - P_2 = l_3 \cot \psi^{AB}$, this is possible only if the aspect ratio obeys

$$AR = \frac{l_2}{l_3} \geq AR_0 = \cot \psi^{AB}, \tag{11.58}$$

where a straightforward computation gives for AR_0 the expression (11.51). Then, from (11.54) we get a solution for

$$t - 1 = \frac{\gamma_3 - \gamma_2}{\gamma_2} \frac{P_2}{l_2} \tag{11.59}$$

i.e. for

$$1 \leq t \leq 1 + \frac{\gamma_3 - \gamma_2}{\gamma_2} \frac{AR - AR_0}{AR} \tag{11.60}$$

We finally compare the present construction with our numerical experiment, using the parameters from Section 11.3.3, i.e. $\gamma_2 = \gamma_3^{-1/2} = a^{1/6} = 2^{-1/6}$ and $l_2/l_3 = 1$ and 3. Then, one obtains $t_{\max}(AR = 3) \simeq 1.347$ and $t_{\max}(AR = 1) \simeq 1.212$, which are both indistinguishable from the numerically obtained threshold within the numerical resolution (see Figure 11.7). The corresponding deformation for aspect ratio 3 is displayed in Figure 11.12.

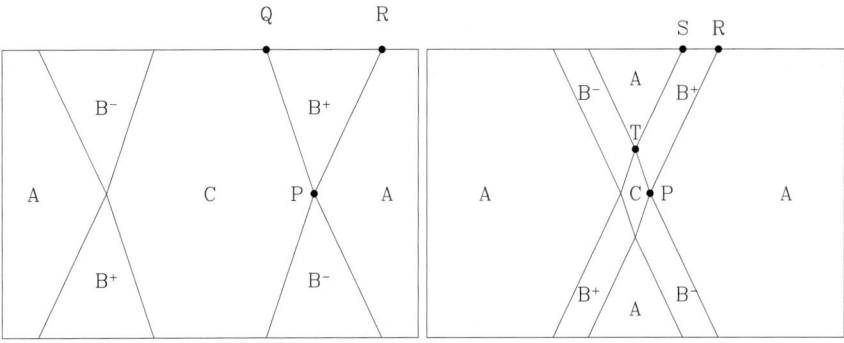

Fig. 11.11. Construction of an energy minimizer u whose gradient takes finitely many values. The left panel shows the case where the first inequality in (11.57) is satisfied, the right panel a case where it is not satisfied (see Section 11.3.4 for details).

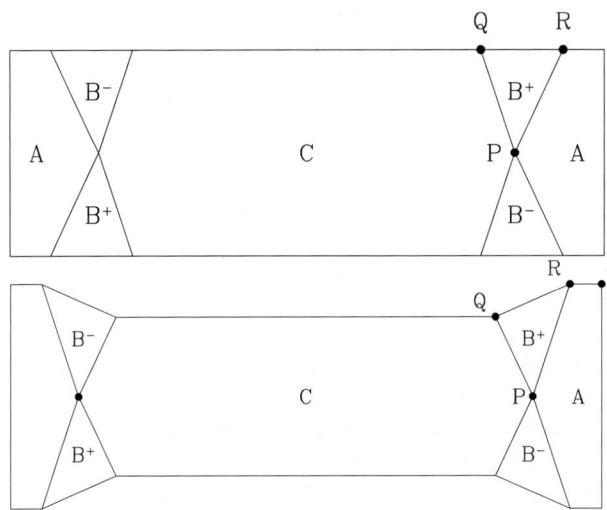

Fig. 11.12. Construction of an energy minimizer u whose gradient takes finitely many values. The upper panel shows the reference configuration, the lower one the deformed configuration (see Section 11.3.4 for details). The figure has been drawn with $l_3/l_2 = 3$, $a = 1/2$.

11.3.5 Discussion and Perspectives

In the previous sections we have presented a complete mathematical and numerical analysis of the predictions of the BTW model for stretching experiments on nematic elastomeric sheets. We now discuss the main results we have obtained, with particular attention to the possibility of comparing with experiment, with the aim not only to assess the general method but also to validate the specific assumptions we have made.

We have shown that, up to a critical stretch, the material behaves as a liquid, in the sense that there is no reaction force to stretching, and the energy is constant. This has been shown both numerically, computing with finite elements based on the quasiconvexified energy, and analytically, by explicitly constructing functions on which the microscopic energy vanishes. The solutions constructed through the two approaches are different, and indeed there is a huge degeneracy: for stretches inside the liquid window there is a continuum of possible zero–energy solutions. This degeneracy has been lifted in the numerical procedure by adding an *ad hoc* regularizing term, which, however, has no physical justification. It is therefore natural to ask whether a more realistic selection mechanism could be obtained to single out the experimentally observed solution. From the point of view mentioned in the introduction, in this problem the robust quantities, which should not be largely affected by small perturbations, are the total energy and its derivative (the force of Figure 11.7). The deformation pattern and the microstructure are instead expected to be significantly affected by small perturbations.

We focus first on the robust quantity, the force, and observe that the general feature of a liquid–like response at small stretches, which is certainly the most interesting experimental result on these materials, is reproduced by the present theoretical approach. However, differences emerge in the details. Indeed, the small stiffness observed in experiments is missed by the BTW model, which gives exactly zero force at small strains.

Non–robust quantities, as for example the local microstructure, give a different picture. For example, at small strains the numerical solution predicts microstructure formation in all of the sample, whereas the analytical solution predicts no microstructure. One standard way to lift degeneracies is to include higher–order singular perturbations, either in the form of higher–order gradient terms or of surface–energy terms. However, this would immediately lead to the conclusion that the explicit construction of Section 11.3.4, with few interfaces, has lower energy than the solution with very fine structures of Section 11.3.3, which is in contrast with the experimental observation of fine–stripe patterns. A different, and probably more specific mechanism needs therefore to be considered.

A correction to the BTW energy which, in principle, can account for all the mentioned effects was proposed by Verwey, Warner, and Terentjev in 1996 [59, 61]. They suggested that in computing the average of Eq. (11.18) one should also average over fluctuations of a in the cross–linking configuration.

By expansion of the product $\operatorname{Tr} U_n^{-2} F U_{n_0}^2 \tilde{F}^T$ and using the definition of U_n one gets

$$\operatorname{Tr} U_n^{-2} \tilde{F} U_{n_0}^2 \tilde{F}^T = |\tilde{F}|^2 + (a-1)|\tilde{F}^T n|^2 + (a^{-1}-1)|\tilde{F} n_0|^2 \\ + (2 - a - a^{-1})(n\tilde{F} n_0)^2 \quad (11.61)$$

which then needs to be averaged over the distribution of a. The special symmetry present in W_{BTW} is only recovered if $\langle 1/a \rangle = 1/\langle a \rangle$. However, it is still possible to eliminate one of the terms in (11.61) by choosing a suitable reference configuration. More precisely, we write

$$\tilde{F} = F b^{1/6} \left[\operatorname{Id} + (b^{-1/2} - 1) n_0 \otimes n_0 \right] \quad (11.62)$$

substitute into (11.61), expand, and choose b so that the coefficient of $|nFn_0|^2$ vanishes, i.e.

$$b = \frac{\langle \frac{1}{a} \rangle - 1}{1 - \langle a \rangle}. \quad (11.63)$$

Minimizing over n, and choosing $b^{1/3} k_B T/2$ as energy units, we get

$$W_{\text{VWT}}(F) = \lambda_1^2(F) + \lambda_2^2(F) + \langle a \rangle \lambda_3^2(F) - \beta |Fn_0|^2 \quad (11.64)$$

where

$$\beta = \frac{\langle a \rangle \langle \frac{1}{a} \rangle - 1}{\langle \frac{1}{a} \rangle - 1} \quad (11.65)$$

is a nonnegative parameter which characterizes the strength of the anisotropy. This energy reduces to the BTW one if $\beta = 0$, i.e. if the distribution of a is a Dirac delta. The energy W_{VWT} is no longer invariant under rotations, and indeed states where the elongation direction is close to n_0 are favored. The consequences of the new term in (11.64) have been analyzed within a restricted geometry in [59, 61], to which we refer for a more detailed discussion. We only observe here that replacing W_{BTW} with W_{VWT} does not change at all the structure of the problem, since no additional gradients are included. However, the explicit computation of the quasiconvex envelope is a much harder problem for the non–isotropic W_{VWT} than it was for W_{BTW}. A full mathematical treatment of W_{VWT}, including computation of the quasiconvex envelope, is, in our opinion, one of the most interesting open questions in the mathematical analysis of nematic elastomers.

Acknowledgments: We thank R.V. Kohn for stimulating discussions, and R. Schäfer for the experimental results displayed in Figures 11.2 and 11.3. This work was partially supported by the EU TMR network *Phase Transitions in Crystalline Solids*, contract FMRX–CT98–0229, and by the NSF through grant DMS0104118. The third author was partially supported by the Max Planck Society.

References

1. G. Alberti and S. Müller, *A new approach to variational problems with multiple scales*, Comm. Pure Appl. Math. **54** (2001), 761–825.
2. L. Ambrosio, C. De Lellis, and C. Mantegazza, *Line energies for gradient vector fields in the plane*, Calc. Var. Partial Diff. Eqs. **9** (1999), 327–355.
3. G. Anzellotti, S. Baldo, and A. Visintin, *Asymptotic behavior of the Landau–Lifshitz model of ferromagnetism*, Appl. Math. Optim. **23** (1991), 171–192.
4. J. Ball and R. D. James, *Fine phase mixtures as minimizers of the energy*, Arch. Rat. Mech. Anal. **100** (1987), 13–52.
5. H. Ben Belgacem, S. Conti, A. DeSimone, and S. Müller, *Energy scaling of compressed elastic films*, Arch. Rat. Mech. Anal. **164** (2002), 1-37.
6. _____, *Rigorous bounds for the Föppl–von Kármán theory of isotropically compressed plates*, J. Nonlinear Sci. **10** (2000), 661–683.
7. H. A. M. van den Berg, *Self-consistent domain theory in soft-ferromagnetic media. ii. basic domain structures in thin film objects*, J. Appl. Phys. **60** (1986), 1104–1113.
8. G. Bertotti, *Hysteresis in magnetism*, Academic Press, San Diego, 1998.
9. P. Bladon, E. M. Terentjev, and M. Warner, *Transitions and instabilities in liquid–crystal elastomers*, Phys. Rev. E **47** (1993), R3838–R3840.
10. A. Braides and A. Defranceschi, *Homogeneization of multiple integrals*, Claredon Press, Oxford, 1998.
11. W. F. Brown, *Micromagnetics*, Wiley, 1963.
12. P. Bryant and H. Suhl, *Thin–film magnetic patterns in an external field*, Appl. Phys. Lett. **54** (1989), 2224.
13. M. Chipot and D. Kinderlehrer, *Equilibrium configurations of crystals*, Arch. Rat. Mech. Anal. **103** (1988), 237–277.
14. S. Conti, G. Dolzmann, and A. DeSimone, *Soft elastic response of stretched sheets of nematic elastomers: a numerical study*, J. Mech. Phys. Solids **50** (2002), 1431–1451.
15. B. Dacorogna, *A relaxation theorem and its application to the equilibrium of gases*, Arch. Rat. Mech. Anal. **77** (1981), 359–386.
16. _____, *Direct methods in the calculus of variations*, Springer, Berlin, 1989.
17. B. Dacorogna and C. Tanteri, *Implicit partial differential equations and the constraints of non linear elasticity*, J. Math. Pure Appl. **81** (2002), 311-341.
18. G. Dal Maso, *An introduction to Γ-convergence*, Birkhäuser, Boston, 1993.
19. E. De Giorgi, *Sulla convergenza di alcune successioni di integrali del tipo dell'area*, Rend. Mat. **8** (1975), 277–294.
20. E. De Giorgi and T. Franzoni, *Su un tipo di convergenza variazionale*, Atti Accad. Naz. Lincei Rend. Cl. Sci. Mat. **58** (1975), 842–850.
21. A. DeSimone and G. Dolzmann, *Macroscopic response of nematic elastomers via relaxation of a class of $SO(3)$-invariant energies*, Arch. Rat. Mech. Anal. **161** (2002), 181-204.
22. A. DeSimone and R.D. James, *A constrained theory of magnetoelasticity*, J. Mech. Phys. Solids **50** (2002), 283-320.
23. A. DeSimone, R.V. Kohn, S. Müller, and F. Otto, *A compactness result in the gradient theory of phase transitions*, Proc. Roy. Soc. Edin. A **131** (2001), 833–844.

24. _____, *Magnetic microstructures – a paradigm of multiscale problems*, ICIAM 99 (J.M. Ball and J.C.R. Hunt, eds.), Oxford Univ. Press, 2000, pp. 175–190.
25. _____, *A reduced theory for thin-film micromagnetics*, Comm. Pure Appl. Math. (to appear).
26. _____, *Repulsive interaction of Néel wall tails*, Mult. Model. and Simul. (in press).
27. A. DeSimone, R.V. Kohn, S. Müller, F. Otto, and R. Schäfer, *Two-dimensional modeling of soft ferromagnetic films*, Proc. Roy. Soc. Lond. A **457** (2001), 2983–2992.
28. L.C. Evans, *Partial differential equations*, American Mathematical Society, Providence, 1998.
29. H. Finkelmann, I. Kundler, E.M. Terentjev, and M. Warner, *Critical stripe-domain instability of nematic elastomers*, J. Phys. II France **7** (1997), 1059–1069.
30. G. Friesecke, R. James, and S. Müller, *Rigorous derivation of nonlinear plate theory and geometric rigidity*, C. R. Acad. Sci. Paris Série I **334** (2002), 173-178.
31. G. Friesecke and F. Theil, *Validity and failure of the Cauchy–Born hypothesis in a 2D mass–spring lattice*, preprint (2001).
32. C.J. Garcia–Cervera and W.E, *Effective dynamics for ferromagnetic thin films*, J. Appl. Phys. **90** (2001), 370–374.
33. P. Gérard, *Microlocal defect measures*, Comm. PDE **16** (1991), 1761–1794.
34. G. Gioia and M. Ortiz, *Delamination of compressed thin films*, Adv. Appl. Mech. **33** (1997), 119–192.
35. L. Golubović and T. C. Lubensky, *Nonlinear elasticity of amorphous solids*, Phys. Rev. Lett. **63** (1989), 1082–1085.
36. A. Hubert and R. Schäfer, *Magnetic domains*, Springer, Berlin, 1998.
37. P. E. Jabin, F. Otto, and B. Perthame, *Line–energy Ginzburg–Landau models: zero–energy states*, Ann. Sc. Normale Pisa (in press).
38. P.E. Jabin and B. Perthame, *Compactness in Ginzburg–Landau energy by kinetic averaging*, Comm. Pure Appl. Math. **54** (2001), 1096–1109.
39. V.V. Jikov, S.M. Kozlov, and O.A. Oleinik, *Homogeneization of differential operators and integral functionals*, Springer, Berlin, 1994.
40. W. Jin and R.V. Kohn, *Singular perturbation and the energy of folds*, J. Nonlinear Sci. **10** (2000), 355–390.
41. W. Jin and P. Sternberg, *Energy estimates of the von Kármán model of thin-film blistering*, J. Math. Phys. **42** (2001), 192–199.
42. J. Kundler and H. Finkelmann, *Strain–induced director reorientation in nematic liquid single crystal elastomers*, Macromol. Chem. Rapid Comm. **16** (1995), 679–686.
43. L.D. Landau and E.M. Lifshitz, *On the theory of the dispersion of magnetic permeability in ferromagnetic bodies*, Phys Z. Sowjetunion **8** (1935), 153–169.
44. C. Le Bris and X. Blanc amd P.–L. Lions, *Convergence de modèles moléculaires vers des modèles de mécanique des milieux continus*, C. R. Acad. Sci. Paris Série I **332** (2001), 949–956.
45. C. Melcher, *The logarithmic tail of Néel walls in thin films*, Preprint MPI–MIS 61 (2001).
46. C.B. Morrey, *Multiple integrals in the calculus of variations*, Springer, Berlin, 1966.

47. S. Müller, *Variational models for microstructure and phase transitions*, in Calculus of Variations and Geometric Evolution Problems, Le ctures given at the 2nd Session of the Centro Internazionale Matematico Estivo, Cetaro 1996 (F. Bethuel, G. Huisken, S. Müller, K. Steffen, S. Hildebrandt, and M. Struwe, eds.), Springer, Berlin, 1999.
48. S. Müller and V. Šverák, *Convex integration with constraints and applications to phase transitions and partial differential equations*, J. Eur. Math. Soc. (JEMS) **1** (1999), 393–442.
49. F. Murat and L. Tartar, *Calcul des variations et homogénéisation*, Les Méthodes de l'Homogénéisation: Théorie et Applications en Physique (D. Bergman et al., ed.), Collect. Dir. Etudes Rech. Electricité de France, vol. 57, Eyrolles, Paris, 1985, pp. 319–369, (translated in [50]).
50. _____, *Calculus of variations and homogenization*, Topics in the Mathematical Modelling of Composite Materials (A. Cherkaev and R. Kohn, eds.), Progress in Nonlinear Differential Equations and Their Applications, Birkhäuser, Boston, 1997, pp. 139–173, (see also the other contributions in this volume).
51. O. Pantz, *Une justification partielle du modèle de plaque en flexion par Γ-convergence*, C. R. Acad. Sci. Paris Série I **332** (2001), 587–592.
52. T. Rivière and S. Serfaty, *Limiting domain wall energy for a problem related to micromagnetics*, Comm. Pure Appl. Math. **54** (2001), 294–338.
53. J. A. Sethian, *Level set methods*, Cambridge University Press, 1996.
54. M. Šilhavý, *Relaxation in a class of $SO(n)$-invariant energies related to nematic elastomers*, preprint (2001).
55. L. Tartar, *Compensated compactness and partial differential equations*, Nonlinear analysis and mechanics: Heriot–Watt Symposium (R. Knops, ed.), vol. IV, Pitman, 1979, pp. 136–212.
56. _____, *H–measures, a new approach for studying homogenization, oscillations and concentration effects in partial differential equations*, Proc. Roy. Soc. Edin. A **115** (1990), 193–230.
57. _____, *Beyond Young measures*, Meccanica **30** (1995), 505–526.
58. R. Tickle, R.D. James, T. Shield, M. Wuttig, and V.V. Kokorin, *Ferromagnetic shape memory in the NiMnGa system*, IEEE Trans. Magn. **35** (1999), 4301–4310.
59. G.C. Verwey, M. Warner, and E. M. Terentjev, *Elastic instability and stripe domain in liquid crystalline elastomers*, J. Phys. II France **6** (1996), 1273–1290.
60. M. Warner, *New elastic behaviour arising from the unusual constitutive relation of nematic solids*, J. Mech. Phys. Sol. **47** (1999), 1355–1377.
61. M. Warner and E.M. Terentjev, *Nematic elastomers – a new state of matter?*, Prog. Polym. Sci. **21** (1996), 853–891.
62. J. Weilepp and H.R. Brand, *Director reorientation in nematic–liquid–single–crystal elastomers by external mechanical stress*, Europhys. Lett. **34** (1996), 495–500.
63. L.C. Young, *Lectures on the calculus of variations and optimal control theory*, Saunders, 1969, reprinted by Chelsea, 1980.
64. E.R. Zubarev, S.A. Kuptsov, T.I. Yuranova, R.V. Talroze, and H. Finkelmann, *Monodomain liquid crystalline networks: reorientation mechanism from uniform to stripe domains*, Liquid crystals **26** (1999), 1531–1540.

Appendix

Color Plates

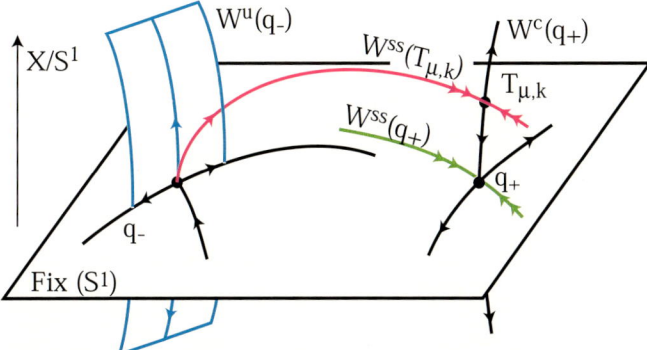

Plate 1. The heteroclinic bifurcation creating a modulated front which invades a Turing pattern

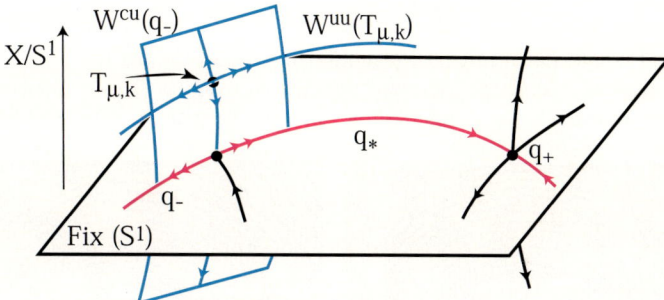

Plate 2. The heteroclinic bifurcation failure of a front leaving a Turing pattern behind

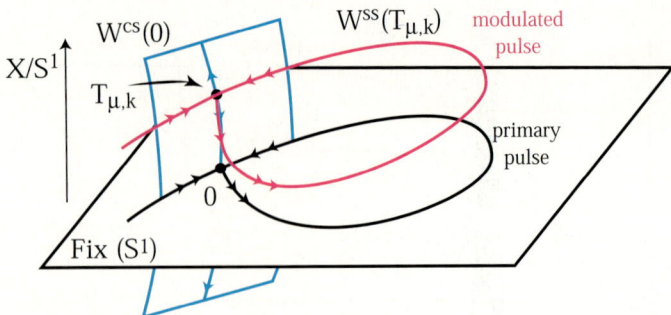

Plate 3. The homoclinic bifurcation creating a modulated pulse travelling through a Turing pattern

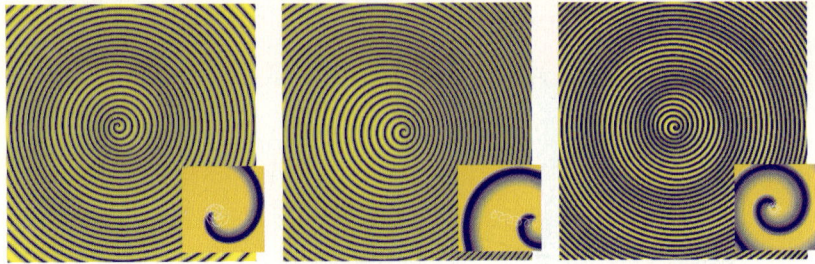

Plate 4. Farfield patterns and tip motion in case of inward meander, drift, and outward meander. The large pictures show the superspiral-shaped deformations of the primary spiral. The small inlets show, on a smaller spatial scale, the tip motion.

Plate 5. Farfield patterns in case of farfield- and core-breakup.

Plate 6. Final state after farfield and core breakup.

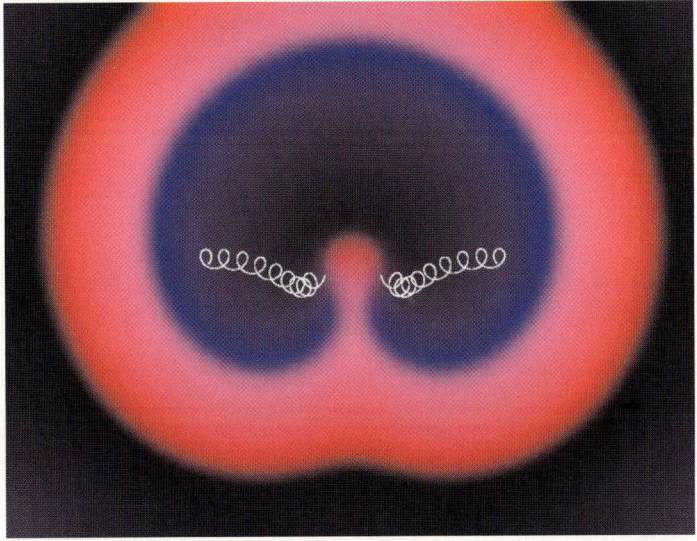

Plate 7. [FiMa00, Fig. 7] Meandering interaction and collision of a pair of planar spiral waves.

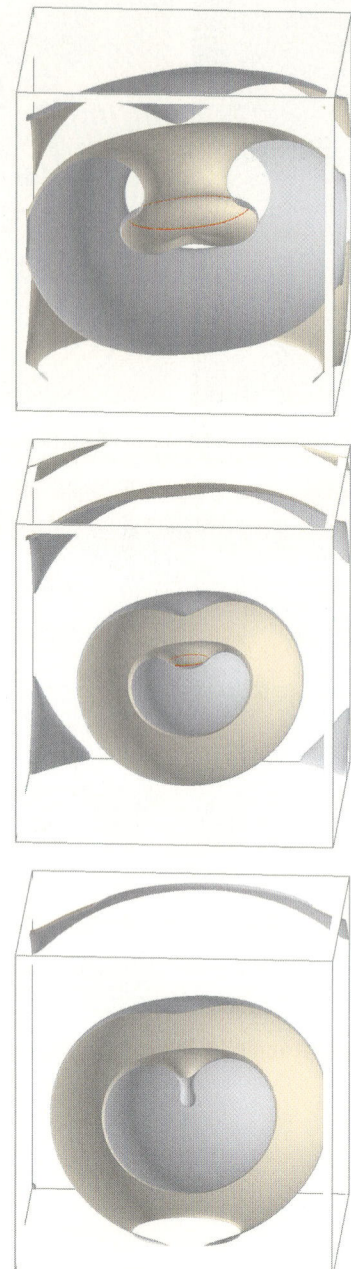

Plate 8. [FiMa00, Fig. 8] Scroll ring annihilation of an untwisted filament. Surfaces $u_2 = 0 \leq u_1$ shown.

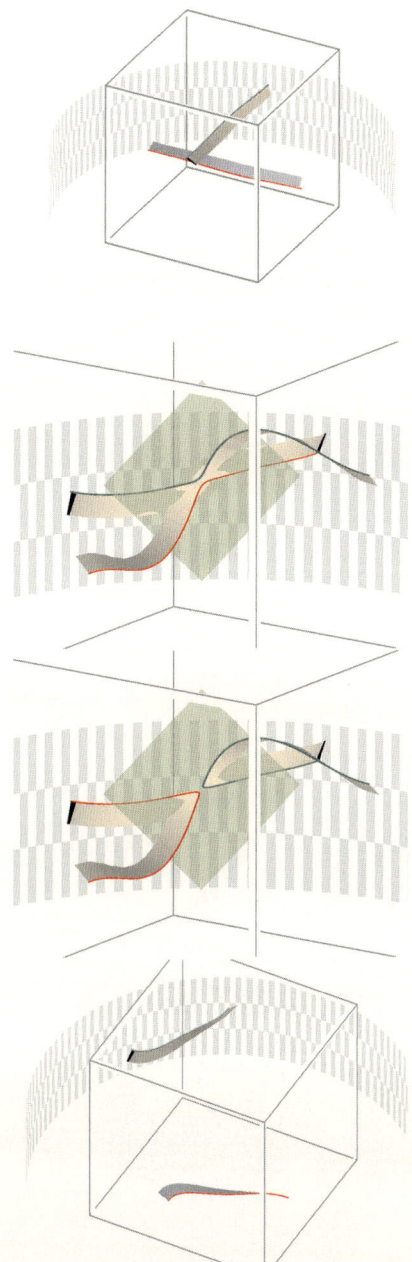

Plate 9. [FiMa00, Fig. 9] Crossover collision of scroll waves including accompanying isochrone bands $u_2 = 0 \leq u_1 \leq \delta$.

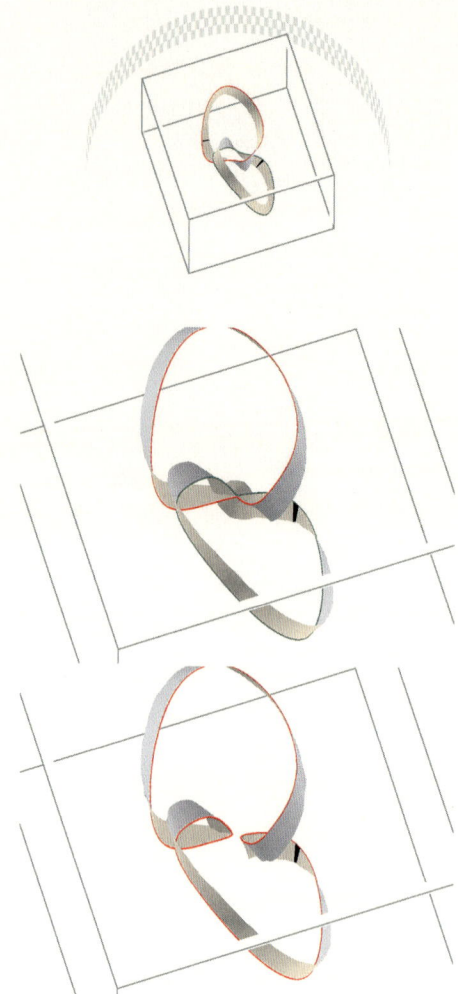

Plate 10. [FiMa00, Fig. 11] Crossover collision of two linked, twisted, circular filaments.

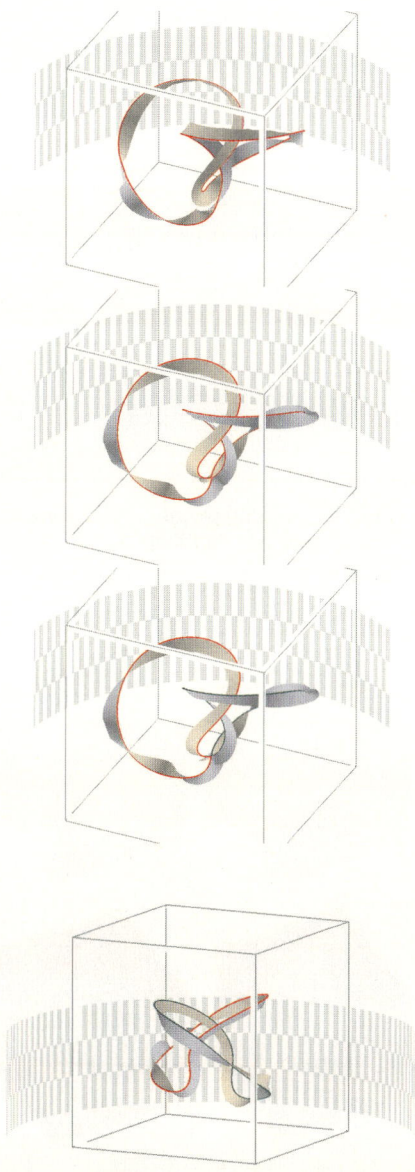

Plate 11. [FiMa00, Fig. 13] Decomposition of the trefoil knot into two linked, twisted circular filaments.

418   Appendix

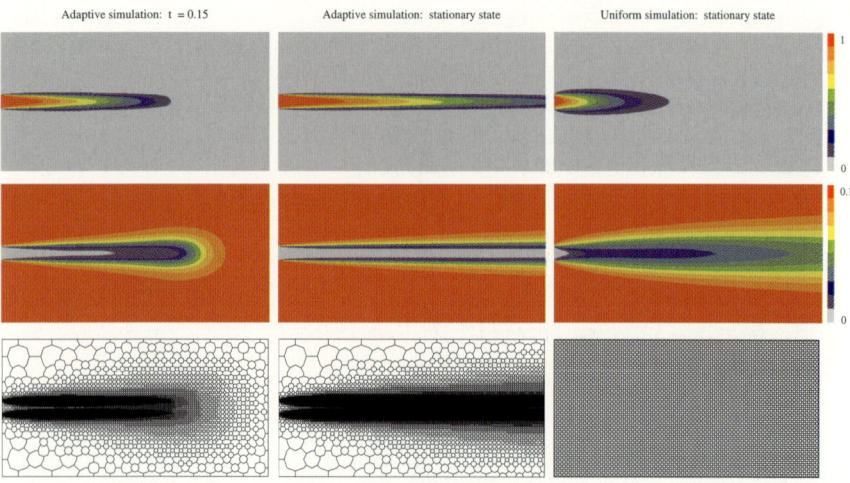

Plate 12. Comparison of the concentrations of substrate (*top*), oxygen (*middle*) and the underlying computational grid (*bottom*) for an adaptive simulation at t = 0.15 (*left*) and t = 5.0 (*middle*) with an uniform calculation at t = 5.0 (*right*).

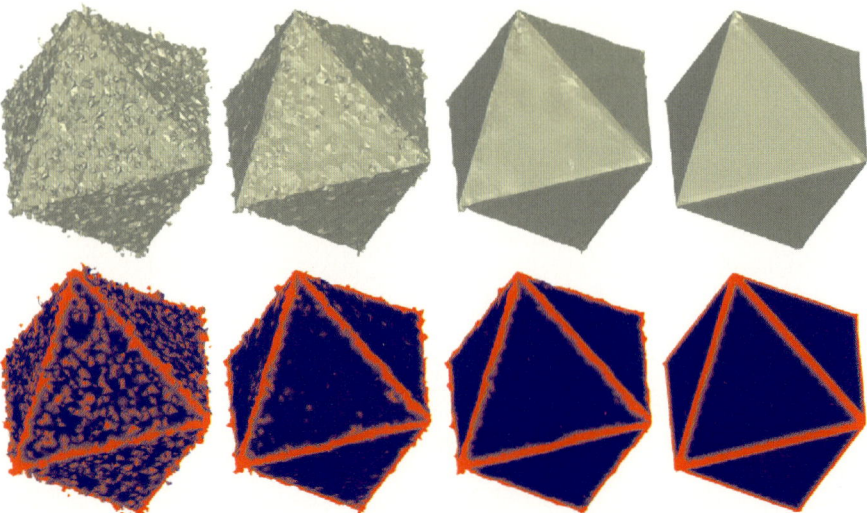

Plate 13. As a test case for the anisotrpic geometric diffusion for still images we consider the function $\phi(x) = |x_1| + |x_2| + |x_3|$ whose level sets are octahedrons. This function was perturbed and then taken as initial data for the anisotropic geometric diffusion method for single images. From left to right an original perturbed level set and the corresponding first, second, and fifth time step of its evolution on a 64^3 grid are depicted. In the bottom row we visualize the dominant curvature on the level sets from the left column. A color ramp from blue to red indicates the dominant curvature value.

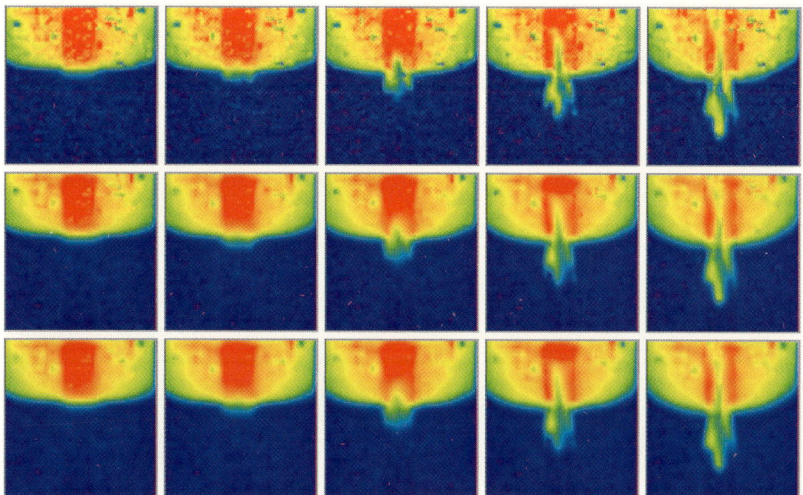

Plate 14. From the experimental data shown in Figure 8.3 vertical slices are extracted. Again from left to right different frames of the sequence and from top to bottom different scales from the evolution are shown.

Printing: Druckhaus Berlin-Mitte
Binding: Buchbinderei Stein & Lehmann, Berlin